T0223973

Lecture Notes in Computer Science 1592

Edited by G. Goos, J. Hartmanis and J. van Leeuwen

Springer

Berlin
Heidelberg
New York
Barcelona
Hong Kong
London
Milan
Paris
Singapore
Tokyo

Jacques Stern (Ed.)

Advances in Cryptology – EUROCRYPT '99

International Conference on the Theory
and Application of Cryptographic Techniques
Prague, Czech Republic, May 2-6, 1999
Proceedings

Springer

Series Editors

Gerhard Goos, Karlsruhe University, Germany
Juris Hartmanis, Cornell University, NY, USA
Jan van Leeuwen, Utrecht University, The Netherlands

Volume Editor

Jacques Stern
Ecole Normale Supérieure
45, rue d'Ulm, F-75230 Paris 05, France
E-mail: Jacques.Stern@ens.fr

Cataloging-in-Publication data applied for

Die Deutsche Bibliothek – CIP-Einheitsaufnahme

Advances in cryptology : proceedings / EUROCRYPT '99, International Conference
on the Theory and Application of Cryptographic Techniques, Prague, Czech
Republic, May 2 - 6, 1999. Jacques Stern (ed.). - Berlin ; Heidelberg ; New
York ; Barcelona ; Hong Kong ; London ; Milan ; Paris ; Singapore ; Tokyo :
Springer, 1999
 (Lecture notes in computer science ; Vol. 1592)
 ISBN 3-540-65889-0

CR Subject Classification (1998): E.3, G.2.1, D.4.6, K.6.5, F.2.1-2, C.2, J.1

ISSN 0302-9743
ISBN 3-540-65889-0 Springer-Verlag Berlin Heidelberg New York

Typesetting: Camera-ready by author
SPIN 10704664   06/3142 – 5 4 3 2 1 0   Printed on acid-free paper

Preface

EUROCRYPT '99, the seventeenth annual Eurocrypt Conference, was sponsored by the International Association for Cryptologic Research (IACR), in cooperation with the Group of Cryptology within the Union of Czech Mathematicians and Physicists. The General Chair, Jaroslav Hruby, was responsible for the overall organization of the conference in the beautiful city of Prague. Let me mention that it was a pleasure to work together: although we were in different locations, we managed to stay in close contact and maintain a smooth organization of the conference.

The Program Committee, consisting of 21 members, considered 120 papers and selected 32 for presentation. In addition, Ross Anderson kindly agreed to chair the traditional rump session for informal short presentations of new results. These proceedings include the revised versions of the 32 papers accepted by the Program Committee. These papers were selected on the basis of originality, quality, and relevance to cryptography. As a result, they should give a proper picture of how the field is evolving. Revisions were not checked and the authors bear full responsibility for the contents of their papers.

The selection of papers was a difficult and challenging task. Each submission was refereed by at least three reviewers and most had four reports or more. I wish to thank the program committee members, who did an excellent job. In addition, I gratefully acknowledge the help of a large number of colleagues who reviewed submissions in their areas of expertise. They are: Michel Abdalla, Josh Benaloh, Charles Bennett, Simon Blackburn, Matt Blaze, Christian Cachin, Jan Camenisch, Ran Canetti, Benny Chor, Galdi Clemente, Jean-Sébastien Coron, Paolo D'Arco, Anand Desai, Uri Feige, Marc Fischlin, Roger Fischlin, Matt Franklin, Steven Galbraith, Rosario Gennaro, Pierre Girard, Dieter Gollmann, Shai Halevi, Helena Handschuh, Yuval Ishai, Markus Jakobsson, Mike Just, Ted Krovetz, Kaoru Kurosawa, Eyal Kushilevitz, Keith Martin, Barbara Masucci, Johannes Merkle, Daniele Micciancio, Victor S. Miller, Fauzan Mirza, Serge Mister, Peter L. Montgomery, Tal Mor, David M'Raïhi, Luke O'Connor, Andrew Odlyzko, Wakaha Ogata, Koji Okada, Pascal Paillier, Pino Persiano, David Pointcheval, Bart Preneel, Tal Rabin, Omer Reingold, Phil Rogaway, Ludovic Rousseau, Berry Schoenmakers, Peter Shor, Jean-Pierre Seifert, Othmar Staffelbach, Ugo Vaccaro, Serge Vaudenay, Ruizhong Wei, Mike Wiener, Rebecca Wright, Xian-Mo Zhang, and Robert Zuccherato. I apologize for any inadvertent omission.

I also wish to thank my PhD students Phong Nguyen, Thomas Pornin, and Guillaume Poupard, who helped me a great deal at various steps of the whole process. Their computer skills and the time and effort they invested were a crucial ingredient of my ability to run the program committee. Thomas ran the electronic submission phase and was able to print all postscript files, including those produced by non-standard word processors. Guillaume opened a private

FTP server and Web site for PC members, and Phong did the editing work, both in paper and in electronic form. I hope I did not distract them too much from their research, but they were kind enough to tell me they had learnt a lot. Thanks also to Joelle Isnard and Nadine Riou, who organized the PC meeting in Paris.

Following the example of CRYPTO '98, EUROCRYPT '99 was the first of the Eurocrypt series with electronic submissions. The electronic submission option was a clear choice for almost all authors, with only 5% of the papers submitted by regular mail. I believe that the time has come to make e-submission mandatory, but it will be the choice of future Crypto and Eurocrypt PC chairs. I wish to thank Joe Kilian, who forwarded us the electronic submission software used for CRYPTO '98 and helped us run it. This software was originally developed by ACM's SIGACT group and I thank the ACM for allowing us to use their system.

Finally, I wish to thank the all authors who submitted papers for making this conference possible by creating the scientific material, and especially the authors of accepted papers. I would also like to thank the publisher, Springer-Verlag, for working within a tight schedule in order to produce these proceedings in due time.

February 1999

Jacques Stern
Program Chair
EUROCRYPT '99

EUROCRYPT '99

May 2 – 6, 1999, Prague, Czech Republic

Sponsored by the

International Association for Cryptologic Research (IACR)

in cooperation with the

Group of Cryptology within the Union of Czech Mathematicians and Physicists

General Chair

Jaroslav Hruby, UCMP, Czech Republic

Program Chair

Jacques Stern, École Normale Supérieure, France

Program Committee

Eli Biham .... Technion, Israel
Mihir Bellare .... University of California, San Diego, USA
Carlo Blundo .... Università di Salerno, Italy
Dan Boneh .... Stanford University, USA
Stefan Brands .... Brands Technologies, Netherlands
Mike Burmester .... Royal Holloway, London, UK
Don Coppersmith .... IBM Research, USA
Claude Crépeau .... McGill University, Canada
Cynthia Dwork .... IBM Almaden Research Center, USA
Joan Feigenbaum .... AT&T Labs - Research, USA
Lars Knudsen .... University of Bergen, Norway
Tsutomu Matsumoto .... Yokohama National University, Japan
Willi Meier .... Fachhochschule Aargau, Switzerland
David Naccache .... Gemplus, France
Jean-Jacques Quisquater .... Université de Louvain, Belgium
Bruce Schneier .... Counterpane Systems, USA
Claus Schnorr .... Universität Frankfurt, Germany
Victor Shoup .... IBM Zurich Research Lab, Switzerland
Paul Van Oorschot .... Entrust Technologies, Canada
Yuliang Zheng .... Monash University, Australia

Table of Contents

Watermarking and Fingerprinting

Elliptic Curves

New Schemes

Block Ciphers

Broadcast and Multicast

Author Index

Cryptanalysis of RSA with
Private Key d Less than $N^{0.292}$

Dan Boneh* and Glenn Durfee**

Computer Science Department, Stanford University, Stanford, CA 94305-9045
{dabo,gdurf}@cs.stanford.edu

Abstract. We show that if the private exponent d used in the RSA public-key cryptosystem is less than $N^{0.292}$ then the system is insecure. This is the first improvement over an old result of Wiener showing that when $d < N^{0.25}$ the RSA system is insecure. We hope our approach can be used to eventually improve the bound to $d < N^{0.5}$.

1 Introduction

To provide fast RSA signature generation one is tempted to use a small private exponent d. Unfortunately, Wiener [10] showed over ten years ago that if one uses $d < N^{0.25}$ then the RSA system can be broken. Since then there have been no improvements to this bound. Verheul and Tilborg [9] showed that as long as $d < N^{0.5}$ it is possible to expose d in less time than an exhaustive search; however, their algorithm requires exponential time as soon as $d > N^{0.25}$.

In this paper we give the first substantial improvement to Wiener's result. We show that as long as $d < N^{0.292}$ one can efficiently break the system. We hope our approach will eventually lead to what we believe is the correct bound, namely $d < N^{0.5}$. Our results are based on the seminal work of Coppersmith [2].

Wiener describes a number of clever techniques for avoiding his attack while still providing fast RSA signature generation. One such suggestion is to use a large value of e. Indeed, Wiener's attack provides no information as soon as $e > N^{1.5}$. In contrast, our approach is effective as long as $e < N^{1.875}$. Consequently, larger values of e must be used to defeat the attack. We discuss this variant in Section 5.

2 Overview of Our Approach

Recall that an RSA public key is a pair $\langle N, e \rangle$ where $N = pq$ is the product of two n-bit primes. For simplicity, we assume $\gcd(p-1, q-1) = 2$. The corresponding private key is a pair $\langle N, d \rangle$ where $e \cdot d \equiv 1 \bmod \frac{\phi(N)}{2}$ where $\phi(N) = N - p - q + 1$.

* Supported by DARPA.
** Supported by Certicom and an NSF Graduate Research Fellowship.

J. Stern (Ed.): EUROCRYPT'99, LNCS 1592, pp. 1–11, 1999.
© Springer-Verlag Berlin Heidelberg 1999

Note that both e and d are less than $\phi(N)$. It follows that there exists an integer k such that

$$ed + k\left(\frac{N+1}{2} - \frac{p+q}{2}\right) = 1. \qquad (1)$$

Writing $s = -\frac{p+q}{2}$ and $A = \frac{N+1}{2}$, we know:

$$k(A + s) \equiv 1 \pmod{e}.$$

Throughout the paper we write $e = N^\alpha$ for some α. Typically, e is of the same order of magnitude as N (e.g. $e > N/10$) and therefore α is very close to 1. As we shall see, when α is much smaller than 1 our results become even stronger.

Suppose the private exponent d satisfies $d < N^\delta$. Wiener's results show that when $\delta < 0.25$ the value of d can be efficiently found given e and N. Our goal is to show that the same holds for larger values of δ. By equation (1) we know that

$$|k| < \frac{2de}{\phi(N)} \le 3de/N < 3e^{1 + \frac{\delta - 1}{\alpha}}.$$

Similarly, we know that

$$|s| < 2N^{0.5} = 2e^{1/2\alpha}.$$

To summarize, taking $\alpha \approx 1$ (which is the common case) and ignoring constants, we end up with the following problem: find integers k and s satisfying

$$k(A + s) \equiv 1 \pmod{e} \quad \text{where} \quad |s| < e^{0.5} \text{ and } |k| < e^\delta. \qquad (2)$$

The problem can be viewed as follows: given an integer A, find an element "close" to A whose inverse modulo e is "small". We refer to this is the *small inverse problem*. Clearly, if for a given value of $\delta < 0.5$ one can efficiently list all the solutions to the small inverse problem, then RSA with private exponent smaller than N^δ is insecure (simply observe that given s modulo e one can factor N immediately, since $e > s$). Currently we can solve the small inverse problem whenever $\delta < 1 - \frac{1}{2}\sqrt{2} \approx 0.292$.

Remark 1. A simple heuristic argument shows that for any $\epsilon > 0$, if k is bounded by $e^{0.5 - \epsilon}$ (i.e. $\delta < 0.5$) then the small inverse problem (equation (2)) is very likely to have a unique solution. The unique solution enables one to break RSA. Therefore, the problem encodes enough information to prove that RSA with $d < N^{0.5}$ is insecure. For $d > N^{0.5}$ we have that $k > N^{0.5}$ and the problem will no longer have a unique solution. Therefore, we believe this approach can be used to show that $d < N^{0.5}$ is insecure, but gives no results for $d > N^{0.5}$.

The next section gives a brief introduction to lattices over $\mathbb{Z}^n$. Our solution to the small inverse problem when α is close to 1 is given in Section 4. In Section 5 we give a solution for arbitrary α. Section 6 describes experimental results with the algorithm.

3   Preliminaries

Let $u_1, \ldots, u_w \in \mathbb{Z}^n$ be linearly independent vectors with $w \leq n$. A lattice L spanned by $\langle u_1, \ldots, u_w \rangle$ is the set of all integer linear combinations of $u_1, \ldots, u_w$. We say that the lattice is full rank if $w = n$. We state a few basic results about lattices and refer to [7] for an introduction.

Let L be a lattice spanned by $\langle u_1, \ldots, u_w \rangle$. We denote by $u_1^*, \ldots, u_w^*$ the vectors obtained by applying the Gram-Schmidt process to the vectors $u_1, \ldots, u_w$. We define the determinant of the lattice L as

$$\det(L) := \prod_{i=1}^{w} \|u_i^*\|.$$

If L is a full rank lattice then the determinant of L is equal to the determinant of the $w \times w$ matrix whose rows are the basis vectors $u_1, \ldots, u_w$.

Fact 1 (LLL). *Let L be a lattice spanned by $\langle u_1, \ldots, u_w \rangle$. Then the LLL algorithm, given $\langle u_1, \ldots, u_w \rangle$, will produce a new basis $\langle b_1, \ldots, b_w \rangle$ of L satisfying:*

1. *$\|b_i^*\|^2 \leq 2\|b_{i+1}^*\|^2$ for all $1 \leq i < w$.*
2. *For all i, if $b_i = b_i^* + \sum_{j=1}^{i-1} \mu_j b_j^*$ then $|\mu_j| \leq \frac{1}{2}$ for all j.*

We note that an LLL-reduced basis satisfies some stronger properties, but those are not relevant to our discussion.

Fact 2. *Let L be a lattice and $b_1, \ldots b_w$ be an LLL-reduced basis of L. Then*

$$\|b_1\| \leq 2^{w/2} \det(L)^{1/w}.$$

Proof. Since $b_1 = b_1^*$ the bound immediately follows from:

$$\det(L) = \prod_i \|b_i^*\| \geq \|b_1\|^w 2^{-w^2/2}.$$

$\square$

In the spirit of a recent result due to Jutla [5] we provide a bound on the norm of other vectors in an LLL reduced basis. For a basis $\langle u_1, \ldots, u_w \rangle$ of a lattice L, define

$$u_{\min}^* := \min_i \|u_i^*\|.$$

Fact 3. *Let L be a lattice spanned by $\langle u_1, \ldots, u_w \rangle$ and let $\langle b_1, \ldots b_w \rangle$ be the result of applying LLL to the given basis. Suppose $u_{\min}^* \geq 1$. Then*

$$\|b_2\| \leq 2^{\frac{w}{2}} \det(L)^{\frac{1}{w-1}}$$

Proof. It is well known that u^*_{min} is a lower bound on the length of the shortest vector in L. Consequently, $\|b_1\| \geq u^*_{min}$. We obtain

$$\det(L) = \prod_i \|b_i^*\| \geq \|b_1^*\| \cdot \|b_2^*\|^{w-1} 2^{-(w-1)^2/2} \geq u^*_{min} \cdot \|b_2^*\|^{w-1} 2^{-(w-1)^2/2}.$$

Hence,

$$\|b_2^*\| \leq 2^{\frac{w-1}{2}} \left[\frac{\det(L)}{u^*_{min}}\right]^{\frac{1}{w-1}} \leq 2^{\frac{w-1}{2}} \det(L)^{\frac{1}{w-1}},$$

which leads to

$$\|b_2\|^2 \leq \|b_2^*\|^2 + \frac{1}{4}\|b_1\|^2 \leq 2^{w-1}\det(L)^{\frac{2}{w-1}} + 2^{w-2}\det(L)^{\frac{2}{w}} \leq 2^w \det(L)^{\frac{2}{w-1}}.$$

Note that $\det(L) \geq 1$ since $u^*_{min} \geq 1$. The bound now follows. $\square$

Similar bounds can be derived for other b_i's. For our purposes the bound on b_2 is sufficient.

4 Solving the Small Inverse Problem

In this section we focus on the case when e is of the same order of magnitude as N, i.e. if $e = N^\alpha$ then α is close to 1. To simplify the exposition, in this section we simply take $\alpha = 1$. In the next section we give the general solution for arbitrary α. When $\alpha = 1$ the small inverse problem is the following: given a polynomial $f(x, y) = x(A + y) - 1$, find (x_0, y_0) satisfying

$$f(x_0, y_0) \equiv 0 \pmod{e} \quad \text{where} \quad |x_0| < e^\delta \text{ and } |y_0| < e^{0.5}.$$

We show that the problem can be solved whenever $\delta < 1 - \frac{1}{2}\sqrt{2} \approx 0.292$. We begin by giving an algorithm that works when $\delta < \frac{7}{6} - \frac{1}{3}\sqrt{7} \approx 0.285$. Our solution is based on a powerful technique due to Coppersmith [2], as presented by Howgrave-Graham [4]. We note that for this particular polynomial our results beat the generic bound given by Coppersmith. For simplicity, let $X = e^\delta$ and $Y = e^{0.5}$.

Given a polynomial $h(x, y) = \sum_{i,j} a_{i,j} x^i y^j$, we define $\|h(x, y)\|^2 := \sum_{i,j} |a_{i,j}^2|$. The main tool we use is stated in the following fact.

Fact 4 (HG98). *Let $h(x, y) \in \mathbb{Z}[x, y]$ be a polynomial which is a sum of at most w monomials. Suppose that*

a. $h(x_0, y_0) = 0 \bmod e^m$ *for some positive integer m where $|x_0| < X$ and $|y_0| < Y$, and*

b. $\|h(xX, yY)\| < e^m/\sqrt{w}$.

Then $h(x_0, y_0) = 0$ holds over the integers.

Proof. Observe that

$$|h(x_0, y_0)| = \left|\sum a_{i,j} x_0^i y_0^j\right| = \left|\sum a_{i,j} X^i Y^j \left(\frac{x_0}{X}\right)^i \left(\frac{y_0}{Y}\right)^j\right| \le$$

$$\sum \left|a_{i,j} X^i Y^j \left(\frac{x_0}{X}\right)^i \left(\frac{y_0}{Y}\right)^j\right| \le \sum |a_{i,j} X^i Y^j| \le$$

$$\sqrt{w} \|h(xX, yY)\| < e^m,$$

but since $h(x_0, y_0) \equiv 0$ modulo e^m we have that $h(x_0, y_0) = 0$.   □

Fact 4 suggests that we should be looking for a polynomial with small norm that has (x_0, y_0) as a root modulo e^m. To do so, given a positive integer m we define the polynomials

$$g_{i,k}(x, y) := x^i f^k(x, y) e^{m-k} \quad \text{and} \quad h_{j,k}(x, y) := y^j f^k(x, y) e^{m-k}.$$

We refer to the $g_{i,k}$ polynomials as x-shifts and the $h_{j,k}$ polynomials as y-shifts. Observe that (x_0, y_0) is a root of all these polynomials modulo e^m for $k = 0, \ldots, m$. We are interested in finding a low-norm integer linear combination of the polynomials $g_{i,k}(xX, yY)$ and $h_{j,k}(xX, yY)$. To do so we form a lattice spanned by the corresponding coefficient vectors. Our goal is to build a lattice that has sufficiently small vectors and then use LLL to find them. By Fact 2 we must show that the lattice spanned by the polynomials has a sufficiently small determinant.

Given an integer m, we build a lattice spanned by the coefficient vectors of the polynomials for $k = 0, \ldots, m$. For each k we use $g_{i,k}(xX, yY)$ for $i = 0, \ldots, m - k$ and use $h_{j,k}(xX, yY)$ for $j = 0, \ldots, t$ for some parameter t that will be determined later. For example, when $m = 3$ and $t = 1$ the lattice is spanned by the rows of the matrix in Figure 1. Since the lattice is spanned by a lower triangular matrix, its determinant is only affected by entries on the diagonal, which we give explicitly. Each "block" of rows corresponds to a certain power of x. The last block is the result of the y-shifts. In the example in Figure 1, $t = 1$, so only linear shifts of y are given. As we shall see, the y-shifts are the main reason for our improved results.

We now turn to calculating the determinant of the above lattice. A routine calculation shows that the determinant of the submatrix corresponding to all x shifts (i.e. ignoring the y-shifts by taking $t = 0$) is

$$\det{}_x = e^{m(m+1)(m+2)/3} \cdot X^{m(m+1)(m+2)/3} \cdot Y^{m(m+1)(m+2)/6}.$$

For example, when $m = 3$ the determinant of the submatrix excluding the bottom block is $e^{20} X^{20} Y^{10}$. Plugging in $X = e^\delta$ and $Y = e^{0.5}$ we obtain

$$\det{}_x = e^{m(m+1)(m+2)(5+4\delta)/12} = e^{\frac{5+4\delta}{12} m^3 + o(m^3)}.$$

It is interesting to note that the dimension of the submatrix is $w = (m+1)(m+2)/2$, and so the wth root of the determinant is $D_x = e^{m(5+4\delta)/6}$. For us to be

	1	x	xy	x^2	x^2y	x^2y^2	x^3	x^3y	x^3y^2	x^3y^3	y	xy^2	x^2y^3	x^3y^4
e^3	e^3													
xe^3		e^3X												
fe^2	−		e^2XY											
x^2e^3				e^3X^2										
xfe^2		−		−	e^2X^2Y									
f^2e	−	−		−		eX^2Y^2								
x^3e^3							e^3X^3							
x^2fe^2					−			e^2X^3Y						
xf^2e		−		−		−	−		eX^3Y^2					
f^3	−	−	−	−		−		−		X^3Y^3				
ye^3											e^3Y			
yfe^2		−										e^2XY^2		
yf^2e		−		−							−		eX^2Y^3	
yf^3		−					−		−		−	−	−	X^3Y^4

Fig. 1. The matrix spanned by $g_{i,k}$ and $h_{j,k}$ for $k = 0..3$, $i = 0..3 - k$, and $j = 0, 1$. The '−' symbols denote non-zero entries whose value we do not care about.

able to use Fact 4, we must have $D_x < e^m$, implying $(5 + 4\delta) < 6$. We obtain $\delta < 0.25$. This is exactly Wiener's result. Consequently, the lattice formed by taking all x-shifts cannot be used to improve on Wiener's result.

To improve on Wiener's result we include the y-shifts into the calculation. For a given value of m and t, the product of the elements on the diagonal of the submatrix corresponding to the y-shifts is:

$$\det{}_y = e^{tm(m+1)/2} \cdot X^{tm(m+1)/2} \cdot Y^{t(m+1)(m+t+1)/2}.$$

Plugging in the values of X and Y, we obtain:

$$\det{}_y = e^{tm(m+1)(1+\delta)/2+t(m+1)(m+t+1)/4} = e^{\frac{3+2\delta}{4}tm^2 + \frac{mt^2}{4} + o(tm^2)}.$$

The determinant of the entire matrix is $\det(L) = \det_x \cdot \det_y$ and its dimension is $w = (m + 1)(m + 2)/2 + t(m + 1)$.

We intend to apply Fact 4 to the shortest vectors in the LLL-reduced basis of L. To do so, we must ensure that the norm of b_1 is less than $e^m/\sqrt{w}$. Combining this with Fact 2, we must solve for the largest value of δ satisfying

$$\det(L) < e^{mw}/\gamma,$$

where $\gamma = (w2^w)^{w/2}$. Since the dimension w is only a function of δ (but not of the public exponent e), γ is a fixed constant, negligible compared to e^{mw}. Manipulating the expressions for the determinant and the dimension to solve for δ requires tedious arithmetic. We provide the exact solution in the full version of this paper. Here, we carry out the computation ignoring low order terms. That is, we write

$$w = \frac{m^2}{2} + tm + o(m^2),$$

$$\det(L) = e^{\frac{5+4\delta}{12}m^3 + \frac{3+2\delta}{4}tm^2 + \frac{mt^2}{4} + o(m^3)}.$$

To satisfy $\det(L) < e^{mw}$ we must have

$$\frac{5+4\delta}{12}m^3 + \frac{3+2\delta}{4}tm^2 + \frac{mt^2}{4} < \frac{1}{2}m^3 + tm^2.$$

This leads to

$$m^2(-1+4\delta) - 3tm(1-2\delta) + 3t^2 < 0$$

For every m the left hand side is minimized at $t = \frac{m(1-2\delta)}{2}$. Plugging this value in leads to:

$$m^2\left[-1+4\delta - \frac{3}{2}(1-2\delta)^2 + \frac{3}{4}(1-2\delta)^2\right] < 0,$$

implying $-7 + 28\delta - 12\delta^2 < 0$. Hence,

$$\delta < \frac{7}{6} - \frac{1}{3}\sqrt{7} \approx 0.285.$$

Hence, for large enough m, whenever $d < N^{0.285-\epsilon}$ for any fixed $\epsilon > 0$ we can find a bivariate polynomial $g_1 \in \mathbb{Z}[x, y]$ such that $g_1(x_0, y_0) = 0$ over the integers. Unfortunately, this is not enough. To obtain another relation, we use Fact 3 to bound the norm of b_2. Observe that since the original basis for L is a triangular matrix, $u^*_{\min}$ is simply the smallest element on the diagonal. This turns out to be the element in the last row of the x-shifts, namely, $u^*_{\min} = X^m Y^m$, which is certainly greater than 1. Hence, Fact 3 applies. Combining Fact 4 and Fact 3 we see that b_2 will yield an additional polynomial g_2 satisfying $g_2(x_0, y_0) = 0$ if

$$\det(L) < e^{m(w-1)}/\gamma'$$

where $\gamma' = (w2^w)^{\frac{w-1}{2}}$. For large enough m, this inequality is guaranteed to hold, since the modifications only effect low order terms. Hence, we obtain another polynomial $g_2 \in \mathbb{Z}[x, y]$ linearly independent of g_1 such that $g_2(x_0, y_0) = 0$ over the integers. We can now attempt to solve for x_0 and y_0 by computing the resultant $h(x) = \text{Res}_y(g_1, g_2)$. Then x_0 must be a root of $h(x)$. By trying all roots x_0 of $h(x)$ we find y_0 using $g_1(x_0, y)$.

Although the polynomials g_1, g_2 are linearly independent, they may not be algebraically independent; they might have a common factor. Indeed, we cannot guarantee that the resultant $h(x)$ is not identically zero. Consequently, we cannot claim our result as a theorem. At the moment it is a heuristic. Our experiments show it is a very good heuristic, as discussed in Section 6. The reason the algorithm works so well is that in our lattice, short vectors produced by LLL appear to behave as independent vectors.

Remark 2. The reader may be wondering why we construct the lattice L using x-shifts and y-shifts of f, but do not explicitly use mixed shifts of the form $x^i y^j f^k$. The reason is that all mixed shifts of f over the monomials used in L are already included in the lattice. That is, any polynomial $x^i y^j f^k e^{m-k}$ can be expressed as an integer linear combination of x-shifts and y-shifts. To see this, observe that for any i, j, we have

$$x^i y^j = \sum_{u=0}^{i} \sum_{v=0}^{u} b_{u,v} x^{u-v} f^v + \sum_{u=1}^{j-i} \sum_{v=0}^{i} c_{u,v} y^u f^v$$

for some integer constants $b_{u,v}$ and $c_{u,v}$. Note that when $j \leq i$ the second summation is vacuous and hence zero. It now follows that

$$x^i y^j f^k e^{m-k} = \sum_{u=0}^{i} \sum_{v=0}^{u} b_{u,v} e^v x^{u-v} f^{v+k} e^{m-v-k} + \sum_{u=1}^{j-i} \sum_{v=0}^{i} c_{u,v} e^v y^u f^{v+k} e^{m-v-k}$$

$$= \sum_{u=0}^{i} \sum_{v=0}^{u} b_{u,v} e^v \cdot g_{u-v,v+k} + \sum_{u=1}^{j-i} \sum_{v=0}^{i} c_{u,v} e^v \cdot h_{u,v+k}$$

Consequently, $x^i y^j f^k e^{m-k}$ is already included in the lattice.

4.1   Improved Determinant Bounds

The results of the last section show that the small inverse problem can be solved when $\delta < 0.285$. The bound is derived from the determinant of the lattice L. It turns out that the lattice L contains a sublattice with a smaller determinant. Working in this sublattice leads to improved results. The idea is to remove some of the rows that enlarge the determinant. We throw away the y-shifts corresponding to low powers of f. Namely, for all r and $i \geq (1 - 2\delta)r$, the polynomials $y^i f^r$ are not included in the lattice. Since these "damaging" y-shifts are taken out, more y-shifts can be included. More precisely, the largest y-shift can now be taken to be $t = m(1 - 2\delta)$ as opposed to $t = \frac{m(1-2\delta)}{2}$ used in the previous section.

The lattice constructed using these ideas is no longer full rank. In particular, the basis vectors no longer form a triangular matrix. As a result, the determinant must be bounded by other means. Nevertheless, an improvement on the bound on the determinant can be established, leading to the result that the small inverse problem can be solved for $\delta < 1 - \frac{1}{2}\sqrt{2} \approx 0.292$. We provide the details in the full version of this paper.

5   Cryptanalysis of Arbitrary e

In his paper, Wiener suggests using large values of e when the exponent d is small. This can be done by adding multiples of $\phi(N)$ to e before making it known as the public key. When $e > N^{1.5}$, Wiener's attack will fail even when d is small. We show that our attack applies even when larger values of e are used.

As described in Section 2, we solve the small inverse problem:

$$k(A + s) \equiv 1 \pmod{e} \quad \text{where} \quad |k| < 2e^{1 + \frac{\delta - 1}{\alpha}} \quad \text{and} \quad |s| < 2e^{1/2\alpha},$$

for arbitrary values of α. We build the exact same lattice used in Section 4. Working through the calculations one sees that the determinant of the lattice in question is

$$\det{}_x(L) = e^{\frac{m^3}{3\alpha}(2\alpha+\delta-\frac{3}{4})+o(m^3)},$$
$$\det{}_y(L) = e^{\frac{tm^2}{2\alpha}(2\alpha+\delta-\frac{1}{2})+\frac{mt^2}{2}\frac{1}{2\alpha}+o(tm^2)}.$$

The dimension is as before. Therefore, to apply Fact 4 we must have

$$\frac{m^3}{3\alpha}(2\alpha+\delta-\frac{3}{4})+\frac{tm^2}{2\alpha}(2\alpha+\delta-\frac{1}{2})+\frac{mt^2}{2}\frac{1}{2\alpha} < \frac{m^3}{2}+tm^2,$$

which leads to

$$m^2(2\alpha+4\delta-3)-3tm(1-2\delta)+3t^2 < 0.$$

As before, the left hand side is minimized at $t_{\min} = \frac{1}{2}m(1-2\delta)$, which leads to

$$m^2[2\alpha+7\delta-\frac{15}{4}-3\delta^2] < 0,$$

and hence

$$\delta < \frac{7}{6}-\frac{1}{3}(1+6\alpha)^{1/2}.$$

Indeed, for $\alpha = 1$, we obtain the results of Section 4. The expression shows that when $\alpha < 1$ our attack becomes even stronger. For instance, if $e \approx N^{2/3}$ then RSA is insecure whenever $d < N^\delta$ for $\delta < \frac{7}{6}-\frac{\sqrt{5}}{3} \approx 0.422$. Note that if $e \approx N^{2/3}$ then d must satisfy $d > N^{1/3}$.

When $\alpha = \frac{15}{8}$ the bound implies that $\delta = 0$. Consequently, the attack becomes totally ineffective whenever $e > N^{1.875}$. This is an improvement over Wiener's bound, which become ineffective as soon as $e > N^{1.5}$.

6 Experiments

We ran some experiments to test our results when $d > N^{0.25}$. Our experiments were carried out using the LLL implementation available in Victor Shoup's NTL library. In all our experiments LLL produced two independent relations $g_1(x,y)$ and $g_2(x,y)$. In every case, the resultant $h(y) := \text{Res}(g_1(x,y), g_2(x,y), x)$ with respect to x was a polynomial of the form $h(y) = (y+p+q)h_1(y)$, with $h_1(y)$ irreducible over $\mathbb{Z}$ (similarly for x). Hence, the unique solution (x_0, y_0) was correctly determined in every trial executed. Below we show the parameters of some attacks executed.

n	δ	m	t	lattice dimension	running time
1000 bits	0.265	5	3	39	45 minutes
3000 bits	0.265	5	3	39	5 hours
10000 bits	0.255	3	1	14	2 hours

These tests were performed under Solaris running on a 400MHz Intel Pentium processor. In each of these tests, d was chosen uniformly at random in the range $\left[\frac{3}{4}N^\delta, N^\delta\right]$ (thus guaranteeing the condition $d > N^{0.25}$). The last row of the table is especially interesting as it is an example in which our attack breaks RSA with a d that is 50 bits longer than Wiener's bound.

7   Conclusions and Open Problems

Our results show that Wiener's bound on low private exponent RSA is not tight. In particular, we were able to improve the bound from $d < N^{0.25}$ to $d < N^{0.285}$. Using an improved analysis of the determinant, we can show $d < N^{0.292}$. Our results also improve Wiener's attack when large values of e are used. We showed that our attack becomes ineffective only once $e > N^{1.875}$. In contrast, Wiener's attack became ineffective as soon as $e > N^{1.5}$.

Unfortunately, we cannot state our attack as a theorem since we cannot prove that it always succeeds. However, experiments that we carried out demonstrate its effectiveness. We were not able to find a single example where the attack fails. This is similar to the situation with many factoring algorithms, where one cannot prove that they work; instead one gives strong heuristic arguments that explain their running time. In our case, the heuristic "assumption" we make is that the two shortest vectors in an LLL reduced basis give rise to algebraically independent polynomials. Our experiments confirm this assumption. We note that a similar assumption is used in the work of Bleichenbacher [1] and Jutla [5].

Our work raises two natural open problems. The first is to make our attack rigorous. More importantly, our work is an application of Coppersmith's techniques to bivariate modular polynomials. It is becoming increasingly important to rigorously prove that these techniques can be applied to bivariate polynomials.

The second open problem is to improve our bounds. A bound of $d < N^{1-\frac{1}{\sqrt{2}}}$ cannot be the final answer. It is too unnatural. We believe the correct bound in $d < N^{1/2}$. We hope our approach eventually will lead to a proof of this stronger bound.

References

1. D. Bleichenbacher, "On the security of the KMOV public key cryptosystem", Proc. of Crypto '97, pp. 235–248.
2. D. Coppersmith, "Small solutions to polynomial equations, and low exponent RSA vulnerabilities", J. of Cryptology, Vol. 10, pp. 233–260, 1997.
3. J. Hastad, "Solving simultaneous modular equations of low degree", SIAM Journal of Computing, vol. 17, pp 336–341, 1988.
4. N. Howgrave-Graham, "Finding small roots of univariate modular equations revisited", Proc. of Cryptography and Coding, LNCS 1355, Springer-Verlag, 1997, pp. 131–142.
5. C. Jutla, "On finding small solutions of modular multivariate polynomial equations", Proc. of Eurocrypt '98, pp. 158–170.

6. A. Lenstra, H. Lenstra, and L. Lovasz. Factoring polynomial with rational coefficients. *Mathematiche Annalen*, 261:515–534, 1982.

7. L. Lovasz, "An algorithmic theory of numbers, graphs and convexity", SIAM lecture series, Vol. 50, 1986.

8. R. Rivest, A. Shamir, L. Adleman, "A method for obtaining digital signatures and public-key cryptosystems", Communications of the ACM, vol. 21, pp. 120–126, 1978.

9. E. Verheul, H. van Tilborg, "Cryptanalysis of less short RSA secret exponents", Applicable Algebra in Engineering, Communication and Computing, Springer-Verlag, vol. 8, pp. 425–435, 1997.

10. M. Wiener, "Cryptanalysis of short RSA secret exponents", IEEE Transactions on Info. Th., Vol. 36, No. 3, 1990, pp. 553–558.

Cryptanalysis of Skipjack Reduced to 31 Rounds Using Impossible Differentials

Eli Biham[1], Alex Biryukov[2], and Adi Shamir[3]

[1] Computer Science Department, Technion – Israel Institute of Technology,
Haifa 32000, Israel,
biham@cs.technion.ac.il,
http://www.cs.technion.ac.il/~biham/
[2] Applied Mathematics Department, Technion – Israel Institute of Technology,
Haifa 32000, Israel,
albi@cs.technion.ac.il.
[3] Department of Applied Mathematics and Computer Science,
Weizmann Institute of Science, Rehovot 76100, Israel,
shamir@wisdom.weizmann.ac.il.

Abstract. In this paper we present a new cryptanalytic technique, based on *impossible differentials*, and use it to show that Skipjack reduced from 32 to 31 rounds can be broken by an attack which is faster than exhaustive search.

Key words: Skipjack, Cryptanalysis, Differential cryptanalysis, Impossible differentials.

1 Introduction

Differential cryptanalysis [6] traditionally considers characteristics or differentials with relatively high probabilities and uses them to distinguish the correct unknown keys from the wrong keys. When a correct key is used to decrypt the last few rounds of many pairs of ciphertexts, it is expected that the difference predicted by the differential appears frequently, while when a wrong key is used the difference occurs less frequently.

In this paper we describe a new variant of differential cryptanalysis in which a differential predicts that particular differences should not occur (i.e., that their probability is exactly zero), and thus the correct key can never decrypt a pair of ciphertexts to that difference. Therefore, if a pair is decrypted to this difference under some trial key, then certainly this trial key is not the correct key. This is a *sieving attack* which finds the correct keys by eliminating all the other keys which lead to contradictions.

We call the differentials with probability zero *Impossible differentials*, and this method of cryptanalysis *Cryptanalysis with impossible differentials*.

We should emphasize that the idea of using impossible events in cryptanalysis is not new. It is well known [7] that the British cryptanalysis of the German Enigma in world war II used several such ideas (for example, a plaintext letter

J. Stern (Ed.): EUROCRYPT'99, LNCS 1592, pp. 12–23, 1999.

could not be encrypted to itself, and thus an incorrectly guessed plaintext could be easily discarded). The first application of impossible events in differential cryptanalysis was mentioned in [6], where zero entries in the difference distribution tables were used to discard wrong pairs before the counting phase. A more recent cryptanalytic attack based on impossible events was described by Biham in 1995 in the cryptanalysis of Ladder-DES, a 4-round Feistel cipher using DES as the F function. This cryptanalysis was published in [3], and was based on the fact that collisions cannot be generated by a permutation. A similar technique was latter used by Knudsen in his description of DEAL [8], a six-round Feistel cipher using DES as the F function. Although the idea of using impossible events of this type was natural in the context of Feistel ciphers with only a few rounds and with permutations as the round function, there was no general methodology for combining impossible events with differential cryptanalytic techniques, and for generating impossible differentials with a large number of rounds.

In this paper we show that cryptanalysis with impossible differentials is very powerful against many ciphers with various structures. We describe an impossible differential of Skipjack [15] which ensures that for all keys there are no pairs of inputs with particular differences with the property that after 24 rounds of encryption the outputs have some other particular differences. This differential can be used to (1) attack Skipjack reduced to 31 rounds (i.e., Skipjack from which only the first or the last round is removed), slightly faster than exhaustive search (using 2^{34} chosen plaintexts and 2^{64} memory), (2) attack shorter variants efficiently, and (3) distinguish whether a black box applies a 24-round variant of Skipjack, or a random permutation. In a related paper [5] we describe the application of this type of cryptanalysis to IDEA [10] and to Khufu [12], which improves the best known attacks on these schemes.

For conventional cryptanalysis of Skipjack with smaller numbers of rounds we refer the reader to [4] and to [9].

The paper is organized as follows: The description of Skipjack is given in Section 2. The 24-round impossible differential of Skipjack is described in Section 3. In Section 4 we describe a simple variant of our attack against Skipjack reduced to 25 and to 26 rounds, and in Section 5 we describe our main attack applied against Skipjack reduced to 31 rounds. Finally, in Section 6 we discuss why the attack is not directly applicable to the full 32-round Skipjack, and summarize the paper. In the Appendix we describe an automated approach for finding impossible differentials.

2   Description of Skipjack

Skipjack is an iterated blockcipher with 32 rounds of two types, called Rule A and Rule B. Each round is described in the form of a linear feedback shift register with additional non linear keyed G permutation. Rule B is basically the inverse of Rule A with minor positioning differences. Skipjack applies eight rounds of Rule A, followed by eight rounds of Rule B, followed by another eight rounds of Rule A, followed by another eight rounds of Rule B. The original definitions of

Rule A	RuleB
$w_1^{k+1} = G^k(w_1^k) \oplus w_4^k \oplus counter^k$	$w_1^{k+1} = w_4^k$
$w_2^{k+1} = G^k(w_1^k)$	$w_2^{k+1} = G^k(w_1^k)$
$w_3^{k+1} = w_2^k$	$w_3^{k+1} = w_1^k \oplus w_2^k \oplus counter^k$
$w_4^{k+1} = w_3^k$	$w_4^{k+1} = w_3^k$

Fig. 1. Rule A and Rule B

Rule A and Rule B are given in Figure 1, where *counter* is the round number (in the range 1 to 32), and where G is a four-round Feistel permutation whose F function is defined as an 8x8-bit S box (called the F table), and each round of G is keyed by eight bits of the key. The key scheduling of Skipjack takes a 10-byte key, and uses four of them at a time to key each G permutation. The first four bytes are used to key the first G permutation, and each additional G permutation is keyed by the next four bytes cyclically, with a cycle of five rounds.

The description becomes simpler if we unroll the rounds, and keep the four elements in the shift register stationary. Figure 2 describes this representation of Skipjack (only the first 16 rounds out of 32 are listed; the next 16 rounds are identical except for the counter values). The unusual structure after round 8 (and after round 24) is the result of simplifying the two consecutive XOR operations at the boundary between Rule A and Rule B rounds.

3   A 24-Round Differential with Probability Zero

We concentrate on the 24 rounds of Skipjack starting from round 5 and ending at round 28 (i.e., without the first four rounds and the last four rounds). For the sake of clarity, we use the original round numbers of the full Skipjack, i.e., from 5 to 28, rather than from 1 to 24. Given any pair with difference only in the second word of the input of round 5, i.e., with a difference of the form $(0, a, 0, 0)$, the difference after round 28 cannot be of the form $(b, 0, 0, 0)$, for any non-zero a and b.

The reason that this differential has probability zero can be explained as a *miss in the middle* combination of two 12-round differentials with probability 1: As Wagner observed in [17], the second input word of round 5 does not affect the fourth word after round 16, and given an input difference $(0, a, 0, 0)$ the difference after 12 rounds is of the form $(c, d, e, 0)$ for some non-zero c, d, and e. On the other hand, we can predict the data after round 16 from the output difference of round 28, i.e., to consider the differentials in the backward direction. Similarly to the 12-round differential with probability 1, there is a backward 12-round differential with probability 1. It has the difference $(b, 0, 0, 0)$ after round 28, and it predicts that the data after round 16 must be of the form $(f, g, 0, h)$ for some non-zero f, g, and h. Combining these two differentials, we conclude

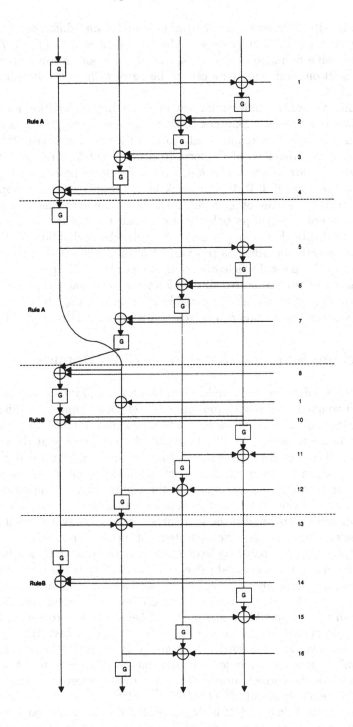

Fig. 2. Skipjack

that any pair with difference $(0, a, 0, 0)$ after round 4 and difference $(b, 0, 0, 0)$ after round 28 must have differences of the form $(c, d, e, 0) = (f, g, 0, h)$ after round 16 for some non-zero c, d, e, f, g, and h. As e and h are non-zero, we get a contradiction, and thus there cannot be pairs with such differences after rounds 4 and 28.

One application of this differential may be to distinguish whether an encryption black box is a 24-round Skipjack (from round 5 to round 28), or a random permutation. Identification requires only to feed the black box with $2^{48}\alpha$ pairs (for some α) with differences of the form $(0, a, 0, 0)$, and to verify whether the output differences are of the form $(b, 0, 0, 0)$. If for some pair the output difference is of the form $(b, 0, 0, 0)$, the black box certainly does not apply this variant of Skipjack. On the other hand, if the black box implements a random permutation, there is only a probability of $e^{-\alpha}$ that none of the $2^{48}\alpha$ pairs has a difference $(b, 0, 0, 0)$. For example, given 2^{52} pairs the probability of the black box to be incorrectly identified as this variant of Skipjack is only $e^{-16} \approx 10^{-7}$. These pairs can be packed efficiently using structures of 2^{16} plaintexts which form 2^{31} pairs. In these structures all the plaintexts are equal except for the second word which ranges over all the possible 2^{16} values. Using these structures, the same distinguishing results can be reached using only $2^{33}\alpha$ encryptions.

4   Attack on Skipjack Reduced to 25-26 Rounds

In this section we describe the simplest cryptanalysis of Skipjack variants, with only one or two additional rounds (on top of the 24-round impossible differential itself). An attack on a 25-round variant of Skipjack from round 5 to round 29 is as follows. Choose structures of 2^{16} plaintexts which differ only at their second word, having all the possible values in it. Such structures propose about 2^{31} pairs of plaintexts. Given 2^{22} such structures (2^{38} plaintexts), collect all those pairs which differ only at the first two words of the ciphertexts; by the structure of Skipjack, only these pairs may result from pairs with a difference $(b, 0, 0, 0)$ after round 28. On average only half of the structures propose such pairs, and thus only about 2^{21} pairs remain. Denote the ciphertexts of such a pair by (C_1, C_2, C_3, C_4) and (C_1^*, C_2^*, C_3, C_4). The pair may have a difference of the form $(b, 0, 0, 0)$ before the last round only if the decrypted values of C_1 and C_1^* by the G permutation in the last round have difference $C_2' = C_2 \oplus C_2^*$. As we know that such a difference is impossible, every key that proposes such a difference is a wrong key. For each pair we try all the 2^{32} possible values of the subkey of the last round, and verify whether the decrypted values by the last G permutation have the difference C_2' (this process can be done efficiently in about 2^{16} steps). It is expected that about 2^{16} values propose this difference, and thus we are guaranteed that these 2^{16} values are not the correct subkey of the last round. After analyzing the 2^{21} pairs, there remain only about $2^{32} \cdot (1 - 2^{-16})^{2^{21}} = 2^{32} \cdot e^{-32} \approx 2^{-14}$ wrong values of the subkey of the last round. It is thus expected that only one value remains, and this value must be the correct subkey. The time complexity of recovering this last 32-bit subkey is about $2^{17} \cdot 2^{21} = 2^{38}$ G permutation computations. Since each encryption consists of about 2^5 applications of G, this time complexity is equivalent to about 2^{33} encryptions. A straightforward implementation of the attack requires an array of 2^{32} bits to keep the information of the already identified wrong keys. A more efficient implementation requires only about 2^{32} G computations on average, which is about 2^{27} encryptions, and using 2^{16} bits of memory.

Essentially the same attack works against a 26-round variant from round 4 to round 29. In this variant, the same subkey is used in the first and last rounds. The attack is as follows: Choose 2^6 structures of 2^{32} plaintexts which differ only in the first two words and get all the 2^{32} values of these two words. Find the pairs which differ only in the first two words of the ciphertexts. It is expected that about $2^6 \cdot 2^{63}/2^{32} = 2^{37}$ pairs remain. Each of these pairs propose one wrong subkey value on average, and thus with a high probability after analysis of all the pairs only the correct first/last subkey remains. The time complexity of this attack when done efficiently is 2^{48}, using an array of 2^{16} bits. The rest of the key bits can be found by exhaustive search of 2^{48} keys, or by more efficient auxiliary techniques.

5 Cryptanalysis of Skipjack Reduced to 31 Rounds

For the cryptanalysis of Skipjack reduced to 31 rounds, we use again the 24-round impossible differential. We first analyze the variant consisting of the first 31 rounds of Skipjack, and then the variant consisting of the last 31 rounds of Skipjack.

Before we describe the full details of the attack, we wish to emphasize several delicate points. We observe that the full 80-bit key is used in the first four rounds (before the differential), and is also used in the last three rounds (after the differential). Therefore, the key-elimination process should discard 80-bit candidate keys. Assuming that the verification of each of the 2^{80} keys costs at least one G computation, and as one G computation is about 31 times faster than one encryption, we end up with an attack whose time complexity is at least $2^{80}/31 \approx 2^{75}$ encryptions. This lower bound is only marginally smaller than exhaustive search, and therefore the attack cannot spend more than a few G operations verifying each key, and cannot try each key more than a few times.

We next observe that if the impossible differential holds in some pair, then the third word of the plaintexts and the third and fourth words of the ciphertexts have zero differences, and the other words have non-zero differences. Given a pair with such differences, and assuming that the differential holds, we get three 16-bit restrictions in rounds 1, 4, and 29. Therefore, we expect that a fraction of 2^{-48} of the keys, i.e., about 2^{32} keys, encrypt the plaintext pair to the input difference of the differential after round 4, and decrypt the ciphertext pair to the output difference of the differential before round 29. Once verified, these keys are discarded. These 2^{32} keys must be discarded with complexity no higher than 2^{32} as we mentioned earlier. Thus, we cannot try all the 2^{80} keys for each pair, but rather, we devise an efficient algorithm to compute the 2^{32} keys.

The general structure of the attack is thus expected to be as follows: we generate a large structure of chosen plaintexts and select the pairs satisfying the required differences. We analyze these pairs, and each of them discards about 2^{32} keys. After the analysis of 2^{48} pairs, about 2^{80} (not necessarily distinct) keys are discarded. We expect that due to collisions, about $1/e$ of the keys remain undiscarded. The analysis of additional pairs decreases the number of undiscarded keys, until after about $2^{48} \ln 2^{80} \approx 2^{48} \cdot 2^6$ pairs only the correct key remains. However, the complexity of such an attack is higher than the complexity of exhaustive search.

Therefore, we analyze only 2^{49} pairs, leaving about $2^{80}/e^2 \approx 2^{77}$ keys undiscarded, and then try the remaining keys exhaustively. We emphasize that the analysis discards keys which cause partial encryption and decryption of a valid pair to match the form of the impossible differential. We thus assume in the attack that the differences proposed by the impossible differential do hold, and discard all keys which confirm this false assumption.

We are now ready to describe the attack. We choose 2^{41} plaintexts whose third words are equal. Given the ciphertexts, we sort (or hash) them by their third and fourth words, and select pairs which collide at these words. It is expected that about $\frac{(2^{41})^2/2}{2^{32}} = 2^{49}$ pairs are selected.

Each selected pair is subjected to the following analysis, consisting of four phases. In the first phase we analyze the first round. We know the two inputs of the G permutation, and its output difference. This G permutation is keyed by 32 bits, and there are about 2^{16} of the possible subkeys that cause the expected difference. This can be done in 2^{16} steps, by guessing the first two bytes of the subkeys, and computing the other two bytes by differential cryptanalytic techniques. As the subkeys of the first and last rounds are the same, we can peel off the last round for each of the possible subkeys.

We then analyze round 4. We know the input and output differences of the G permutation in round 4. Due to the complementation properties [4][1] of the G permutation, we can assume that the inputs are fixed to some arbitrary pair of values, and find about 2^{16} candidate subkeys corresponding to these values. The complexity of this analysis is 2^{16}. We can then complete all the possible combinations of inputs and subkeys using the complementation properties. The analysis of round 29 is similar. We now observe that the same subkey is used in round 4 and in round 29. The possible subkeys of rounds 4 and 29 are kept efficiently by using the complementation property, and thus we cannot directly search for two equal subkey values. Instead, we observe that the XOR value of the first two subkey bytes with the other two subkey bytes is independent of complementation, and we use this XOR value as the common value which is used to join the two lists of subkeys of both rounds. By a proper complementation we get a list of about 2^{16} tuples of the subkey, the input of round 4 and the output of round 29. The complexity of this analysis is about 2^{16} steps. This list can still

---

[1] The G permutation of Skipjack has $2^{16} - 1$ complementation properties: Let $O = G_K(I)$, and let $d = (d_0, d_1)$ be any 16-bit value. Then $O \oplus d = G_{K \oplus (d_1, d_0, d_1, d_0)}(I \oplus d)$.

be subjected to the complementation property to get all the (about 2^{32}) possible combinations.

The third phase joins the two lists, into a list of about 2^{32} entries of the form $(cv_0, \ldots, cv_5, X_3, X_{30})$ where $cv_0, \ldots, cv_5$ are the six key bytes used in rounds 1, 4, and 29, X_3 is the feedback of the XOR operation in round 3 (i.e., the output of the third G permutation), X_{30} is the feedback in round 30 (i.e., the input of the 30'th G permutation, which is the same in both members of the pair if $cv_0, \ldots, cv_5$ are correct). For each of these values we can now encrypt the first half of round 2 (using cv_4 and cv_5) and decrypt the second half of round 3 (using X_3, cv_0, and cv_1). We can view the second half of round 2 and the first half of round 3 as one permutation, which we call G', which has an additional feedback (the third plaintext word) in its middle. We are left now with only two equalities involving $cv_6, \ldots, cv_9$ which should hold, as we know the input and output of round 30, and we know the two outputs of G'. There is only one solution of $cv_6, \ldots, cv_9$ on average, and given the solution we find a key which encrypts the plaintexts to the input difference of the impossible differential after round 4, and decrypts the ciphertexts to the impossible difference before round 29. Therefore, we find a key which is certainly wrong, and thus should be discarded.

In total we find about 2^{32} such keys during the analysis of each pair. By analyzing 2^{49} pairs selected from the 2^{41} chosen plaintexts, we find a total of $2^{49} \cdot 2^{32} = 2^{81}$ keys, but some of them are found more than once. It is expected that a fraction of $(1-2^{-80})^{2^{81}} = 1/e^2 \approx 1/8$ of the keys are not discarded. These keys are then tested by trial encryptions in the fourth phase.

To complete the description of the attack we should describe two delicate implementation details: The first detail describes how to find the subkey $cv_6, \ldots, cv_9$ using one table lookup. The inputs and outputs of G and G' consist of 80 bits, and for each choice of the 80-bit query there is on average only one solution for the subkey. Therefore, we could keep a table of 2^{80} entries, each storing the solution(s) for a specific query. But the size of this table and the time of its precomputation are larger than the complexities we can afford. Instead, we observe that the complementation property of the G permutation [4] enables us to fix one of the input words (say to zero) by XORing the other input, the two outputs, and the proposed subkeys (excluding the intermediate feedback of G') by the original value of this input. We can, therefore, reduce the size of the table to 2^{64}, and the precomputation time to 2^{64} as well. Each entry of the table contains on average one 32-bit subkey. The size of the table can be halved by keeping only the first 16 bits of the subkey, observing that the second half can then be easily computed given the first half.

The second delicate implementation detail is related to the way we keep the list of discarded keys. The simplest way is to keep the list in a table of 2^{80} binary entries whose values are initialized to 0, and are set to 1 when the corresponding keys are discarded. But again, this table is too large (although its initialization and update times are still considerably faster than the rest of the attack). Instead, we observe that we can perform the attack iteratively (while caching the results of phase 2), where in each iteration we analyze only the keys

Rounds	Chosen Plaintexts	Steps
25 (5–29)	2^{38}	2^{27}
26 (4–29)	2^{38}	2^{49}
28 (1–28)	2^{34}	2^{77}
29 (1–29)	2^{34}	2^{77}
30 (1–30)	2^{34}	2^{77}
31 (1–31)	2^{41}	2^{78}
31 (2–32)	2^{34}	2^{78}

Table 1. Complexities of Chosen Plaintext Attacks Against Reduced-Round Skipjack

whose first two bytes cv_0 and cv_1 are fixed to the index of the iteration. This modification can be performed easily as the attack guesses these two bytes in its first phase, and each guess leads to independent computations. We thus perform exactly the same attack with a different order of instructions. As the first 16 bits of the keys are now fixed in each iteration, the number of required entries in the table is reduced to 2^{64}.

The complexities of phases 1 and 2 are about 2^{16} for each pair, and $2^{49} \cdot 2^{16} = 2^{65}$ in total for all the pairs. The complexity of phase 3 is as follows: For each pair, and for each value in the joined list, we compute two halves of a G permutation and solve for $cv_6, \ldots, cv_9$ given the inputs and outputs of the third G and of G'. Assuming that this solution costs about one computation of a G permutation, the total complexity of phase 3 is $2^{49} \cdot 2^{32}(2 \cdot \frac{1}{2} + 1) = 2^{82}$ computations of a G permutation, which is equivalent to $2^{82}/31 \approx 2^{77}$ encryptions. The complexity of phase 4 is about $2^{80}/8 = 2^{77}$ encryption. Therefore, the total complexity of the attack is about 2^{78} encryptions, which is four times faster than exhaustive search. The average time complexity of the attack is about 2^{77}, which is also four times faster than the average case of exhaustive search.

An attack on the reduced variant consisting of rounds 2 to 32 requires fewer chosen plaintexts, and the same complexity. Given four structures of 2^{32} chosen plaintexts with words 3 and 4 fixed, we can select the $\frac{4 \cdot (2^{32})^2/2}{2^{16}} = 2^{49}$ required pairs, and apply the same attack to these pairs (exchanging rounds 1 and 32, rounds 2 and 31, etc.). This attack can also be applied as a chosen ciphertext attack against the variant consisting of rounds 1 to 31 using 2^{34} chosen ciphertext blocks.

6   Discussion and Conclusions

The best complexities of our attack when applied to reduced-round variants of Skipjack are summarized in Table 1.

This attack cannot be directly used against the full 32 rounds of Skipjack because each pair may discard only about 2^{16} keys. However, the analysis of phases 1 and 2 (which in the case of the full Skipjack also includes the analysis of the last round) cannot be reduced below 2^{32} G computations. Therefore, the complexity of the attack is lower bounded by $2^{16}/32 = 2^{11}$ times the number of discarded keys (instead of being a few times smaller than the number of discarded keys), and thus the time required to eliminate all but the correct key is longer than exhaustive search.

Note that the above attacks against Skipjack are independent of the choice of the G permutation or the F table. Also note that if in addition to the 5-round cycle of the key schedule, Skipjack had 5-round groups of rules (instead of 8-round groups of rules), i.e., had consecutive groups of five rounds of Rule A followed by five rounds of Rule B, followed by five Rule A and five Rule B rounds, etc, then it would have a 27-round impossible differential.

We are aware of several impossible differentials of various blockciphers, such as a 9-round impossible differential of Feal [16,13], 7-round impossible differential of DES [14], 20-round impossible differential of CAST-256 [1], 18-round impossible differential of Khufu [12], and 2.5-round impossible differential of IDEA [10]. In a related paper [5] we use these impossible differentials to cryptanalyze IDEA with up to 4.5 rounds, and to cryptanalyze Khufu with up to 20 rounds. Both attacks analyze more rounds than any other published attack against these ciphers.

There are many modifications and extensions of the ideas presented in this paper. For example, cryptanalysis with impossible differentials can be used with low-probability (rather than zero-probability) differentials, can be used with conditional characteristics [2] (or differentials), and can be combined with linear [11] (rather than differential) cryptanalysis.

Designers of new blockciphers try to show that their schemes are resistant to differential cryptanalysis by providing an upper bound on the probability of characteristics and differentials in their schemes. One of the interesting consequences of the new attack is that even a rigorously proven upper bound of this type is insufficient, and that designers also have to consider lower bounds in order to prove resistance against attacks based on impossible or low-probability differential properties.

A *Shrinking*: An Automated Technique for Finding Global Impossible Differentials

In Section 3 we used the miss in the middle approach to find the 24-round impossible differential of Skipjack. In this appendix we describe an automated approach for finding all the impossible differentials which are based on the global structure of the cipher. The simplest way to automate the search is to encrypt many pairs of plaintexts under various keys, and to conclude that every differential proposed by the encrypted plaintexts (i.e., any differential formed by a plaintext difference and the corresponding ciphertext difference) is not an impossible differential. Therefore, by elimination, only differentials that never occur in our trials may be impossible.

The main problem is that the space of differentials is too large. The problem can be greatly simplified when considering wordwise truncated differentials whose differences distinguish only between zero and arbitrary non-zero differences in the various words (e.g., Skipjack divides the blocks into four words, and thus there are only 16 possible truncated plaintext differences, and 16 possible truncated ciphertext differences, yielding 256 truncated differentials). By selecting various plaintext pairs and computing the ciphertext differences, we can easily discard most differentials which are not impossible. However, when long blocks are divided into many small words, we may never encounter an input pair whose outputs are almost identical, except for a single word.

To overcome this problem we analyze scaled down variants of the cipher, which preserve its global structure but change its local details (including the size of words and the definition of the various functions and permutations). In many cases, including the impossible differential used against Skipjack in this paper, the particular implementation of the G permutation, the F table, and the key schedule do not affect the impossible differentials. In such cases, we can replace the local operations in the cipher by other operations, maintaining the global structure. Moreover, we can also reduce the word size to a smaller word size, together with reducing the size of the local operations without affecting the impossible differentials. We therefore replace the word size by a few bits (typically three, since any invertible function with fewer bits is affine), and replace the large functions by appropriate smaller functions.[2] Impossible differentials resulting from the global structure of the cipher remain impossible even in the scaled down variant. As the block size of the new variant is small (e.g., 12 bits in the case of Skipjack), we can easily encrypt all the 2^{12} plaintexts and calculate all their differences (by exhaustive computation of all the 2^{23} pairs of plaintexts and ciphertexts). By repeating this process for several random independent choices of the local functions, and taking the intersection of the resulting impossible differentials, we can get with high probability all the impossible differentials which are a consequence of the global structure of the cipher.[3] We call this technique *shrinking*.

Using this approach we searched for the wordwise truncated impossible differentials of Skipjack with various numbers of rounds. We found a large number of impossible differentials with fewer than 24 rounds (some of them with more than one non-zero word difference in the plaintext or the ciphertext), and confirmed that the longest impossible differential based on the global structure of Skipjack has 24 rounds. The most notable shorter impossible differentials of Skipjack are (1) the two 23-round impossible differentials (rounds 5–27) which are $(0, a, 0, 0) \not\rightarrow (b, 0, 0, 0)$ and $(0, a, 0, 0) \not\rightarrow (0, b, 0, 0)$ (where a and b are non-zero), and (2) the two 22-round impossible differentials (rounds 5–26) which are

---

[2] The new functions should preserve the main character of the original functions. For example, large permutations should be replaced by smaller permutations, linear functions by smaller linear functions, etc.

[3] This technique can also find wordwise truncated differentials with probability 1 which are based on the global structure of the cipher.

$(0, a, 0, 0) \not\rightarrow (0, b, 0, 0)$, and the more useful $(0, a, 0, 0) \not\rightarrow (x, 0, y, 0)$, where x and y can have any value.

References

1. Carlisle M. Adams, *The CAST-256 Encryption Algorithm*, AES submission, available at http://www.entrust.com/resources/pdf/cast-256.pdf.
2. Ishai Ben-Aroya, Eli Biham, *Differential Cryptanalysis of Lucifer*, Lecture Notes in Computer Science, Advances in Cryptology, proceedings of CRYPTO'93, pp. 187–199, 1993.
3. Eli Biham, *Cryptanalysis of Ladder-DES*, proceedings of Fast Software Encryption, Haifa, Lecture Notes in Computer Science, pp. 134–138, 1997.
4. Eli Biham, Alex Biryukov, Orr Dunkelman, Eran Richardson, Adi Shamir, *Initial Observations on Skipjack: Cryptanalysis of Skipjack-3XOR*, proceedings of Selected Areas in Cryptography, SAC'1998, Lecture Notes in Computer Science, 1998.
5. Eli Biham, Alex Biryukov, Adi Shamir, *Miss in the Middle Attacks on IDEA, Khufu and Khafre*, proceedings of Fast Software Encryption, Rome, Lecture Notes in Computer Science, 1999.
6. Eli Biham, Adi Shamir, *Differential Cryptanalysis of the Data Encryption Standard*, Springer-Verlag, 1993.
7. Cipher A. Deavours, Louis Kruh, *Machine Cryptography and Modern Cryptanalysis*, Artech House, 1985.
8. Lars Ramkilde Knudsen, *DEAL - A 128-bit Block Cipher*, AES submission, available at http://www.ii.uib.no/~larsr/papers/deal.ps, 1998.
9. Lars Ramkilde Knudsen, Matt Robshaw, David Wagner, private communication, 1998.
10. Xuejia Lai, James L. Massey, Sean Murphy, *Markov Ciphers and Differential Cryptanalysis*, Lecture Notes in Computer Science, Advances in Cryptology, proceedings of EUROCRYPT'91, pp. 17–38, 1991.
11. Mitsuru Matsui, *Linear Cryptanalysis Method for DES Cipher*, Lecture Notes in Computer Science, Advances in Cryptology, proceedings of EUROCRYPT'93, pp. 386–397, 1993.
12. Ralph C. Merkle, *Fast Software Encryption Functions*, Lecture Notes in Computer Science, Advances in Cryptology, proceedings of CRYPTO'90, pp. 476–501, 1990.
13. Shoji Miyaguchi, Akira Shiraishi, Akihiro Shimizu, *Fast Data Encryption Algorithm FEAL-8*, Review of electrical communications laboratories, Vol. 36, No. 4, pp. 433–437, 1988.
14. National Bureau of Standards, *Data Encryption Standard*, U.S. Department of Commerce, FIPS pub. 46, January 1977.
15. *Skipjack and KEA Algorithm Specifications*, Version 2.0, 29 May 1998. Available at the National Institute of Standards and Technology's web page, http://csrc.nist.gov/encryption/skipjack-kea.htm.
16. Akihiro Shimizu, Shoji Miyaguchi, *Fast Data Encryption Algorithm FEAL*, Lecture Notes in Computer Science, Advances in Cryptology, proceedings of EUROCRYPT'87, pp. 267–278, 1987.
17. David Wagner, *Further Attacks on 16 Rounds of SkipJack*, private communication, July 1998.

Software Performance of Universal Hash Functions

Wim Nevelsteen and Bart Preneel[*]

Katholieke Universiteit Leuven, Dept. Electrical Engineering-ESAT
Kardinaal Mercierlaan 94, B–3001 Heverlee, Belgium
wnevelst@eps.agfa.be, bart.preneel@esat.kuleuven.ac.be

Abstract. This paper compares the parameters sizes and software performance of several recent constructions for universal hash functions: bucket hashing, polynomial hashing, Toeplitz hashing, division hashing, evaluation hashing, and MMH hashing. An objective comparison between these widely varying approaches is achieved by defining constructions that offer a comparable security level. It is also demonstrated how the security of these constructions compares favorably to existing MAC algorithms, the security of which is less understood.

1 Introduction

In many commercial applications, protecting the integrity of information is even more important than protecting its secrecy. Digital signatures, introduced in 1976 by Diffie and Hellman [13], are the main tool for protecting the integrity of information. They are essential to build a worldwide trust infrastructure. However, there are still a significant number of applications for which digital signature are *not* cost-effective:

- For applications with *short* messages, the limitation is that signing and verifying is too demanding for processors in low-cost smart cards. On a more modern processor[1], the combined time of signing and verifying a digital signature using RSA, DSA or ECDSA typically exceeds 30 milliseconds.
- For applications with *long* messages (several Megabytes), the speed of signing is limited by the speed of present-day hash functions, which is about 100 Mbit/s.
- Finally, the overhead of a digital signature varies between 25 to 128 bytes, and the keys and system parameters require between 80 and a few hundred bytes of storage.

For the reasons indicated above, many applications use conventional MAC (Message Authentication Code) algorithms to provide data integrity and data origin authentication. MACs do not provide non-repudiation of origin, unlike

---

[*] F.W.O. postdoctoral researcher, sponsored by the Fund for Scientific Research – Flanders (Belgium).

[1] Throughout this paper performance numbers will be given for a 200 MHz Pentium.

J. Stern (Ed.): EUROCRYPT'99, LNCS 1592, pp. 24–41, 1999.

digital signatures, that can be used in a setting where the parties do not trust each other. Moreover, MACs rely on shared symmetric keys, which requires additional key management functions. Banks have been using MACs since the late seventies [36,37] for message authentication. Recent applications in which MACs have been introduced include electronic purses (such as Proton and Mondex) and credit/debit applications (e.g., the EMV specifications). MACs are also being deployed for securing the Internet (e.g., IP security). For all these applications MACs are preferred over digital signatures because they are two to three orders of magnitude faster, and MAC results are shorter (typically between 4 ... 16 bytes). On present day machines, software implementations of MACs can achieve speeds from 50 ... 250 Mbit/s, and MACs require very little resources on inexpensive 8-bit smart cards and on the currently deployed Point of Sale (POS) terminals. During the last five years, our understanding of MACs has improved considerably, through development of security proofs (Bellare et al. [3,5,6]) and new attacks (Knudsen [23] and Preneel and van Oorschot [30,31]).

An important disadvantage of both digital signatures and MAC algorithms is that their security is only computational. That implies that an opponent with sufficient computing power can in principle forge a message. A second problem is that shortcut attacks might exist, which means that forging a message can be much easier than expected. This problem can partially be solved by developing security proofs; such a proof can reduce the security of a MAC or a digital signature scheme to another primitive, such as a pseudo-random function or to a problem that is believed to be difficult, such as factoring the product of two large primes. However it seems wise to anticipate further progress in cryptanalysis of specific primitives. In the nineties we have witnessed the development of differential attacks [8], linear attacks [26], and of the use of optimization techniques as in [14]. The ultimate solution to this problem is unconditional security.

The idea of unconditionally secure authentication (and the so-called *authentication codes*) dates back to the early seventies, when Simmons was developing for Sandia National Laboratories a system for the verification of treaty compliance, such as the comprehensive nuclear test-ban treaty between the USA and the USSR [37]. The motivation for his research was that apparently the NSA refused to export strong conventional cryptographic mechanisms to the USSR. The first construction of authentication codes appeared in a 1974 paper by Gilbert et al. [18]. Subsequently their theory has been developed further by Simmons, analogous to Shannon's theory of secrecy systems [34]. An overview of the theory of authentication codes can be found in the work of Simmons [36] and Stinson [38]. In the seventies and the eighties, the research on authentication codes in the cryptographic community focussed mainly on the properties of authentication codes that meet certain bounds (such as perfect authentication codes, cf. §2.1). While this work illustrates that combinatorial mathematics and information theory provides powerful tools to develop an understanding of cryptographic primitives, it was widely believed that this work was of purely academic interest only.

This is the more surprising because Carter and Wegman developed already in the late seventies efficient authentication codes under the name of *strongly universal hash functions* [12,40]. They show that this is an interesting combinatorial tool that can be applied to other problems as well (such as interactive proof systems, pseudo-random number generation, and probabilistic algorithms). Carter and Wegman make the following key observations: i) long messages can be authenticated efficiently using short keys if the number of bits in the authentication tag is increased slightly compared to 'perfect' schemes; ii) if a message is hashed to a short authentication tag, weaker properties are sufficient for the first stage of the compression; iii) under certain conditions, the hash function can remain the same for many plaintexts, provided that the hash result is encrypted using a one-time pad. Mehlhorn and Vishkin propose more efficient constructions in [28]. At Crypto'82, Brassard pointed out that combining this primitive with a pseudo-random string generator will result in efficient computationally secure message authentication with short keys [11].

In the beginning of the nineties, the two 'independent' research threads are brought together. Stinson improves the work by Wegman and Carter, and establishes an explicit link between authentication codes and strongly universal hash functions [39]. A second important development is that Johansson, Kabatianskii, and Smeets establish a relation between authentication codes and codes correcting independent errors [22]. This provides a better understanding of the existing constructions and their limitations.

During the last five years, progress has been made both in theory and practice of universal hash functions. Krawczyk has proposed universal hash functions that are linear with respect to bitwise xor [24,25]. This property makes it easier to reuse the authentication code (with the same key): one encrypts the m-bit hash result for each new message using a one-time pad. This approach leads to simple and efficient constructions based on polynomials and Linear Feedback Shift Registers (LFSRs). Other constructions based on polynomials over finite fields are proposed and analyzed by Shoup [35]. Shoup [35] and Afanassiev et al. [1] study efficient software implementations of this primitive. Another line of research has been to improve the speed at the cost of an increased key size and size of the authentication tag. Rogaway has introduced bucket hashing in [33]; a slower variant with shorter keys was proposed by Johansson in [21]. Halevi and Krawczyk have developed an extremely fast scheme (MMH) which makes optimal used of the multiply and accumulate instruction of the Pentium MMX processor [19]. Recently Black et al. have further improved the performance on high end processors with the UMAC construction [9].

While it is clear that authentication codes (or universal hash functions) have a large potential for certain applications, they are not widely known to application developers. Some of the reasons might be that the research is too new, and that it is difficult to choose among the many schemes. For example, Halevi and Krawczyk write "An exact comparison is not possible since the data available on the most efficient implementations of other functions are based on different platforms" [19, p. 174]. The latter problem makes it more difficult to introduce them into standards. For the time being, there is also a lack of public domain implementations, that can demonstrate the benefits of this approach.

This paper intends to solve part of these problems by providing an objective comparison of performance and parameter sizes for the most promising constructions. For three related universal hash functions, similar work has been done by Shoup [35]. Atici and Stinson [2] provide an overview of the general parameters of several schemes, but do not discuss the performance.

The remainder of this paper is organized as follows. §2 introduces the most important definitions, and §3 presents the constructions that will be compared in this paper. The comparison of implementation speeds and memory requirements of the different schemes is presented in §4, and §5 contains some concluding remarks.

2   Definitions and Background

This section presents the model for authentication without secrecy. Next universal hash functions and strongly universal hash functions are introduced, and it is explained how they can be combined.

2.1   Authentication Codes

As usually in cryptography, the main players are the sender Alice, who wants to send some information to the receiver Bob; the opponent of Alice and Bob is the active eavesdropper Eve. Here, Alice and Bob are not concerned about the secrecy of the information. In order to detect the actions of Eve, Alice attaches to the plaintext an authentication tag that is a function of a shared secret key and of the plaintext. Bob recomputes the tag and accepts the plaintext as authentic if the tag is the same. As in the Vernam scheme, the secret key can be used only once.

Eve can perform three types of attacks: (i) Eve can create a new plaintext and send it to Bob, pretending that it came from Alice (*impersonation attack*); (ii) Eve can wait until she observes a plaintext and replace it by a different plaintext (*substitution attack*); (iii) Eve can choose freely between both strategies (*deception attack*). The probability of success (when the strategy of Eve is optimal) will be denoted with P_i, P_s, and P_d respectively. A first result that follows from Kerckhoffs' assumption (namely that the strategy to choose the key is known by Eve) is that $P_d = \max(P_i, P_s)$ [27].

In the following the length (in bits) of the plaintext, authentication tag, and key is denoted with m, n, and k respectively. The combinatorial bounds state that P_i and P_s are at least $1/2^n$. In the following we will consider only schemes for which $P_i = 1/2^n$. Another important bound is the square root bound; it states that $P_d \geq 1/2^{k/2}$. This is a corollary of the 'authentication channel capacity theorem' which states that an authentication code can only be secure if the authentication tag reveals a significant amount of information on the secret key (see Massey [27] for details).

Stinson proves that if $P_i = P_s = 1/2^m$, the number of plaintexts is at most a linear function of the number of keys [39]. This shows that schemes of this type require large keys for large messages, which makes them impractical. On the other hand, Kabatianskii et al. [22] showed that if P_s exceeds P_i by an arbitrarily small amount, the number of plaintexts grows exponentially with the number of keys. This research developed from exploring connections to the rich theory of error-correcting codes, and connects to the work of Wegman and Carter [12,40]. The disadvantage of $P_s > 1/2^n$ is that for a given security level (say, $P_d = 1/2^{64}$), slightly more than 64 bits are required for the authentication tag. While [22] shows efficient constructions that require only a single extra bit, in practice one can afford to send one or more extra bytes.

2.2   Universal Hash Functions

A universal hash function is a mapping from a finite set A with size a to a finite set B with size b. For a given hash function h and for a pair (x, x') with $x \neq x'$ the following function is defined: $\delta_h(x, x') = 1$ if $h(x) = h(x')$, and 0 otherwise. For a finite set of hash functions H (in the following this will be denoted with a *family* of hash functions), $\delta_H(x, x')$ is defined as $\sum_{h \in H} \delta_h(x, x')$, or $\delta_H(x, x')$ counts the number of functions in H for which x and x' collide. When a random choice of h is made, then for any two distinct inputs x and x', the probability that these two inputs yield a collision equals $\delta_H(x, x')/|H|$. For a universal hash function, the goal is to minimize this probability together with the size of H.

Definition 1. *Let ϵ be any positive real number. An ϵ-almost universal family (or ϵ-AU family) H of hash functions from a set A to a set B is a family of functions from A to B such that for any distinct elements $x, x' \in A$*

$$| \{h \in H : h(x) = h(x')\} | = \delta_H(x, x') \leq \epsilon \cdot |H| \ .$$

This definition states that for any two distinct inputs the probability for a collision is at most ϵ. In [12] the case $\epsilon = 1/b$ is called universal; the smallest possible value for ϵ is $(a - b)/(b(a - 1))$.

Definition 2. *Let ϵ be any positive real number. An ϵ-almost strongly universal family (or ϵ-ASU family) H of hash functions from a set A to a set B is a family of functions from A to B such that*

- *for every $x \in A$ and for every $y \in B$, $| \{h \in H : h(x) = y\} | = |H| /b$,*
- *for every $x_1, x_2 \in A$ ($x_1 \neq x_2$) and for every $y_1, y_2 \in B$ ($y_1 \neq y_2$),*
  $$| \{h \in H : h(x_1) = y_1, h(x_2) = y_2\} | \leq \epsilon \cdot |H| /b \ .$$

The first condition states that the probability that a given input x is mapped to a given output y equals $1/b$. The second condition implies that if x_1 is mapped to y_1, then the conditional probability that x_2 (different from x_1) is mapped to y_2 is upper bounded by ϵ. The lowest possible value for ϵ equals $1/b$ and this family has been called strongly universal functions in [40]. For this family the first condition in the definition follows from the second one [39].

If an Abelian group can be defined in the set B using the operation $\oplus$ (bitwise exclusive-or), Krawczyk defines the following variant [24] (the terminology is from [33]):

Definition 3. *Let ϵ be any positive real number. An ϵ-almost XOR universal family (or ϵ-AXU family) H of hash functions from a set A to a set B is a family of functions from A to B such that for any distinct elements $x, x' \in A$ and for any $b \in B$*

$$| \{h \in H : h(x) \oplus h(x') = b\} | \leq \epsilon \cdot |H| \ .$$

It follows directly from the definition that ϵ-ASU families of hash functions are equivalent to authentication codes with $P_i = 1/b$ and $P_s = \epsilon$ [39,40].

Theorem 4. *There exists an ϵ-ASU family H of hash functions from A to B iff there exists an authentication code with a plaintexts, b authenticators and $k = |H|$ keys, such that $P_i = 1/b$ and $P_s \leq \epsilon$.*

A similar result has been proved by Krawczyk for ϵ-AXU families [24,40].

Theorem 5. *There exists an ϵ-AXU family H of hash functions from A to B iff there exists an authentication code with a plaintexts, b authenticators and $k = |H| \cdot b$ keys, such that $P_i = 1/b$ and $P_s \leq \epsilon$.*

The construction consists of hashing an input using a hash function from H followed by encrypting the result by xoring a random element of B (which corresponds to a one-time pad).

Rogaway [33] and Shoup [35] show how the one-time pad can be replaced by a finite pseudo-random function (respectively permutation). In addition, they develop models for the use of counters and random tags. If the keys are generated using a finite pseudo-random function, the unconditional security is lost, but one has achieved a clear separation between compression (in a combinatorial way) and the final cryptographic step. This makes it easier to analyze and understand the resulting scheme.

2.3   Composition Constructions

The following propositions show how universal hash functions can be combined in different ways in order to increase their domains, reduce ϵ, or decrease the range. Several of these results were applied by Wegman and Carter [40].

Proposition 6 (Cartesian Product [39]). *If there exists an ϵ-AU family H of hash functions from A to B, then, for any integer $i \geq 1$, there exists an ϵ-AU family H^i of hash functions from A^i to B^i with $|H^i| = |H|$.*

Proposition 7 (Concatenation [33]). *If there exists an ϵ_1-AU family H_1 of hash functions from A to B and an ϵ_2-AU family H_2 of hash functions from A to C, then there exists an ϵ-AU family H of hash functions from A to $B \times C$, where $H = H_1 \times H_2$, $|H| = |H_1| \cdot |H_2|$, and $\epsilon = \epsilon_1 \epsilon_2$.*

Proposition 8 (Composition 1 [39]). *If there exists an ϵ_1-AU family H_1 of hash functions from A to B and an ϵ_2-AU family H_2 of hash functions from B to C, then there exists an ϵ-AU family H of hash functions from A to C, where $H = H_1 \times H_2$, $|H| = |H_1| \cdot |H_2|$, and $\epsilon = \epsilon_1 + \epsilon_2 - \epsilon_1\epsilon_2 \leq \epsilon_1 + \epsilon_2$.*

Proposition 9 (Composition 2 [39]). *If there exists an ϵ_1-AU family H_1 of hash functions from A to B and an ϵ_2-ASU family H_2 of hash functions from B to C, then there exists an ϵ-ASU family H of hash functions from A to C, where $H = H_1 \times H_2$, $|H| = |H_1| \cdot |H_2|$, and $\epsilon = \epsilon_1 + \epsilon_2 - \epsilon_1\epsilon_2 \leq \epsilon_1 + \epsilon_2$.*

Proposition 10 (Composition 3 [39]). *If there exists an ϵ_1-AU family H_1 of hash functions from A to B and an ϵ_2-AXU family H_2 of hash functions from B to C, then there exists an ϵ-AXU family H of hash functions from A to C, where $H = H_1 \times H_2$, $|H| = |H_1| \cdot |H_2|$, and $\epsilon = \epsilon_1 + \epsilon_2 - \epsilon_1\epsilon_2 \leq \epsilon_1 + \epsilon_2$.*

The most important results are Proposition 9 and Proposition 10, as they allow to use more efficient (in terms of key size and computation) ϵ-AU universal hash functions in the first stage of the compression.

3 Constructions

The schemes that are discussed here are: bucket hashing, bucket hashing with a short key, fast polynomial evaluation, Toeplitz hashing, evaluation hash function, the division hash function, and MMH.

3.1 Bucket Hashing

The first hashing technique we consider is bucket hashing, which is an ϵ-AU introduced by Rogaway [33]. Fix a *word size* $w \geq 1$. For $M \geq N$ the hash functions of the family $B[w, M, N]$ are defined as mappings from $A = \{0, 1\}^{wM}$ to $B = \{0, 1\}^{wN}$. Each $h \in B[w, M, N]$ is specified by a list of length M, each entry of which contains three integers in the interval $[0, N - 1]$. Denote this list by $h = h_1 \ldots h_M$, where $h_i = \{h_i^1, h_i^2, h_i^3, \}$. The hash family $B[w, M, N]$ is the set of all possible lists h subjected to the constraints that no two of the 3-element sets in the list are the same, i.e., $h_i \neq h_j, \forall i \neq j$.

For a given hash function $h = h_1 \ldots h_M$ and a given input $X = x_1 \ldots x_M$ the hash result $h(X)$ is computed as follows. First, for each $j \in \{1, \ldots, N\}$, initialize y_j to 0^w. Then for each $i \in \{1, \ldots, M\}$ and $k \in h_i$, replace y_k by $y_k \oplus x_i$. When this operation is completed, set $h(X) := y_1 \| y_2 \| \ldots \| y_N$.

The name bucket hashing is derived from the following interpretation of the computation. We start with N empty buckets y_1 through y_N. Each word of the input is thrown into three buckets; the ith word x_i is thrown in the buckets h_i^1, h_i^2, and h_i^3. Then, the xor of the content in each of the buckets is computed, and the hash function output is the concatenation of the final content of the buckets.

The bucket hash family is ϵ-AU with the collision probability given by a complicated expression in the number N of buckets (see Rogaway, [33, p. 35]). It is important to note that the number N of buckets increases very fast if ϵ decreases. For example, for $\epsilon = 2^{-28}$, $N = 100$ buckets are needed, but for $\epsilon = 2^{-34}$ already 197 buckets are needed.

Table 1 indicates the performance and parameter sizes for an input block of 4 Kbyte and a word length w equal to 32. The Assembly code is hand optimized, and makes optimal use of the two parallel pipes of the Pentium. Several alternatives have been compared, but it was decided not to use self-modifying code, as this poses problems in most applications. For some of these results, we have combined several bucket hash functions using the rules from §2.3 (details are omitted due to space constraints). The memory in the table below corresponds to the processed key and the hash result; the memory to store the input is not included.

Note that for this hash function only the speed was measured under DOS, while for the other schemes (that use a finite field arithmetic library), the speed was measured under Windows '95. Timing measurements under DOS tend to be a little better.

Table 1. Characteristics of bucket-hashing for a block of 4 Kbyte

ϵ	2^{-16}	2^{-32}	2^{-48}	2^{-64}
Parameters	$M = 256$	$M = 1024$	$M = 1024$	$M = 1024$
	$N = 24$	$N = 160$	$N = 62$	$N = 160$
Speed (Mbit/s)	543	341	147	138
Key (bits)	3521	22493	36582	44986
Hash result (bytes)	384	640	496	1280
Memory (bytes)	1152	3712	6640	7424

We conclude that bucket-hashing is a very fast technique, but it requires a long key and a large memory. The hash result becomes very large for small values of ϵ.

3.2   Bucket Hashing with Small Key Size

The bucket-hashing approach from §3.1 gives rise to ϵ-AU hash functions that are very fast to compute, at the cost of a very large key and a long hash result. To overcome these disadvantages Johansson proposed bucket hashing with small key size [21].

Let $N = 2^{s/L}$. Each hash function $h \in B'[w, M, N]$ is specified by a list of length M, where each entry contains L integers in the interval $[0, N - 1]$. Next L arrays are introduced, each containing N buckets. Each word from the input is thrown in one bucket of each array, based on the list that describes the hash function h. Next, each array is compressed to s/L words, using a fixed primitive element $\gamma \in \mathrm{GF}(2^{s/L})$. The hash result is equal to the concatenation of the L compressed arrays, each containing s/L words.

Table 2 indicates the performance and parameter sizes for an input block of 4 Kbyte and a word length w equal to 32. Again the Assembly code is hand optimized for the Pentium. It is not possible to use exactly the same values for ϵ as for bucket hashing, because the constraints on the parameters (for example, L has to divide s). For each value of ϵ, one has to determine the optimal value for L. Too large values of L imply that the input has to be thrown in too many buckets; too small values of L imply that N becomes too large.

Table 2. Characteristics of bucket-hashing with small key for a block of 4 Kbyte

ϵ	2^{-18}	2^{-32}	2^{-46}	2^{-62}
Parameters	$s = 28$	$s = 42$	$s = 224$	$s = 72$
	$L = 4$	$L = 6$	$L = 7$	$L = 12$
	$N = 128$	$N = 128$	$N = 256$	$N = 256$
Speed (Mbit/s)	128	93	75	58
Key (bits)	28	42	56	72
Hash result (bytes)	112	168	224	288
Memory (bytes)	11264	16896	25088	43008

We conclude that bucket-hashing with small key size results indeed in very small keys, at the cost of a factor 2 to 4 in performance (depending on the value of ϵ). However, the memory requirements are still large, and the hash results are a little shorter.

3.3 Hash Family Based on Fast Polynomial Evaluation

The next family of hash functions has been proposed by Bierbrauer et al. [7]; it is based on polynomial evaluation over a finite field. Let $q = 2^r$, $Q = 2^m = 2^{r+s}$, $n = 1 + 2^s$, and π be a linear mapping from GF(Q) onto GF(q), where $Q = q_0^m$, $q = q_0^r$, and q_0 a prime power. Let $f_a(x) = a_0 + a_1 x + \ldots + a_{n-1}x^{n-1}$, where $x, y, a_0, a_1, \ldots a_{n-1} \in$ GF(Q), $z \in$ GF(q) and

$$H = \{h_{x,y,z} : h_{x,y,z}(a) = h_{x,y,z}(a_0, a_1, \ldots, a_{n-1}) = \pi(y \cdot f_a(x)) + z\} \ .$$

It is shown in [7] that the hash family in the construction above is ϵ-ASU with $\epsilon \leq 2/2^r$. For $q_0 = 2$, the function is also ϵ-AXU (for other values, a different group operation has to be used for the difference). The main step in the hash function construction is the evaluation of a polynomial in some point determined by the key. Afanassiev, Gehrmann and Smeets [1] have developed a very fast construction to evaluate a polynomial $f_a(x)$ in an element α (the MinWal procedure). This procedure makes use of Minimal W-nomials. Before evaluating the polynomial $f_a(x)$ in α, $f_a(x)$ is first reduced modulo the minimal W-nomial $\tau_{\alpha,w}(x)$. The minimal W-nomial $\tau_{\alpha,w}(x)$ is a multiple of the minimal polynomial of α with the lowest degree and with less than W non-zero terms.

Table 3 indicates the performance and parameter sizes for an input blocks of 4, 64, and 256 Kbyte. Most of the code has been written in C++ (compiled with

Borland C++ 5.0). The most critical step, the reduction modulo the minimal W-nomial has been written in Assembly language. Calculations are performed in GF(2^{32}). The maximal input length for one instance is equal to 256 Kbyte; of course Proposition 6 (cf. §2.3) can be used for large inputs, and ϵ can be reduced using Proposition 7. We can show that for our software, the optimal value for $W = 5$. Finding a minimal 5-nomial requires about 40 seconds using sub-optimal code. Note that this operation has be done only for the set-up phase. If one adds the (pre-computed) 5-nomials to the key, one needs about 42 bits per additional 5-nomial.

Table 3. Characteristics of hashing based on fast polynomial evaluation

ϵ	2^{-15}	2^{-30}	2^{-45}	2^{-60}
Speed 4 Kbyte (Mbit/s)	9	5	3	2
Speed 64 Kbyte (Mbit/s)	104	56	34	25
Speed 256 Kbyte (Mbit/s)	207	87	50	38
Key (bits)	80	160	240	320
Hash result (bytes)	2	4	6	8
Memory (bytes)	30	60	90	120

For large inputs (256 Kbyte or more), the polynomial evaluation hash function is rather fast and the keys sizes are reasonable. The two main advantages are the very small memory requirements, both for the computation and for the storage of the hash result.

3.4   Hash Family Using Toeplitz Matrices

The next hash family is the Toeplitz construction proposed by Krawczyk [25]. Toeplitz matrices are matrices with constant values on the left-to-right diagonals. A Toeplitz matrix of dimension $n \times m$ can be used to hash messages of length m to hash results of length n by vector-matrix multiplication. The Toeplitz construction uses matrices generated by sequences of length $n + m - 1$ drawn from δ-biased distributions. δ-biased distributions, introduced by Naor and Naor [29], are a tool for replacing truly random sequences by more compact and easier to generate sequences. The lower δ, the more random the sequence is.

Krawczyk proves that the family of hash functions associated with a family of Toeplitz-matrices corresponding to sequences selected from a δ-biased distribution is ϵ-AXU with $\epsilon = 2^{-n} + \delta$ [25]. He proposes to use the LFSR construction due to Alon et al. to construct a δ-biased distribution. This construction associates with r random bits a δ-biased sequence of length l with $\delta = l/2^{r/2}$.

Table 4 indicates the performance and parameter sizes for an input block of 4 Kbyte and a word length w equal to 32. As pointed out by Krawczyk [25], this construction is more suited for hardware, and is not very fast in software. In this case, the compiled C++ code could not be improved manually. For this version of the code, the complete matrix has been stored to improve the performance.

Table 4. Characteristics of Toeplitz hashing for a block of 4 Kbyte

ϵ	2^{-16}	2^{-32}	2^{-48}	2^{-64}
Parameters	$n = 17$	$n = 33$	$n = 44$	$n = 65$
	$r = 68$	$r = 88$	$r = 120$	$r = 142$
Speed (Mbit/s)	65	33	21	16
Key (bits)	68	88	120	142
Hash result (bytes)	68	132	176	260
Memory (bytes)	2176	4224	5632	8320

3.5   Evaluation Hash Function

The evaluation hash function was proposed by Mehlhorn and Vishkin in 1984 [28]. It is one of the variants analyzed by Shoup in [35]. The input (of length $\leq tn$) is viewed as a polynomial $M(x)$ of degree $< t$ over GF(2^n). The key is a random element $\alpha \in$ GF(2^n), and the hash result is equal to $M(\alpha) \cdot \alpha \in GF(2^n)$. This family of hash functions is ϵ-AXU with $\epsilon = t/2^n$.

We have written an implementation for $n = 64$, where GF(2^{64}) was represented as GF(2)$[x]/f(x)$, with $f(x) = x^{64} + x^4 + x^3 + x + 1$. The evaluation of the polynomial is performed using Horner's rule, and with a precomputation of the mapping $\beta \mapsto \alpha \cdot \beta$ with $\beta \in$ GF(2^n). As in [35], two options have been considered, that provide a time-memory trade-off.

For this construction ϵ grows with the number of n-bit blocks in the input. The fastest method achieves a speed of approximately 240 Mbit/s in optimized Assembly language (122 Mbit/s in C++), and requires about 16 Kbyte of memory. The second method is about a factor of 7 slower (18 Mbit/s in C++), but requires only 2 Kbyte of memory. Shoup's implementation in C is a little slower than our Assembly version, but faster than our C++ code; the latter can probably be explained by better optimization in C versus C++, and maybe by the overhead of the operating system (Linux versus Windows '95).

3.6   Division Hash Function

The division hash function was proposed by Krawczyk [24], inspired by an earlier scheme by M.O. Rabin. It represents the input as a polynomial $M(x)$ of degree less than tn over GF(2). The hash key is a random irreducible polynomial $p(x)$ of degree n over GF(2). The hash result is $m(x) \cdot x^n \bmod p(x)$. Since the total number of irreducible polynomials of degree n is roughly equal to $2^n/n$, it follows that this family of hash functions is ϵ-AXU with $\epsilon = tn/2^n$.

Again, we have written an implementation for $n = 64$. The main step is the reduction, which can be optimized by using a precomputation of the mapping $g(x) \mapsto g(x) \cdot x^{64} \bmod p(x)$, with $\deg g(x) < 64$. Again, following [35], two options were considered, that provide a time-memory trade-off. For the key generation, see [35].

For this construction $\epsilon = t/2^{58}$, with t the number of 8-byte blocks in the input (for the same value of n, the security level is 6 bits smaller compared to the evaluation hash function). The slower implementation uses 2 Kbyte of memory and runs at 14 Mbit/s in C++. Our fastest implementation uses 8 Kbyte of memory and achieves a speed of approximately 115 Mbit/s in C++, which is still slower than the evaluation hash function (in contrast to the conclusions of Shoup [35]). Therefore it was decided not to write optimized Assembly language.

Shoup generalizes this construction to polynomials over $GF(2^k)$, where k divides n [35]. The main conclusion is that for this variant the key generation is faster, but the precomputation is a little slower. For $n = 64$, $\epsilon = t/2^{58}$ (for the same value of n, the security is 3 bits better than the simple division hash, but 3 bits worse than the evaluation hash), and the performance is identical to that of the division hash.

3.7   MMH Hashing

Halevi and Krawczyk propose MMH (Multilinear Modular Hashing) in [19]. This hash function consists of a (modified) inner product between message and key modulo a prime p (close to 2^w, with w the word length; below $w = 32$.) MMH is an ϵ-AXU, but with xor replaced by subtraction modulo p. The core hash function maps 32 32-bit message words and 32 32-bit key words to a 32-bit result. The key size is 1024 bits and $\epsilon = 1.5/2^{30}$. For larger messages, a tree construction can be used based on Proposition 6 and Proposition 10; the value of ϵ and the key length have to be multiplied by the height of the tree.

This algorithm is very fast on the Pentium Pro, which has a multiply and accumulate instruction (and on other machines with this feature). On a 32-bit machine, MMH requires only 2 instructions per byte for a 32-bit result. We have not (yet) implemented MMH, but include the impressive speed given in [19] for a 200 MHz Pentium Pro (optimized Assembly language): 1.2 Gbit/s for $\epsilon = 1.5/2^{30}$, and 500 Mbit/s for $\epsilon = 1.125/2^{59}$ (for large messages, if the data resides in cache). Note that this does not take into account the final addition of the key. The memory size of the implementation is not mentioned, but 1 Kbyte is probably sufficient.

The Pentium does not have this 'multimedia' instruction, and therefore the speed is reduced to about 350 Mbit/s for $\epsilon = 1.5/2^{30}$. However, one can use the floating point co-processor; this requires that one reduces the key words from 32 bits to 27 bits to avoid overflow. This results in about 500 Mbit/s for $\epsilon = 1.5/2^{25}$, and 260 Mbit/s for the double length variant with $\epsilon \approx 1.1/2^{49}$.

4   Comparing the Hash Functions

In §3, the properties of the different constructions have been listed. However, this information does not allow to compare the different schemes. As pointed out in §2.2, for message authentication, an ϵ-ASU or an ϵ-AXU combined with an encryption are required. For this purpose, Table 5 defines six algorithms that provide a comparable functionality. Note that all these functions are ϵ-AXU

Table 5. Six schemes for message authentication and a comparison of their performance ('+' denotes composition)

Scheme	Definition
A	bucket hash(AU) + evaluation hash (AXU)
B	bucket hash/short key (AU) + evaluation hash (AXU)
C	Toeplitz hash (AXU) + evaluation hash (AXU)
D	fast polynomial evaluation (AXU)
E	evaluation hash (AXU)
F	MMH (AXU) + evaluation hash (AXU)

Scheme	A	B	C	D	E	F
ϵ	2^{-32}	2^{-32}	2^{-32}	2^{-30}	2^{-49}	$1.1 \cdot 2^{-49}$
Speed (Mbit/s)	323	89	33	87	240	$250^\dagger$
Key (bits)	45,114	170	216	160	128	1243
Hash result (bytes)	8	8	8	4	8	8
Memory (Kbyte)	64	26	12	0.03	8	$8.5^\dagger$

† estimated

(some functions need a group operation other than exor such as scheme D with $q_0 \neq 2$).

The six algorithms from Table 5 are applied to an input of 256 Kbyte with as goal $\epsilon \approx 2^{-32}$. Note that it is not possible to compare these schemes with exactly the same parameters, because the value of ϵ for the best performance is typically related to the word size of the processor. Messages of 256 Kbyte offer a fair basis of comparison, because for shorter messages the performance varies more with the message size. By introducing an unambiguous padding rule, one can also process shorter inputs with the same code. The constructions can be extended easily to larger message lengths, either by extending the basic construction or by using trees. The full version of this paper will provide an extended comparison for different values of ϵ and input sizes.

All parameters are chosen to optimize for speed (rather than for memory), and the critical part of the code has been written in Assembly language. For schemes A, B, C, and F the input is divided into blocks and Proposition 6 of §2.3 is applied. This has the advantage that the description of the hash function fits in the cache memory. The second hashing step for these schemes uses the evaluation hash with $n = 64$. The results are summarized in Table 5.

Scheme A: the input is divided into 32 blocks of 8 Kbyte; each block is hashed using the same bucket hash function with $N = 160$, which results in an intermediate string of 20 480 bytes.

Scheme B: the input is divided into 64 blocks of 4 Kbyte; each block is hashed using the same bucket hash function with short key ($s = 42$, $L = 6$, $N = 128$), which results in an intermediate string of 10 752 bytes.

Scheme C: the input is divided into 64 blocks of 4 Kbyte; each block is hashed using a 33×1024 Toeplitz matrix, based on a δ-biased sequence of length 1056 generated using an 88-bit LFSR. The length of the intermediate string is 8 448 bytes.

Scheme D: the input is hashed twice using the polynomial evaluation hash function with $\epsilon = 2^{-15}$, resulting in a combined value of 2^{-30}; the value of $W = 5$. The performance is slightly key dependent; therefore an average over a number of keys has been computed.

Scheme E: this is simply the evaluation hash function with $t = 32\,768$. Note that the resulting value of ϵ is too small. However, choosing a smaller value of n that is not a multiple of 32 induces a performance penalty.

Scheme F: the input is divided into 2048 blocks of 128 bytes; each block is hashed twice using MMH. The length of the intermediate string is $16\,384$ bytes. It is not possible to obtain a value of ϵ closer to 2^{-32} in an efficient way.

Note that for bucket hashing and its variant the speed was measured under DOS, while for the other schemes (that use a finite field arithmetic library), the speed was measured under Windows '95. Timing measurements under DOS tend to be a little better.

The main conclusion is that scheme A, E and F are the fastest schemes. Scheme A offers the best performance for $\epsilon = 2^{-32}$. However, if the application needs a smaller value of ϵ ($\approx 2^{-49}$), scheme E and F are faster. Moreover, the key size and memory size for scheme A are large. If the key is generated using a pseudo-random function, or if the expanded key has to be decrypted before it can be used, this will introduce a performance penalty (for example, 13.7 msec if 3-DES is used, which runs at 13.8 Mbit/s and 0.43 msec for SEAL-3, which runs at 440 Mbit/s [10]). If memory requirements (both for the hash function and for the result) are an issue, scheme D is the best solution. It is about 4 times slower than scheme A, and requires less memory than scheme B. Note however that the other schemes can reduce the memory requirement (for the hash function) at the cost of a reduced speed. Scheme E and F offer a reasonable compromise between performance and memory requirements; scheme F needs a larger key and a powerful multiplier.

Recently Black et al. [9] have proposed UMAC, that uses a different type of inner product. UMAC is faster than MMH on processors with a fast multiplication (Pentium II, PowerPC 604). They report a performance of 3.4 Gbit/s on a 233 MHz Pentium II. The value of $\epsilon = 2^{-30}$, and the key size is about $32\,768$ bits (but slower versions with a shorter key are possible). It is also suggested to replace the encryption at the end by a pseudo-random function that takes a nonce as second input.

We provide a comparison with MAC algorithms based on [10]. The performance of HMAC [3] and MDx-MAC [30] depends on the underlying hash function (MDx-MAC is a few percent slower than HMAC). For MD5 [32], SHA-1 [17], RIPEMD-160, and RIPEMD-128 [15] the speeds are respectively 228 Mbit/s, 122 Mbit/s, 101 Mbit/s, and 173 Mbit/s (note however that the security of MD5 as a hash-function is questionable; this has no immediate impact to its use in HMAC and MDx-MAC, but it is prudent to plan for its replacement). For CBC-MAC [6,20], the performance corresponds approximately to that of the underlying block cipher. For DES [16] this is 37.5 Mbit/s; for other block ciphers, this varies between 20 and 100 Mbit/s. XOR-MAC [5] is about 25% slower.

5   Concluding Remarks

The main advantages of universal hash functions are that their security is unconditional, and that their speed is comparable to or better than that of currently used MAC algorithms. In addition, they are easy to implement and easy to parallelize. Finally, they are often incremental [4] (this means that after small updates to the input, the output can be recomputed quickly). If they are used with a pseudo-random string generator, the unconditional security is lost, but what remains is a scheme that is easy to understand (the only cryptographic requirement is concentrated in one primitive).

Applications where universal hash functions can be used are the protection of high speed telecommunication networks, video streams, and for the integrity protection of file systems.

Many banking systems currently use unique MAC keys per transaction: for each transaction a new MAC key is derived from a master key. Therefore it seems natural to replace the MAC algorithms by universal hash functions, but with the following caveats: for short messages, the performance advantage of universal hash functions is limited. Moreover, constructions based on universal hash functions often give away part of their key bits (as an example, an input consisting of zero bits is often mapped to a hash result of zero bits). This is not a problem for the authentication, because the hash result is encrypted using a one-time pad. The opponent cannot exploit this property to forge messages, but he can find easily the output bits of the pseudo-random string generator. Therefore, any cryptographic weakness in the pseudo-random string generator may compromise the master keys, and it would be advisable to invest some of the time gained by using a universal hash function in strengthening the pseudo-random string generator. The security of the MAC algorithms depends on a cryptographic assumption, and thus it might well be possible that one finds a way to forge messages. However, for none of the state-of-the art MAC algorithms, an attack is known that can recover one or more key bits by observing a single text-MAC pair. Therefore an opponent will not be able to learn the MAC keys, and mounting an attack on the pseudo-random string generator will probably be more difficult (note that there is no proof of this). In summary, universal hash functions solve in an elegant and very efficient way the authentication problem, but put a higher requirement on the pseudo-random string generator, while MAC algorithms divide the (conjectured) cryptographic strength between the MAC algorithm and the pseudo-random string generator.

Acknowledgments. We would like to thank Antoon Bosselaers, Hugo Krawczyk, and Shai Halevi for helpful discussions and the anonymous referees for their constructive comments.

References

1. V. Afanassiev, C. Gehrmann, B. Smeets, "Fast message authentication using efficient polynomial evaluation," *Fast Software Encryption, LNCS 1267*, E. Biham, Ed., Springer-Verlag, 1997, pp. 190–204.
2. M. Atici, D.R. Stinson, "Universal hashing and multiple authentication," *Proc. Crypto'96, LNCS 1109*, N. Koblitz, Ed., Springer-Verlag, 1996, pp. 16–30.
3. M. Bellare, R. Canetti, H. Krawczyk, "Keying hash functions for message authentication," *Proc. Crypto'96, LNCS 1109*, N. Koblitz, Ed., Springer-Verlag, 1996, pp. 1–15. Full version: http:// www.research.ibm.com/security/.
4. M. Bellare, O. Goldreich, S. Goldwasser, "Incremental cryptography: the case of hashing and signing," *Proc. Crypto'94, LNCS 839*, Y. Desmedt, Ed., Springer-Verlag, 1994, pp. 216–233.
5. M. Bellare, R. Guérin, P. Rogaway, "XOR MACs: new methods for message authentication using block ciphers," *Proc. Crypto'95, LNCS 963*, D. Coppersmith, Ed., Springer-Verlag, 1995, pp. 15–28.
6. M. Bellare, J. Kilian, P. Rogaway, "The security of cipher block chaining," *Proc. Crypto'94, LNCS 839*, Y. Desmedt, Ed., Springer-Verlag, 1994, pp. 341–358.
7. J. Bierbrauer, T. Johansson, G. Kabatianskii, B. Smeets, "On families of hash functions via geometric codes and concatenation," *Proc. Crypto'93, LNCS 773*, D. Stinson, Ed., Springer-Verlag, 1994, pp. 331–342.
8. E. Biham, A. Shamir, *"Differential Cryptanalysis of the Data Encryption Standard,"* Springer-Verlag, 1993.
9. J. Black, S. Halevi, H. Krawczyk, T. Krovetz, P. Rogaway, "UMAC: fast and secure message authentication," preprint, 1999.
10. A. Bosselaers, "Fast implementations on the Pentium," http://www.esat.kuleuven.ac.be/~bosselae/fast.html.
11. G. Brassard, "On computationally secure authentication tags requiring short secret shared keys," *Proc. Crypto'82*, D. Chaum, R.L. Rivest, and A.T. Sherman, Eds., Plenum Press, New York, 1983, pp. 79–86.
12. J.L. Carter, M.N. Wegman, "Universal classes of hash functions," *Journal of Computer and System Sciences*, Vol. 18, 1979, pp. 143–154.
13. W. Diffie, M.E. Hellman, "New directions in cryptography," *IEEE Trans. on Information Theory*, Vol. IT–22, No. 6, 1976, pp. 644–654.
14. H. Dobbertin, "RIPEMD with two-round compress function is not collisionfree," *Journal of Cryptology*, Vol. 10, No. 1, 1997, pp. 51–69.
15. H. Dobbertin, A. Bosselaers, B. Preneel, "RIPEMD-160: a strengthened version of RIPEMD," *Fast Software Encryption, LNCS 1039*, D. Gollmann, Ed., Springer-Verlag, 1996, pp. 71–82. See also http://www.esat.kuleuven.ac.be/~bosselae/ripemd160.
16. FIPS 46, *"Data Encryption Standard,"* Federal Information Processing Standard, National Bureau of Standards, U.S. Department of Commerce, Washington D.C., January 1977 (revised as FIPS 46-1:1988; FIPS 46-2:1993).
17. FIPS 180-1, *"Secure Hash Standard,"* Federal Information Processing Standard (FIPS), Publication 180-1, National Institute of Standards and Technology, US Department of Commerce, Washington D.C., April 17, 1995.
18. E. Gilbert, F. MacWilliams, N. Sloane, "Codes which detect deception," *Bell System Technical Journal*, Vol. 53, No. 3, 1974, pp. 405–424.
19. S. Halevi, H. Krawczyk, "MMH: Software message authentication in the Gbit/second rates," *Fast Software Encryption, LNCS 1267*, E. Biham, Ed., Springer-Verlag, 1997, pp. 172–189.

20. ISO/IEC 9797, *"Information technology – Data cryptographic techniques – Data integrity mechanisms using a cryptographic check function employing a block cipher algorithm,"* ISO/IEC, 1994.

21. T. Johansson, "Bucket hashing with a small key size," *Proc. Eurocrypt'97, LNCS 1233,* W. Fumy, Ed., Springer-Verlag, 1997, pp. 149–162.

22. G.A. Kabatianskii, T. Johansson, B. Smeets, "On the cardinality of systematic A-codes via error correcting codes," *IEEE Trans. on Information Theory,* Vol. IT–42, No. 2, 1996, pp. 566–578.

23. L. Knudsen, "Chosen-text attack on CBC-MAC," *Electronics Letters,* Vol. 33, No. 1, 1997, pp. 48–49.

24. H. Krawczyk, "LFSR-based hashing and authentication," *Proc. Crypto'94, LNCS 839,* Y. Desmedt, Ed., Springer-Verlag, 1994, pp. 129–139.

25. H. Krawczyk, "New hash functions for message authentication," *Proc. Eurocrypt'95, LNCS 921,* L.C. Guillou and J.-J. Quisquater, Eds., Springer-Verlag, 1995, pp. 301–310.

26. M. Matsui, "The first experimental cryptanalysis of the Data Encryption Standard," *Proc. Crypto'94, LNCS 839,* Y. Desmedt, Ed., Springer-Verlag, 1994, pp. 1–11.

27. J.L. Massey, "An introduction to contemporary cryptology," in *"Contemporary Cryptology: The Science of Information Integrity,"* G.J. Simmons, Ed., IEEE Press, 1991, pp. 3–39.

28. K. Mehlhorn, U. Vishkin, "Randomized and deterministic simulations of PRAMs by parallel machines with restricted granularity of parallel memories," *Acta Informatica,* Vol. 21, Fasc. 4, 1984, pp. 339–374.

29. J. Naor, M. Naor, "Small bias probability spaces: efficient construction and applications," *Siam Journal on Computing,* Vol. 22, No. 4, 1993, pp. 838–856.

30. B. Preneel, P.C. van Oorschot, "MDx-MAC and building fast MACs from hash functions," *Proc. Crypto'95, LNCS 963,* D. Coppersmith, Ed., Springer-Verlag, 1995, pp. 1–14.

31. B. Preneel, P.C. van Oorschot, "On the security of two MAC algorithms," *Proc. Eurocrypt'96, LNCS 1070,* U. Maurer, Ed., Springer-Verlag, 1996, pp. 19–32.

32. R.L. Rivest, "The MD5 message-digest algorithm," *Request for Comments (RFC) 1321,* Internet Activities Board, Internet Privacy Task Force, April 1992.

33. P. Rogaway, "Bucket hashing and its application to fast message authentication," *Proc. Crypto'95, LNCS 963,* D. Coppersmith, Ed., Springer-Verlag, 1995, pp. 29–42. Full version http://www.cs.ucdavis.edu/~rogaway/papers.

34. C.E. Shannon, "Communication theory of secrecy systems," *Bell System Technical Journal,* Vol. 28, 1949, pp. 656–715.

35. V. Shoup, "On fast and provably secure message authentication based on universal hashing, *Proc. Crypto'96, LNCS 1109,* N. Koblitz, Ed., Springer-Verlag, 1996, pp. 313–328.

36. G.J. Simmons, "A survey of information authentication," in *"Contemporary Cryptology: The Science of Information Integrity,"* G.J. Simmons, Ed., IEEE Press, 1991, pp. 381–419.

37. G.J. Simmons, "How to insure that data acquired to verify treat compliance are trustworthy," in *"Contemporary Cryptology: The Science of Information Integrity,"* G.J. Simmons, Ed., IEEE Press, 1991, pp. 615–630.

38. D.R. Stinson, "The combinatorics of authentication and secrecy codes," *Journal of Cryptology,* Vol. 2, No. 1, 1990, pp. 23–49.

39. D.R. Stinson, "Universal hashing and authentication codes," *Designs, Codes, and Cryptography*, Vol. 4, No. 4, 1994, pp. 369–380.
40. M.N. Wegman, J.L. Carter, "New hash functions and their use in authentication and set equality," *Journal of Computer and System Sciences*, Vol. 22, No. 3, 1981, pp. 265–279.

Lower Bounds for Oblivious Transfer Reductions

Yevgeniy Dodis[1] and Silvio Micali[2]

[1] Laboratory for Computer Science, Massachusetts Institute of Technology, USA
yevgen@theory.lcs.mit.edu
[2] Laboratory for Computer Science, Massachusetts Institute of Technology, USA
silvio@tiac.net

Abstract. We prove the first *general* and *non-trivial* lower bound for the number of times a 1-out-of-n Oblivious Transfer of strings of length ℓ should be invoked so as to obtain, by an information-theoretically secure reduction, a 1-out-of-N Oblivious Transfer of strings of length L. Our bound is tight in many significant cases.

We also prove the first non-trivial lower bound for the number of random bits needed to implement such a reduction whenever the receiver sends no messages to the sender. This bound is also tight in many significant cases.

1 Introduction

THE OBLIVIOUS TRANSFER. The *Oblivious Transfer* (OT) is a fundamental primitive in secure protocol design, which has been defined in many different ways and contexts (e.g. [17], [10], [9]) and has found enormously many applications (e.g. [2], [17], [9], [13], [7], [16], [1], [14], [11]).

The OT is a protocol typically involving two players, the *sender* and the *receiver*, and several parameters. In the most used form, the $\binom{N}{1}$-OT_2^L, the sender has N binary secrets of length L, and the receiver gets exactly one of these strings, the one he chooses, but no information about any other secret (even if he cheats), while the sender (even if she cheats) gets no information about the secret learned by the receiver.

Also important is the notion of a *weak* Oblivious Transfer, a relaxation of the traditional OT. The only difference in a weak $\binom{N}{1}$-OT_2^L is that a cheating receiver is allowed to obtain partial information about several secrets, but at most L bits of information overall.

REDUCTIONS BETWEEN DIFFERENT OTs. Protocol reductions facilitate protocol design because they enable one to take advantage of implementing cryptographically only a few, carefully chosen, primitives. Information-theoretic reductions are even more attractive, because they guarantee that the security of a complex construction *automatically coincides* with that of the chosen primitive, once the latter is implemented cryptographically.

J. Stern (Ed.): EUROCRYPT'99, LNCS 1592, pp. 42–55, 1999.
© Springer-Verlag Berlin Heidelberg 1999

But to be really useful, reductions must be efficient. In particular, because even the best cryptographic implementation of a chosen primitive may be expensive to run, it is crucial that reductions call such primitives as few times as possible.

Because of the importance of OT, numerous *reductions* from "more complex" to "simpler" OT appear in the literature (e.g. [5], [8], [3], [6]). Particular attention has been devoted to reducing $\binom{N}{1}$-OT_2^L to $\binom{n}{1}$-OT_2^ℓ, where $N \geq n$ and $L \geq \ell$, both in the weak and in the strong case. Typically, these reductions are information-theoretically secure if the simpler OT is assumed to be so secure.

An important class of OT reductions are the ones in which the receiver sends no messages to the sender. Such reductions are called *natural*, both because all known OT reductions are of this type (e.g. [5], [6], [3]), and because they immediately imply that the sender gets no information about the receiver's index.

So far, researchers have been focusing on improving the *upper bounds* of these reductions, that is, the number of times one calls $\binom{n}{1}$-OT_2^ℓ in order to construct $\binom{N}{1}$-OT_2^L. However, little is known about the corresponding *lower bounds*. Indeed,

What is the minimum number of times that the given $\binom{n}{1}$-OT_2^ℓ must be invoked so as to obtain the desired $\binom{N}{1}$-OT_2^L?

Lower bounds were previously addressed in the context of very *specific* reduction techniques, and for very *specific* OTs. For instance, in [5] simple lower bounds are derived for reductions of $\binom{2}{1}$-OT_2^L to $\binom{2}{1}$-OT_2^1 that use *zigzag* functions.

Another natural resource of a reduction of $\binom{N}{1}$-OT_2^L to $\binom{n}{1}$-OT_2^ℓ is the amount of *needed randomness*. That is, an OT protocol is necessary probabilistic, but

What is the minimum number of random bits needed in a information-theoretically secure reduction of $\binom{N}{1}$-OT_2^L to $\binom{n}{1}$-OT_2^ℓ?

To the best of our knowledge, no significant results have ever been obtained about this crucial aspect.

OUR RESULTS. In this paper we provide the first *general* lower bounds for such information-theoretic OT reductions, and prove that these bounds are *tight* in significant cases. Namely, we prove that

- In any information-theoretically secure reduction of (even weak!) $\binom{N}{1}$-OT_2^L to $\binom{n}{1}$-OT_2^ℓ, the latter protocol must be invoked at least $\frac{L}{\ell} \cdot \frac{N-1}{n-1}$ times.
- The lower bound is tight for weak $\binom{N}{1}$-OT_2^L.
- The lower bound is tight for ("strong") $\binom{N}{1}$-OT_2^L when $L = \ell$.

We also prove the first general lower bound for the amount of randomness needed in a natural OT reduction. Namely,

- In any natural reduction of (even weak) $\binom{N}{1}$-OT_2^L to $\binom{n}{1}$-OT_2^ℓ, the sender must flip at least $\frac{L(N-n)}{n-1}$ coins.

- The lower bound is tight for weak $\binom{N}{1}$-OT_2^L.
- The lower bound is tight for ("strong") $\binom{N}{1}$-OT_2^L when $L = \ell$.

We note that, in a natural reduction, the amount of randomness used by the sender necessarily coincides with the total amount of randomness needed by both parties.

We point out the interesting special case when $n = 2$ and $\ell = 1$, i.e. reducing $\binom{N}{1}$-OT_2^L to $\binom{2}{1}$-OT_2, the simplest possible 1-out-2 Oblivious Transfer. We obtain that we need at least $L(N-1)$ invocations of $\binom{2}{1}$-OT_2 and, for a natural OT reduction, at least $L(N-2)$ random bits.

LOWER BOUNDS VIA INFORMATION THEORY. No general lower bound for OT reduction would be provable without very precisely and generally defining what such a reduction is. Fortunately, one such definition was successfully given by Brassard, Crépeau, and Sántha [5] based on information theory, and in particular the notion of *mutual information*. This framework is very useful since it allows one to define precisely such intuitive (but hard to capture formally) notions as "learn at most k bits of information" or "learn no information other than ...".

We point out, however, that information theory is much more useful than merely defining the problem. Indeed, we shall demonstrate that its powerful machinery is essential in *solving* our problem, for example, in proving our $\frac{L}{\ell} \cdot \frac{N-1}{n-1}$ lower bound on the number of invocations. Only the trivial bound of $\frac{L}{\ell}$ appears to be provable without information theory. But getting the additional $\frac{N-1}{n-1}$ factor in the lower bound (which is essential when $L = \ell$) requires explicit or implicit use of information theory.

We believe and hope that information theory will prove useful for other types of lower bounds in protocol problems.

2   Preliminaries

2.1   Information Theory Background

Let X, Y, Z by random variables over domains $\mathcal{X}, \mathcal{Y}, \mathcal{Z}$. Let us denote by $P_X(x)$, $P_{X|Z}(x|z)$, $P_{X,Y}(x,y)$ the probability distribution of X, conditional probability distribution of X given Z, and joint distribution of X and Y respectively.

Definition 1.

- *The* entropy $\mathbf{H}(X) = -\sum_x P_X(x) \log_2 P_X(x)$.
  The entropy of a random variable X tells how many truly random bits one can extract from X, i.e. how much "uncertainty" is in X.
- *The* conditional entropy $\mathbf{H}(X|Z)$ *is the average over z of the entropy of the variable X_z distributed according to $P_{X|Z}(x|z)$ (denoted $\mathbf{H}(X|Z=z)$), i.e.*

$$\mathbf{H}(X|Z) = \sum_z P_Z(z) \mathbf{H}(X|Z=z) = -\sum_z P_Z(z) \sum_x P_{X|Z}(x|z) \log_2 P_{X|Z}(x|z)$$

$\mathbf{H}(X|Z)$ *measures how much uncertainty X still has when one knows Z.*

- *The* joint entropy *of X and Y is the entropy of the joint variable (X, Y),* i.e.

$$\mathbf{H}(X, Y) = -\sum_{x,y} P_{X,Y}(x, y) \log_2 P_{X,Y}(x, y)$$

- *The* mutual information *between X and Y is* $\mathbf{I}(X; Y) = \mathbf{H}(X) - \mathbf{H}(X|Y)$.
- *The* mutual information between X and Y given Z is $\mathbf{I}(X; Y|Z) = \mathbf{H}(X|Z) - \mathbf{H}(X|Y, Z)$.

  The mutual information between X and Y (given Z) tells how much common information is between X and Y (given Z), i.e. by how much the uncertainty of X (given Z) decreases after one learns Y.

The following easily verified lemma summarizes some of the properties we will need.

Lemma 2.

1. $\mathbf{H}(X, Y) = \mathbf{H}(X) + \mathbf{H}(Y|X) = \mathbf{H}(Y) + \mathbf{H}(X|Y)$.
2. $\mathbf{I}(X; Y) = \mathbf{I}(Y; X) = \mathbf{H}(Y) - \mathbf{H}(Y|X) = \mathbf{H}(X) - \mathbf{H}(X|Y) = \mathbf{H}(X) + \mathbf{H}(Y) - \mathbf{H}(X, Y)$.
3. $\mathbf{I}(X, Z; Y) = \mathbf{I}(X; Y) + \mathbf{I}(Z; Y|X)$.
4. $\mathbf{H}(X|Y) = 0$ *iff X is a deterministic function of Y.*
5. $\mathbf{H}(X|Y) \le \mathbf{H}(X)$ *with equality iff X and Y are independent.*
   (Thus, $\mathbf{I}(X; Y) \ge 0$ with equality iff X and Y are independent.)
6. $\mathbf{I}(X; Y) \le \mathbf{H}(X) \le \log_2 |\mathcal{X}|$.
7. $\mathbf{I}(X; Y) \le \mathbf{I}(X; Y|Z) + \mathbf{H}(Z)$.
8. $\mathbf{H}(U_n) = n$, *where U_n is the uniform distribution over n-bit strings.*

2.2 Information-Theoretically Secure OT Reductions

Assuming some familiarity with the notions of an interactive Turing machine (ITM) [12], let us semi-formally define (1) protocols with an ideal $\binom{n}{1}$-OT$_2^\ell$ and then (2) information-theoretically secure reduction of $\binom{N}{1}$-OT$_2^L$ to $\binom{n}{1}$-OT$_2^\ell$.

Despite the difference in presentation, the following definition is a *simplification* of that of [5]. (For instance, we simplify it by ignoring the additional condition of *awareness* that is not going to affect our lower bound in any way.)

PROTOCOLS WITH IDEAL $\binom{n}{1}$-OT$_2^\ell$. Let us denote by a *n-sender* a probabilistic ITM having n special registers, and by a *n-receiver* is probabilistic ITM having a single special register. Let A be a n-sender and B a n-receiver. We say that (A, B) is a *protocol with ideal* $\binom{n}{1}$-OT$_2^\ell$ if, letting a be a private input for A and b be a private input for B, the computation of (A, B) proceeds as that of pair of ITMs, except that it consists of three (rather than the usual two) types of rounds: sender-rounds, receiver-rounds and OT-rounds, where by convention the first round always is a sender-round and the last is a receiver-round. In a sender-round, only A is active, and it sends a message to B (that will become an input to B at the start of the next receiver-round). In a receiver-round, only B is active and, except for the last round, it sends a message to A (this message will become an input to A at the start of the next sender-round). In an OT round,

(1) A places for each $j \in [1, n]$ an ℓ-bit string σ_j in its jth special register, and
(2) B places an integer $i \in [1, n]$ in its special register, and
(3) σ_i will become a distinguished input to B at the start of the next receiver-round. A will obtain no information about i.

At the end of any execution of (A, B), B computes a distinguished string called B's *output*.

MESSAGES AND VIEWS. Let (A, B) be a protocol with ideal $\binom{n}{1}$-OT$_2^\ell$. Then, in an execution of (A, B), we refer to the messages that A sends in a sender-round as A's *ordinary messages*, and to the strings that A writes in its special registers in an OT-round as A's *potential OT messages*. For each OT-round, only one of the n potential messages will be received by B, and we shall refer to all such received messages as B's *actual OT messages*. Recalling that both A and B are probabilistic, in a random execution of (A, B) where the private input of A is a and the private input of B is b, let us denote by VIEW$_A[A(a), B(b)]$ the random variable consisting of

(1) a, (2) A's coin tosses, and (3) the ordinary messages received by A;

and let us denote by VIEW$_B[A(a), B(b)]$ the random variable consisting of

(1) b, (2) B's coin tosses, and (3) all messages (both the ordinary and the actual OT ones) received by B.

REDUCTION OF $\binom{N}{1}$-OT$_2^L$ TO $\binom{n}{1}$-OT$_2^\ell$. Denote by $\mathcal{W}$ the set of all N-long sequences of L-bit stings and, given $w \in \mathcal{W}$, let w_i be the ith string of w. Denote by W the random variable that selects an element of $\mathcal{W}$ with uniform probability; by I the random variable selecting an integer in $[1, N]$ with uniform probability; and let A be an n-sender and B be an n-receiver. We say that (A, B) is an *information-theoretically secure* reduction of $\binom{N}{1}$-OT$_2^L$ to $\binom{n}{1}$-OT$_2^\ell$ if the following three properties are satisfied:

(P1) (Correctness) $\forall w \in \mathcal{W}$ and $\forall i \in [1, N]$, and $\forall$ execution of (A, B) where A's private input is w and B's private input is i,

$$B\text{'s output is } w_i;$$

(P2) (Receiver Privacy) $\forall$ sender A' and $\forall$ string a',

$$\mathbf{I}(\text{VIEW}_{A'}[A'(a'), B(I)] \; ; \; I) = 0; \tag{1}$$

(P3) (Sender Privacy) $\forall$ receiver B' and string b', $\exists$ a random variable $\tilde{I} \in [1, N]$ *independent* of W s.t.

$$\mathbf{I}(W \; ; \; \text{VIEW}_{B'}[A(W), B'(b')] \mid W_{\tilde{I}}) = 0. \tag{2}$$

In the context of a reduction of $\binom{N}{1}$-OT$_2^L$ to $\binom{n}{1}$-OT$_2^\ell$, we shall sometimes say that we are given $\binom{n}{1}$-OT$_2^\ell$ as a black-box.

The Correctness Property states that when A and B are honest, B will always obtain the string he wants. The Receiver Privacy Property states that no malicious sender A' can learn any information about the index of the honest receiver B. Finally, the Sender Privacy Property states that a malicious receiver B' can learn information about *at most one* of N strings of the sender A. Moreover, the index $\tilde{I}$ of this single string cannot depend on W (e.g. we don't want B' to learn the first string in W that starts with 10). In other words, both A and B do not gain anything by not following the protocol.

REDUCTION OF WEAK $\binom{N}{1}$-OT$_2^L$ TO $\binom{n}{1}$-OT$_2^\ell$. We call (A, B) an *information-theoretically secure* reduction of weak $\binom{N}{1}$-OT$_2^L$ to $\binom{n}{1}$-OT$_2^\ell$ if all the properties of the reduction of $\binom{N}{1}$-OT$_2^L$ to $\binom{n}{1}$-OT$_2^\ell$ hold except (Sender Privacy) is relaxed to the following:

(P3') (Weak Sender Privacy) $\forall$ receiver B' and string b'

$$\mathbf{I}(W \; ; \; \text{VIEW}_{B'}[A(W), B'(b')]) \leq L. \tag{3}$$

This property says that we allow a malicious receiver B' to obtain partial information about possibly *several* strings, provided he learns *no more than L bits* of information overall. To emphasize the difference, we will sometimes refer to the (regular) reduction between $\binom{N}{1}$-OT$_2^L$ and $\binom{n}{1}$-OT$_2^\ell$ as reducing *strong* $\binom{N}{1}$-OT$_2^L$ to $\binom{n}{1}$-OT$_2^\ell$. To justify this terminology, we show

Lemma 3. *If (A, B) is a reduction of (strong) $\binom{N}{1}$-OT$_2^L$ to $\binom{n}{1}$-OT$_2^\ell$, then it is a reduction of weak $\binom{N}{1}$-OT$_2^L$ to $\binom{n}{1}$-OT$_2^\ell$.*

Proof. By Lemma 2 (equations 7 and 6),

$$\mathbf{I}(W; \text{VIEW}_{B'}[A(W), B'(b')]) \leq \mathbf{I}(W; \text{VIEW}_{B'}[A(W), B'(b')] \mid W_{\tilde{I}}) + \mathbf{H}(W_{\tilde{I}})$$
$$= \mathbf{H}(W_{\tilde{I}}) \leq |W_{\tilde{I}}| = L$$

3 Lower Bounds

To simplify our notation, we do not worry about "floors" and "ceilings" in the rest of the paper, assuming that $(N - 1)$ is divisible by $(n - 1)$ and that L is divisible by ℓ (handling the the general case presents no significant difficulties). We will also refer to the sender as Alice and to the receiver as Bob.

Let α be the number of OT-rounds (invocations of $\binom{n}{1}$-OT$_2^\ell$) needed to reduce (weak) $\binom{N}{1}$-OT$_2^L$ to $\binom{n}{1}$-OT$_2^\ell$. Since we concentrate on the worst possible number of OT-rounds, we can assume w.l.o.g. that α is a fixed number and that the sender and receiver always perform exactly α OT-steps. We start with a sharp lower bound on α.

3.1   Lower Bound on the Number of Invocations of $\binom{n}{1}$-OT_2^ℓ

Theorem 4. *Any information-theoretically secure reduction of* **weak**[1] $\binom{N}{1}$-OT_2^L *to* $\binom{n}{1}$-OT_2^ℓ *must have*

$$\alpha \geq \frac{L}{\ell} \cdot \frac{N-1}{n-1} \tag{4}$$

Proof. Let us first give the *informal* intuition behind the proof. We know by the (weak) sender privacy condition that Bob can learn at most L (out of total NL) bits of information about W. However, if in each of the OT rounds Bob was somehow able to obtain *all* n strings that Alice put as her local inputs to this OT round (rather than getting only one of them), Bob should be able to learn all (NL bits) of W. Indeed, if Bob could not cannot learn some W_i with certainty, Alice will know that Bob's index is not i (if it was i, honest Bob should be able to get W_i with probability 1 by the correctness property). But this would contradict the receiver privacy condition as Alice learns some information about Bob's index. Hence, $\alpha n\ell - n\ell = \alpha\ell(n-1)$ bits that Bob did *not* get from the OT rounds, "contain information" about the remaining at least $NL - L = L(N-1)$ bits of W that Bob did not learn. The bound follows.

Let us now turn this intuition into a formal proof. Let P, $P = (Alice, Bob)$, be an information-theoretically secure reduction of $\binom{N}{1}$-OT_2^L to $\binom{n}{1}$-OT_2^ℓ that uses α invocations to $\binom{n}{1}$-OT_2^ℓ. First, we need the following simple lemma.

Local Lemma: $\forall$ input $w = w_1, \ldots, w_N$, $\forall$ random tape R_A for Alice, $\forall$ distinct $i, i' \in [1, N]$ and $\forall$ random tape tape R_B' for Bob, there exists a tape R_B for Bob such that the sequence of messages, M, received by $Alice(w, R_A)$ from $Bob(i', R_B')$ coincides with the sequence of messages that $Alice(w, R_A)$ receives from $Bob(i, R_B)$.

Proof: Assume that R_B does not exist. Then, executing with $Bob(i', R_B')$, we get that $Alice(w, R_A)$ will determine for sure that Bob's index is not i. Thus, when Bob's index is i', with non-zero probability over Bob's random string, $Alice(w, R_A)$ would obtain information about Bob's index (that it is not i), contradicting the receiver privacy condition. ∎

To derive our lower bound for α, we define the following two notions: that of a special execution of P and that of a pseudo-execution of P.

SPECIAL EXECUTION. A *special execution of* P is an execution of P in which Alice's input is a sequence of N randomly selected strings of length L, Alice's tape consists of randomly and independently selected bits, Bob's index is 1, and Bob's tape is the all-zero string, $\mathbf{0}$. In other words, we fix the behavior of Bob by fixing his index and the random string. With respect to a special execution of P, define the following random variables:

- W — Alice's N L-bit strings, $W = W_1, \ldots, W_N$;
- R — Alice's random tape;

---

[1] Since we are proving a lower bound, it clearly applies to (strong) $\binom{N}{1}$-OT_2^L as well.

- M_s — the ordinary messages sent by sender Alice;
- M_r — the ordinary messages sent by receiver Bob;
- V — Alice's potential messages (an $\alpha n l$-bit string, that is, for each of the α invocations of $\binom{n}{1}$-OT$_2^\ell$, the n ℓ-bit strings that are Alice's local inputs in the invocation).
- V_r — the actual messages received by Bob in the OT-rounds, (an αl-bit string, that is, for each of the α invocations of $\binom{n}{1}$-OT$_2^\ell$, the ℓ-bit string that Bob received depending on his local index during that invocation).

PSEUDO-EXECUTION. Let $\bar{M}_s$ be a sequence of messages, let $\bar{V}$ be a sequence of α sequences of n strings of length l each, let $\bar{i}$ be an index in $[1, N]$, and let $\bar{R}_B$ be a bit-sequence. A *pseudo-execution of* P with inputs $\bar{M}_s$, $\bar{V}$, $\bar{i}$, and $\bar{R}_B$, denoted by $\bar{P}(\bar{M}_s, \bar{V}, \bar{i}, \bar{R}_B)$, is the process of running Bob with index $\bar{i}$ and coin tosses $\bar{R}_B$, letting the kth message from the sender be the kth string of $\bar{M}_s$, and by letting the sender's input to the jth invocation of $\binom{n}{1}$-OT$_2^\ell$ to be the jth n-tuple of l-bit strings in $\bar{V}$. In other words, we pretend to be Alice and see what Bob will do in this situation on some particular index and random string.

Our lower bound for α immediately follows from the following two claims.

Local Claim 1: $\mathbf{I}((V, M_s) ; W) = NL$.

Proof: By the definition of mutual information, we have

$$\mathbf{I}((V, M_s) ; W) = \mathbf{H}(W) - \mathbf{H}(W \mid (V, M_s)).$$

Because W is randomly selected, $\mathbf{H}(W) = NL$. Therefore, to establish our claim we must prove that $\mathbf{H}(W \mid (V, M_s)) = 0$. We do that by showing that W is computable from V and M_s by means of the following algorithm.

1. Run $\bar{P}(V, M_s, 1, \mathbf{0})$ and let M_r be the resulting "ordinary messages sent by Bob".
   (*Comment:* Bob's view and Bob's messages sent in this pseudo-execution are distributed exactly as in a special execution.)
2. For $i = 1 \ldots N$ compute W_i as follows:
   - Find a string R_i such that, when executing $\bar{P}(V, M_s, i, R_i)$, the sequence of messages sent by Bob equals M_r.
     (*Comment:* The *existence* of at least one such R_i follows from the Local Lemma with $i' = 1$, $R'_B = \mathbf{0}$, $w = W$ and $R_A = R$. Further notice that, because M_r, W and R totally determine Alice's behavior, the messages and "potential" messages that $Alice(W, R)$ sends to $Bob(1, \mathbf{0})$ and to $Bob(i, R_i)$ are exactly V and M_s in both cases. Hence, *any* R_i that produces M_r in the pseudo-execution $\bar{P}(V, M_s, i, R_i)$, implies that $Alice(W, R)$ would produce messages M_s and "potential" messages V when communicating with $Bob(i, R_i)$.)
   - Let W_i be Bob's output in $\bar{P}(V, M_s, i, R_i)$.
     (*Comment:* By the correctness property of our reduction, $Bob(i, R_i)$ would correctly output W_i when talking to $Alice(W, R)$. And as we noticed, $Alice(W, R)$ would produce M_s and V when communicating with

$Bob(i, R_i)$, so running pseudo-execution $\bar{P}(V, M_s, i, R_i)$ indeed makes Bob to produce the correct W_i).

$\blacksquare$

Local Claim 2: $\mathbf{I}((V, M_s) ; W) \leq L + \alpha\ell(n - 1)$.

Proof: By Lemma 2 (equation 3), we have

$$\mathbf{I}((V, M_s) ; W) = \mathbf{I}((V_r, M_s) ; W) + \mathbf{I}((V \backslash V_r) ; W \mid (V_r, M_s)).$$

Now, because P implements **weak** $\binom{N}{1}$-OT_2^L, and because (V_r, M_s) consists of Bob's view in a (special) execution of P, we have by (P3') that $\mathbf{I}((V_r, M_s) ; W) \leq L$. Also, by Lemma 2 (equations 5 and 6),

$$\mathbf{I}((V \backslash V_r) ; W \mid (V_r, M_s)) \leq |V \backslash V_r| = \alpha\ell(n - 1).$$

The claim follows. $\blacksquare$

By combining Local Claims 1 and 2, we have $NL \leq L + \alpha\ell(n - 1)$, from which the desired lower bound for α immediately follows.

3.2   Lower Bound on the Number of Random Bits

Let us now prove the lower bound on the number of random bits needed by the sender in a natural reduction.

Theorem 5. *In any informationally-theoretic natural reduction of* **weak**[2] $\binom{N}{1}$-OT_2^L *to* $\binom{n}{1}$-OT_2^ℓ *the sender must flip at least* $\frac{L(N-n)}{n-1}$ *random coins.*

Proof. Let P, $P = (Alice, Bob)$, be an information-theoretically secure natural reduction from weak $\binom{N}{1}$-OT_2^L to $\binom{n}{1}$-OT_2^ℓ. As before, let W be the random input of Alice, R be her random tape, M_s be her ordinary messages sent to Bob and V be her "potential" messages. We notice that since the reduction is natural, the distribution of V and M_s does not depend on Bob's index and his random string. Let V_j, $j = 1 \dots n$, be an α-tuple consisting of string number j taken from each of the α invocations of $\binom{n}{1}$-OT_2^ℓ. We see that V is the disjoint union of $V_1, \dots, V_n$.

As before, we proceed by expanding the mutual information between W and (V, M_s) in two different ways.

$$\mathbf{I}((V, M_s); W) = \mathbf{H}(W) - \mathbf{H}(W \mid (V, M_s)) = NL - 0 = NL \qquad (5)$$

Here we used the fact that W is determined from V and M_s. Indeed, since V and M_s do not depend on Bob's input or random string, Alice should make sure that honest Bob can retrieve any W_i with probability 1 (if his input is i).

On the other hand, it is a possible behavior for a (malicious) Bob to read string number j in all the OT-rounds, i.e. to obtain V_j. By the weak sender

---

[2] Again, same result applies to (strong) $\binom{N}{1}$-OT_2^L as well.

privacy condition, $I((V_j, M_s); W) \leq L$, and, therefore, for any $j \in [1, n]$ we have (using Lemma 2, equations 5 and 6)

$$\mathbf{I}((V, M_s); W) = \mathbf{I}((V_j, M_s); W) + \mathbf{I}(V \backslash V_j; W \mid (V_j, M_s)) \leq L + \mathbf{H}(V \backslash V_j \mid V_j)$$

Combining this with Equation (5), we get

$$\mathbf{H}(V \backslash V_j \mid V_j) \geq L(N - 1), \qquad \forall j \in [1, n] \tag{6}$$

Since V is a disjoint union of V_j's, we get from the above equation (for $j = n$) and Lemma 2 (equations 1 and 5) that $L(N-1) \leq \mathbf{H}(V \backslash V_n \mid V_n) \leq \sum_{j=1}^{n-1} \mathbf{H}(V_j \mid V_n)$. Hence, there is an index $j \in [1, n-1]$ s.t. $\mathbf{H}(V_j) \geq \mathbf{H}(V_j \mid V_n) \geq \frac{L(N-1)}{n-1}$. W.l.o.g. assume $j = 1$, i.e. $\mathbf{H}(V_1) \geq \frac{L(N-1)}{n-1}$. Since for a fixed W, the only randomness of V came from R, we have by Equation (6) and Lemma 2 (equation 1)

$$|R| \geq \mathbf{H}(V \mid W) = \mathbf{H}(V, W) - \mathbf{H}(W) = \mathbf{H}(V_1) + \mathbf{H}(V \backslash V_1 \mid V_1) - NL$$
$$\geq \frac{L(N-1)}{n-1} + L(N-1) - LN = \frac{L(N-n)}{n-1}$$

Here $\mathbf{H}(V, W) = \mathbf{H}(V)$ as W is a function of V, and then we use (6) for $j = 1$ and our assumption on $\mathbf{H}(V_1)$. This completes the lower bound proof.

4 Upper Bounds

Though this paper focuses on proving lower bounds, we need to touch briefly upon upper bounds to demonstrate the tightness of Theorems 4 and 5. This is done by means of a *single* natural reduction of weak $\binom{N}{1}$-OT_2^L to $\binom{n}{1}$-OT_2^ℓ that *simultaneously* achieves both the lower bounds for the number of invocations of $\binom{n}{1}$-OT_2^ℓ and the number of random bits needed by the sender. This protocol is a simple generalization of the one given by Brassard, Crépeau and Sántha [5] for the case $L = \ell$, $n = 2$. For completeness purposes, we also include the proof that this protocol works. Though a similar proof could be derived from [5], the one included here is more direct because our definition of a reduction is slightly simpler.[3] Note that the same protocol also proves that our lower bounds are tight for reduction of (strong) $\binom{N}{1}$-OT_2^ℓ to $\binom{n}{1}$-OT_2^ℓ.

Theorem 6. *There exists a natural information-theoretically secure reduction of weak $\binom{N}{1}$-OT_2^L to $\binom{n}{1}$-OT_2^ℓ such that*

- *it uses $\frac{L}{\ell} \cdot \frac{N-1}{n-1}$ invocations of $\binom{n}{1}$-OT_2^ℓ.*
- *the sender uses $\frac{L(N-n)}{n-1}$ random bits.*

---

[3] You might notice, we embed the security of $\binom{n}{1}$-OT_2^ℓ into the definition of our reduction. Without doing so, one would have to argue about "nested mutual information".

Moreover, for $L = \ell$, the reduction actually is a reduction of (strong) $\binom{N}{1}$-OT_2^ℓ to $\binom{n}{1}$-OT_2^ℓ.

Proof. We start with $L = \ell$, i.e. a reduction of (strong) $\binom{N}{1}$-OT_2^ℓ to $\binom{n}{1}$-OT_2^ℓ, making $\alpha = \frac{N-1}{n-1}$ invocations and using $\frac{\ell(N-n)}{n-1}$ random bits for Alice. Let $w = w_1, \ldots, w_N$ be Alice's N strings of length ℓ each, and let i be Bob's index.

Protocol $P(w, i)$:

1. Alice chooses $(\alpha - 1)$ random ℓ-bit strings $x_1, \ldots, x_{\alpha-1}$ using $\ell(\alpha - 1) = \frac{\ell(N-n)}{n-1}$ random bits. Set $x_0 = 0^\ell$, $x_\alpha = w_N$.
2. Perform α invocations of the $\binom{n}{1}$-OT_2^ℓ where transfer $j = 0 \ldots (\alpha - 1)$ implements
   $$\binom{n}{1}\text{-}OT_2^\ell\ [w_{j(n-1)+1} \oplus x_j, w_{j(n-1)+2} \oplus x_j, \ldots, w_{(j+1)(n-1)} \oplus x_j, x_{j+1} \oplus x_j].$$
   Let z_j be the value Bob reads from the jth invocation, described next.
3. Let $j_0 \in \{0 \ldots (\alpha - 1)\}$ be the index of the box which has the XOR-ed value of w_i $(= \lfloor \frac{i-1}{n-1} \rfloor$, if $i \neq N$, and $= (\alpha - 1)$, otherwise). Bob reads the value $z_{j_0} = w_i \oplus x_{j_0}$ from box number j_0 and values $z_j = x_{j+1} \oplus x_j$ for all $j \neq j_0$.
4. Bob outputs $\bigoplus_{j=0}^{j_0} z_j$.

We now prove that the above protocol indeed implements strong $\binom{N}{1}$-OT_2^ℓ. The Correctness Property (P1) is clear since $(w_i \oplus x_{j_0}) \oplus (x_{j_0} \oplus x_{j_0-1}) \oplus \ldots \oplus (x_2 \oplus x_1) \oplus x_1 = w_i$. The Receiver Privacy (P2) is clear as well since the scheme is natural and, as we just saw, Bob can recover any w_i. We now show the main condition (P3).

Let $W = W_1, \ldots, W_N$ be chosen at random as well as Alice's random string $R = X_1, \ldots, X_{\alpha-1}$. Let V be the random variable containing all (αn) values of the $\binom{n}{1}$-OT_2^ℓ boxes. We can assume w.l.o.g. that in each of the α OT boxes, Bob indeed read one entire ℓ-bit string that he chose (he can not learn more and it "does not hurt" to learn as much as possible). Thus, define V_r to be the α-tuple of ℓ-bit strings that Bob read, i.e. everything that Bob learned from the protocol. Let $t_0, \ldots, t_{\alpha-1}$, where $t_j \in [1, n]$, be the (random variables denoting the) indices of α strings that Bob read.

Let j_0 be the smallest number such that $t_{j_0} \neq n$, if it exists. Otherwise, $j_0 = \alpha - 1$. Thus, Bob learned $X_1, X_1 \oplus X_2, \ldots, X_{j_0-2} \oplus X_{j_0-1}$ and some $W_i \oplus X_{j_0-1}$. Clearly, this enables him to reconstruct W_i (the exceptional case of all $t_j = n$ falls here as well giving Bob W_N). We let $\tilde{I} = i$. First of all, $\tilde{I}$ is *independent* from W. Indeed, Bob choose to read index t_{j_0} in the j_0^{th} invocation of $\binom{n}{1}$-OT_2^ℓ only based on his random coins and $X_1, X_1 \oplus X_2, \ldots, X_{j_0-2} \oplus X_{j_0-1}$, which does not depend on W. Thus, it suffices to show that $I(V_r; W \mid W_{\tilde{I}}) = 0$. But we already observed that $W_{\tilde{I}}$ is determined from V_r. Hence, using Lemma 2 (equations 4 and 3),

$$I(V_r\ ;\ W) = I((V_r, W_{\tilde{I}})\ ;\ W) = I(W_{\tilde{I}}\ ;\ W) + I(V_r\ ;\ W \mid W_{\tilde{I}})$$
$$= \ell + I(V_r\ ;\ W \mid W_{\tilde{I}})$$

Thus, we only need to show that $\mathbf{I}(V_r; W) = \ell$, i.e. to establish the weak property (P3'). Intuitively, Bob always learns some $W_{\bar{\jmath}}$, i.e. ℓ bits of information. So if we show that he does not learn more than ℓ bits of information, we know that the only thing he learned was that *one* string $W_{\bar{\jmath}}$. We proceed by showing a sequence of easy claims.

Local Claim 1: W is a function of V, i.e.

$$\mathbf{H}(W \mid V) = 0 \tag{7}$$

Proof: We already saw from correctness that V determines each string W_i. ∎

Local Claim 2:

$$\mathbf{H}(V \backslash V_r \mid V_r) = \ell(N - 1) \tag{8}$$

Proof: We show that all (αn) ℓ-bit strings of V are totally independent when W and R are randomly chosen. Let us view each such string in V as an $(N + \alpha - 1)$-dimensional vector over $\mathbb{Z}_2$ by taking the characteristic vector of the equation defining this string. Since all W_i and X_j are chosen randomly, our strings are independent iff the corresponding vectors are *linearly independent*. Assume that some linear combination of vectors in V is zero. This combination cannot include a vector depending on some W_i as there is only one such vector in V. And the remaining vectors $X_1, X_1 \oplus X_2, \ldots, X_{\alpha-2} \oplus X_{\alpha-1}$ are clearly linearly independent. And since our disjoint split of V into V_r and $V \backslash V_r$ does not depend on $V \backslash V_r$, we get that $V \backslash V_r$ is independent of V_r, so by Lemma 2 (equation 5), $\mathbf{H}(V \backslash V_r \mid V_r) = |V \backslash V_r| = \ell(n - 1)\alpha = \ell(N - 1)$. ∎

Local Claim 3: $V \backslash V_r$ is determined from W and V_r, i.e.

$$\mathbf{H}(V \backslash V_r \mid (V_r, W)) = 0 \tag{9}$$

Proof: The knowledge of W and any string $W_i \oplus X_{\alpha-1}$ in the last $\binom{n}{1}$-OT_2^ℓ box (which we have from V_r) determines $X_{\alpha-1}$. Knowing $X_{\alpha-1}$, W and any string of the form $z \oplus X_{\alpha-2}$ from the next to last $\binom{n}{1}$-OT_2^ℓ box (which we have from V_r where z is either some W_i or $X_{\alpha-1}$) enables one to deduce $X_{\alpha-2}$. Continuing this way, we determine X_1 from the first $\binom{n}{1}$-OT_2^ℓ box which allows us to reconstruct the whole $V \backslash V_r$. ∎

Combining Local Claims 1,2,3 and using Lemma 2 (equations 1, 2 and 3),

$$\ell N = \mathbf{H}(W) = \mathbf{H}(W) - \mathbf{H}(W \mid V) = \mathbf{I}(V; W) = \mathbf{I}(V_r; W) + \mathbf{I}(V \backslash V_r; W \mid V_r)$$
$$= \mathbf{I}(V_r; W) + \mathbf{H}(V \backslash V_r \mid V_r) - \mathbf{H}(V \backslash V_r \mid (V_r, W)) = \mathbf{I}(V_r; W) + \ell(N - 1)$$

Hence, $\mathbf{I}(V_r; W) = \ell$ indeed. This completes the proof of correctness when $L = \ell$.

For $\ell < L$ we give a trivial protocol that sacrifices the strong property (P3) leaving only (P3'). The protocol simply splits each of the strings of the database into L/ℓ disjoint parts of length ℓ each, and performs the previous protocol implementing $\binom{N}{1}$-OT_2^ℓ using $\binom{n}{1}$-OT_2^ℓ. It uses $\frac{L}{\ell} \cdot \frac{N-1}{n-1}$ invocations of $\binom{n}{1}$-OT_2^ℓ and $\frac{L}{\ell} \cdot \frac{\ell(N-n)}{n-1} = \frac{L(N-n)}{n-1}$ random bits as claimed. The correctness is clear except Alice's privacy. We clearly loose the strong property (P3) as Bob can learn up to L/ℓ different blocks of length ℓ from different strings. However, weak property (P3') still holds as the L/ℓ groups of boxes are totally independent, and from each of them Bob can learn at most ℓ bits about W, i.e. a total of at most $\ell \cdot \frac{L}{\ell} = L$ bits.

Acknowledgments

We would like to thank Amos Lapidoth for his critical comments about this paper. Special thanks to Madhu Sudan and Peter Winkler for useful remarks and suggestions.

References

1. M. Bellare, S. Micali. Non-interactive Oblivious Transfer and Applications. In *Advances in Cryptology: Proceedings of Crypto '90*, pp. 547-559, Springer-Verlag, 1990.
2. M. Blum. How to Exchange (Secret) Keys. In *ACM Transactions of Computer Systems*, vol. 1, No. 2, pp. 175–193, May 1983.
3. G. Brassard, C. Crépeau. Oblivious Transfers and Privacy Amplification. In *Advances in Cryptology: Proceedings of Eurocrypt '97*, Springer-Verlag, pp. 334-347, 1997.
4. G. Brassard, C. Crépeau, J. Robert. Information theoretic reductions among disclosure problems. In *27th Symp. of Found. of Computer Sci.*, pp. 168-173, IEEE, 1986.
5. G. Brassard, C. Crépeau, M. Sántha. Oblivious Transfers and Intersecting Codes. In *IEEE Transaction on Information Theory, special issue in coding and complexity*, Volume 42, Number 6, pp. 1769-1780, 1996.
6. C. Crépeau. Equivalence between two flavors of oblivious transfers. In *Advances in Cryptology: Proceedings of Crypto '87*, volume 293 of Lecture Notes in Computer Science, pp. 350-354, Springer-Verlag, 1988.
7. C. Crépeau. A zero-knowledge poker protocol that achieves confidentiality of the players' strategy or how to achieve an electronic poker face. In *Advances in Cryptology: Proceedings of Crypto '86*, pp. 239–247. Springer-Verlag, 1987.
8. C. Crépeau, J. Kilian. Weakening security assumptions and oblivious transfer. In *Advances in Cryptology: Proceedings of Crypto '88*, volume 403 of Lecture Notes in Computer Science, pp. 2-7, Springer-Verlag, 1990.
9. S. Even, O. Goldreich, A. Lempel. A Randomized Protocol for Signing Contracts. In *Advances of Cryptology: Proceedings of Crypto '83*, Plenum Press, New York, pp. 205-210, 1983.
10. M. Fisher, S. Micali, C. Rackoff. A Secure Protocol for the Oblivious Transfer. In *Journal of Cryptology*, vol. 9, No. 3 pp. 191–195, 1996.
11. O. Goldreich, S. Micali, A. Wigderson. How to play any mental game, or: A completeness theorem for protocols with honest majority. In *Proceedings of 19th Annual Symp. on Theory of Computing*, 218-229, 1987.

12. S. Goldwasser, S. Micali, C. Rackoff. The knowledge complexity of interactive proof-systems. In *SIAM Journal on Computing*, 18:186-208, 1989.
13. J. Kilian. Founding Cryptography on Oblivious Transfer. In *Proceedings of 20th Annual Symp. on Theory of Computing*, pp. 20–31, 1988.
14. J. Kilian, S. Micali, R. Ostrovsky. Minimum Resource Zero-Knowledge Proofs. In *Proceedings of 30th Annual Symp. on Foundations of Computer Science*, pp. 474–479, 1989.
15. S. Micali, P. Rogaway. Secure Computation. In *Advances in Cryptology: Crypto' 91 Proceedings*, pp. 392-404, Springer-Verlag, 1992.
16. H. Nurmi, A. Salomaa, L. Santean. Secret ballot elections in computer networks. In *Computer and Security*, volume 10, No. 6, pp. 553-560, 1991.
17. M. Rabin. How to Exchange Secrets by Oblivious Transfer. In *Technical Memo TR-81*, Aiken Computation Laboratory, Harvard University, 1981.

On the (Im)possibility of Basing Oblivious Transfer and Bit Commitment on Weakened Security Assumptions

Ivan Damgård[1], Joe Kilian[2], and Louis Salvail[1]

[1] BRICS, Basic Research in Computer Science, center of the Danish National Research Foundation, Department of Computer Science,University of Århus, Ny Munkegade, DK-8000 Århus C, Denmark.
{ivan,salvail}@daimi.aau.dk
[2] NEC Research Institute, 4 Independence Way, Princeton, NJ 08540.
joe@research.nj.nec.com

Abstract. We consider the problem of basing Oblivious Transfer (OT) and Bit Commitment (BC), with information theoretic security, on seemingly weaker primitives. We introduce a general model for describing such primitives, called *Weak Generic Transfer* (WGT). This model includes as important special cases *Weak Oblivious Transfer* (WOT), where both the sender and receiver may learn too much about the other party's input, and a new, more realistic model of noisy channels, called *unfair noisy channels*. An unfair noisy channel has a known range of possible noise levels; protocols must work for any level within this range against adversaries who know the actual noise level.

We give a precise characterization for when one can base OT on WOT. When the deviation of the WOT from the ideal is above a certain threshold, we show that no information-theoretic reductions from OT (even against passive adversaries) and BC exist; when the deviation is below this threshold, we give a reduction from OT (and hence BC) that is information-theoretically secure against active adversaries.

For unfair noisy channels we show a similar threshold phenomenon for bit commitment. If the upper bound on the noise is above a threshold (given as a function of the lower bound) then no information-theoretic reduction from OT (even against passive adversaries) or BC exist; when it is below this threshold we give a reduction from BC. As a partial result, we give a reduction from OT to UNC for smaller noise intervals.

1 Introduction

A 1 out of 2 Oblivious transfer (1-2 OT) protocol is one by which a sender with 2 bits b_0, b_1 as input can interact with a receiver with a bit c as input. Ideally, the sender should learn nothing new from the protocol, whereas the receiver should learn b_c and nothing more. Several variants of OT exist, but it does not matter much which one we consider, as they are almost all equivalent (see e.g. [8]).

A bit commitment scheme is a pair of protocols Commit and Open executed by two parties, a commiter, C, and a receiver, R. First, C and R execute Commit,

J. Stern (Ed.): EUROCRYPT'99, LNCS 1592, pp. 56–73, 1999.
© Springer-Verlag Berlin Heidelberg 1999

where C has a bit b as input; R either accepts that a commitment has taken place or rejects. Ideally, the receiver should learn no information about b from this. Later, they may execute Open, after which R returns *accept* 1, *accept* 0 or *reject*. We require our protocols to be *correct*, *private* and *binding*:

Correctness: If both parties follow the protocol, R should always accept with the same value (b) that C wished to commit to.

Privacy: Committing to b reveals nothing to the receiver about b.

Binding: C cannot cause R to accept a commitment, and then be able to execute Open so that R accepts a 1 and also be able to execute Open so that R accepts a 0.

We have described the ideal requirements here. However, usually when building such protocols, one accepts an error that can be made negligibly small as a function of some security parameter k.

A great deal of work has gone into how to implement oblivious transfer and bit commitment based on seemingly weaker primitives. For example, a binary symmetric channel (BSC) is one that allows a sender S to send a bit b_S to a receiver R, such that a bit b_R will be received, which is not necessarily equal to b_S. There is a constant probability $0 < \delta < 1/2$, called the *noise level* of the channel such that each time the channel is invoked, $Pr(b_S \neq b_R) = \delta$. Another, essentially equivalent formulation has S and R receiving random bits b_S and b_R that are individually unbiased but correlated so that $Pr(b_S \neq b_R) = \delta$. Another equivalent formulation has a random bit b transmitted to both parties through independent noisy channels. One motivation for the last two formulations is that one might want to implement noisy channels by a very weak broadcast source, such as a satellite.

Crépeau and Kilian [9] showed that a BSC can be used to implement 1–2 OT with unconditional (information theoretic) security; the efficiency of this reduction was later improved by Crépeau [6], who also directly implemented a bit commitment scheme (indirectly, bit commitment can be based on 1–2 OT).

The reductions given in [9,6] rely on the fact that δ is known exactly by each party. However, in real life it may be possible for one party to surreptitiously alter the noise level of the channel. If the noise is induced by a communications channel then it may be possible to alter the mechanism (say by heating it up or cooling it down), or change the way it uses the mechanism, to change the noise rate. For example, suppose the channel consists of two pieces of optical fibre with a repeater station in between, a very common case in practice. If one party has access to the data received by the repeater station, then he can send and receive a cleaner signal than the other party expects him to. In the case of a noisy broadcast channel, an adversary might send a jamming signal or buy a more sensitive antenna. Note that while it may be hard to hide the fact that one has made a channel noisier, one can always hide the fact that one has made it less noisy, simply by deliberately garbling ones own messages and pretending to hear a more garbled version than one has actually heard.

Such "unfair advantages" are not always devastating for applications to cryptography: Maurer [15] shows that secure key exchange between two parties with access to a random but noisy signal is possible, even in the presence of an enemy who can receive the signal much more reliably than the honest players. However, this scenario is a game for two parties who trust each other and want to be protected "against the rest of the world." It is natural to ask if we can still make do with unfair channels in case of games between two *mutually distrusting parties*. Unfortunately, the protocols of [9,6] break down in this scenario.

1.1   Our Results

In this paper we propose a general model for two-party primitives where a cheating player can get more information than an honest one; we call this model *Weak Generic Transfer* (WGT). We then consider a number of important subcases and show when they can and cannot be used as a basis for bit commitment and oblivious transfer.

We consider a family of *Weak Oblivious Transfers*, which are 1-2 OT protocols with the following faults: with probability (at most) p a cheating sender will learn which bit the receiver chose to receive, and with probability q a cheating receiver will learn both of the sender's input bits. Note that the honest participants only receive what they are supposed to receive; this extra information cannot be relied on. We call such a protocol a (p, q)-WOT. We give tight results for when one can reduce oblivious transfer to (p, q)-WOT.

In the statement of our results, when we use "reduction" we mean a reduction that is information-theoretically secure against unbounded adversaries, where deviations from the ideal are negligible in a given security parameter.

Theorem 1. *1-2 OT and BC can be reduced to (p, q)-WOT iff $p + q < 1$.*

We also consider a still noisier model, denoted (p, q, ϵ)-WOT, in which with probability *at most* ϵ an honest receiver receives $\bar{b}_c$ instead of b_c (i.e., the incorrect value); a cheating receiver is under no such handicap. In this case, we prove positive and negative results that are no longer tight.

Theorem 2. *1-2 OT can be reduced to (p, q, ϵ)-WOT, for the case of passive adversaries, if $p + q + 2\epsilon < .45$. No reductions from 1-2 OT or BC exist if $p + q + 2\epsilon \geq 1$.*

Passive adversaries, also known as "honest but curious" adversaries follow the protocol, but then try to use their view of the protocol execution to violate the security conditions.

Both theorems comprise a constructive result and an impossibility result. The constructive result of Theorem 1 generalizes a theorem of [9], which solves the special cases where either p or q is 0 (or negligible in the security parameter). Brassard/Crépeau [2] and Cachin [4] consider a more general model of WOT, where the extra information that an adversary learns is only specified by a general information measure, but here again the weakness is one-sided: only the receiver learns extra information. Prior to this work, few nontrivial impossibility results of this type were known (see [14] for one such result). These impossibility results hold even if security against passive cheating is required and the honest players are allowed infinite computing power.

We note that one motivation for the study of these imperfect protocols is that they provide easier to achieve steps for other reductions. For example, our reduction from 1-2 OT to unfair noisy channels first reduces (p, q, ϵ)-WOT to unfair noisy channels.

We finally consider *unfair noisy channels* (UNC). These channels have parameters γ and δ, where $\gamma, \delta \leq 1/2$. The noise level p of this channel is guaranteed to fall into the interval $[\gamma, \delta]$. The protocol must work for any p in this range; however the value of p is *not* known to the honest players (but may be set within this range by the adversary).

Theorem 3. *For* $\delta > 2\gamma(1 - \gamma)$, *neither 1-2 OT nor BC may be reduced to* (γ, δ)-*UNC. For* $\delta < 2\gamma(1 - \gamma)$ *BC may be reduced to* (γ, δ)-*UNC. Finally, 1-2 OT may be reduced to* (γ, δ)-*UNC if* $\alpha^3\beta^3(1 - \zeta(1 - \alpha)) > \frac{0.775 + \epsilon(\delta)}{1 - \epsilon(\delta)}$, *where* $\epsilon(\delta) = \frac{\delta^2}{\delta^2 + (1-\delta)^2}$, $\alpha = \frac{1-\delta-\gamma}{1-2\gamma}$, $\beta = \frac{1-\gamma}{1-\delta}$, *and* $\zeta = \frac{1-\gamma}{\delta}$.

1.2   Techniques Used

All of our impossibility results rely on a general simulation technique that allows us to leverage the result that it is impossible to implement 1-2 OT (information-theoretically) given only a clear channel.

Our upper bounds for (p, q)-WOT and (p, q, ϵ)-WOT use some reductions first used in [9]. The reduction from bit commitment to (γ, δ)-UNC is based on the interactive hashing technique of [16]. The precise hashing method of [16] doesn't work for our application; instead we use families of universal hash functions [10]. Hash functions are ubiquitous in cryptography; two classic results on achieving privacy with universal hash functions are [13] and [1]. For the specifics of our analysis we use bounds on their behaviour implied by the results of [17].

Guide to the Paper In Section 2 we give the general scenario for weak generic transfer. In Section 3 we show impossibility results for reducing 1-2 OT and bit commitment to (p, q)-WOT, (p, q, ϵ)-WOT and (γ, δ)-UNC. In Section 4 we give reductions from 1-2 OT (and hence bit commitment) to (p, q)-WOT and (p, q, ϵ)-WOT. In Section 5 we give a reduction from bit commitment to (γ, δ)-UNC. In Section 6 we in give a reduction from 1-2 OT to (γ, δ)-UNC.

2   The General Scenario: Weak Generic Transfer

In order to show more clearly the basic properties we study, we start with a general scenario that includes as special cases those primitives we later study in greater detail.

First, we describe a specification for standard two party primitives, and then show how to augment these specifications to model interesting deviations from the ideal behaviour of the protocol.

Initially, our scenario includes two players A, B that start with private inputs x_A, x_B, respectively chosen from domains X_A and X_B (the precise nature of these domains has no impact on the following discussion). A specification for a standard two-party primitive is a function output that maps (x_A, x_B) to a probability measure D on $Y_A \times Y_B$. When the primitive is executed with inputs (x_A, x_B), $D = \text{output}(x_A, x_B)$ is computed, (y_A, y_B) is chosen according to D, y_A is sent to A and y_B is sent to B. This framework is powerful enough to model primitives such as OT, 1–2 OT and binary symmetric channels.

To model the possibility that a primitive might inadvertently leak information, we modify D to be a distribution $(Y_A \times Z_A) \times (Y_B \times Z_B)$; $((y_A, z_A), (y_B, z_B))$ are sampled from D. If A is honest, then A receives only y_A, but if A is corrupt, it also receives z_A; B behaves symmetrically.

We can model passive ("honest but curious") adversaries by simply specifying that an adversary $Q \in \{A, B\}$ follows the protocol, ignoring z_Q, and then later learns what it can from the values of z_Q that is obtained. An active adversary may immediately use this extra information in planning its next move.

We have modeled deviations from privacy; we now model deviations in behaviour. Instead of having a single function output, we have a (possibly infinite) set S of functions {output}, which contains a "default" output_0. When the protocol is executed, the adversary has the option of choosing output from S; the protocol then behaves as before. If there is no adversary, then the default output_0 is used. We say that S specifies a *Weak Generic Transfer* (WGT). We will assume throughout that A and B have access to the WGT as a black box and can execute it independently as many times as they wish.

A WGT may consist of a protocol where for instance a noisy channel is used several times, and the protocol instructs one player to send the same bit each time. An active cheater may choose not to do so, and so he may behave in a way that is not consistent with any legal input value; in this case we say that he inputs "?." We cannot require in general that a WGT prevents such behaviour: this would require that the cheater was committed to his input, and it is not clear a priori that a WGT implies bit commitment (indeed some WGT's don't, as we shall see). The best we can ask for in case of active cheating is therefore that the following is satisfied:

– For any active cheating strategy followed by A (B), there exists an input value x such that A (B) learns nothing more than what is implied by the view of a passively cheating player with input x.

If this is satisfied, we say that the WGT is *secure against active cheating* (but note that the only security we ask for here is that active cheating does not give any advantage over passive cheating).

It should be clear that (γ, δ)-UNC, (p, q)-WOT and (p, q, ϵ)-WOT are special cases of WGT. Note that only for the case of (γ, δ)-UNC may an adversary choose between more than one output distribution function.

3   Impossibility Results

The basic question we can ask is now: given a WGT, can we build OT or BC based on it? It is easy to characterize a class of WGT's where the answer is no. We first consider the case where there is only noiseless communication between A and B, and consider any interactive protocol between them, of the following form:

- A starts with input x_A, B starts with input x_B.
- The players exchange a finite number of messages, and the protocol specifies at each stage a probabilistic algorithm for computing the next message, given the input, and all message and random coins seen so far.
- The *view* of a player ($View_A/View_B$) is as usual defined to be the player's input and random coins, along with all messages received from the other player. At the end, A and B compute their results, y_A and y_B from their views by some function, i.e., $y_Q = f_Q(View_Q)$.

It is clear that any such protocol can be seen as a WGT by letting $z_A = View_A$ and $z_B = View_B$; this method for producing y_A, y_B, z_A, z_B from x_A, x_B defines a probability measure $D(x_A, x_B)$, and we define just one output distribution function output which always return $D(x_A, x_B)$. Honest players will ignore anything except for the result specified (Y_A, Y_B), but a passively cheating player may use its entire view to compute extra information.

It is well known (and easy to see) that in a two-player scenario with only noiseless communication, OT and BC with information theoretic security is not possible, even if only passive cheating is assumed, and players are allowed infinite computing power. Hence, OT and BC are not reducible to the above WGT. We call such a WGT *trivial*.

We now show how to "implement" (p, q, ϵ)-WOT in this manner, where $2\epsilon = 1 - p - q$. Consider the following protocol, in which A has input (b_0, b_1) and B has input c.

Protocol SimNoisyWOT$[p, q]((b_0, b_1), c)$

1. With probability q, A announces (b_0, b_1), B computes b_c and the protocol terminates; otherwise, A announces "pass".
2. If A passes, then with probability $p/(1 - q)$, B sends c to A who replies with b_c; otherwise, B chooses b_c at random.

By a straightforward case analysis, B learns both b_0 and b_1 with probability q, A learns c with probability p and B receives an incorrect value of b_c with probability $(1 - p - q)/2 = \epsilon$. Aside from easily simulated messages, such as "pass", the view of each side corresponds to the view it could obtain from an actual run of a (p, q, ϵ)-WOT primitive. Now, suppose we had an 1–2 OT protocol based on a (p, q, ϵ)-WOT primitive. If we replaced each execution of the (p, q, ϵ)-WOT by an execution of the SimNoisyWOT$[p, q]$ primitive, then the view of each party, taken in isolation, would be unchanged. Since the security of 1–2 OT (at least against passive adversaries) is defined solely in terms of properties of Player

A's view and properties of Player B's view, the resulting protocol would remain secure against passive adversaries. This would give a "mental 1-2 OT" protocol, information-theoretically secure against passive adversaries, a contradiction. Similarly, there is no information-theoretically secure (against both parties) mental bit commitment protocol, even if both parties are guaranteed to follow the Commit protocol; we can in a very similar way derive a contradiction.

The above argument implies the following lemma.

Lemma 4. *There is no reduction from 1-2 OT or BC to (p, q, ϵ)-WOT when $p + q + 2\epsilon \geq 1$, even if only security against passive adversaries is required.*

Remark: The simulation argument was for $p+q+2\epsilon = 1$. If $p+q+2\epsilon > 1$, choose $\epsilon' = (1-p-q)/2 < \epsilon$; the impossibility argument works for (p, q, ϵ')-WOT. Note that a (p, q, ϵ')-WOT primitive also meets the requirements of a (p, q, ϵ)-WOT primitive, since its error rate cannot be higher than ϵ, so a protocol that works for (p, q, ϵ)-WOT must work for (p, q, ϵ')-WOT as well.

We now consider the case of the noisy channel. Consider the following purely mental protocol, in which A has input b.

Protocol SimUNC$[\gamma](b)$

1. A and B pick b_A and b_B respectively, such that $Pr(b_A = 1) = Pr(b_B = 1) = \gamma$.
2. A sends $b' = b \oplus b_A$ to B. B computes $b^* = b' \oplus b_B$, denoting b^* as the received bit, while no output is defined for A.

Consider a WGT which between honest players A and B is a BSC with noise level δ, but where if A or B cheats passively, then some extra information is available and allows to reduce the noise level to γ, seen from the cheater's point of view. Let us call this a (γ, δ)-PassiveUNC. It is similar to but not the same as a (γ, δ)-UNC. It immediately follows from the above protocol that a (γ, δ)-PassiveUNC is trivial if $\delta = 2\gamma(1 - \gamma)$, and in fact in general if $\delta \geq 2\gamma(1 - \gamma)$. And so there is no reduction of 1-2 OT or BC to (γ, δ)-PassiveUNC in this case, not even if only passive security is required.

Now, suppose we have a reduction from 1-2 OT to a (γ, δ)-UNC, where $\delta = 2\gamma(1 - \gamma)$, one secure against active attacks. We compare the following two cases: In case 1 the reduction is attacked by an adversary using the following active cheating strategy for a player $Q \in \{A, B\}$: Q sets the noise level for the UNC to be γ always, and then does the following: Whenever Q is supposed to send a bit through the channel, Q first flips it with probability γ and then sends it. Similarly, whenever Q receives a bit from the channel, Q flips it with probability γ and acts as if that was the bit actually received. In any other cases, Q follows the algorithm specified by the reduction. Case 2: we execute the algorithm of the reduction substituting the (γ, δ)-UNC by a (γ, δ)-PassiveUNC, and the adversary executes a passive attack.

There is no difference between the cases from the honest player's point of view. Observe that in case 1, the adversary following the strategy for Q knows as much about every bit sent and received by his opponent as a passive adversary knows in case 2. So since the reduction is secure in case 1, it must be secure in case 2, and we have a contradiction. Essentially the same argument works for bit commitment. So we have proved:

Lemma 5. *There is no reduction from 1–2 OT or BC to* (γ, δ)*-UNC when* $\delta \geq 2\gamma(1 - \gamma)$.

This motivates the following interesting and open question: *Which non-trivial cryptographic primitives (if any) can be implemented based on a WGT assuming only that it is non trivial?*

4   Reducing 1–2 OT to (p, q)-WOT and (p, q, ϵ)-WOT

We now look at the possibility of building a 1–2 OT or commitments from a WOT. A reduction that accomplishes such a task can be thought of as a program that gets the noise levels of a UNC or the error probabilities of a WOT and a security parameter value k as input and then instructs at each point in time one of the players to either send a message in the clear to the other player, or send a bit through the noisy channel. Any information known to the player at the time can be used, together with any number of random bits, to compute the next message to send. We make no assumption on the amount of computation required.

4.1   Preliminaries

For a reduction of 1–2 OT to UNC, let $I_c(k, \delta, \gamma), I_{b_c}(k, \delta, \gamma), I_{b_{1-c}}(k, \delta, \gamma)$ be the expected information that the sender obtains about c, the receiver obtains about b_c, and the receiver obtains about b_{1-c} respectively. We will say that *the reduction works* for values δ, γ, if $\lim_{k\to\infty} I_c(k, \delta, \gamma) = \lim_{k\to\infty} I_{b_{1-c}}(k, \delta, \gamma) = 0$ and $\lim_{k\to\infty} I_{b_c}(k, \delta, \gamma) = 1$ For a reduction of 1–2 OT to (p, q)-WOT, we use the same definitions, but with (δ, γ) replaced by (p, q).

For a reduction of bit commitment to UNC, let $I_b(k, \delta, \gamma)$ be the expected information the receiver obtains about b in the *Commit* protocol, and let $p(k, \delta, \gamma)$ be the probability that the binding condition fails. We will say that *the reduction works* for values δ and γ, if $\lim_{k\to\infty} I_b(k, \delta, \gamma) = \lim_{k\to\infty} p(k, \delta, \gamma) = 0$

We refer to [3] for a more sophisticated definition of 1–2 OT; our protocols meet this definition as well.

The set of pairs for which a reduction works will be called the *range* of the reduction. We will say that a reduction *works efficiently* in a point in its range, if the required convergence in k is exponential, and that the number of calls to the WOT or UNC is polynomial in k. This is usually required for a reduction to be useful in practice, but note that our impossibility results hold even if efficiency is not required.

4.2   Some Useful Reductions

We use the following two known reductions for basing 1-2 OT on (p, q)-WOT. The first is designed to reduce the chance the sender (A) learns too much, while the second is targeted against the chance of the receiver (B). Both reductions are assumed to be given as a black-box a protocol $\mathcal{W}$ implementing (p, q)-WOT and work with security parameter k. S-Reduce is taken from [9], while R-Reduce is more or less folklore.

Protocol S-Reduce$(k, \mathcal{W})$

1. Let (b_0, b_1) resp. c be the input of the sender, resp. the receiver.
2. $\mathcal{W}$ is executed k times, with inputs $(b_{0i}, b_{1i}), i = 1..k$ from the sender and $c_i, i = 1..k$ from the receiver. Here, the b_{0i}'s are uniformly chosen, such that $b_0 = \oplus_{i=1}^{k} b_{0i}$, $b_{1i} = b_{0i} \oplus b_0 \oplus b_1$ and the c_i's are uniformly chosen such that $c = \oplus_{i=1}^{k} c_i$.
3. The receiver computes his output bit as the xor of all bits received in the k executions of $\mathcal{W}$.

Protocol R-Reduce$(k, \mathcal{W})$

1. Let (b_0, b_1) resp. c be the input of the sender, resp. the receiver.
2. $\mathcal{W}$ is executed k times, with inputs $(b_{0i}, b_{1i}), i = 1..k$ from the sender and $c_i, i = 1..k$ from the receiver. Here, $c_i = c$, $b_{01} \oplus ... \oplus b_{0k} = b_0$ and $b_{11} \oplus ... \oplus b_{1k} = b_1$
3. The receiver computes the XOR of all bits received.

Lemma 6. *When given k and a (p, q)-WOT $\mathcal{W}$ as input, S-Reduce$(k, \mathcal{W})$ implements a $(p^k, 1 - (1 - q)^k)$-WOT, and R-Reduce$(k, \mathcal{W})$ implements a $(1 - (1-p)^k, q^k)$-WOT protocol. Both protocol produce a WOT secure against active cheating if the given WOT has this property.*

Proof (sketch). First, it follows by inspection that the protocols allow the players to compute the correct output. As for the error probabilities, note that for S-Reduce a bad sender will learn c iff he learns all c_i's, which happens with probability p^k. On the other hand, a bad receiver can learn both b_0 and b_1 if he learns just one pair (b_{0i}, b_{1i}), and this happens with probability $1 - (1 - q)^k$. The case of R-Reduce is similar, but with the chances of sender and receiver reversed. The last claim follows easily: In S-Reduce, security of W means that none of the parties can gain anything from inputing ? to W. And if indeed no ? is input to any W instance, R always behaves consistently with some input c, namely the value $c = \oplus_{i=1}^{k} c_i$. S can behave inconsistently by choosing bad values for his bits, but this will not give him more information on c. The case of R-Reduce is similar.   □

4.3   A Reduction to (p, q)-WOT

Lemma 7 shows that the lower bound given by Lemma 4 is tight when $\epsilon = 0$. Lemmata 4 and 7 imply Theorem 1.

Lemma 7. *There exists a reduction for building 1-2 OT from a (p, q)-WOT, the range of which is $\{p, q \mid p + q < 1\}$. It works efficiently for all points in its range.*

Proof. Suppose we start with a (p, q)-WOT $\mathcal{W}$, and apply first R-Reduce$(t, \mathcal{W})$ and then S-Reduce$(t', \mathcal{W})$. We call this combination RS-Reduce. It follows easily that RS-Reduce produces a $((1 - (1 - p)^t)^{t'}, 1 - (1 - q^t)^{t'})$-WOT. Of course, we can also apply S-Reduce first, and obtain a $(1 - (1 - p^t)^{t'}, (1 - (1 - q)^t)^{t'})$-WOT. We call this combination SR-Reduce.

The strategy for our reduction is to apply repeatedly SR-Reduce or RS-Reduce, in order to reduce as quickly as possible the sum of the error probabilities. When given errors (p, q), we apply RS-Reduce if $p \leq q$, and SR-Reduce otherwise. This will be called one step in the reduction.

To analyse the effect of one step, define $x = q, y = 1 - p$ when $p \leq q$, and $x = p, y = 1 - q$ otherwise. It follows that the difference between the sum of the errors before and after the transformation is

$$f(t, t', x, y) = (1 - y^t)^{t'} + 1 - (1 - x^t)^{t'} - (1 - y + x) = (1 - y^t)^{t'} + y - ((1 - x^t)^{t'} + x)$$

The constraints we have on p, q imply that $1/2 < y < 1$ and $1 - y \leq x < y$. And we see that the progress we make is the difference between the values of the function $g_{t,t'}(z) = (1 - z^t)^{t'} + z$ evaluated at points x and y. The trick is now to choose, given x and y, values of t and t' such that the above difference becomes numerically "large". Note that since we are subtracting the sum before the step from the sum after, the difference is hopefully negative.

In any situation where the error probability sum before a step is greater than 0.2, one of the following three cases apply:

$y \leq 0.8$: This is a case where the smallest of p, q is at least 0.2, so p, q are both "large." In this case, we choose $t = t' = 2$. By direct inspection of $g_{2,2}(x)$, one finds that for any x, y obeying the restrictions, $(g_{2,2}(y) - g_{2,2}(x))/(y - x) \leq -0.1$. Since $y - x = 1 - (p + q)$, this shows that taking one step in this case multiplies the distance from $p + q$ to 1 by a factor of at least 1.1.

$y > 0.8, x > .4$: In this case, $p + q$ is also "large," but this time because one probability is small and the other is large. In this case, we choose $t = 2$ and $t' = 1$. Again, by direct inspection, one can verify that $(g_{2,1}(y) - g_{2,1}(x))/(y - x) \leq -0.2$. By the same argument as before, we see that in this case, the distance from $p + q$ to 1 is multiplied by at least 1.2 by taking one step.

$y > 0.8, x \leq .4$: In this case, both p and q and hence $p + q$ are "small." We choose $t = t' = 2$. Observe that for the large y, $g_{2,2}(y)$ approaches y as y approaches 1, while for the small x, $g_{2,2}(x)$ approaches $1 + x$ as x approaches 0. As a result, $(g_{2,2}(y) - g_{2,2}(x))/(1 - y + x) \simeq -1$ for small x and large y, and is in fact less than -0.2 for the range specified. However, $1 - y + x = p + q$, so we see that in this case, taking one step reduces $p + q$ to at most 0.8 of its previous value.

As soon as we have an error probability sum which is at most 0.2, we start doing steps where we always have $t = t' = 4$. In this case one finds that if the error sum was s before a step it is at most s^2 afterwards.

The overall strategy is now as follows: we first do whatever number of steps is necessary to bring the error probability sum below 0.2. We then do $\log_2(k)$ steps, where k is the security parameter. It follows from the above that the resulting error probability sum is exponentially small in k, at most 0.2^k. The number of calls we make to the WOT is exponential in the number of steps, but since we only take a logarithmic number of steps, the total number of calls is polynomial in k.   □

The above argument only considers p, q as being constants. However, even if we have a case where $p + q$ is a function of some parameter n and converges polynomially to 1 in n, e.g. $p(n) + q(n) = 1 - 1/n$, the reduction in the proof still works in time polynomial in n.

4.4   A Reduction to (p, q, ϵ)-WOT

Lemma 4 shows that no reduction of 1-2-OT to (p, q, ϵ)-WOT exists if $p + q + 2\epsilon \geq 1$ and this, even in the case of passive adversaries. We show that if $p + q + 2\epsilon < 0.45$, such a reduction does exist. We adapt SR-Reduce to deal with transmission errors. We then characterize triplets (p, q, ϵ) for which 1-2 OT is reducible to (p, q, ϵ)-WOT. The reduction we consider assumes only passive adversaries.

The following error detection phase accepts parameter $l > 0$ and, given a (p, q, ϵ)-WOT $\mathcal{W}$, produces a (p', q', ϵ')-WOT $\mathcal{W}'$ such that $\epsilon' < \epsilon$. As usual b_0 and b_1 denote the two bits to be transmitted and c is the selection bit.

Protocol ErRed$(l, \mathcal{W})$

1. A chooses $q_0, q_1 \in_R \{0, 1\}$ and B chooses $s \in_R \{0, 1\}$,
2. A sends l times the bits (q_0, q_1) through the (p, q, ϵ)-WOT $\mathcal{W}$ and B selects the bit q_s l times,
3. If B did not receive the same bits $\hat{q}_s$ l times then A and B go to Step 1.
4. B announces $y = 0$ if $s = c$ and $y = 1$ otherwise.
5. A announces r_0 and r_1 such that $b_y = r_0 \oplus q_0$ and $b_{1-y} = r_1 \oplus q_1$, allowing B to compute $b_c = \hat{q}_s \oplus r_s$.

We are now ready to describe a reduction of 1-2 OT to (p, q, ϵ)-WOT which basically inserts calls to ErRed into SR-Reduce (and RS-Reduce). Given positive integers l_0, k, l_1, k', l_2 and a (p, q, ϵ)-WOT $\mathcal{W}_0$, protocol SR_ϵ produces a new (p', q', ϵ')-WOT $\mathcal{W}$:

Protocol $\mathsf{SR}_\epsilon(l_0, k, l_1, k', l_2, \mathcal{W}_0)$

- $\mathcal{W} \leftarrow \mathsf{ErRed}(l_2, \mathsf{R\text{-}Reduce}(k', \mathsf{ErRed}(l_1, \mathsf{S\text{-}Reduce}(k, \mathsf{ErRed}(l_0, \mathcal{W}_0)))))$.

$RS_\epsilon(l_0, k, l_1, k', l_2)$ is defined the same way except the calls to S-Reduce and R-Reduce are swapped. Similarly to Lemma 6, one can characterize exactly the transformation taking place in a call to $SR_\epsilon(l_0, k, l, k', l')$ for any parameters l_0, k, l, k', and l'. In particular, $SR_\epsilon(l_0, k, l, k', l')$ transforms a (p, q, ϵ)-WOT into a (p', q', ϵ')-WOT where $p' = 1 - (1 - (1 - (1-p)^{l_0})^k)^{l_1 \cdot k' \cdot l_2}$, $q' = 1 - (1 - (1 - (1 - q)^{l_0 \cdot k \cdot l_1})^{k'})^{l_2}$, and ϵ' is of similar but slightly more complicated form. A brute force analysis, using linear programming, shows that SR_ϵ can be tuned to work at 45% the optimum (the sketch of the proof can be found in [11]).

Lemma 8. *1–2 OT can be implemented given any (p, q, ϵ)-WOT that satisfies $p + q + 2\epsilon \leq 0.45$.*

The above bound is not tight especially whenever one of $p + q$ and ϵ is small. In particular, SR_ϵ works for all $(p, q, 0)$ such that $p + q < 1$ and for all $(0, 0, \epsilon)$ such that $\epsilon < \frac{1}{2}$. A natural question arises: Is it possible to use a different method for choosing parameters l_0, k, l_1, k' and l_2 such that reduction SR_ϵ works also for $p + q + 2\epsilon \gg 0.45$? The following lemma suggests that if one wants to get closer to the bound $p + q + 2\epsilon = 1$, one has to find a different reduction.

Lemma 9. *There exists triplets (p, q, ϵ) that satisfy $p + q + 2\epsilon \leq 0.70$ such that SR_ϵ and RS_ϵ does not work for any value of parameters l_0, k, l_1, k' and l_2.*

Proof (sketch). Let $p = q = 0.2$ and $\epsilon = 0.15$ be the parameters of a WOT that satisfies $p + q + 2\epsilon = 0.7$. It can be shown that whatever the parameters l_0, k, l_1, k' and l_2 are, the reduction always generates an intermediary simulatable triplet. □

Lemma 9 suggests that introducing noise in a WOT might lead to a primitive that is strictly weaker than 1–2 OT even for non-simulatable but noisy WOT. However, the gap between the bounds could be narrowed down by finding a better simulation and/or a new reduction. It is unknown to us if such a gap necessarily exists.

5 Reducing Bit Commitment to (γ, δ)-UNC

5.1 Preliminaries

Our commitment protocol makes extensive use of t-universal hash functions, first introduced in [10]; we use the following slightly stronger notion that has become more or less standard. Given a domain D and a range R, a t-universal family of hash functions is a distribution $\mathcal{H}$ on a set of functions $\{h_i\}$ such that for any distinct $X_1, \ldots, X_t \in D$, if h is chosen according to $\mathcal{H}$, the induced distribution on $(h(X_1), \ldots, h(X_t))$ is uniform over R^t. For our application, $D = \{0, 1\}^k$, $R = \{0, 1\}^l$, for some k, l. For any k and l, there exists a t-universal family of hash functions whose functions may be represented using $poly(k, t)$ bits, and for which the operations of sampling h from the distribution and computing $h(X)$ may be performed in $poly(k, t)$ time. Hence, we speak of one party "sending" a function, abstracting all representational details.

68   Ivan Damgård, Joe Kilian, and Louis Salvail

Given two bit-sequences X and X', let $d(X, X')$ denote their Hamming distance, i.e., the number of places where they differ. We use distance as shorthand for Hamming distance.

There is a huge body of literature on universal hash functions and their use in cryptography. Despite superficial differences, our method is motivated by that of [16].

5.2   What We Need To Achieve

Note that it suffices to produce a protocol for committing to $x = r$ for a random bit r; as a standard trick one can commit to $y = b$ by revealing $b' = b \oplus r$ and defining $y = x \oplus b'$. For the rest of the discussion, we analyze the case of committing to random values. We also allow the receiver to reject even though the commiter followed the protocol, but only with probability negligible in the security parameter, k.

5.3   The Protocol

On a high level, our (weak) commitment protocol consists of the following steps. First, C sends string X over the noisy channel to C. R queries C about the value of $h_i(X)$ for $i = 1, 2$. Finally, C chooses a hash function h and designates $h(X)$ as the random committed value. To reveal a bit, C sends X to R. R accepts if X is close to the received value and is consistent with the queried hash values.

Protocol Commit(γ, δ, k)

Define d_0 by $\gamma(1 - d_0) + (1 - \gamma)d_0 = \delta$ and let $d_1 = (d_0 + \gamma)/2$, $d = (d_1 + \gamma)/2$ and $\ell = \lfloor \lg(\sum_{i=1}^{\lfloor dk \rfloor} \binom{k}{i}) \rfloor$. We let $d^* = (d_1 + d)/2$, and define ℓ^* as ℓ, where we replace dk by d^*k. Note that, by a standard argument, it follows that $\ell - \ell^* > ck$ for some constant c as k grows sufficiently large.

Let $\mathcal{H}, \mathcal{H}_1$ and $\mathcal{H}_2$ be canonical $64k$-universal families of hash functions from $\{0,1\}^k$ to $\{0,1\}, \{0,1\}^{\ell^*}$ and $\{0,1\}^{\ell-\ell^*}$, respectively.
1. C uniformly chooses $X = x_1, \ldots, x_k \in \{0,1\}^k$ and sends X to R over the (γ, δ) channel. Denote by $X' = x'_1, \ldots, x'_k$ the string received by R.
2. For $i = 1$ to 2
    R chooses $h_i \leftarrow \mathcal{H}_i$ and sends h_i to C.
    C sends $y_i = h_i(X)$ to R.
3. C chooses $h \leftarrow \mathcal{H}$ and sends h to C. The committed bit is defined as $h(X)$.

Protocol Open(γ, δ, k)

Let $X, X', y_1, y_2, h, h_1, h_2, d_0, d_1$ and d be as in the execution of Commit for the bit to be opened, and let $\delta' = \gamma(1 - d_1) + (1 - \gamma)d_1$.
1. C sends X to R.
2. R rejects if $y_i \neq h_i(X)$ for any i or if $d(X, X') \geq \delta'k$ locations. Otherwise, R accepts the committed value of $h(X)$.

5.4   Analysis of the Protocols

We first observe that the protocol behaves correctly if both parties are honest. For the rest of this discussion, "negligible" means smaller than $1/k^c$, for any c, as k grows sufficiently large and "almost always" means with probability negligibly close to 1. Proofs of some of the Lemmata below are sketched in the appendix.

Lemma 10. *If C and R both correctly execute* Commit *and* Open, *then R accepts the value $r = h(X)$ almost always (where h and X are as generated by C during* Commit.

We next show that the commiter has only a negligible probability of breaking the commitment scheme.

Lemma 11. *Regardless of C's strategy for generating $X, (h_1, y_1), (h_2, y_2)$ during* Commit, *there will almost certainly be at most one string, denoted X^*, that C can send R with a nonnegligible probability of acceptance.*

Hence, C is committed to $h(X^*)$. Note that C can ensure that $h(X^*)$ is not random, but this does not constitute a break of the commitment scheme. In the reduction from a standard commitment protocol to a random-bit commitment protocol, C and only C benefits from the randomness of the committed bit.

Before proving the lemma, we first define a set of viable X^* that C can reasonably send during the Open protocol.

Definition 12. *Given $X, (h_1, y_1), (h_2, y_2)$. We say that X^* is viable if it differs from X in at most d^*k places and $y_i = h_i(X^*)$ for $i = 1, 2$.*

Proposition 13. *If X^* is not viable, then C will accept X^* with negligible probability, where the probability is taken over the behavior of the noisy channel.*

Proof (of Lemma 11). (Sketch) We can view the process of generating (h_i, y_i) as progessively constraining and shrinking the viable set S. Initially, the viable set S consists of those strings whose distance from X is at most d^*k .

We use the following result by Rompel [17]

Lemma 14. *([17]) Let $X_1, ..., X_n$ be a set of t-wise independent random variables taking 0/1 values. Let $X = \sum_i X_i$, and $\mu = E(X)$. Then for any $A > 0$, we have that $Pr(|X - \mu| > A) < O((\frac{t\mu + t^2}{A^2})^{t/2})$.*

Fix any string $y \in \{0, 1\}^{\ell^*}$, and define X_i as $X_i = 1$ iff the i'th viable string is mapped to y by h_1; $X_i = 0$ otherwise. Then clearly, $\mu = 1$. We apply the above lemma with $t = 4k$ and $A = t^2$, and we say that y is bad if its preimage under h_1 has more than t^3 viable strings in it. The lemma can be used because the X_i's by construction are $64k > t$-wise independent. It immediately implies that the probability that y is bad is at most $2^{-t/2}$. The probability that ANY y is bad is at most $2^{\ell^*} \leq 2^k$ times larger, so since we have chosen $t = 4k$, even this last probability becomes exponentially small (in k). So except with exponentially small probability, at most $t^3 = 64k$ viable strings remain. The final constraint added is the value of $h_2(X)$. Since h_2 is $64k$-universal and $\ell - \ell^* > ck$, we can view this constraints as assigning at least ck random bits to each string $X^* \in S$. In order for two strings $X_1^*, X_2^* \in S$ to both remain viable, they must both receive the same bit sequence; the probability of this occurring for any such pair is negligible.

Finally, we show that after Commit, R can predict r with only a small advantage.

Lemma 15. *At the conclusion of* Commit, *the expected amount of information R holds about $h(X)$ is exponentially small in k.*

6 Reducing 1–2 OT to (γ, δ)-UNC

We first reduce 1–2 OT to (γ, δ)-PassiveUNC by a reduction secure against passive adversaries. The reduction is a straightforward adaptation of a reduction of Crépeau and Kilian [9] that builds 1–2 OT from a BSC. The same procedure is then shown to reduce 1–2 OT to (γ, δ)-UNC. Bit commitments can finally be used to tolerate active adversary for the same price.

In the appendix, reduction WOTfromPassiveUNC is described. Given a (δ, γ)-PassiveUNC, it produces a $(p(\delta, \gamma), q(\delta, \gamma), \epsilon(\delta))$-WOT $\mathcal{W}$ that can be used in reductions SR_ϵ and RS_ϵ. Using lemma 8, 1–2 OT can be obtained from any (δ, γ)-PassiveUNC that satisfies $p(\delta, \gamma) + q(\delta, \gamma) + 2\epsilon(\delta) \leq 0.45$. Unlike the bit commitment case, we were not able to show that as soon as the unfairness of the PassiveUNC is not simulatable then 1–2 OT is possible. Nevertheless, the next lemma gives a partial answer leaving a "grey" area of values for γ, δ where neither the impossibility result, nor our reduction applies. Due to space limitations, we refer the reader to [11] for the proof of next lemma.

Lemma 16. *There exists a reduction secure against passive cheating of 1–2 OT to (γ, δ)-PassiveUNC such that $\alpha^3 \beta^3 (1 - \zeta(1 - \alpha)) > \frac{0.775 + \epsilon(\delta)}{1 - \epsilon(\delta)}$ where $\epsilon(\delta) = \frac{\delta^2}{\delta^2 + (1-\delta)^2}, \alpha = \frac{1-\delta-\gamma}{1-2\gamma}, \beta = \frac{1-\gamma}{1-\delta}$, and $\zeta = \frac{1-\gamma}{\delta}$.*

To give a numerical example, when $\delta = .075$, one can reduce 1–2 OT to (γ, δ)-PassiveUNC for $\gamma \approx .06$; no such reduction is possible for $\gamma < .039$.

The reduction is also secure when a (γ, δ)-UNC is used instead of a (γ, δ)-PassiveUNC. The following straightforward lemma establishes this fact.

Lemma 17. *Any secure reduction to (γ, δ)-PassiveUNC against passive adversaries is also a secure reduction to (γ, δ)-UNC given it produces the correct output for any noise level in the interval $[\gamma \ldots \delta]$.*

Intuitively, the adversary maximizes the information he gets by reducing the noise level of a (δ, γ)-UNC to γ. In this case, the information obtained is the same as if a (γ, δ)-PassiveUNC was used. If in addition, the reduction to PassiveUNC produces the right output for any noise level in $[\gamma \ldots \delta]$ then a (γ, δ)-UNC can be used instead. It is easy to verify that WOTfromPassiveUNC is a reduction working for any noise level in the interval $[\gamma \ldots \delta]$.

Any reduction of 1–2 OT to UNC that is secure against passive cheating can handle the case of active cheating as well by proceeding along the same lines as [5]. To briefly sketch the construction, we first note that once A and B get a bit commitment scheme, they can prove in ZK that they follow the protocol honestly [12]. They can also use the bit commitment scheme in a cut and choose manner for showing that the bits sent and received through the channel are used according the protocol description. The result being that from an UNC and a bit commitment scheme, a committed UNC is built. From this point, general ZK techniques are used to make sure that no active cheating occurs. Using lemma 17 and the above argument leads to the following corollary:

Corollary 18. *Lemma 16 applies against active adversaries for both the (γ, δ)-PassiveUNC and the (γ, δ)-UNC.*

References

1. C. H. BENNETT, G. BRASSARD, C. CRÉPEAU, AND U. M. MAURER. "Generalized privacy amplification". *IEEE Trans. Info. Theory*, vol. 41, no.6, pp. 1915–1923, 1995.
2. G. BRASSARD AND C. CRÉPEAU. "Oblivious transfers and privacy amplification". *EUROCRYPT '97*, LNCS series, vol. 1223, pp. 334-347, 1997.
3. G. BRASSARD, C. CRÉPEAU, AND M. SÁNTHA. "Oblivious Transfer and Intersecting Codes". *IEEE Trans. Info. Theory*, vol. 42, No. 6, pp.1769–1780, 1996.
4. C. CACHIN. "On the Foundations of Oblivious Transfer", *EUROCRYPT '98*, LNCS series, vol. 1403, pp. 361-374.
5. C. CRÉPEAU. "Verifiable disclosure of secrets and applications". *EUROCRYPT '89*, LNCS series, vol. 434, pp.150-154, 1998.
6. C. CRÉPEAU. "Efficient Cryptographic Protocols based on Noisy Channels", *EUROCRYPT '97*, LNCS series, vol.1233, pp.306-317.
7. C. CRÉPEAU. Private communication, 1998.
8. C. CRÉPEAU. "Equivalence between two flavours of oblivious transfer", *CRYPTO '87*, LNCS series, pp.350-354, 1987.
9. C. CRÉPEAU AND J. KILIAN. "Achieving Oblivious Transfer using Weakened Security Assumptions", *FOCS 88*, pp.42-52, 1988.
10. L. CARTER AND M. WEGMAN. "Universal Classes of Hash Functions". *JCSS*, 18, pp. 143–154, 1979.
11. I. DAMGÅRD, J. KILIAN, AND L. SALVAIL, "On the (Im)possibility of basing Oblivious Transfer and Bit Commitment on Weakened Security Assumptions", BRICS report available from http://www.brics.dk/Publications/, 1998.
12. O. GOLDREICH, S. MICALI, AND A. WIGDERSON, "Proofs That Yield Nothing but Their Validity or All Languages in NP Have Zero-Knowledge Proof Systems", J. Assoc. Comput. Mach., vol. 38, pp. 691–729, 1991.
13. J. HÅSTAD, R. IMPAGLIAZZO, L. A. LEVIN, AND M. LUBY. "Construction of a pseudo-random generator from any one-way function". Technical Report TR-91-068, International Computer Science Institute, Berkeley, CA, 1991.

14. J. KILIAN. "A general completeness theorems for 2-party games". *STOC '91*, pp. 553–560, 1991.
15. U.M. MAURER. "Secret Key Agreement by Public Discussion from Common Information", *IEEE Trans. Info. Theory*, vol. 39, p.733-742, 1993.
16. M. NAOR, R. OSTROVSKY, R. VENKATESAN, AND M. YUNG. "Perfect zero-knowledge arguments for NP using any one-way permutation". *Journal of Cryptology*, vol. 11, 1998.
17. J. ROMPEL. *Techniques for Computing with Low-Independence Randomness*. PhD-thesis, MIT, 1990.

A   WOT from PassiveUNC and Proofs from Section 5

In this appendix, we first give the reduction of WOT to PassiveUNC and second, we provide the proofs from section 5. Protocol WOTfromPassiveUNC$(b_0, b_1)(c)$

1. A picks $x, y \in_R \{0, 1\}$,
2. A sends (xx, yy) through the PassiveUNC(γ, δ) and B receives $(\hat{x}\hat{x}', \hat{y}\hat{y}')$,
3. If B receives $(\hat{x} \oplus \hat{x}', \hat{y} \oplus \hat{y}') \notin \{(0, 1), (1, 0)\}$ then they go to step 1.
4. B announces w such that
   - $w = 0$ if $((\hat{x} \oplus \hat{x}' = 0) \wedge (c = 0)) \vee ((\hat{y} \oplus \hat{y}' = 0) \wedge (c = 1))$
   - $w = 1$ if $((\hat{x} \oplus \hat{x}' = 0) \wedge (c = 1)) \vee ((\hat{y} \oplus \hat{y}' = 0) \wedge (c = 0))$
5. A announces
   - $(a, b) = (x \oplus b_0, y \oplus b_1)$ if $w = 0$,
   - $(a, b) = (y \oplus b_0, x \oplus b_1)$ if $w = 1$,
6. B computes
   - $b_0 = a \oplus \hat{x}$ if $c = 0$ and $w = 0$,
   - $b_0 = a \oplus \hat{y}$ if $c = 0$ and $w = 1$,
   - $b_1 = b \oplus \hat{y}$ if $c = 1$ and $w = 0$,
   - $b_1 = b \oplus \hat{x}$ if $c = 1$ and $w = 1$.

Proof Sketch of Lemma 10   By inspection, if R accepts it will always recover $r = h(X)$ (assuming C is good), and R will only reject if X and X' differ in at least $\delta' k$ places. Now, since $\delta < 2\gamma(1 - \gamma)$, $d_0 < d_1 < \gamma$ and hence $\delta' > \delta$. However, for each i, $x_i' \neq x_i$ with independent probability at most δ, so by a standard Chernoff bound, the probability that $x_i' \neq x_i$ in $\delta' k$ is negligible in k. Note that by the universality of $\mathcal{H}$, r is distributed uniformly over $\{0, 1\}$. $\square$

Proof Sketch of Proposition 13   Clearly, R will reject if $y_i \neq h_i(X^*)$. Suppose that X^* differs from X in $d^* k$ places and the channel flips each bit with probability at least γ. Then X^* and X' will differ in at least $\delta^* k$ expected places, where $\delta^* = \gamma(1 - d^*) + (1 - \gamma)d^* > \delta'$. By a standard Chernoff bound, they will almost always differ in more than $\delta' k$ places, causing R to reject. $\square$

Proof Sketch of Lemma 15   First, we conceptually give R the value of $d(X', X)$; this can only help R. Let set S denote those Xs of the given distance; after receiving X', each $X \in S$ is equally likely. We first observe that for some constant $c_1 > 0$, $H(X', X) - dn \geq c_1 n$ almost always; it follows that for some constant $c_2 > 0$, $|S|/2^\ell \geq 2^{c_2 k}$.

Now, after receiving X', R can obtain $h_1(X), h_2(X)$. Note that we cannot assume these functions are chosen randomly. Still, conceptually, we can view R as flipping its random coins (if it uses any) and then constructing a (quite shallow) decision tree. Each interior vertex v is labelled with a hash function h to be sent to C; the edges from v to its children correspond to the possible answers C might give (2^{ℓ^*} possibilities in the first level, $2^{\ell-\ell^*}$ in the second).

For every vertex v in the tree we define the set S_v as those $X \in S$ that are consistent with the sequence of hash functions and answers given on the path from the root vertex to v. We can view Step 2 of Commit as a traversal from the root of the tree to a leaf l.

By a simple probability argument, the probability that a given leaf l is reached is $|S_l|/|S|$, and the conditional distribution on X is uniform over S_l. Since the tree has only 2^ℓ leaves, the probability of reaching a leaf l with $|S_l| < 2^{c_2 k/2}$ is at most $2^{-c_2 k/2}$; we can safely ignore this event. On the other hand, in every case where $|S_l| \geq 2^{c_2 k/2}$, it follows immediately from the privacy amplification result in [1] that R's expected information about $h(X)$ is exponentially small.

□

Conditional Oblivious Transfer and Timed-Release Encryption

Giovanni Di Crescenzo[1], Rafail Ostrovsky[2], and
Sivaramakrishnan Rajagopalan[2]

[1] Computer Science Department, University of California San Diego,
La Jolla, CA, 92093-0114
giovanni@cs.ucsd.edu
Work done while at Bellcore.
[2] Bell Communications Research,
445 South Street, Morristown, NJ, 07960-6438, USA
{rafail,sraj}@bellcore.com

Abstract. We consider the problem of sending messages "into the future." Previous constructions for this task were either based on heuristic assumptions or did not provide anonymity to the sender of the message. In the public-key setting, we present an efficient and secure *timed-release encryption scheme* using a "time server" which inputs the current time into the system. The server has to only interact with the receiver and never learns the sender's identity. The scheme's computational and communicational cost per request are only logarithmic in the time parameter. The construction of our scheme is based on a novel cryptographic primitive: a variant of oblivious transfer which we call *conditional oblivious transfer*. We define this primitive (which may be of independent interest) and show an efficient construction for an instance of this new primitive based on the quadratic residuosity assumption.

1 Introduction

Time is a critical aspect of many applications in distributed computer systems and networks. Among other things, time is used to co-ordinate remote actions, to guarantee and monitor services, and to create linear order in some distributed transactions systems. Roughly speaking, applications of time in distributed systems fall in two categories: those that use relative time between events (e.g. one hour from the last reboot) and those that use absolute time (e.g. 0900 hours, May 2, 1999 GMT). We can concentrate on the second category as relative timing can be implemented using absolute time but not vice-versa. While the existence of a common view of current time is essential in systems that use absolute time, it is generally hard to implement in a distributed system – either the local clock is assumed to be an acceptable approximation of the universal time or there is a central time "server" that is available whenever needed and local clocks are periodically synchronized [24]. In some cases, the trustworthiness of the time server may be an issue as adversarial programs may try to change the value of

J. Stern (Ed.): EUROCRYPT'99, LNCS 1592, pp. 74–89, 1999.
© Springer-Verlag Berlin Heidelberg 1999

the local clock or spoof a network clock to their advantage and this may have an unacceptable negative effect on the system. Such problems can be solved using authentication mechanisms as long as the applications only need the current time (or to be more accurate, the "latest time"). There are many applications that depend on a common assumption of an absolute time that is in the future, where, say, the opening of a document before a specified time is unacceptable. An example is the Internet programming contest where teams located all over the world need to be given access to the challenge problems at a certain time. Another example may be in trading stocks: suppose one wants to send an e-mail message from their laptop computer to a broker to sell a stock at a particular time in the future. The broker should not be able to see the message before that time and gain an unfair advantage, and yet one cannot rely on a service such as e-mail to work correctly and expeditiously if the message was sent exactly on release-time. The essence of the problem is this: the message has to be sent early but the broker should not be able to read the message before the release-time. Also, it would be preferable from a security perspective if the time server never learns the identity of the sender of the message.

The act of encrypting a document so that it cannot be read before a release-time has been called "sending information in to the future" or timed-release cryptography by May [22]. Rivest, Shamir and Wagner [27] give a number of applications for timed-release cryptography: electronic actions, sealed bids and chess moves, release of documents (like memoirs) over time, payment schedules, press releases, etc. Bellare and Goldwasser [2] propose that timed release may also be used in key escrow: they suggest that the delayed release of escrowed keys may be a suitable deterrent in some contexts to the possible abuse of escrow.

Prior techniques: There are two main approaches suggested in the literature: the first one is based on so-called "time-lock puzzles," and the second one is based on trusted "time-servers." Time-lock puzzles were first suggested by Merkle [23] and extended by Rivest et al. [27]. The idea is that the time to recover a secret is given by the minimum computational effort needed by any machine, serial or parallel, to perform some computation which enables one to recover the secret. This approach has the interesting feature of not requiring any third party or trust assumption. On the other hand, it escapes a formal proof that a certain lock is valid for a certain time. The time-lock puzzle presented by Bellare and Goldwasser in [2] is based on the heuristic assumption that exhaustive search on the key space is the fastest method to recover the key of, say, 40-bit DES. In [27], Rivest et al. point out that this only works on average: for instance, for a particular key, exhaustive search may find the key well before the assigned time; they then propose a time-lock puzzle based on the hardness of factoring which does not have this problem, although it still uses a different heuristic assumption about the minimum time needed to perform some number-theoretic calculations. A major disadvantage of time-lock puzzles from our point of view is that they can only solve the relative time problem.

A "time-server" is a trusted third party that is expected to allow release of the message at the appointed time only. May [22] suggests that the third party be a trusted escrow agent that stores the message and releases it at release-time. This does not scale well since the agent has to store all escrowed messages until their release-times. Moreover, no anonymity is guaranteed: the server knows the message, the release-time, and the identity of the two parties. In Rivest et al. [27] it was suggested that the server simply use the iterates of a one-way function (or a public-key sequence) and publish the next iterate value after one unit of time. A sender wishing to release a document at time t gets the server to encrypt the document (or a key to an encrypted version) with a secret key that the server will only publish at time t. This scheme has the advantage that the server does not have to remember anything except the seed to the sequence of one-way function iterates. This scheme requires the server to generate and publish a large number of keys, so that the overall computation and storage costs are linear in the time parameter. Furthermore, the receivers are anonymous but the sender is not anonymous in this scheme and the time server knows the release-time requested.

The Model. The general discussion above shows that it is necessary and advantageous to have a system in which the current absolute time is available with the facility of posting messages that can be read at a future time. For completeness, we first lay out some basic considerations. To begin with, time needs to be defined. In computers, as in real life, time is simply defined to be the output of a certain clock (such as the Denver Atomic Clock). We can assume the existence of such an entity which we will call the Time Server (or simply, server) that defines the time. Naturally, the server outputs (periodically or upon request) messages that indicate the current time (down to whatever granularity is needed or possible). The server is endowed with a universally known public key. However, the server is not required to remember any other keys, such as keys of senders and receivers (in this respect, our model is different from that of Rivest et al. [27]).

What we are concerned with here is timed release of electronic documents: it is straightforward to see that using encryption, we can reduce this problem to timed release of the decryption key (rather than the document itself which is assumed to be delivered ahead in encrypted form). Now, if the sender of a document is available at the time of its release, this problem is trivial since the sender can herself provide the decryption key at release-time. Thus, we can assume that the sender of the document is only present prior to, but not at, the release time. Furthermore, as May [22] suggested, if we have a trusted escrow agent that supplies the decryption key at the appointed time, again the problem is solved (but the solution does not scale well for the server). Next, the problem is also trivial if the receiver can be trusted by the sender to not decrypt the document before the release-time. Therefore, we assume that the the sender does not trust the receiver to not try to decrypt the document before the release-time. Finally, it may be that the sender is remote and hence may not be able to communicate directly with the server (this is not possible in [27]). Hence, we will also assume that the sender and server cannot interact at any time. However, the receiver can interact with the server and we will be interested in keeping this interaction to the minimum as well.

Putting all this together, the problem of timed-release encryption that we address in this paper can be restated as follows: how can a sender, without talking to the server, create a document with a release-time (defined using the notion of time as marked by the server) such that a receiver can decrypt this document only after the release-time has passed by interacting with the server and such that the server does not learn the identity of the sender?

Our results: We present a formal definition for the cryptographic task of a *timed-release encryption scheme*, and a solution to this task. Also, we introduce a new variant of the oblivious transfer protocol, which we call *conditional oblivious transfer*. We present a formal definition for this variant, and a construction for an instance of it. The properties of this construction will be crucial for the design of our timed-release encryption scheme.

Conditional Oblivious Transfer. The Oblivious Transfer protocol was introduced by Rabin [26]. Informally, it can be described as follows: it is a game between two polynomial time parties Alice and Bob; Alice wants to send a message to Bob in such a way that with probability $1/2$ Bob will receive the same message Alice wanted to send, and with probability $1/2$ Bob will receive nothing. Moreover, Alice does not know which of the two events really happened. There are other equivalent formulations of Oblivious Transfer (see, e.g., [8,9,15,3,21]). This primitive has found numerous applications (see, e.g., [19,16,29,17]).

In this paper, we consider a variant of Oblivious Transfer, which we call Conditional Oblivious Transfer. In this variant, Bob and Alice have private inputs and share a public *predicate* that is evaluated over the private inputs and is computable in polynomial time. The conditional oblivious transfer of (for simplicity), say, a bit b from Alice to Bob has the following requirements. If the predicate holds, then Bob successfully receives the bit Alice wanted to send him and if the predicate does not hold, then no matter how Bob plays, he will have no information about the bit Alice wanted to send him. Furthermore, no efficient strategy can help Alice during the protocol in computing the actual value of the predicate. Of course, such a game can be easily implemented as an instance of secure function evaluation [29,17,4,10,16], however, we are interested here in more efficient implementations of this particular game. To the best of our knowledge, such a variant has not been considered previously in the literature.

Timed-Release Encryption. The setting is as follows. There are three participants: the sender, the receiver and the server. First, the sender transmits to the receiver an encrypted messages and a release-time. Then, the receiver can engage in a conversation with a server. Our timed-release encryption scheme uses, in a crucial way, a protocol for conditional oblivious transfer. In particular, the server and receiver engage in a conditional oblivious transfer such that if the release-time is not less than the current time as defined by the server, then the receiver gets the message. Otherwise, the receiver gets nothing. Furthermore, the server does not learn any information about the release-time or the identity of the sender. In particular, the server does not learn whether the release-time is less than, equal to, or greater than the current time. Our protocol has minimal round-complexity: an execution of the scheme consists of a single message from the sender to the receiver and one request-answer interaction between the receiver and the time server. Moreover, we present an implementation of our scheme, using efficient primitives and cryptosystems as building blocks, that only require communication and computation logarithmic in the size of the time parameter. Finally, we note that the trust placed on the server can be further decreased if more servers are available.

2   Notations and Definitions

In this section we present notations and definitions needed for this paper. We start with basic notations, then we define conditional oblivious transfer and timed-release encryption schemes. For the necessary number-theoretic background on quadratic residuosity and Blum integers, we refer the reader to [1,11].

2.1   Basic Notations and Model

An algorithm is a Turing machine. An *efficient* algorithm is an algorithm running in probabilistic polynomial time. An interactive Turing machine is a probabilistic algorithm with an additional communication tape. A pair of interactive Turing machines is an *interactive protocol*. The notation $x \xleftarrow{D} S$ denotes the *probabilistic experiment* of choosing element x from set S according to distribution D; we only write $x \leftarrow S$ in the case D is the uniform distribution over S. The notation $y \leftarrow A(x)$, where A is an algorithm, denotes the probabilistic experiment of obtaining y when running algorithm A on input x, where the probability space is given by the random coins (if any) of algorithm A. Similarly, the notation $t \leftarrow (A(x), B(y))(z)$ denotes the probabilistic experiment of running interactive protocol (A,B), where x is A's input, y is B's input, z is an input common to A and B, and t is the transcript of the communication between A and B during such execution. By $\mathrm{Prob}[R_1; \ldots ; R_n : E]$ we denote the probability of event E, after the execution of probabilistic experiments $R_1, \ldots , R_n$. Let $a \oplus b$ denote the string obtained as the bitwise logical xor of strings a and b. Let $a \circ b$ denote the string obtained by concatenating strings a and b. A language is a subset of $\{0,1\}^*$.

The predicate GE. Given two sequences of k bits $t_1, \ldots , t_k$, and $d_1, \ldots , d_k$, define predicate GE as follows: $\mathrm{GE}(t_1, \ldots , t_k, d_1, \ldots , d_k) = 1$ if and only if $(t_1 \circ \cdots \circ t_k) \geq (d_1 \circ \cdots \circ d_k)$, when strings $t_1 \circ \cdots \circ t_k$ and $d_1 \circ \cdots \circ d_k$ are interpreted as integers.

The public random string model. In the sequel, we define two cryptographic protocols: conditional oblivious transfer, and timed-release encryption, in a setting that is well known as the "public random string" model. In this model, the parties share a public and uniformly distributed string. It was introduced by Blum, De Santis, Feldman, Micali and Persiano in [6,5], and was motivated by the goal of reducing the ingredients needed for the implementation of zero-knowledge proofs. This model has been well studied in Cryptography since then as a minimal setting for obtaining non-interactive zero-knowledge proofs and several other cryptographic protocols.

2.2 Conditional Oblivious Transfer: Definition

We now give the formal definition of Conditional Oblivious Transfer.

Definition 1. Let Alice and Bob be two probabilistic Turing machines running in time polynomial in some security parameter n. Also, let x_A (x_B) be Alice's (respectively, Bob's) private input, let b be the private bit Alice wants to send to Bob, and let $q(\cdot, \cdot)$ be a predicate computable in polynomial time. We say that (Alice,Bob) is a CONDITIONAL OBLIVIOUS TRANSFER protocol for predicate q if there exists a constant a such that:

1. *Transfer Validity.* If $q(x_A, x_B) = 1$ then for each $b \in \{0, 1\}$, it holds that

$$\text{Prob}\left[\sigma \leftarrow \{0,1\}^{n^a}; tr \leftarrow (\text{Alice}(x_A, b), \text{Bob}(x_B))(\sigma) : \text{Bob}(\sigma, x_B, tr) = b\right] = 1.$$

2. *Security against Bob.* If $q(x_A, x_B) = 0$ then for any Bob', the random variables X_0 and X_1 are equally distributed, where, for $b \in \{0, 1\}$,

$$X_b = [\sigma \leftarrow \{0,1\}^{n^a}; tr \leftarrow (\text{Alice}(x_A, b), \text{Bob}'(x_B))(\sigma) : (\sigma, tr)]$$

3. *Security against Alice.* For any efficient Alice', there exists an efficient simulator M such that for any constant c, and any sufficiently large n, it holds that $|p_0 - p_1| \leq n^{-c}$, where, p_0 and p_1 are equal to, respectively,

$$\text{Prob}\left[\sigma \leftarrow \{0,1\}^{n^a}; tr \leftarrow (\text{Alice}'(x_A), \text{Bob}(x_B))(\sigma): \text{Alice}'(\sigma, x_A, tr) = q(x_A, x_B)\right]$$

$$\text{Prob}\left[\sigma \leftarrow \{0,1\}^{n^a} : \text{M}(\sigma, x_A) = q(x_A, x_B)\right].$$

Notice that here we are defining the security against a possibly cheating Bob even if he is infinitely powerful (requirement 2), similar to [25]. In the sequel, we will also consider security with respect to a honest-but-curious Bob, meaning that Bob follows his program, but at the end may arbitrarily try to distinguish random variables X_0 and X_1. We also note that a definition suitable for the public-key setting can be easily obtained by the above one. In particular each party would use its private string too and would be given access also to the public keys of the other parties; namely, Alice and M would be given Bob's public key, and Bob would be given Alice's public key. Finally, we note that the above definition can be extended in a natural way to the case of a message containing more than one bit.

2.3 Timed-Release Encryption: Definition

First, we give an informal description. A timed release encryption scheme is a protocol between three polynomial time parties: a sender S, a receiver R and a server V. Time (a positive integer) is represented as a k-bit string and is entirely managed by V. Each message sent in an "encrypted" form from S to R will be associated with a release-time $d = (d_1, \ldots, d_k)$, where $d_i \in \{0, 1\}$, for $i = 1, \ldots, 2^k$. R can check if the message it got from S is "well-formed". If the message is "well-formed" then R is guaranteed to be able to decrypt some message after the release-time d. R is allowed to interact with V, while S never needs to interact with V. Also, for any efficient strategy of R, R can not decrypt before the release-time. Finally, V just acts as a time server; i.e., the conversation V sees reveals no information about the actual message, its release-time or which sender/receiver pair is involved in the current conversation. Time is managed by V by answering timing requests from R; namely, first R sends a message to V, then V answers. V's answer contains some information allowing R to decrypt if and only if the current time is greater than the release-time.

By (p_r, s_r) (resp., (p_v, s_v)) we will denote a pair of R's (resp., V's) public/secret keys; by σ a sufficiently long public random string; by $m \in \{0,1\}^*$ a message, by t and d the current time according to V's clock and the release-time of message m, both being represented as a k-bit string. We now present our formal definition of timed release encryption scheme.

Definition 2. Let S,R,V be three probabilistic Turing machines running in time polynomial in some security parameter n. Let $\perp$ denote a special symbol that may be output by any of S,R,V, on any input, meaning that such input is not of the specified form. We say that (S,R,V) is a TIMED-RELEASE ENCRYPTION SCHEME if there exists a constant a such that:

0. *Correctness.* For any $m \in \{0,1\}^*$ and any $d \in \{0,1\}^k$,

$$\text{Prob} [\, \sigma \leftarrow \{0,1\}^{n^a}; (p_v, s_v) \leftarrow V(\sigma); (p_r, s_r) \leftarrow R(\sigma);$$
$$(enc, d) \leftarrow S(p_r, p_v, m, d); req \leftarrow R(p_r, s_r, p_v, enc, d);$$
$$ans \leftarrow V(p_v, s_v, t, req, \sigma):$$
$$(t < d) \vee (R(p_r, s_r, ans) = m)] = 1.$$

1. *Security against S.* For any probabilistic polynomial time S', any constant c, and any sufficiently large n,

$$\text{Prob} [\, \sigma \leftarrow \{0,1\}^{n^a}; (p_v, s_v) \leftarrow V(\sigma); (p_r, s_r) \leftarrow R(\sigma);$$
$$(enc, d) \leftarrow S'(p_r, p_v); req \leftarrow R(p_r, s_r, p_v, enc, d);$$
$$ans \leftarrow V(p_v, s_v, t, req, \sigma):$$
$$(t < d) \vee (req = \perp) \vee (R(p_r, s_r, ans) \neq \perp)] \geq 1 - n^{-c}.$$

2. *Security against R.* For any probabilistic polynomial time R'=(R'$_1$,R'$_2$,R'$_3$,R'$_4$), any constant c, and any sufficiently large n,

$$\text{Prob} [\, \sigma \leftarrow \{0,1\}^{n^a}; (p_v, s_v) \leftarrow V(\sigma); (p_r, s_r) \leftarrow R'_1(\sigma, p_v);$$
$$(m_0, m_1, aux) \leftarrow R'_2(\sigma, p_v, p_r, s_r); i \leftarrow \{0,1\}; (enc, d) \leftarrow S(p_r, p_v, m_i, d);$$
$$req \leftarrow R'_3(p_r, s_r, p_v, enc, d); ans \leftarrow V(p_v, s_v, t, req, \sigma):$$
$$(t < d) \wedge R'_4(p_r, s_r, aux, ans) = i] \leq 1/2 + n^{-c}.$$

3. *Security against V.* For any probabilistic polynomial time $V'=(V'_1,V'_2,V'_3)$, any constant c, and any sufficiently large n,

$$\text{Prob} \left[\sigma \leftarrow \{0,1\}^{n^a}; (p_r, s_r) \leftarrow R(\sigma); (p_v, s_v) \leftarrow V'_1(\sigma, p_r); \right.$$
$$((m_0, d_0), (m_1, d_1), aux) \leftarrow V'_2(\sigma, p_r); i \leftarrow \{0, 1\};$$
$$(enc, d_i) \leftarrow S(p_r, p_v, m_i, d_i); req \leftarrow R(p_r, s_r, p_v, enc, d_i) :$$
$$\left. V'_3(p_v, s_v, t, req, aux, \sigma) = i \right] \leq 1/2 + n^{-c}.$$

Notes on Definition 2 :

- The validity of the encryption scheme is defined even with respect to malicious senders (requirement 1): even if S is malicious, and tries to send a message for which he claims the release-time to be d, then R can always decrypt after time d.
- The security against malicious R (requirement 2), and V (requirement 3) have been defined in the sense of semantic security [18] against chosen message attack. Extensions to chosen ciphertext attack can be similarly formalized.
- A scheme satisfying the above definition also protects the sender's anonymity: namely, the sender does not need to use his public or private keys when talking to the receiver, and never talks to the server.

3   A Conditional Oblivious Transfer Protocol for GE

In this section we show a conditional oblivious transfer protocol for predicate GE. Our result is the following

Theorem 3. Let GE be the predicate defined in Section 2. The protocol (Alice,Bob) presented in Section 3 is a conditional oblivious transfer protocol for GE, for which requirement 1 of Definition 1 holds with respect to any honest-but-curious and infinitely powerful Bob and requirement 3 of Definition 1 holds with respect to any probabilistic polynomial time Alice under the hardness of deciding quadratic residuosity modulo Blum integers.

The rest of this section is devoted to the proof of Theorem 3.

3.1   Our Conditional Oblivious Transfer Protocol

We first give an informal description of the ideas in our protocol and then give the formal description.

An informal description. We will use as a subprotocol (A,B) a simple variation of the oblivious transfer protocol given in [13,12]. based on quadratic residuosity modulo Blum integers. For lack of space, we omit the description of such protocol but give the properties necessary for our construction here. By NQR-COT-Send($b, (x,y)$) we denote the algorithm A on input a bit b and (x,y), where x is a Blum integer, $y \in Z_x^{+1}$. By NQR-COT-Receive($mes, (x,p,q,y)$)

we denote the algorithm B using the factors p, q of x to decode a message mes sent by A using (x, y), where the result of such decoding will be either b or $\perp$ (indicating an invalid message). We recall that in protocol (A,B), algorithm B will receive bit b sent by A if y is a quadratic non residue and the actual value of b will remain information-theoretically hidden with respect to a honest-but-curious B otherwise. Moreover, no efficient strategy allows A to guess whether B actually received the right value for b or not.

Informally, our COT protocol for the predicate GE works as follows. At the beginning of the protocol, Alice has a k-bit string $t = (t_1, \ldots, t_k)$ as her secret key and Bob has a k-bit string $d = (d_1, \ldots, d_k)$ as his secret key. Moreover, let b be the bit that Alice wants to send to Bob. First of all, Bob computes a Blum integer x, and a k-tuple $(D_1, \ldots, D_k)$ as his public key, where $D_i = r_i^2 y^{d_i} \bmod x$, where the d_i's are part of Bob's secret key. Similarly, Alice computes her public key as integers $T_1, \ldots, T_k \in Z_x^{+1}$, where T_i is a quadratic non residue if and only if $t_i = 1$. Now, one would like to use properly computed products of the T_i's and D_i's to play the role of integer y in the above mentioned protocol (A,B), where the products are computed according to the boolean expression that represents predicate GE over bit strings. A protocol implementing this would require $\Theta(k^2)$ modular multiplications. We show below a protocol which only requires $8k$ modular multiplications.

First of all Alice splits bit b into bit a and bit $a \oplus b$, for random a, and sends a using (x, T_1) and $a \oplus b$ using $(x, D_1 T_1 \bmod x)$ as inputs for the subprotocol (A,B) (notice that this allows Bob to receive b if and only if $t_1 > d_1$). Then, Alice will send a random bit c using $(x, -T_1 D_1 \bmod x)$ as input (this allows Bob to receive c if and only if $t_1 = d_1$). The gain in having the latter step is that it allows Alice to run the same scheme recursively on the tuple $(T_2, \ldots, T_k, , D_2, \ldots, D_k)$, using as input $b \oplus c$. Notice that if $t < d$ Bob will be able to compute only bits with distribution uniform and independent from b. In this protocol Alice only performs 8 modular multiplications for each i (this is because A's algorithm only requires 2 modular multiplications). We now proceed with the formal description of our scheme (Alice,Bob).

THE ALGORITHM ALICE: On input $b, t_1, \ldots, t_k \in \{0, 1\}$, Alice does the following:

1. **Receive:** $x, D_1, \ldots, D_k$ from Bob and set $b_1 = b$.
2. For $i = 1, \ldots, k$,
    uniformly choose $a_i, c_i \in \{0, 1\}, r_i \in Z_x^*$ and compute $T_i = r_i^2(-1)^{t_i} \bmod x$;
    if $i = k$ then set $c_i = b_i$;
    compute $mes_{i1} = \text{NQR-COT-Send}(a_i, (x, T_i))$;
    compute $mes_{i2} = \text{NQR-COT-Send}(a_i \oplus b_i, (x, D_i T_i \bmod x))$;
    compute $mes_{i3} = \text{NQR-COT-Send}(c_i, (x, -D_i T_i \bmod x))$;
    set $b_{i+1} = b_i \oplus c_i$;
    set $p_A = (T_1, \ldots, T_k)$ and $mes = ((mes_{11}, mes_{12}, mes_{13}), \ldots, (mes_{k1}, mes_{k2}, mes_{k3}))$;
    send: (p_A, mes) to Bob.

THE ALGORITHM BOB: On input a sufficiently long string σ and $d_1, \ldots, d_k \in \{0, 1\}$, Bob does the following

1. Uniformly choose two n-bit primes p, q such that $p \equiv q \equiv 3 \bmod 4$ and set $x = pq$; for $i = 1, \ldots, k$, uniformly choose $r_i \in Z_x^*$ and compute $D_i = r_i^2 (-1)^{d_i} \bmod x$; let $p_B = (x, D_1, \ldots, D_k)$ and **send:** p_B to Alice.
2. **Receive:** $((T_1, \ldots, T_k), (mes_{11}, mes_{12}, mes_{13}, \ldots, mes_{k1}, mes_{k2}, mes_{k3}))$ by Alice.
3. For $i = 1, \ldots, k$,
    compute $a_i = \text{NQR-COT-Receive}(mes_{i1}, (x, p, q, T_i))$;
    compute $e_i = \text{NQR-COT-Receive}(mes_{i2}, (x, p, q, D_i T_i \bmod x))$;
    if $a_i \neq \bot$ and $e_i \neq \bot$ then
        output: $a_i \oplus e_i \oplus c_{i-1} \oplus \cdots \oplus c_1$ and halt;
    else compute $c_i = \text{NQR-COT-Receive}(mes_{i3}, (x, p, q, -D_i T_i \bmod x))$;
    if $i = k$ and $c_i \neq \bot$ then **output:** c_i and halt;
    output: $\bot$.

3.2 Conditional Oblivious Transfer: The Proof

We need to prove three properties: transfer validity, security against Alice and security against Bob.

Transfer validity. Assume predicate q is true; i.e., $t_1 \circ \cdots \circ t_k \geq d_1 \circ \cdots \circ d_k$. Notice that if $t_1 > d_1$ then T_1 and $D_1 T_1 \bmod x$ are quadratic non residue and by the validity property of NQR-COT-Receive, Bob can compute a_1, e_1. and therefore b as $a_1 \oplus e_1$. Now, assume that $t_j = d_j$, for $j = 1, \ldots, i-1$ and $t_i > d_i$, for some $i \in \{1, \ldots, t\}$. Then, since $t_j = d_j$, the integer $-T_j D_j \bmod x$ is a quadratic non residue modulo x and Bob can compute c_j, for each $j = 1, \ldots, i-1$, by the validity property of NQR-COT-Receive; moreover, since $t_i > d_i$, Bob can compute both a_i and e_i from Alice's message. Since $e_i = a_i \oplus b \oplus c_1 \oplus \cdots \oplus c_{i-1}$, Bob can compute b as $a_i \oplus e_i \oplus c_1 \oplus \cdots \oplus c_{i-1}$. Finally, the case of $t_j = d_j$, for $j = 1, \ldots, k$, is similarly shown, by just observing that c_k is set equal to $b \oplus c_1 \oplus \cdots \oplus c_{k-1}$.

Security against Bob. To prove security against any honest-but-curious algorithm Bob', first, assume that x is a Blum integer, $D_1, \ldots, D_k \in Z_x^{+1}$, and predicate q is false; i.e., $t_1 \circ \cdots \circ t_k < d_1 \circ \cdots \circ d_k$. Consequently, for some $i \in \{1, \ldots, k\}$ it must be that $t_j = d_j$, for $j = 1, \ldots, i-1$, and $t_i < d_i$. Note that according to Alice's algorithm, b can be written as $a_1 \oplus e_1$ or as $c_1 \oplus \cdots \oplus c_{l-1} \oplus a_l \oplus e_l$, for some l. Then, since $T_j D_j$, for $j = 1, \ldots, i-1$, is a quadratic residue modulo x, for each j, it holds that at most c_j and a_j can be computed from mes_{j1}, mes_{j3}, but e_j is information-theoretically hidden given mes_{j2}; from the properties of NQR-COT-Send. Notice that both a_j and c_j are independently and uniformly distributed bits. Then, since $t_i < d_i$, Bob' has no information about either a_i or c_i; this guarantees that even for any $i' > i$ such that $t_{i'} > d_{i'}$, Bob' will obtain $a_{i'}$ and $a_{i'} \oplus b_{i'}$, but not b since $b_{i'} = b \oplus c_1 \oplus \cdots \oplus c_{i'-1}$. Moreover, even for such values i', the values received by Bob' are again independently and uniformly distributed bits. Hence, for any b, Bob' only sees uniform and independent bits. Therefore, the two variables X_0, X_1 are equally distributed.

Security against Alice. Notice that Alice's role in the protocol consists of a single message to Bob. Therefore, if after the protocol, Alice has a non-negligible advantage over any efficient simulator M in deciding the predicate q, then she has the same advantage when she is given only Bob's public message p_B before running the protocol. Therefore, there exists an efficient M that has the same advantage in deciding predicate q as Alice. Finally, using a standard hybrid argument, M has a non-negligible advantage in deciding the quadratic residuosity modulo the Blum integer x of one of the D_i's, and therefore any $y \in Z_x^{+1}$.

This concludes the proof of Theorem 3.

4   A Timed-Release Encryption Scheme

In this section, we present our construction of a timed-release encryption scheme which can be viewed as a transformation from any ordinary encryption scheme into a timed-release one. It uses as additional tools, a non-malleable encryption scheme and a conditional oblivious transfer protocol for the predicate GE. Our scheme can be based on several different intractability assumptions, according to what goal one is interested in (i.e., generality of the assumptions, efficiency in terms of communication, and efficiency in terms of computation). We discuss all these variants after our formal description and proof. Our result is the following

Theorem 4. The scheme (S,R,V) defined in Section 4.1 is a timed-release encryption scheme.

4.1   Description of Our Scheme

We start with an informal description of the ideas needed for our scheme. A first idea for the construction of our scheme would be to use the conditional oblivious transfer designed in Section 3 as follows. Assume the receiver has obtained the release-time $d = (d_1, \ldots, d_k)$ of the message from the sender. Since the server has the current time $t = (t_1, \ldots, t_k)$, the server and the receiver can execute the conditional oblivious transfer protocol for predicate GE, where the server plays as Alice on input t and the receiver plays as Bob on input d. Additionally, the receiver, by running this protocol should get the information required to decrypt and compute the message. The properties of conditional oblivious transfer guarantee that the receiver will be able to receive some private information if and only if the time of the receiver's request was not earlier than the release-time. First, we have to decide what secret information should be sent from the server to the receiver in the event that the release-time is past. This can be as follows: the sender will first encrypt the message using the receiver's public key, and then encrypt this encryption using the server's public key. Let z_0 be the resulting message. Then the private information sent by the server to the receiver could be the decryption of z_0 under the server's public key. Note this is the encryption of the message under the receiver's key and therefore this would give the receiver a way to compute the message. Moreover, the sender does not get any information about the message since he only sees an encryption of it.

A second issue is about the release-time. So far, we have assumed that the receiver encrypts the same release-time that he obtains from the sender, and the server uses those encryptions for the conditional oblivious transfer. However, a malicious receiver could simply replace the release-time with an earlier one and obtain the message earlier. Now, a first approach to prevent this is the following: the server will compute the bit-by-bit encryption of the release-time needed for the conditional oblivious transfer and send it to the receiver, together with a further encryption of it under the server's public key. Let z_1 be the resulting message. The idea would be that the receiver will be required to send z_1 to the server so that the server can verify that he is using the right encryptions. Still, the receiver can compute a faked encryption z_1' and repeat the same attack as before. However, now we can have the sender encrypt under the server's key the concatenation of the encryption of the release-time (under the receiver's key) and the encryption of the message (under the receiver's key). In other words, z_0 and z_1 are actually merged into a single encryption z. Now, the only attack the receiver can try is to modify z into something which may be decrypted as encryptions of the same message and a different release-time. However, this can be prevented by requiring that the encryption scheme of the server is non-malleable. the preceding discussion gives us our timed-release encryption scheme described formally below.

A formal description of our scheme. Let (nm-G, nm-E, nm-D) be a non-malleable encryption scheme, and denote by $(nm\text{-}pk, nm\text{-}sk)$ the pair of public and secret keys output by nm-G. Also, let (Alice,Bob) denote the conditional oblivious transfer protocol for predicate GE given in Section 3. We now describe the three algorithms S, R, V; in each algorithm, when necessary, we differentiate between the key-generation phase and the encryption/decryption phase. Also, we assume wlog that the message m to be encrypted is a single bit.

THE ALGORITHM S:

Key-Generation Phase: no instruction required.

Encryption Phase:

1. Let (m, d) be the pair message/release-time input to S, where $m \in \{0, 1\}$;
2. let (x) be R's public key;
3. uniformly choose $r \in Z_x^*$ and compute $c_m = r^2(-1)^m \bmod x$;
4. let $d = d_1, \ldots, d_k$, where $d_i \in \{0, 1\}$, for $i = 1, \ldots, k$;
5. for $i = 1, \ldots, k$, uniformly choose $r_{2i} \in Z_x^*$ and compute $D_i = r_{2i}^2(-1)^{d_i} \bmod x$;
6. let $c_d = (D_1, \ldots, D_k)$;
7. compute $cc = nm\text{-}E(nm\text{-}pk, c_d \circ c_m)$ and **output:** (cc, c_d, d).

THE ALGORITHM R:

Key-Generation Phase:

1. Uniformly choose two $n/2$-bit primes p, q such that $p \equiv q \equiv 3 \bmod 4$ and set $x = pq$;
2. let L be the language $\{x \mid x$ is a Blum integer $\}$;

3. using σ, p, q, compute a non-interactive zero-knowledge proof Π for L;
4. **output:** (x, Π).

Decryption Phase:

1. Let (cc, c_d, d) be the triple received by S;
2. let $d = d_1, \ldots, d_k$, where $d_i \in \{0, 1\}$, for $i = 1, \ldots, k$;
3. for $i = 1, \ldots, k$,
       using p, q, set $d'_i = 1$ if $(x, D_i) \in NQR$ or $d'_i = 0$ otherwise;
       if $d'_i \neq d_i$ then **output:** $\perp$ and halt.
4. run step 1 of algorithm Bob, by sending $(x, D_1, \ldots, D_k)$ to V;
5. send cc, Π to V;
6. run step 2 of algorithm Bob, by receiving $(T_1, \ldots, T_k), mes$ from V;
7. run step 3 of algorithm Bob, by decoding mes as c_m;
8. if $c_m \neq \perp$ then compute $m = D(sk, pk, c_m)$ and **output:** m else **output:** $\perp$.

THE ALGORITHM V:

Key-Generation Phase:

1. Run algorithm nm-G to generate a pair $(nm\text{-}pk, nm\text{-}sk)$;
2. output: $nm\text{-}pk$.

Timing service phase:

1. run step 1 of algorithm Alice, by receiving $(x, D_1, \ldots, D_k)$ from R;
2. receive cc, Π from R;
3. verify that the proof Π is accepting;
4. compute $(c'_d, c'_m) = nm\text{-}D(nm\text{-}sk, nm\text{-}pk, cc)$;
5. if $c_d \neq c'_d$ or the above verification is not satisfied then **output** $\perp$ to R and **halt**;
6. let $t = (t_1, \ldots, t_k)$ be the current time, where $t_i \in \{0, 1\}$, for $i = 1, \ldots, k$;
7. for $i = 1, \ldots, k$, uniformly choose $r_i \in Z_x^*$ and compute $T_i = r_i^2(-1)^{t_i} \bmod x$;
8. run step 2 of algorithm Alice, by computing mes;
9. **output:** $(T_1, \ldots, T_m), mes$.

Round complexity: In the above description, we use the specific conditional oblivious transfer protocol of Section 3, based on quadratic residuosity, since this protocol shows that the entire timed-release can be implemented with minimum interaction. Notice that the sender does not interact at all with the server. Moreover, the sender only sends one message to the receiver in order to encrypt a message, after the receiver has published his public key. Finally the interaction between receiver and server is one round (after both parties have published their own public keys).

Efficiency: In the above description, we can use a generic non-malleable encryption scheme. A practical implementation would use, for instance, the scheme by Cramer and Shoup [7], that is based on the hardness of the decision Diffie-Hellman problem. Recall that the scheme in [7] requires about 5 exponentiations from its parties. The rest of the communication between sender and receiver is based on computing an encryption of the message m and release-time d, which requires at most k modular products (which is less expensive than one exponentiation, since k is the number of bits to encode time, and therefore a very small constant). Then, the interaction between server and receiver requires only $8nk$

modular products (which is about $8k$ n-bit exponentiations). We observe that the communication complexity is $12nk + n \log t$ and the storage complexity is $6n + n \log t$, where t is the soundness parameter required for the non-interactive zero-knowledge proof.

Complexity Assumptions: We remark that by using the non-malleable encryption scheme in [14], and implementing the conditional oblivious transfer protocol using well-known results on private two-party computation [28,17,16], our scheme can be implemented using any one-way trapdoor permutation.

5 Timed-Release Encryption: The Proof

We would like to prove Theorem 4. First of all observe that S,R,V run in probabilistic polynomial time; in particular the non-interactive zero-knowledge proof Π can be efficiently computed and verified using the protocol in [11]. Now we need to prove four properties: correctness, security against S, security against R and security against V. The correctness requirement directly follows from the properties of the conditional oblivious transfer and the encryption schemes used as subprotocols. We now concentrate on the remaining three properties.

Security against S. We need to show that for any probabilistic polynomial time S', if R does not output $\perp$ and $t \geq d$ then R can compute the message m sent by S'. Notice that if R does not output $\perp$ then the release-time has a right format; then, since $t \geq d$, by the transfer validity property of the conditional transfer protocol used, R will always receive c_m and then compute m with probability 1.

Security against R. Assume that S and V are honest. Consider the following experiment for any probabilistic polynomial time algorithm $R'=(R'_1, R'_2, R'_3, R'_4)$. Let $((x, \Pi), (p, q))$ be the pair computed by R'_1 on input σ, p_v and let (m_0, m_1) be the two messages returned by R'_2 on input σ, x, p, q. Let b a uniformly chosen bit, and let $((cc, c_d), d)$ be the output of S on input message m_b, the public key pk by R' and the public key $nm\text{-}pk$ of V. Now, let $req = (x, \Pi, cc', c'_d)$ be the request made by R'_3 to V, and let ans be V's reply at some time $t < d$. We now want to show that for any t such that $t < d$, the probability p that $R'_4(x, \Pi, p, q, ans) = b$ is at most $1/2 + n^{-c}$, for any constant c and all sufficiently large n. We divide the proof in three cases.

Case 1: $cc' = cc$ and $c'_d = c_d$. Assume that there exists a t such that $t < d$ and the above probability p is at least $1/2 + n^{-c}$, for some constant c and infinitely many n. Now, we explain how to turn R' into an algorithm B' that can break the scheme (nm-G, nm-E, nm-D). The idea is of course to simulate an entire execution of the protocol, and then use R' in order to break the mentioned scheme. Specifically, B' uses R'_1 in order to generate the pair of public/private keys. Now, given the two messages m_0, m_1 output by R'_2 as candidates to be encrypted using the timed release encryption scheme, B' will compute the two messages $nm\text{-}m_0, nm\text{-}m_1$ that are the candidates to be encrypted under the non-malleable scheme. These two messages are computed by encrypting the messages m_0, m_1, respectively, using the public key output by R'_1. Now, a bit b is uniformly chosen in the attack experiment associated to the non-malleable scheme, and $nm\text{-}m_b$ is encrypted using such encryption scheme (this is the encryption cc). This automatically chooses message m_b in the experiment associated with the timed release scheme. Now, B$'$ uses R$_3'$ to send a request (x, Π, cc', c_d') to V; recall that we are assuming that $cc = cc'$ and $c_d = c_d'$, therefore, B$'$ will now simulate the server using the assumed time t. Notice that he does not need to know the string c_m that is part of the decryption of cc' since when $t < d$, by Property 2 of the conditional oblivious transfer (Alice,Bob), the receiver is only obtaining transfers of uniformly distributed bits, which are therefore easy to simulate. Finally, B$'$ runs algorithm R$_4'$ on the (simulated) answer obtained by V. Now, notice that since the simulation of V is perfect, the probability that B$'$ breaks the non-malleable encryption scheme is the same as p, which contradicts the security of scheme $(nm\text{-}G, nm\text{-}E, nm\text{-}D)$.

Case $cc' = cc$ and $c_d' \neq c_d$. This case cannot happen, since V decrypts cc' as (c_d, c_m) and can see that $c_d' \neq c_d$, and therefore outputs $\perp$ and halts.

Case $cc' \neq cc$. This case contradicts the non-malleability of the scheme used by V. This is because given history c_d about plaintext $pl = (c_d, c_m)$, and ciphertext cc, R$'$ is able to compute a ciphertext cc' of some related plaintext $pl' = (c_d, c_m')$, i.e., such that c_m' is a valid encryption of m under the key of R$'$. The fact that c_m' is a valid encryption of m under such key is guaranteed by our original contradiction assumption that R$'$ successfully breaks the timed release encryption scheme.

Security against V. We see that the server V receives a tuple (x, Π, cc, c_d), and he can decrypt cc as (c_m, c_d') and check that $c_d' = c_d$. Namely, he obtains encryptions of the message m and the release-time d under the receiver's key. The semantic security of the encryption scheme used guarantees that the server does not obtain any additional information about m, d. Moreover, notice that the tuple (x, Π, cc, c_d) is independent from the sender's identity and the receiver's identity. Therefore, V does not obtain any information about the sender's or the receiver's identity either.

This concludes the proof of Theorem 4.

References

1. E. Bach and J. Shallit, *Algorithmic Number Theory*, MIT Press, 1996.
2. M. Bellare and S. Goldwasser, *Encapsulated Key-Escrow*, MIT Tech. Report 688, April 1996.
3. G. Brassard, C. Crépeau, and J.-M. Robert, *Information Theoretic Reductions among Disclosure Problems*, in Proc. of FOCS 86.
4. M. Ben-or, S. Goldwasser, and A. Wigderson, *Completeness Theorems for Non-Cryptographic Fault-Tolerant Distributed Computation*, in Proc. of STOC 88.
5. M. Blum, A. De Santis, S. Micali, and G. Persiano, *Non-Interactive Zero-Knowledge*, SIAM Journal of Computing, vol. 20, no. 6, Dec 1991, pp. 1084–1118.
6. M. Blum, P. Feldman, and S. Micali, *Non-Interactive Zero-Knowledge and Applications*, in Proc. of STOC 88.

7. R. Cramer and V. Shoup, *A Practical Cryptosystem Provably Secure under Chosen Ciphertext Attack,* in Proc. of CRYPTO 98.
8. C. Crépeau, *Equivalence between Two Flavors of Oblivious Transfer,* in Proc. of CRYPTO 87.
9. C. Crépeau and J. Kilian, *Achieving Oblivious Transfer Using Weakened Security Assumptions,* in Proc. of FOCS 1988.
10. D. Chaum, C. Crepeau, and I. Damgard, *Multiparty Unconditionally Secure Protocols,* in Proc. of STOC 88.
11. A. De Santis, G. Di Crescenzo, and G. Persiano, *The Knowledge Complexity of Quadratic Residuosity Languages,* Theoretical Computer Science, vol. 132, (1994), pp. 291–317.
12. A. De Santis, G. Di Crescenzo, and G. Persiano, *Zero-Knowledge Arguments and Public-Key Cryptography,* Information and Computation, vol. 121, (1995), pp. 23–40.
13. A. De Santis and G. Persiano, *Public Randomness in Public-Key Cryptography,* in Proc. of EUROCRYPT 92.
14. D. Dolev, C. Dwork, and M. Naor, *Non-Malleable Cryptography,* in Proc. of STOC 91.
15. S. Even, O. Goldreich and A. Lempel, *A Randomized Protocol for Signing Contracts,* Communications of ACM, vol. 28, 1985, pp. 637-647.
16. O. Goldreich, *Secure Multi-Party Computation,* 1998. First draft available at http://theory.lcs.mit.edu/~oded
17. O. Goldreich, S. Micali, and A. Wigderson, *How to Play any Mental Game,* in Proc. of STOC 87.
18. S. Goldwasser and S. Micali, *Probabilistic Encryption,* in Journal of Computer and System Sciences. vol. 28 (1984), n. 2, pp. 270–299.
19. J. Kilian, *Basing Cryptography on Oblivious Transfer ,* in Proc. of STOC 88.
20. J. Kilian, S. Micali and R. Ostrovsky *Minimum-Resource Zero-Knowledge Proofs,* in Proc. of FOCS 89.
21. E. Kushilevitz, S. Micali, and R. Ostrovsky, *Reducibility and Completeness in Multi-Party Private Computations,* Proc. of FOCS 94 (full version joint with J. Kilian to appear in SICOMP).
22. T. May, *Timed-Release Crypto,* Manuscript.
23. R.C. Merkle, *Secure Communications over insecure channels* Communications of the ACM, 21:291-299, April 1978.
24. R. Ostrovsky and B. Patt-Shamir, *Optimal and Efficient Clock Synchronization Under Drifting Clocks,* in Proc. of PODC 99, to appear.
25. R. Ostrovsky, R. Venkatesan, and M. Yung, *Fair Games Against an All-Powerful Adversary,* in Proc. of SEQUENCES 91, Positano, Italy. Final version in AMS DIMACS Series in Discrete Mathematics and Theoretical Computer Science, vol. 13, pp. 155–169, 1993.
26. M. Rabin, *How to Exchange Secrets by Oblivious Transfer,* TR-81 Aiken Computation Laboratory, Harvard, 1981.
27. R. Rivest, A. Shamir, and D. Wagner, *Time-Lock Puzzles and Timed-Release Crypto,* manuscript at http://theory.lcs.mit.edu/ rivest.
28. A.C. Yao, *Protocols for Secure Computations,* in Proc. of FOCS 82.
29. A.C. Yao, *How to Generate and Exchange Secrets,* in Proc. of FOCS 86.

An Efficient *Threshold* Public Key Cryptosystem Secure Against Adaptive Chosen Ciphertext Attack

(Extended Abstract)

Ran Canetti[1] and Shafi Goldwasser[2]*

[1] IBM T. J. Watson Research Center, Yorktown Height, NY, 10598,
canetti@watson.ibm.com
[2] Laboratory for Computer Science, Massachusetts Institute of Technology,
545 Technology Square, Cambridge, MA 02139,
shafi@theory.lcs.mit.edu

Abstract. This paper proposes a simple *threshold* Public-Key Cryptosystem (PKC) which is secure against adaptive chosen ciphertext attack, under the Decisional Diffie-Hellman (DDH) intractability assumption.

Previously, it was shown how to design non-interactive threshold PKC secure under chosen ciphertext attack, in the random-oracle model and under the DDH intractability assumption [25]. The random-oracle was used both in the proof of security and to eliminate interaction. General completeness results for multi-party computations [6,13] enable in principle converting any single server PKC secure against CCA (e.g., [19,17]) into a threshold one, but the conversions are inefficient and require much interaction among the servers for each ciphertext decrypted.

The recent work by Cramer and Shoup [17] on single server PKC secure against adaptive CCA is the starting point for the new proposal.

1 Introduction

A threshold public-key cryptosystem (PKC) [18] extends the idea of a PKC as follows: instead of a single party holding the private decryption key, there are n *decryption servers*, each of which hold a piece of the private decryption key. When a user receives a ciphertext c to be decrypted, she sends c to each decryption server, receives a piece of information from each, and recovers the cleartext from the collected pieces.

Semantic security of encryption schemes [28] can be easily extended to the threshold PKC case. A threshold PKC is called *t-secure* if a coalition of t curious but honest servers cannot distinguish between ciphertexts of different messages, yet sufficiently many servers can jointly reconstruct the cleartext. A threshold

---

* Supported by DARPA grant DABT63-96-C-0018.

J. Stern (Ed.): EUROCRYPT'99, LNCS 1592, pp. 90–106, 1999.

PKC is called *t-robust* if it meets these requirements even when up to t servers are maliciously faulty.

Secure and robust threshold PKC's can be designed, under general assumptions such as the existence of trapdoor permutations and using multi-party computation completeness theorems [6,26], to convert any centralized semantically secure PKC into a threshold one. More efficient threshold PKC's have been designed based on the RSA and DH intractability assumptions [24,18,35]. All of these proposals require interaction among the servers and the user, in order to achieve robustness for a linear fraction of faults. The general conversions require interaction to achieve both security and robustness. In the work of [24] the presence of a trusted dealer, which distributes verification data for each pair of server and user in a pre-processing stage, is proposed as a way to eliminate interaction and yet achieve robustness for linear number of faults (they actually address RSA signatures but the work can be easily reformulated for RSA decryption).

Stronger notions of security of centralized encryption schemes, namely security against 'Lunch-time Attacks' and 'chosen ciphertext attacks' (CCA) were defined, constructed, and studied in [33,38,19,17,4]. These notions capture additional security concerns when using encryption within a general security application. CCA security of *threshold* PKC has been recently defined in [25]. In principal the Dolev-Dwork-Naor PKC secure against CCA (using non-interactive zero knowledge) can be converted, using multi-party completeness theorems, into a threshold PKC secure against CCA if trapdoor functions exist, but the resulting scheme is inefficient and requires much interaction among the servers. *Efficient* CCA-secure threshold PKC schemes were proposed in [25], in the *random oracle model* under the DDH intractability assumption. The use of random oracles was essential for proving security against CCA. Once the random oracle was present it was also used to eliminate interaction to achieve robustness of the scheme against a linear number of faulty servers. Our goal is to design an efficient threshold PKC secure against CCA *not* in the random oracle model.

A threshold decryption service has several applications. Let us sketch a few. One application (suggested in [25]) is for distributing the escrow service in a *key recovery* mechanism and allowing it to decrypt only specific messages rather than entirely recover the key. Another attractive application is for having public encryption keys associated with an *organization*. Here messages directed to the organization are encrypted with the organization's public key; the organization's decryption servers now direct the decrypted plaintext to an appropriate organization member. Another application is for a decryption service that 'sits on the net' and offers decryption services for customers who do not have their own certified public keys. This service can also be part of an 'electronic vault' application (e.g., [23]). Here it may be important that the decryption be done so that no one except some specified party, not even the decryption servers themselves, will learn anything about the plaintext. (Our security requirements from a threshold PKC take these scenarios into consideration, in an explicit way.)

1.1 New Results

In this paper we present a new threshold PKC, which is provably secure against CCA based on the DDH intractability assumption. Our scheme makes no use of random oracles. The scheme achieves security against a coalition of t honest but curious servers upto $t < \frac{n}{2}$.

The starting point for our scheme is the recent attractive result of Cramer and Shoup [17] which proposed (using techniques reminiscent of those of [25]) an efficient *centralized* PKC secure against adaptive CCA, under the DDH intractability assumption.

The idea of the Cramer-Shoup scheme is that the ciphertext carries with it a *tag*, which the decryption algorithm checks for validity before computing the cleartext. If the tag is valid then the cleartext is output, else the decrypting algorithm outputs 'invalid'. Simplistically stated, unless the legal encryption algorithm was used to produce the ciphertext, it is computationally hard to come up with anything but an invalid tag, and thus it is safe for the server to decrypt ciphertexts carrying a valid tag.

Differently from previous PKC's proved secure against lunch-time attacks and CCA [33,19], this scheme is **not** publicly verifiable. That is, deciding whether the tag is valid or not requires the knowledge of the private key. In particular, this knowledge enables computing from the ciphertext a *tag'* which should equal *tag* when the ciphertext is valid.

We now turn our attention to trying to make Cramer-Shoup into a threshold PKC system. First note that if one is willing to increase the size of the ciphertext (and of the public encryption key) proportionally to the number of servers then achieving threshold CCA security is very simple: Let each server have a separate public key of the Cramer-Shoup scheme, and modify the encryption algorithm to that it first generates a Shamir secret sharing $m_1...m_n$ of the message m, and then each m_i is separately encrypted using the public key of the ith server. Each server decrypts its share as usual and hands it to the decrypting user.

However, we are interested in schemes where the ciphertext and the encryption key are small, and in particular independent of the number of servers. A straightforward approach would thus be to distribute the private key among all the decryption servers. When a ciphertext arrives, the servers distributively compute whether the tag is valid or not and if it is valid each server outputs a piece of the cleartext. The user then uses the pieces to recover the cleartext. The basic problem of this approach is: *how to distributively implement the check that the tags are valid?* General completeness results for multi-party computation indicate that this is of course possible in principle, but requires interaction between servers for every ciphertext received. More efficient, DDH based protocols seem to require interaction as well.

The Main Idea: Avoiding the Validity Check The new idea is to first modify the PKC scheme so as to avoid an explicit validity check. Instead, the decrypting algorithm (still in the standard PKC case) will output the cleartext if the ciphertext is valid, and otherwise a value indistinguishable from random[1]. Thus, when the ciphertext is invalid (as defined in [17]) the user will get essentially 'random garbage' computationally unrelated to the ciphertext. Such a modified scheme (which we label M-CS) enjoys a very similar security proof to the original scheme, but it is now possible to turn it into a threshold PKC scheme avoiding the distributed validity check.

In our threshold PKC scheme, each of the n servers will output a piece of information with the following property $\mathcal{P}$:

• if the ciphertext was *valid*, then the cleartext can be recovered from the pieces sent by the decryption servers; but

• if the ciphertext was *invalid*, then the collection of all the pieces is indistinguishable from random garbage.

How is this achieved? Let *tag, tag'* be as discussed above. We come up with a function f such that (1) $f(tag, tag') = 1$ if *tag=tag'* ; (2) $f(tag, tag') = rval$ if *tag≠tag'* (where *rval* is indistinguishable from random); and (3) f is easy to distribute in the sense that it is easy to compute a share of $f(tag, tag')$ from a share of the secret key. Condition (3) is necessary for threshold PKC whereas any function with input/output behavior as specified in conditions (1)-(2) suffices for M-CS . Using such f, each server will compute from the ciphertext and its share of the private key, a share of $f(tag, tag')$ and send to the user a share of *cleartext* · $f(tag, tag)$.[2] The user will combine the shares to obtain *cleartext* · $f(tag, tag')$. This choice of f guarantees property $\mathcal{P}$.

We propose to use $f(tag, tag') = (tag/tag')^s$ where s is a random exponent. In order to implement f, at system startup the servers will agree on a sequence of random numbers s shared between them using some secret sharing method such as polynomial secret sharing, and will use these numbers for f as ciphertexts arrive.

Where Does the Randomness in Decoding Come from? The idea described above requires that for *each* ciphertext, the servers will use a new random number that is shared among them using a secret sharing method such as polynomial secret sharing. How are these numbers chosen and shared? We suggest the following method.

A straightforward implementation would be that before the start of the system the servers agree using standard methods (e.g [39,6,20,35,22]) on m random numbers $r_1, ...r_m$ each of which is shared using a polynomial secret sharing among the n players. These are used for decrypting m ciphertexts, after which time a new set of random numbers will be chosen. This means, that each server must store in local memory m shares of m random numbers in addition to his secret

---

[1] This value does not have to be random. It would actually be sufficient to output, in case of invalid tag, a value which is unrelated to the ciphertext.

[2] This is a slight over simplification for purpose of exposition in the introduction. In the actual scheme the server sends a share of *mask* · $f(tag, tag')$ where *mask* · *cleartext* is part of the ciphertext. Receiving *mask* will enables the user to compute the cleartext. See exact details within.

key. (Alternatively, the servers may generate these random numbers every time a ciphertext appears. However, this method requires interaction among servers at the time of decryption, and is thus not recommended.)

This implementation may encounter synchronization problems when the servers receive the ciphertexts in different orders. We suggest solutions within.

We do not know how to keep the memory requirements of the servers independent of the number of decryptions to be performed, without interaction among the servers. This is left as an interesting open problem. (See more details in Section 3.1.)

Robustness Suppose now that some of the decryption servers are maliciously faulty. To achieve t-robustness we propose several variants of our basic scheme, all of which use [39,20] style polynomial secret sharing as a building block. Our solutions use standard tools which have been used in the literature to address robustness of threshold signature and encryption schemes such as the prover proving in zero-knowledge to the user that the share provided is proper; we come up with efficient instantiations of such tools tailored to the tasks at hand. We stress that in all methods the public encryption key and the encryption algorithm are identical to those of Cramer-Shoup, and in particular are independent of the number of servers. We sketch these methods, all of which achieve t-robustness for up to $t < \frac{n}{3}$ malicious server faults.

• A first method is fully non-interactive, and is efficient when $t = O(\sqrt{n})$. Practically speaking, when, say, $n = 7$ and $t = 2$ this method is quite efficient.

• A second method requires a simple four-round interactive proof between the user and the decryption servers (no interaction between the decryption servers themselves is necessary). Here each server proves to the user that the piece of decryption information provided is correct. The interactive protocol can be avoided when sufficient number of decryption servers do not act in a faulty fashion. The user first runs a local detection-algorithm to see if she can use the pieces of information she received from the servers to decrypt the ciphertext. Only when the user detects that too many pieces were faulty, should she carry out the interactive-proofs to determine which pieces were faulty and should be discarded. Here the decrypting user needs some verification information for each of the servers. Thus the size of the public file grows by a factor of n. Yet, it is stressed that the encryption algorithm remains identical to that of Cramer-Shoup, and the public key needed for encryption remains small.

• A third method uses the technique of check-vectors introduced by [37] for VSS implementation and used by [24] to achieve robustness of threshold RSA signatures. The idea of [24] was that at the time of key generation, a trusted dealer generates additional *verification data* for every pair of user-server, and gives part of this data to the user and part to the server. At the time of signature verification, the user uses her verification data to verify non-interactively that each piece of RSA signature she received from each server is non-faulty. A slight modification of the idea of [24] can be applied to our scheme as well to make it non-interactive and t-robust for $t < \frac{n}{3}$. It will however require each potential decrypting user to have some secret information, and increase the size of each server's key proportionally to the number of potential decrypting users. Thus, this variant is adequate when the number of decrypting users is small, or a 'decryption gateway' is available. (For lack of space, this method is deleted from the proceedings version. It is described in [12].)

Remark: The question of whether it is possible to achieve robustness efficiently against a linear number of faults without interaction (either among the servers or between the servers and the user) or a trusted dealer is an interesting open problem for threshold PKC regardless of which security is desirable, be it semantic security or CCA.

1.2 Additional Contributions of This Paper

A New Definition of Security for Threshold PKC. Another contribution of our work is proposing an alternative definition of security for threshold PKCs. than the definition of [17]. (The definition of [17] is stated in the random-oracle model; yet it can be readily transformed to protocols that do not use the random oracle.) An attractive feature of the new definition (which follows a standard methodology for defining security of cryptographic protocols [27,31,1,9]) is that it is geared towards defining the security of the threshold PKS as a component within larger systems. In particular, on top of guaranteeing CCA security it addresses issues like non-malleability [19], plaintext awareness [5,4] and security against dynamic adversaries [3,10].

Remote Key Encryption. One of the by-products of our method is yet another variant of our PKC (for the single or multiple server case) such that the user can send the ciphertext to a decryption server(or several servers) on line and receive information which allows the user to recover the cleartext. Yet, neither the servers nor anyone else listening on line can get any information about the cleartext. This functionality has been introduced and (very different) constructions were given in [7]. This variant is secure against lunch-time attacks only.

Proactiveness. Our techniques can be 'proactivized' (i.e., modified to withstand mobile faults, as suggested in [34,11]) in standard ways [29]. See more discussion in [12].

2 Security of Threshold Cryptosystems

We present a measure of security of threshold PKCs. Our formalization is geared towards capturing the security requirements that emerge when using the system as a "service" in a complex and unpredictable environment. In a nutshell, the definition here requires that the system behaves like an "ideal encryption service" under *any usage pattern*. Indeed, this requirement incorporates known security measures like CCA security, non-malleability, plaintext awareness, and security against dynamic adversaries.

The definition here takes a different approach than that of [25], where threshold CCA-security is regarded as a natural extension of the standard definition of CCA-security to the context of threshold cryptosystems. In particular, security according to the definition here implies security according to the definition of [25]. (The converse does not hold in general.)

For lack of space we only sketch the definition in this extended abstract. See [12] for full details on the definition and the relations with that of [25].

Outline of our definition. Following the approach used for defining security of general cryptographic protocols [27,31,1,9], we proceed in three steps. First we formalize the model of computation and specify a syntax for threshold PKCs. Next we specify an idealized model where a threshold PKC is replaced with an "ideal encryption service". Finally, we say that a threshold PKC is secure if it *emulates* the ideal service in a way described below.

THE COMPUTATIONAL MODEL AND THRESHOLD PKCS. There are n decryption servers $S_1...s_n$, an encrypting user E and a decrypting user U. A threshold PKC consists of:

A key generation module, that given the security parameter generates a public key, pk, known to all parties, and some secret key, sk_i, known to each server S_i;

An encryption algorithm (run by E) that, given pk, a message m to be encrypted, and random input ρ, outputs a ciphertext $c = \text{ENC}_{pk}(m, \rho)$;

A server decryption module that, when operating within server S_i and given sk_i and a ciphertext c, possibly interacts with D and other servers and eventually generates a decryption share μ_i;

A user decryption module (run by D) that, given a ciphertext c, interacts with the servers, and eventually outputs $m = \text{DEC}_{pk}(c, \rho_i, \mu_1...\mu_n)$.

A run of the system consists of an invocation of the key generation module (at the end of which pk is made public and the secret keys are given to the corresponding servers), followed by an interaction among the parties via some standard model of distributed communication. (For simplicity assume ideally secure and authenticated communication links). The interaction is orchestrated by an adversary A who can invoke E and D on cleartexts and ciphertexts of its choice; in addition, A can corrupt D and up to t servers. (The corruptions are either static or dynamic. Corrupting D gives A access to the decrypted data.) We augment the model by allowing the adversary to freely interact with an additional entity, called an environment machine Z. This (Turing) machine models the external environment, and in particular provides the adversary with arbitrary and interactive 'auxiliary input' throughout the run of the system. In particular, Z learns all the information learned by A (and, in general, can have additional information that is unknown to A.)

We let the global output $\text{EXEC}_{\tau,A,Z}$ of a run of a threshold PKC τ with adversary A and environment Z be the concatenation of the output of all the parties, the adversary, and the environment machine. In particular, the global output registers all the encryption requests made to E, all the decryption requests made to D and each S_i, and the resulting ciphertexts and cleartexts.

THE IDEAL ENCRYPTION MODEL. The ideal model consists of replacing the four modules of a threshold PKC with a trusted service T, parameterized by a threshold t, and a security parameter k. First T receives a description of a distribution Γ from the adversary (who is now called an ideal model adversary, S).[3] Next, the trusted party provides the following services:

Ideal Encryption, where E hands T a message m to encrypt. In response, E receives a receipt c, chosen from distribution Γ *independently of* m.

Ideal Decryption, where The servers can hand a receipt c to T. Once t servers have handed c to T, and if c was previously generated by T, then T hands D the message m that corresponds to c. Otherwise T ignores the request.

A run of the system in the ideal model is orchestrated by the adversary in the same way as described above.

Let the ideal global output IDEAL$_{t,S,Z}$ be defined analogously to EXEC$_{\tau,A,Z}$ with respect to parties running in the ideal encryption model with ideal-model adversary S, where t is the trusted party's threshold.

SECURITY OF THRESHOLD PKCs. A threshold PKC τ is called t-secure if it *emulates* the ideal encryption service, in the following way. For any adversary A there should exist an ideal model adversary S such that for any environment Z the global outputs IDEAL$_{t,S,Z}$ and EXEC$_{\tau,A,Z}$ are computationally indistinguishable (when regarded as distribution ensembles). We stress that the environment Z is the same in the real-life and ideal executions; that is, S can "mimic" the behavior of A in *any* environment.

Replacing an ideal service. The quantification over all environments Z provides a powerful guarantee. In particular, it captures the interaction of *any application protocol* with the PKC in question. Consequently, this definition can be used to show the following attractive property of PKCs. Consider an arbitrary, multi-party 'application protocol' π where, in addition to communicating over the specified communication channels, the parties have access to an ideal encryption service similar to the one described above. Let τ be a PKC that meets the above definition, and let π^τ be the protocol where each call to the ideal service is replaced, in the natural way, with an invocation of the corresponding module of τ. Then π^τ *emulates* π, where the notion of emulation is similar to the one used above. See more details in [12].

3   A Threshold Cryptosystem

Our threshold cryptosystem is based on the Cramer-Shoup cryptosystem [17]. We first briefly review the (basic variant of the) Cramer-Shoup scheme, denoted

---

[3] Typically, Γ will be the distribution of an encryption of a random message in the domain, under a randomly chosen public key.

CS, and modify it as a step towards constructing the distributed scheme. Next we present the basic scheme and its extensions.

The Cramer-Shoup scheme. Given security parameter k, the secret key is $(p, g_1, g_2, x_1, x_2, y_1, y_2, z, H)$ where p is a k-bit prime, g_1, g_2 are generators of a subgroup of $\mathbb{Z}_p$ of a large prime order q, function H is a hash function chosen from a collision-resistant hash function family and $x_1, x_2, y_1, y_2, z \xleftarrow{R} \mathbb{Z}_q$.[4] The public key is (p, g_1, g_2, c, d, h) where $c = g_1^{x_1} g_2^{x_2}$, $d = g_1^{y_1} g_2^{y_2}$, and $h = g_1^z$.

It is assumed that messages are encoded as elements in $\mathbb{Z}_q$. To encrypt a message m choose $r \xleftarrow{R} \mathbb{Z}_q$ and let $\text{ENC}(m, r) = (g_1^r, g_2^r, h^r m, c^r d^{r\alpha})$, where $\alpha = H(g_1^r, g_2^r, h^r m)$. Decrypting a ciphertext (u_1, u_2, e, v) proceeds as follows. First compute $v' = u_1^{x_1 + y_1 \alpha} \cdot u_2^{x_2 + y_2 \alpha}$. Next, perform a validity check: if $v \neq v'$ then output an error message, denoted '?'. Otherwise, output $m = e/u_1^z$. Security of this scheme against CCA is proven, based on the decisional Diffie-Hellman assumption (DDH), in [17].

Towards a threshold scheme. We first observe that this scheme can be slightly modified as follows, without losing in security. If the decryptor finds $v \neq v'$ then instead of outputting '?' it outputs a random value in $\mathbb{Z}_q$. In a sense, the modified scheme is even "more secure" since the adversary does not get notified by the decryptor whether a ciphertext is valid.

Next, modify this scheme further, as follows. The decryption algorithm now does not explicitly check validity. Instead, given (u_1, u_2, e, v) it outputs $e/u_1^z \cdot (v'/v)^s$, where v' is computed as before and $s \xleftarrow{R} \mathbb{Z}_q$. (Note that now the decryption algorithm is *randomized*.) To see the validity of this modification, notice that if $v = v'$ then $(v/v')^s = 1$ for all s, and the correct value is output. If $v \neq v'$ then the decryption algorithm outputs a uniformly distributed value in $\mathbb{Z}_q$, independent of m, as in the previous scheme. Call this scheme M-CS.

Claim. If scheme CS is secure against CCA then so is scheme M-CS.

Proof. Correctness of M-CS (i.e., correct decryption of correctly generated ciphertexts) clearly holds. To show security against CCA, consider an adversary $\mathcal{A}$ that wins in the 'CCA-game' (see [17]) against M-CS with probability non-negligibly more than one half. Construct the following adversary, $\mathcal{A}'$ that operates against CS. $\mathcal{A}'$ runs $\mathcal{A}$, with the exception that whenever $\mathcal{A}'$ receives an answer '?' from the decryption oracle it chooses $r \xleftarrow{R} \mathbb{Z}_q$ and gives r to $\mathcal{A}$. Finally $\mathcal{A}'$ outputs whatever $\mathcal{A}$ does. The view of $\mathcal{A}'$ is distributed identically to its view in an interaction with M-CS, thus it predicts the bit b chosen by the encryption oracle of $\mathcal{A}'$ with probability non-negligibly more than one half.

Verifying Validity of Ciphertexts. An apparent disadvantage of M-CS is that even a legitimate user of the decryption algorithm does not learn whether a

---

[4] In fact, H can be a target-collision-resistant hash function. The notation $e \xleftarrow{R} D$ means that element e is drawn uniformly at random from domain D.

ciphertext was valid. However, this information may be obtained in several ways:
First, when applying the decryption algorithm twice to an invalid ciphertext,
two independent random numbers are output, but if the ciphertext is valid then
both applications output the same cleartext. Alternatively, valid cleartexts can
be assumed to have a pre-defined format (say, a leading sequence of zeros). The
output of the decryption algorithm on an invalid ciphertext, being a random
number, has the right format with probability that can be made negligibly small.

On Remotely Keyed Encryption: As a side remark, one can trivially change CS
and M-CS to qualify as a remotely-keyed-encryption scheme [7] secured against
lunch-time attacks. Simply, drop d from the public key, and let $ENC(m, r) =
(g_1^r, g_2^r, h^r m, c^r)$.[5] Then the user sends to be decrypted remotely only (g_1^r, g_2^r, c^r),
dropping the third component of the ciphertext. To decrypt, the server who gets
(u_1, u_2, v) computes $v' = u_1^{x_1} u_2^{x_2}$ and sends $p = \frac{(v'/v)^s}{u_1^z}$ back to the user. The
user sets $m = e \cdot p$. Clearly, the server got no information about m. A similar
modification can be applied to the threshold PKC coming up in the next section,
to obtain a remotely keyed threshold PKC secure against lunch time attacks.

3.1   An Threshold Cryptosystem for Passive Server Faults

The basic threshold scheme, denoted T-CS, distributes scheme M-CS in a straightforward way. Let p, q, g_1, g_2 be as in the original scheme, and let t be the threshold. The scheme requires an additional parameter, L, specifying the number of
decryption performed before the public and secret keys need to be refreshed. We
first describe the scheme for the case where all serves follow their protocol.

KEY GENERATION. For simplicity we assume a trusted dealer for this stage. This
simplifying assumption can be replaced by an interactive protocol performed by
the servers. This can be done using general multi-party computation techniques
[26,6,13] or more efficiently using techniques from [35]. Say that a polynomial
$P(\xi) = \sum_{i=0}^d a_i \xi^i \pmod q$ is a random polynomial for a if $a_0 = a$ and $a_1...a_d \xleftarrow{R}
\mathbb{Z}_q$. The dealer generates:

- $x_1, x_2, y_1, y_2, z \xleftarrow{R} \mathbb{Z}_q$ as in the original CS, and random degree t polynomials
  $P^{x_1}(), P^{x_2}(), P^{y_1}(), P^{y_2}(), P^z()$ for x_1, x_2, y_1, y_2, z, respectively.
- L values $s_1...s_L \xleftarrow{R} \mathbb{Z}_q$ and random degree t polynomials $P^{s_1}()...P^{s_L}()$ for them.
- L random degree $2t$ polynomials $P^{o_1}()...P^{o_L}()$ for the value 0.[6]

Let $x_{j,i} = P^{x_j}(i)$. Let $y_{j,i}, z_i, s_{l,i}, o_{l,i}$ be defined similarly. The secret key of server
S_i is now set to $sk_i = (p, q, g_1, g_2, x_{1,i}, x_{2,i}, y_{1,i}, y_{2,i}, z_i, s_{1,i}...s_{L,i}, o_{1,i}...o_{L,i})$. The
public key is identical to that of CS: $pk = (p, q, g_1, g_2, c, d, h)$ where $c = g_1^{x_1} g_2^{x_2}$,
$d = g_1^{y_1} g_2^{y_2}$, and $h = g_1^z$.

---

[5] This simplification was suggested in [17].

[6] Looking ahead, we note that these values are needed to make sure that the partial
decryptions are computed based on a *random* degree $2t$ polynomial. More specifically,
these shares make sure that polynomial $Q()$ defined in Equation (2) below is a *random*
degree $2t$ polynomial for the appropriate value.

ENCRYPTION is identical to CS: $\text{ENC}_{\mathsf{pk}}(m, r) = (g_1^r, g_2^r, h^r m, c^r d^{r\alpha})$, where $\alpha = H(g_1^r, g_2^r, h^r m)$.

DECRYPTION. Each server S_i proceeds as follows, to decrypt the lth ciphertext, (u_1, u_2, e, v). First it computes a share v_i' of v', by letting $v_i' = u_1^{x_{1,i} + y_{1,i}\alpha} u_2^{x_{2,i} + y_{2,i}\alpha}$. Then it computes a share $u_1^{z_i}$ of u_1^z and a share $g_1^{o_{l,i}}$ of the value '1'. Next it computes and outputs the partial decryption:[7]

$$f_i = u_1^{z_i} \cdot (v/v_i')^{s_{l,i}} \cdot g_1^{o_{l,i}}.$$

The user module collects the partial decryptions $f_1 \ldots f_n$ and computes the value $f_0 = \Pi_{i=1}^n f_i^{\lambda_i}$, where the λ_i's are the appropriate Lagrange interpolation coefficients; that is, the λ_i's satisfy that for any degree $2t$ polynomial $P()$ over $\mathbb{Z}_q$ we have $P(0) = \sum_{i=1}^n \lambda_i P(i)$. Next, the user outputs $m = e/f_0$.

Theorem 1. *If the DDH assumption holds then* T-CS *is a t-secure threshold cryptosystem for any $t < \frac{n}{2}$, provided that even corrupted servers follow their protocol.*

Proof. See proof in [12]. Here we only verify that the output of the user's decryption module is identical to the output of the decryption module in M-CS. Each partial decryption f_i can be written as follows:

$$f_i = u_1^{z_i - s_{l,i}(x_{1,i} + y_{1,i}\alpha)} u_2^{-s_{l,i}(x_{2,i} + y_{2,i}\alpha)} v^{s_{l,i}} g_1^{o_{l,i}} \qquad (1)$$

Let $r_1 = \log_{g_1} u_1$ (i.e., r_1 satisfies $g_1^{r_1} = u_1$), let $r_2 = \log_{g_2} u_2$, and let $r_3 = \log_{g_1} v$. Then we have

$$f_i = g_1^{r_1 \cdot z_i - r_1 \cdot s_{l,i} x_{1,i} - r_1 \alpha \cdot s_{l,i} y_{1,i} - r_2 \cdot s_{l,i} x_{2,i} - r_2 \alpha \cdot s_{l,i} y_{2,i} + r_3 \cdot s_{l,i} + o_{l,i}}.$$

Consequently, $f_i = g_1^{Q(i)}$ where $Q()$ is the degree $2t$ polynomial:

$$\begin{aligned} Q() = {} & r_1 P^z() - r_1 P^{s_l}() P^{x_1}() - r_1 \alpha P^{s_l}() P^{y_1}() - r_2 P^{s_l}() P^{x_2}() \\ & - r_2 \alpha P^{s_l}() P^{y_2}() + r_3 P^{s_l}() + P^{o_l}() \end{aligned} \qquad (2)$$

It follows that

$$f_0 = g_1^{Q(0)} = g_1^{r_1 \cdot z - r_1 \cdot s_l x_1 - r_1 \alpha \cdot s_l y_1 - r_2 \cdot s_l x_2 - r_2 \alpha \cdot s_l y_2 + r_3 \cdot s_l + 0} = u_1^z \cdot (v/v')^{s_l}$$

therefore $e/f_0 = m \cdot (v'/v)^{s_l}$.

How to synchronize the s's. The above scheme may encounter synchronization problems when the servers receive the ciphertexts in different orders, and consequently associate shares of different s's with the same ciphertext. A way for solving this problem is to have the servers agree on a bivariate polynomial $H(x, y)$

---

[7] Once the partial decryption is generated, the server *erases* the shares $o_{l,i}, s_{l,i}$. This provision is important for proving security of the scheme against *dynamic* adversaries that may corrupt parties during the course of the computation.

of degree t in x and degree L in y, where each server P_i holds the degree-m univariate polynomial $H_i(y) = H(i, y)$. The value $s_{i,c}$ associated with the ciphertext c is computed as $s_{i,c} = H(i, h(c))$ where $h()$ is a collision-resistant hash function that outputs numbers in $\mathbb{Z}_q$. It now holds that the first L ciphertexts will be associated with L independent s's, regardless of the relative order of arrival at the servers. (Ciphertext c will be associated with the value $s_c = H(0, h(c))$. Using bivariate polynomials for related purposes is common in the literature. The use here is similar to the one in [32].

This method does not reduce the memory requirements from the servers, since each H_i has $L + 1$ coefficients. Furthermore, our proof of security against *dynamic* adversaries does not go through when this method is used. (Security against static adversaries remains unchanged.)

In an alternative method (that allows the proof against dynamic adversaries to go through) the servers use a universal hash function h (not cryptographic, just avoiding collisions with high probability) to map the ciphertext c to an index i. Once an s_l has been utilized, erase it from the list.[8] Note that universal hash functions suffice here, as it is in the interest of the encrypting party to prevent collisions in hashed ciphertexts. However, only a fraction of the s's are used before collisions become frequent.

On pseudorandomly generated s's. The need to prepare in advance the s's and the o's (i.e., the shared random values and the shares of the value 0) is a drawback of our scheme. It raises an interesting open problem of whether it is possible to construct a non-interactive and efficient implementation of a threshold pseudorandom function (TPRF), namely a PRF family $\{f_k\}$ where the secret key k is shared by a number of servers so that the servers can jointly evaluate the function, yet the function remains pseudorandom to an adversary who may control a coalition of some of the servers. For our scheme, we would need in addition that the shares of the servers of $f_k(c)$ would correspond to the values of a degree-t polynomial whose free term is $f_k(c)$. If such function family would exist, then instead of pre-sharing the random s's, each server S_i will, given a ciphertext c, set s_i to be the ith share of $f_k(c)$. (The shares of the value '0' can be pseudorandomly generated using similar methods.)

In fact, a threshold pseudorandom generator (TPRG) will suffice for us and could possibly be easier to implement. In a TPRG suitable for our purpose, the seed to the generator would be shared among the servers. Each server would compute a point on a degree t random polynomial whose free term is the ith output block of the generator.

---

[8] In the event that a c' arrives s.t. $h(c') = i$ for an s_i that was previously used and erased, the server alerts the user to replace c' with c'' (a perturbed c') and $s_{h(c'')}$ is used instead.

3.2   Achieving Robustness

This section deals with protecting against actively faulty decryption servers. Notice that since scheme T-CS is non-interactive then actively faulty servers cannot help the adversary in compromising the secrecy of encrypted messages that were not explicitly decrypted by the non-corrupted servers. The only damage that actively faulty servers can cause is denial of service. This is a lesser concern than secrecy, and in particular can usually be dealt with using external methods, such as notifying a higher-layer protocol or an operator. Still, we describe three methods for dealing with such active faults, as sketched in the Introduction.

Local error correcting. The first method uses the fact that, as long as $t < \frac{n}{3}$, the correct value f_0 is uniquely determined. This holds even if up to t of the f_i's are arbitrary elements in $\mathbb{Z}_q$: Let $Q()$ be the polynomial defined in Equation (2). Then at least $n - t$ of the partial decryptions $f_1...f_n$ satisfy $f_i = g_1^{Q(i)}$. Furthermore, there exists only a single degree $2t$ polynomial that agrees with $n - t$ of the f_i's.

We describe below a method for finding $f_0 = g_1^{Q(0)}$. This method is efficient only when $t = O(\sqrt{n})$. We do not know how to efficiently find f_0 for larger values of t; this 'error correction in the exponent' is an interesting and general open problem with various applications for cryptographic protocols. In particular, standard error correction algorithms for Reed-Solomon codes [30,41], which work when the perturbed $Q(i)$'s are explicitly given, do not seem to work here.

Our simplistic method for finding the value $f_0 = g_1^{Q(0)}$ proceeds as follows. We first pick at random a set $G = \{f_{i_1}...f_{i_d}\}$ of $d = 2t + 1$ f_i's, and check its validity using the appropriate Lagrange coefficients. That is, let $\lambda_1^x...\lambda_d^x$ be such that $P(x) = \sum_{k=1}^{d} \lambda_k^x P(i_k)$ for all polynomials $P()$ of degree $2t$. (These λ_k^x's are specific for x and for the set G.) Then, for each $j = 1..n$ we test whether $f_j = \Pi_{k=1}^{d} f_{i_k}^{\lambda_k^j}$. Say that s is valid if the test fails for at most t f_j's. We are guaranteed by the uniqueness of $Q()$ that if G is valid then letting $f_0 = \Pi_{k=1}^{d} f_{i_k}^{\lambda_k^0}$ yields the correct value. Furthermore, if S_i is uncorrupted (and thus $f_i = g_1^{Q(i)}$) for all $i \in G$ then G is valid.

We thus repeatedly choose random sets of size $2t + 1$ and check for validity. Each trial succeeds with probability $\Omega(e^{-2t^2/n})$. Thus when $t = O(\sqrt{n})$ we are guaranteed that a valid set G is found within a constant number of trials. (A similar argument is used in [2]). When n is small — as would be the case for practical applications — this method is quite efficient.

Interactive proofs of validity of partial decryptions. This method calls for the decrypting user to perform a (four-move) Zero Knowledge interaction with each of the servers to verify the validity of the partial decryptions. While making sure that neither corrupted servers nor a corrupted user gather more information (or, rather, more computational ability) than in the basic scheme (T-CS), these interactions guarantee that the user will almost never accept an invalid partial decryption as valid. Once the interactions are done, the user interpolates f_0 as in the basic scheme, based on the acceded partial decryptions. We remark that the user need not always perform these interactions. It can first locally check validity (using terminology from the previous method) of the entire set $\{f_1...f_n\}$ and interact with the servers only if $\{f_1...f_n\}$ is found invalid.

We use standard techniques for discrete-log based ZK proofs of membership and knowledge [21,14,15,40,16,8]. First the following verification information is added to the public key. (We stress that this information is not needed for encrypting messages; it is used only by the decrypting users.) For each server S_i and each $l = 1..L$ we add:

$$g_1^{z_i}, \; g_1^{s_{l,i}}, \; g_2^{s_{l,i}}, \; g_2^{o_{l,i}}.$$

Now, given the lth ciphertext (u_1, u_2, e, v) server S_i sends to the decrypting user, along with the partial decryption f_i (computed as in T-CS), also the values $\tilde{u}_1 = u_1^{s_{l,i}}$ and $\tilde{u}_2 = u_2^{s_{l,i}}$. Next, S_i and the user U engage in the following interaction, whose purpose can be informally described as follows. Recall that $\alpha = H(u_1, u_2, e)$. Server S_i proves to U that:

1. $\log_{u_1} \tilde{u}_1 = \log_{g_1} g_1^{s_{l,i}}$ and $\log_{u_2} \tilde{u}_2 = \log_{g_2} g_2^{s_{l,i}}$.
2. S_i "knows" values w_1, w_2, w_3, w_4, w_5 and $x_{1,i}, x_{2,i}, y_{1,i}, y_{2,i}$, such that:
   (a) $w_1 \cdot w_2 \cdot w_3 \cdot w_4 \cdot w_5 = f_i$
   (b) $w_1^{-1} = \tilde{u}_1^{x_{1,i}} \tilde{u}_2^{x_{2,i}}$ and $w_2^{-1} = \tilde{u}_1^{y_{1,i}\alpha} \tilde{u}_2^{y_{2,i}\alpha}$
   (c) $\log_{u_1} w_3 = \log_{g_1} g_1^{z_i}$
   (d) $\log_v w_4 = \log_{g_1} g_1^{s_{l,i}}$
   (e) $\log_{g_1} w_5 = \log_{g_2} g_2^{o_{l,i}}$.

The proof proceeds as follows. We describe the proof in two parts. These parts are performed in parallel. (In fact, U can use the same challenge for both proofs.) First, to prove item (1) a standard ZK proof of equality of discrete-logs [16] is performed:

1. U commits to a challenge $c \xleftarrow{R} \mathbb{Z}_q$.[9]
2. S_i chooses $r_1, r_2 \xleftarrow{R} \mathbb{Z}_q$, and sends $b_1 = u_1^{r_1}, b_2 = g_1^{r_1}, b_3 = u_2^{r_2}, b_4 = g_2^{r_2}$ to U.
3. U de-commits to c
4. S_i sends $a_1 = r_1 + cs_{l,i}$ and $a_2 = r_2 + cs_{l,i}$ to U.
5. U accepts if $u_1^{a_1} = \tilde{u}_1^{cb_1}$ and $g_1^{a_1} = g_1^{s_{l,i}cb_2}$ and $u_2^{a_2} = \tilde{u}_2^{cb_3}$ and $g_2^{a_2} = g_2^{s_{l,i}cb_4}$.

The above interaction consists of two [16] proofs, that use the same challenge. It can be seen that using the same challenge does not significantly increase the probability of error for the user.

Next, to show item (2) above, server S_i and user U engage in the following interaction (which is a combination of the above proof of equality of discrete logs, and a proof of "knowledge of a representation" from [14,15] (we use the formulation of [8]).

___

[9] Specifically, we use the Pedersen commitment scheme [36]. Here the parties may use two predetermined generators g, h of the subgroup of size q in $\mathbb{Z}_p^*$. The user commits to c by sending $g^c h^s$ for a randomly chosen s in $\mathbb{Z}_q$.

1. U commits to a challenge $c \overset{R}{\leftarrow} \mathbb{Z}_q$, as before.
2. S_i chooses $r_1...r_7 \overset{R}{\leftarrow} \mathbb{Z}_q$ and sends $b_1 = \tilde{u}_1^{r_1}\tilde{u}_1^{r_2}$, $b_2 = \tilde{u}_2^{r_3\alpha}\tilde{u}_2^{r_4\alpha}$, $b_3 = u_1^{r_5}$, $b_4 = g_1^{r_5}$, $b_5 = v^{r_6}$, $b_6 = g_1^{r_6}$, $b_7 = g_1^{r_7}$, $b_8 = g_2^{r_7}$ to U.
3. U de-commits to c.
4. S_i sends $a_1 = r_1 + x_{1,i}c$, $a_2 = r_2 + x_{2,i}c$, $a_3 = r_3 + y_{1,i}c$, $a_4 = r_4 + y_{2,i}c$, $a_5 = r_5 + z_ic$, $a_6 = r_6 + s_{l,i}c$, $a_7 = r_6 + o_{l,i}c$ to U.
5. U accepts f_i if $g_1^{a_5} = g_1^{z_i c b_4}$ and $g_1^{a_6} = g_1^{s_{l,i} c b_6}$ and $g_2^{a_7} = g_2^{o_{l,i} c b_8}$ and

$$\tilde{u}_1^{a_1}\tilde{u}_2^{a_2}\tilde{u}_1^{a_3\alpha}\tilde{u}_2^{a_4\alpha}u_1^{a_5}v^{a_6}g_1^{a_7} = f_i^c b_1 b_2 b_3 b_5 b_7. \tag{3}$$

The above interaction combines three [16] proofs of equality of discrete logarithms with two [14,15] proofs of knowledge of representation. In addition to using the same challenge, here the verifier's acceptance conditions of the five proofs are combined in a single product (3). This allows the verifier to check the validity of the product f_i without knowing the individual w_i's. Correctness of this interaction is based on the fact that if S_i 'knows' representations $w_1^{-1} = \tilde{u}_1^{x_{1,i}}\tilde{u}_2^{x_{2,i}}$ and $w_2^{-1} = \tilde{u}_1^{y_{1,i}\alpha}\tilde{u}_2^{y_{2,i}\alpha}$ then the values $x_{1,i}, x_{2,i}, y_{1,i}, y_{2,i}$ must be the ones from S_i's secret key (otherwise a knowledge extractor for S_i can be used to find the index of g_2 w.r.t. g_1).

User U decides that f_i is valid if it accepted f_i in both of the above interactions. Finally, U proceeds to compute f_0 and m based on the valid f_i's, as in the basic scheme. Let I-CS denote this interactive variant of T-CS.

Theorem 2. *If the DDH assumption holds then* I-CS *is a t-robust threshold cryptosystem for any* $t < \frac{n}{2}$.

The proof combines the simulation technique from the proof of Theorem 1 with the proofs of the protocols of [16,14,15]. We omit details from this version. (Here we only withstand static adversaries.) We remark that the protocols described here do not withstand *asynchronously concurrent* interactions between a corrupted user and the servers. This problem can be solved once general mechanisms for efficiently dealing with the concurrency problem are provided.

Acknowledgments

We thank Rosario Gennaro, Oded Goldreich, Shai Halevi, Tal Rabin, Omer Reingold, Ronitt Rubinfeld, Victor Shoup and Madhu Sudan for helpful discussions. The first author expresses special thanks to Rosario and Tal for very instructive tutorials. We also thank Moni Naor and Benny Pinkas for pointing out the bivariate-polynomial solution to the synchronization problem among the servers.

References

1. D. Beaver, "Secure Multi-party Protocols and Zero-Knowledge Proof Systems Tolerating a Faulty Minority", J. Cryptology (1991) 4: 75-122.

2. D. Beaver and J. Feigenbaum, "Hiding instances in multi-oracle queries", *STACS*, 1990.

3. D. Beaver and S. Haber, "Cryptographic Protocols Provably secure Against Dynamic Adversaries", *Eurocrypt*, 1992.

4. M. Bellare, A. Desai, D. Pointcheval and P. Rogaway, "Relations among notions of security for public-key encryption schemes", *CRYPTO '98*, 1998, pp. 26-40.

5. M. Bellare and P. Rogaway, "Optimal Asymmetric Encryption", *Eurocrypt '94 (LNCS 950)*, 92-111, 1995.

6. M. Ben-Or, S. Goldwasser and A. Wigderson, "Completeness Theorems for Non-Cryptographic Fault-Tolerant Distributed Computation", *20th STOC*, 1988, pp. 1-10.

7. M. Blaze, J. Feigenbaum and M. Naor, "A formal treatment of remotely keyed encryption", *Eurocrypt '98*, 1998, pp. 251-165.

8. S. Brands, "An efficient off-line electronic cash system based on the representation problem", CWI TR CS-R9323, 1993.

9. R. Canetti, "Security and composition of multi-party protocols", Available at the Theory of Cryptography Library, http://philby.ucsd.edu, 1998.

10. R. Canetti, U. Feige, O. Goldreich and M. Naor, "Adaptively Secure Computation", *28th STOC*, 1996. Fuller version in MIT-LCS-TR #682, 1996.

11. R. Canetti and A. Herzberg, "Maintaining security in the presence of transient faults", *CRYPTO'94*, 1994.

12. R. Canetti and S. Goldwasser, "A threshold public-key cryptosystem secure against chosen ciphertext attacks", Available at the Theory of Cryptography Library, http://philby.ucsd.edu, 1999.

13. D. Chaum, C. Crepeau, and I. Damgard, "Multi-party Unconditionally Secure Protocols", *20th STOC*, 1998, pp. 11–19.

14. D. Chaum, E. Everetse and J. van der Graaf, "An improved protocol for demonstrating possession of discrete logarithms and some generalizations", *Eurocrypt '87,*, LNCS 304, 1987, pp. 127-141.

15. D.Chaum, A. Fiat and M. Naor, "Untraceable electronic cash", *CRYPTO '88*, LNCS 403, 1988, pp. 319-327.

16. D. Chaum and T. Pedersen, "Wallet databases with observers", *CRYPTO '92*, 1992, pp. 89-105.

17. R. Cramer and V. Shoup, "A paractical public-key cryptosystem provably secure against adaptive chosen ciphertext attack", *CRYPTO '98*, 1998.

18. Y. Desmedt and Y. Frankel, "Threshold cryptosystems", *Crypto '89 (LNCS 435)*, 1989, pages 307–315.

19. D. Dolev, C. Dwork and M. Naor, Non-malleable cryptography, *23rd STOC*, 1991.

20. P. Feldman, "A practical scheme for non-interactive Verifiable Secret Sharing", *28th FOCS*, 1987, pp. 427-437.

21. A. Fiat and A. Shamir, "How to prove yourself: Practical solutions to identification and signature problems", *CRYPTO'86 (LNCS 263)*, 186-194, 1986.

22. R. Gennaro, S. Jarecki, H. Krawczyk, and T. Rabin, "Secure Distributed Key Generation for Discrete-Log Based Cryptosystems", *these proceedings*.

23. J. Garay, R. Gennaro, C. Jutla and T. Rabin, "Secure Distributed Storage and Retrieval" Proceedings of 11th International Workshop on Distributed Algorithms (WDAG97) Lecture Notes in Computer Science 1320, pp. 275-289, 1997.

24. R. Gennaro, S. Jarecki, H. Krawczyk, and T. Rabin, "Robust and efficient sharing of RSA functions, *CRYPTO '96*, 1996, pp. 157-172.

25. R. Gennaro and V. Shoup, "Securing threshold cryptosystems against chosen ciphertext attack", *Eurocrypt '98*, 1998, pp. 1-16.
26. O. Goldreich, S. Micali and A. Wigderson, "How to Play any Mental Game", *19th STOC*, 1987, pp. 218-229.
27. S. Goldwasser, and L. Levin, "Fair Computation of General Functions in Presence of Immoral Majority", *CRYPTO, 1990.*
28. S. Goldwasser and S. Micali, "Probabilistic encryption", *JCSS*, Vol. 28, No 2, April 1984, pp. 270-299.
29. A. Herzberg, S. Jarecki, H. Krawczyk and M. Yung "Proactive Secret Sharing or: How to Cope with Perpetual Leakage", *CRYPTO '95*, LNCS 963, 1995. pp. 339–352.
30. F. J. Macwiliams and N. J. A. Sloane, *"The Theory of Error Correcting Codes"*, North-Holland, 1977.
31. S. Micali and P. Rogaway, "Secure Computation", unpublished manuscript, 1992. Preliminary version in *CRYPTO 91*.
32. M. Naor, B. Pinkas, and O. Reibgold "Distributed Pseudo-random Functions and KDCs", *these proceedings.*
33. M. Naor and M. Yung, "Public key cryptosystems provably secure against chosen ciphertext attacks", *22nd STOC*, 427-437, 1990.
34. R. Ostrovsky and M. Yung. "How to withstand mobile virus attacks". *10th PODC*, 1991, pp. 51–59.
35. T. Pedersen, "A threshold cryptosystem without a trusted party", *Eurocrypt '91*, 1991, pp. 522-526.
36. T. Pedersen. Distributed provers with applications to undeniable signatures. *Eurocrypt '91*, 1991.
37. T. Rabin and M. Ben-Or, "Verifiable Secret Sharing and Multi-party Protocols with Honest Majority", *21st STOC*, 1989, pp. 73-85.
38. C. Rackoff and D. Simon, "Non-interactive zero-knowledge proof of knowledge and chosen ciphertext attack", *CRYPTO '91*, 1991.
39. A. Shamir. "How to Share a Secret", *Communications of the ACM*, 22:612–613, 1979.
40. C. Schnorr, "Efficient signature generation by smart cards", *J. Cryptology* 4:161-174, 1991.
41. M. Sudan, "Algorithmic issues in coding theorey", *17th Conf. on Foundations of Software Technology and Theoretical Computer Science*, Kharapur, India, 1997. Available on-line at **theory.lcs.mit.edu/~madhu/**

Proving in Zero-Knowledge that a Number Is the Product of Two Safe Primes

Jan Camenisch[1]* and Markus Michels[2]**

[1] BRICS
Department of Computer Science, University of Aarhus,
Ny Munkegade, DK – 8000 Århus C, Denmark
camenisch@daimi.au.dk
[2] Entrust Technologies Europe, r3 security engineering ag,
Glatt Tower, CH – 8301 Glattzentrum, Switzerland
Markus.Michels@entrust.com

Abstract. We present the first efficient statistical zero-knowledge protocols to prove statements such as:
- A committed number is a prime.
- A committed (or revealed) number is the product of two safe primes, i.e., primes p and q such that $(p-1)/2$ and $(q-1)/2$ are prime.
- A given integer has large multiplicative order modulo a composite number that consists of two safe prime factors.

The main building blocks of our protocols are statistical zero-knowledge proofs of knowledge that are of independent interest. We show how to prove the correct computation of a modular addition, a modular multiplication, and a modular exponentiation, where all values including the modulus are committed to but *not* publicly known. Apart from the validity of the equations, no other information about the modulus (e.g., a generator whose order equals the modulus) or any other operand is exposed. Our techniques can be generalized to prove that any multivariate modular polynomial equation is satisfied, where only commitments to the variables of the polynomial and to the modulus need to be known. This improves previous results, where the modulus is publicly known. We show how these building blocks allow to prove statements such as those listed earlier.

1 Introduction

The problem of proving that a number n is the product of two primes p and q of special form arises in many recently proposed cryptographic schemes (e.g., [7,8,20,21]) whose security is based on both the infeasibility of computing discrete logarithms and of computing roots in groups of unknown order. Such schemes typically involve a designated entity that knows the group's order and hence

---

* BRICS - Basic Research in Computer Science, Center of the Danish National Research Foundation.

** Part of this work was done while this author was with Ubilab, UBS, Switzerland.

J. Stern (Ed.): EUROCRYPT'99, LNCS 1592, pp. 107–122, 1999.
© Springer-Verlag Berlin Heidelberg 1999

is able to compute roots. Although the other involved entities must not learn the group's order, nevertheless, they want to be assured that it is large and not smooth, i.e., that computing discrete logarithms is infeasible to the designated entity as well. An example of groups used in such schemes are subgroups of $\mathbb{Z}_n^*$. Here, it suffices that the designated entity proves n to be the product of two safe primes, i.e., primes p and q such that $(p-1)/2$ and $(q-1)/2$ are prime. More precisely, if n is the product of two safe primes p and q and $a^2 \not\equiv 1 \pmod{n}$ and $\gcd(a^2 - 1, n) = 1$ holds for some a (which the verifier can check easily), then a has multiplicative order $(p-1)(q-1)/4$ or $(p-1)(q-1)/2$ [21]. Another example are elliptic curves over $\mathbb{Z}_n$. In this case, n is required to be the product of two primes p and q such that $(p+1)/2$ and $(q+1)/2$ are prime [25]. Finally, standards such as X9.31 require the modulus to be the product of two primes p and q, where $(p-1)/2$, $(p+1)/2$, $(q-1)/2$, and $(q+1)/2$ have a large prime factor that is between 100 and 120 bit [39][1]. Previously, the only way known to prove such properties was applying inefficient general zero-knowledge proof techniques (e.g., [23,5,16]).

In this paper we describe an efficient protocol for proving that a committed integer is in fact the modular addition of two committed integer modulo another committed integer without revealing any other information whatsoever. Then, we provide similar protocols for modular multiplication, modular exponentiation, and, more general, for any multivariate polynomial equation. Previously known protocols allow only to prove that algebraic relations modulo a *publicly known* integer hold [4,9,16,18]. Furthermore, we present an efficient zero-knowledge argument of primality of a committed number and, as a consequence, a zero-knowledge argument that an RSA modulus n consists of two safe primes. The additional advantage of this method is that only a commitment to n but not n itself must be publicly known. If the number n is publicly known, however, more efficient protocols can be obtained by combining our techniques with known results which are described in the next paragraph.

A number of protocols for proving properties of composite numbers are found in literature. Van de Graaf and Peralta [37] provide an efficient proof that a given integer n is of the form $n = p^r q^s$, where r and s are odd, p and q are primes and $p \equiv q \equiv 3 \pmod 4$. A protocol due to Boyar et al. [2] allows to prove that a number n is square-free, i.e., there is no prime p with $p|n$ such that $p^2|n$. Hence, if both properties are proved, it follows that n is the product of two primes p and q, where $p \equiv q \equiv 3 \pmod 4$. This result was recently strengthened by Gennaro et al. [22] who present a proof system for showing that a number n (satisfying certain side-conditions) is the product of quasi-safe primes, i.e., primes p and

---

[1] However, it is unnecessary to explicitly add this requirement to the RSA key generation. For randomly chosen large primes, the probability that $(p-1)/2$, $(p+1)/2$, $(q-1)/2$, and $(q+1)/2$ have a large prime factor is overwhelming. This is sufficient protection against the Pollard $p-1$ and Williams $p+1$ factoring methods [32,38]. Moreover, an efficient proof that an arbitrarily generated RSA modulus is not weak without revealing its factors seems to be hard to obtain as various conditions have to be checked (e.g., see [1]).

q for which $(p-1)/2$ and $(q-1)/2$ is a prime *power*. However, their protocol can not guarantee that $(p-1)/2$ and $(q-1)/2$ are indeed primes which is what we are aiming for. Finally, Chan et al. [11] and Mao [29] provide protocols for showing that a committed number consists of two large factors, and, recently, Liskov & Silverman describe a proof that a number is a product of two nearly equal primes [28].

2 Tools

In the following we assume a group $G = \langle g \rangle$ of large known order Q and a second generator h whose discrete logarithm to the base g is not known. We define the discrete logarithm of y to the base g to be any integer x such that $y = g^x$ holds, in particular discrete logarithms are allowed to be negative. Computing discrete logarithms is assumed to be infeasible.

2.1 Commitment Schemes

Our schemes use commitment schemes that allow to prove algebraic properties of the committed value. There are two kinds of commitment schemes. The first kind hides the committed value information theoretically from the verifier (unconditionally hiding) but is only conditionally binding, i.e., a computationally unbounded prover can change his mind. The second kind is only computationally hiding but unconditionally binding. Depending on the kind of the commitment scheme employed, our schemes will be statistical zero-knowledge arguments (proofs of knowledge) or computational zero-knowledge proof systems. Cramer and Damgård [16] describe a class of commitment schemes allowing to prove algebraic properties of the committed value. It includes RSA-based as well as discrete-logarithm-based schemes of both kinds. For easier description of our protocols, we will use a particular commitment scheme which is due to Pedersen [31]: A value $a \in \mathbb{Z}_Q$ is committed to by $c_a := g^a h^r$, where r is randomly chosen from $\mathbb{Z}_Q$. This scheme is unconditionally hiding and computationally binding, i.e., a prover able to compute $\log_g h$ can change his mind. Therefore our protocol will be statistical zero-knowledge proofs of knowledge (or arguments). However, our protocols can easily be adapted to work for all the commitment scheme exposed in [16].

2.2 Various Proof-Protocols Found in Literature

We review various zero-knowledge protocols for proving knowledge of and about discrete logarithms and introduce our notation for such protocols.

Proving the knowledge of a discrete logarithm x of a group element y to a base g [13,35]. The prover chooses a random $r \in_R \mathbb{Z}_Q$ and computes $t := g^r$ and sends t to the verifier. The verifier picks a random challenge $c \in_R \{0,1\}^k$ and sends it to the prover. The prover computes $s := r - cx \pmod{Q}$ and sends

s to the verifier. The verifier accepts, if $g^s y^c = t$ holds. This protocol is an *honest-verifier zero-knowledge proof of knowledge* if $k = \Theta(\text{poly}(\log Q))$ and a *zero-knowledge proof of knowledge* if $k = \mathcal{O}(\log\log(Q))$ and when serially repeated $\Theta(\text{poly}(\log Q))$ times. This holds for all other protocols described in this section (when not mentioned otherwise). Adopting the notation in [8], we denote this protocol by $PK\{(\alpha) : y = g^\alpha\}$, where PK stands for "proof of knowledge".

Proving the knowledge of a representation of an element y to the bases $g_1, \ldots, g_l$ [3,12], i.e., proving the knowledge of integers $x_1, \ldots, x_l$ such that $y = \prod_{i=1}^l g_i^{x_i}$. This protocol is an extension of the previous one with respect to multiple bases. The prover chooses random integers $r_1, \ldots, r_l \in_R \mathbb{Z}_Q$, computes $t := \prod_{i=1}^l g_i^{r_i}$, and sends the verifier t. The verifier returns her a randomly picked challenge $c \in_R \{0,1\}^k$. The prover computes $s_i := r_i - cx_i \pmod Q$ for $i = 1, \ldots, l$ and sends the verifier all s_i's, who accepts, if $t = y^c \prod_{i=1}^l g_i^{s_i}$ holds. This protocol is denoted by $PK\{(\alpha_1, \ldots, \alpha_l) : y = \prod_{i=1}^l g_i^{\alpha_i}\}$.

Proving the equality of the discrete logarithms of elements y_1 and y_2 to the bases g and h, respectively [14]. Let $y_1 = g^x$ and $y_2 = h^x$. The prover chooses a random $r \in \mathbb{Z}_Q^*$, computes $t_1 := g^r, t_2 := h^r$, and sends t_1, t_2 to the verifier. The verifier picks a random challenge $c \in \{0,1\}^k$ and sends it to the prover. The prover computes $s := r - cx \pmod Q$ and sends s to the verifier. The verifier accepts, if $g^s y_1^c = t_1$ and $h^s y_2^c = t_2$ holds. This protocol is denoted by $PK\{(\alpha) : y_1 = g^\alpha \wedge y_2 = h^\alpha\}$. Note that this method allows also to prove that one discrete log is the square of another one (modulo the group order), e.g., $PK\{(\alpha) : y_1 = g^\alpha \wedge y_2 = y_1^\alpha\}$.

Proving the knowledge of (at least) one out of the discrete logarithms of the elements y_1 and y_2 to the base g (proof of OR) [17,34]. W.l.o.g., we assume that the prover knows $x = \log_g y_1$. Then $r_1, s_2 \in_R \mathbb{Z}_Q^*, c_2 \in_R \{0,1\}^k$ and computes $t_1 := g^{r_1}, t_2 := g^{s_2} y_2^{c_2}$ and sends t_1 and t_2 to the verifier. The verifier picks a random challenge $c \in \{0,1\}^k$ and sends it to the prover. The prover computes $c_1 := c \oplus c_2$ and $s_1 := r_1 - c_1 x \pmod Q$ (where $\oplus$ denotes the bit-wise XOR operation) and sends s_1, s_2, c_1, and c_2 to the verifier. The verifier accepts, if $c_1 \oplus c_2 = c$ and $t_i = g^{s_i} y_i^{c_i}$ holds for $i \in \{1,2\}$. This protocol is denoted by $PK\{(\alpha, \beta) : y_1 = g^\alpha \vee y_2 = g^\beta\}$. This approach can be extended to an efficient system for proving arbitrary monotone statements built with $\wedge$'s and $\vee$'s [17,34].

Proving the knowledge of a discrete logarithm that lies in a given range, that is, $2^{\ell_1} - 2^{\ell_2} < \log_g y < 2^{\ell_1} + 2^{\ell_2}$, for some parameters ℓ_1 and ℓ_2. (The parameter 2^{ℓ_1} acts as an offset and can also chosen to be zero.) In principle, this statement can be proved by first committing to every bit of $x = \log_g y$ and then showing that the committed values are either a 0 or a 1 and constitute the binary representation of x. This method is linear in the number of bits of x. A more efficient but only statistical zero-knowledge protocol can be obtained from the basic protocol proving the knowledge of $\log_g y$ by restricting the verifier to binary challenges and by requiring the prover's response s to satisfy $2^{\ell_1} - 2^{\epsilon\ell_2+1} < s < 2^{\ell_1} + 2^{\epsilon\ell_2+1}$, where $\epsilon > 1$ is a security parameter. Now, when considering how the knowledge extractor can compute an $x = \log_g y$ from two accepting protocol views with the same first message, it can be concluded that the prover must know an $x = \log_g y$ such that $2^{\ell_1} - 2^{\epsilon\ell_2+2} < x < 2^{\ell_1} + 2^{\epsilon\ell_2+2}$ holds [11]. We denote this protocol by

$$PK\{(\alpha) : y = g^\alpha \wedge 2^{\ell_1} - 2^{\ddot{\ell}_2} < \alpha < 2^{\ell_1} + 2^{\ddot{\ell}_2}\},$$

where $\ddot{\ell}_2$ denotes $\epsilon\ell_2+2$ (we will stick to that notation for the rest of the paper). For more details on this protocol we refer to [6,11]. Finally, the restriction to binary challenges can be dropped if the order of the group is not known to the prover (e.g., if a subgroup of an RSA-ring is used) and when believing in the non-standard strong RSA-assumption[2] [18,19]. Although we describe our protocols in the following in the setting where the group's order is known to the prover, all protocols can easily be adapted to the case where the prover does not know the group's order using the techniques from [18,19].

All described protocols can be combined in natural ways. First of all, one can use multiple bases instead of a single one in any of the preceding protocols. Then, executing any number of instances of these protocols in parallel and choosing the same challenges for all of them in each round corresponds to the ∧-composition of the statements the single protocols prove. Using this approach, it is even possible to compose instances according to any monotone formula [17,34]. In the following we will use of such compositions without having explained the technical details involved for which we refer to [4,9,10,17,34].

3   Secret Computations with a Secret Modulus

The goal of this section is to provide an efficient protocol to prove that $a^b \equiv d$ (mod n) holds for some committed integers without revealing the verifier any further information (i.e., the protocol is zero-knowledge). A step towards this goal are protocols to prove that a committed integer is the addition or the multiplication of two other committed integers modulo a third committed integer n.

The algebraic setting is as follows. Let ℓ be an integer such that $-2^\ell < a, b, d, n < 2^\ell$ holds and $\epsilon > 1$ be security parameter (cf. Section 2). Furthermore, we assume that a group G of order $Q > 2^{2\epsilon\ell+5}$ ($= 2^{2\ddot{\ell}+1}$) and two generators g and h are available such that $\log_g h$ is not known. This group could for instance be chosen by the prover in which case she would have to prove that she has chosen it correctly. Finally, let the prover's commitments to a, b, d, and n be $c_a := g^a h^{r_1}$, $c_b := g^b h^{r_2}$, $c_d := g^d h^{r_3}$, and $c_n := g^n h^{r_4}$, where r_1, r_2, r_3, and r_4 are randomly chosen elements of $\mathbb{Z}_Q$.

---

[2]   The strong RSA assumption states that there exists a probabilistic polynomial-time algorithm G that on input $1^{|n|}$ outputs an RSA-modulus n and an element $z \in \mathbb{Z}_n^*$ such that it is infeasible to find integers $e \notin \{-1, 1\}$ and u such that $z \equiv u^e$ (mod n).

3.1 Secret Modular Addition and Multiplication

We assume that the verifier already obtained the commitments c_a, c_b, c_d, and c_n. Then the prover can convince the verifier that $a + b \equiv d \pmod{n}$ holds by sequentially running the protocol denoted[3] by

$$S_+ := PK\{(\alpha, \beta, \gamma, \delta, \varepsilon, \zeta, \eta, \vartheta, \pi, \lambda): \quad c_a = g^\alpha h^\beta \wedge (-2^{\ddot{\ell}} < \alpha < 2^{\ddot{\ell}}) \wedge$$
$$c_b = g^\gamma h^\delta \wedge (-2^{\ddot{\ell}} < \gamma < 2^{\ddot{\ell}}) \wedge c_d = g^\varepsilon h^\zeta \wedge (-2^{\ddot{\ell}} < \varepsilon < 2^{\ddot{\ell}}) \wedge$$
$$c_n = g^\eta h^\vartheta \wedge (-2^{\ddot{\ell}} < \eta < 2^{\ddot{\ell}}) \wedge \frac{c_d}{c_a c_b} = c_n^\pi h^\lambda \wedge (-2^{\ddot{\ell}} < \pi < 2^{\ddot{\ell}})\}$$

k times. Alternatively, she can convince the verifier that $ab \equiv d \pmod{n}$ holds by running the protocol

$$S_* := PK\{(\alpha, \beta, \gamma, \delta, \varepsilon, \zeta, \eta, \vartheta, \varrho, \varsigma): \quad c_a = g^\alpha h^\beta \wedge (-2^{\ddot{\ell}} < \alpha < 2^{\ddot{\ell}}) \wedge$$
$$c_b = g^\gamma h^\delta \wedge (-2^{\ddot{\ell}} < \gamma < 2^{\ddot{\ell}}) \wedge c_d = g^\varepsilon h^\zeta \wedge (-2^{\ddot{\ell}} < \varepsilon < 2^{\ddot{\ell}}) \wedge$$
$$c_n = g^\eta h^\vartheta \wedge (-2^{\ddot{\ell}} < \eta < 2^{\ddot{\ell}}) \wedge c_d = c_b^\alpha c_n^\varrho h^\varsigma \wedge (-2^{\ddot{\ell}} < \varrho < 2^{\ddot{\ell}})\}$$

k times with him.

Remark. In some applications the prover might be required to show that n has some minimal size. This can by showing that η lies in the range $2^{\ell_1} - 2^{\ell_2} < \eta < 2^{\ell_1} + 2^{\ell_2}$ instead of $-2^{\ddot{\ell}} < \eta < 2^{\ddot{\ell}}$ for some appropriate values of ℓ_1 and ℓ_2 (cf. Section 2.2).

Theorem 1. *Let a, b, d, and n be integers that are committed to by the prover as described above and assume computing discrete logarithms in G is infeasible. Then the protocol S_+ is a statistical zero-knowledge argument that $a + b \equiv d \pmod{n}$ holds. Furthermore, the protocol S_* is a statistical zero-knowledge argument that $ab \equiv d \pmod{n}$ holds. The soundness error probability for both protocols is 2^{-k}.*

Proof. The statistical zero-knowledge claims follows from the statistical zero-knowledgeness of the building blocks.

Let us argue why the modular relations among the committed integers hold. First, we consider what the clauses prove that S_+ and S_* have in common. Running the prover with either protocol (and using standard techniques), the knowledge extractor can compute integers $\hat{a}$, $\hat{b}$, $\hat{d}$, $\hat{n}$, $\hat{r}_1$, $\hat{r}_2$, $\hat{r}_3$, and $\hat{r}_4$ such that $c_a = g^{\hat{a}} h^{\hat{r}_1}$, $c_b = g^{\hat{b}} h^{\hat{r}_2}$, $c_d = g^{\hat{d}} h^{\hat{r}_3}$, and $c_n = g^{\hat{n}} h^{\hat{r}_4}$ holds. Moreover, $-2^{\ddot{\ell}} < \hat{a} < 2^{\ddot{\ell}}$, $-2^{\ddot{\ell}} < \hat{b} < 2^{\ddot{\ell}}$, $-2^{\ddot{\ell}} < \hat{d} < 2^{\ddot{\ell}}$, and $-2^{\ddot{\ell}} < \hat{n} < 2^{\ddot{\ell}}$ holds for these integers.

When running the prover with S_+, the knowledge extractor can further compute integers $\hat{r}_5 \in \mathbb{Z}_Q$ and $\hat{u}$ with $-2^{\ddot{\ell}} < \hat{u} < 2^{\ddot{\ell}}$ such that $c_d/(c_a c_b) = c_n^{\hat{u}} h^{\hat{r}_5}$ holds. Therefore, we have $g^{\hat{d}-\hat{a}-\hat{b}} h^{\hat{r}_3-\hat{r}_1-\hat{r}_2} = g^{\hat{n}\hat{u}} h^{\hat{u}\hat{r}_4+\hat{r}_5}$ and hence, provided

---

[3] Recall that $\ddot{\ell}$ denotes $\epsilon\ell + 2$.

$\log_g h$ is not known, we must have $\hat{d} \equiv \hat{a} + \hat{b} + \hat{u}\hat{n} \pmod{Q}$. Thus we have $\hat{d} = \hat{a} + \hat{b} + \hat{u}\hat{n} + \bar{w}Q$ for some integer $\bar{w}$. Since $2^{2\ddot{\ell}+1} < Q$ and due to the constraints on $\hat{a}$, $\hat{b}$, $\hat{d}$, $\hat{n}$, and $\hat{u}$, we can conclude that the integer $\bar{w}$ must be 0 and so $\hat{d} \equiv \hat{a} + \hat{b} \pmod{\hat{n}}$ must hold.

Now consider the case when running the prover with S_*. In this case the knowledge-extractor can additionally compute integers $\hat{r}_6 \in \mathbb{Z}_Q$ and $\hat{v}$ with $-2^{\ddot{\ell}} < \hat{v} < 2^{\ddot{\ell}}$ such that $c_d = c_b^{\hat{a}} c_n^{\hat{v}} h^{\hat{r}_6}$ and thus $g^{\hat{d}} h^{\hat{r}_3} = g^{\hat{a}\hat{b}+\hat{v}\hat{n}} h^{\hat{a}\hat{r}_2 + \hat{v}\hat{r}_4 + \hat{r}_6}$ holds. Again, assuming that $\log_g h$ is not known, we have $\hat{d} \equiv \hat{a}\hat{b} + \hat{v}\hat{n} \pmod{Q}$. As before, due to $2^{2\ddot{\ell}+1} < Q$ and the constraints on $\hat{a}$, $\hat{b}$, $\hat{d}$, $\hat{n}$, and $\hat{v}$ we can conclude that $\hat{d} \equiv \hat{a}\hat{b} \pmod{\hat{n}}$ must hold for the committed values.   □

3.2   Secret Modular Exponentiation

We now extend the ideas from the previous paragraph to a method for proving that $a^b \equiv d \pmod{n}$ holds. Using the same approach as above, i.e., having the prover to provide a commitment to an integer $\tilde{a}$ that equals a^b (in $\mathbb{Z}$) and proving this, would required that G has order about $2^{b\ell}$ and thus such a protocol would become rather inefficient. A more efficient protocol is obtained by constructing a^b (mod n) step by step according to the square & multiply algorithm[4], committing to all intermediary results, and then prove that everything is consistent. This protocol is exposed in the following. We assume that an upper-bound $\ell_b \leq \ell$ on the length of b is publicly known.

1. Apart from her commitments c_a, c_b, c_d, and c_n to a, $b = \sum_{i=0}^{\ell_b-1} b_i 2^i$, d, and n, the prover must commit to all the bits of b: let $c_{b_i} := g^{b_i} h^{\tilde{r}_i}$ with $\tilde{r}_i \in_R \mathbb{Z}_Q$ for $i \in \{0, \dots, \ell_b - 1\}$. Furthermore she needs to provide commitments to the intermediary results of the square & multiply algorithm: let $c_{v_i} := g^{(a^{2^i} \bmod n)} h^{\hat{r}_i}$, $(i = 1, \dots, \ell_b - 1)$, be her commitments to the powers of a, i.e., $a^{2^i} \pmod{n}$, where $\hat{r}_i \in_R \mathbb{Z}_Q$, and let $c_{u_i} := g^{u_i} h^{\bar{r}_i}$, $(i = 0, \dots, \ell_b - 2)$, where $u_i := u_{i-1}(a^{2^i})^{b_i} \pmod{n}$, $(i = 1, \dots, \ell_b - 2)$, $u_0 = a^{b_0} \pmod{n}$, and $\bar{r}_i \in_R \mathbb{Z}_Q$. The prover sends the verifier all these commitments.
2. To prove that $a^b \equiv d \pmod{n}$ holds, they carry out the following protocol k times.

$$S_\uparrow := PK\Big\{ (\alpha, \beta, \xi, \chi, \gamma, \delta, \varepsilon, \zeta, \eta, (\lambda_i, \mu_i, \nu_i, \xi_i, \varsigma_i, \tau_i, \vartheta_i, \varphi_i, \psi_i)_{i=1}^{\ell_b-1}, (\pi_i, \varrho_i)_{i=1}^{\ell_b-2}) :$$

$$c_a = g^\alpha h^\beta \ \wedge \ -2^{\ddot{\ell}} < \alpha < 2^{\ddot{\ell}} \ \wedge \tag{1}$$

$$c_d = g^\gamma h^\delta \ \wedge \ -2^{\ddot{\ell}} < \gamma < 2^{\ddot{\ell}} \ \wedge \tag{2}$$

$$c_n = g^\varepsilon h^\zeta \ \wedge \ -2^{\ddot{\ell}} < \varepsilon < 2^{\ddot{\ell}} \ \wedge \tag{3}$$

$$\Big(\prod_{i=0}^{\ell_b-1} c_{b_i}^{2^i} \Big) / c_b = h^\eta \ \wedge \tag{4}$$

---

[4] In practice a more enhanced exponentiation algorithm might be used (see, e.g., [15]), but one should keep in mind that it must not leak additional information about the exponent.

$$c_{v_1} = g^{\lambda_1} h^{\mu_1} \;\wedge\; \ldots \;\wedge\; c_{v_{\ell_b - 1}} = g^{\lambda_{\ell_b - 1}} h^{\mu_{\ell_b - 1}} \;\wedge \tag{5}$$

$$c_{v_1} = c_a^{\alpha} c_n^{\nu_1} h^{\xi_1} \;\wedge\; c_{v_2} = c_{v_1}^{\lambda_1} c_n^{\nu_2} h^{\xi_2} \;\wedge\; \ldots \;\wedge\; c_{v_{\ell_b - 1}} = c_{v_{\ell_b - 2}}^{\lambda_{\ell_b - 2}} c_n^{\nu_{\ell_b - 1}} h^{\xi_{\ell_b - 1}} \;\wedge \tag{6}$$

$$-2^{\ddot{\ell}} < \lambda_1 < 2^{\ddot{\ell}} \;\wedge \ldots \wedge\; -2^{\ddot{\ell}} < \lambda_{\ell_b - 1} < 2^{\ddot{\ell}} \;\wedge \tag{7}$$

$$-2^{\ddot{\ell}} < \nu_1 < 2^{\ddot{\ell}} \;\wedge \ldots \wedge\; -2^{\ddot{\ell}} < \nu_{\ell_b - 1} < 2^{\ddot{\ell}} \;\wedge \tag{8}$$

$$c_{u_1} = g^{\tau_1} h^{\varrho_1} \;\wedge\; \ldots \;\wedge\; c_{u_{\ell_b - 2}} = g^{\pi_{\ell_b - 2}} h^{\varrho_{\ell_b - 2}} \;\wedge \tag{9}$$

$$-2^{\ddot{\ell}} < \pi_1 < 2^{\ddot{\ell}} \;\wedge \ldots \wedge\; -2^{\ddot{\ell}} < \pi_{\ell_b - 2} < 2^{\ddot{\ell}} \;\wedge \tag{10}$$

$$\left((c_{b_0} = h^{\varsigma_0} \;\wedge\; c_{u_0}/g = h^{\tau_0}) \;\vee\; (c_{b_0}/g = h^{\vartheta_0} \;\wedge\; c_{u_0}/c_a = h^{\psi_0}) \right) \;\wedge \tag{11}$$

$$\left((c_{b_1} = h^{\varsigma_1} \;\wedge\; c_{u_1}/c_{u_0} = h^{\tau_1}) \;\vee \right. \tag{12}$$

$$\left. (c_{b_1}/g = h^{\vartheta_1} \;\wedge\; c_{u_1} = c_{u_0}^{\lambda_1} c_n^{\varphi_1} h^{\psi_1} \;\wedge\; -2^{\ddot{\ell}} < \varphi_1 < 2^{\ddot{\ell}}) \right) \;\wedge\; \ldots \;\wedge$$

$$\left((c_{b_{\ell_b - 2}} = h^{\varsigma_{\ell_b - 2}} \;\wedge\; c_{u_{\ell_b - 2}}/c_{u_{\ell_b - 3}} = h^{\tau_i}) \;\vee \right. \tag{13}$$

$$\left. (c_{b_{\ell_b - 2}}/g = h^{\vartheta_{\ell_b - 2}} \;\wedge\; c_{u_{\ell_b - 2}} = c_{u_{\ell_b - 3}}^{\lambda_{\ell_b - 2}} c_n^{\varphi_{\ell_b - 2}} h^{\psi_{\ell_b - 2}} \;\wedge\; -2^{\ddot{\ell}} < \varphi_{\ell_b - 2} < 2^{\ddot{\ell}}) \right) \;\wedge$$

$$\left((c_{b_{\ell_b - 1}} = h^{\varsigma_{\ell_b - 1}} \;\wedge\; c_d/c_{u_{\ell_b - 2}} = h^{\tau_i}) \;\vee \right. \tag{14}$$

$$\left. (c_{b_{\ell_b - 1}}/g = h^{\vartheta_{\ell_b - 1}} \;\wedge\; c_d = c_{u_{\ell_b - 2}}^{\lambda_{\ell_b - 1}} c_n^{\varphi_{\ell_b - 1}} h^{\psi_{\ell_b - 1}} \;\wedge\; -2^{\ddot{\ell}} < \varphi_{\ell_b - 1} < 2^{\ddot{\ell}}) \right) \}$$

Let us now explain why this protocol proves that $a^b \equiv d \pmod{n}$ holds and consider the clauses of sub-protocol $S_\uparrow$. What the Clauses 1–3 prove should be clear. The Clause 4 shows that the c_{b_i}'s indeed commit to the bits of the integer committed to in c_b (that these are indeed bits is shown in the Clauses 11–14). From this it can further be concluded that c_b commits to a value smaller than 2^{ℓ_b}. The Clauses 5–8 prove that the c_{v_i}'s indeed contain $a^{2^i} \pmod{n}$ (cf. Section 3.1). Finally, the Clauses 9–14 show that c_{u_i}'s commit to the intermediary results of the square & multiply algorithm and that c_d commits to the result: The Clauses 9 and 10 show that the c_{u_i}'s commit to integers that lie in $\{-2^{\ddot{\ell}}+1, \ldots, 2^{\ddot{\ell}}-1\}$ (for c_{u_0} this follows from Clause 11). Then, Clause 11 proves that either c_{b_0} commits to a 0 and c_{u_0} commits to a 1 or c_{b_0} commits to a 1 and c_{u_0} commits to the same integer as c_a. The Clauses 12 and 13 show that for each $i = 1, \ldots, \ell_b - 2$ either c_{b_i} commits to a 0 and c_{u_i} commits to same integer as $c_{u_{i-1}}$ or c_{b_i} commits to a 1 and c_{u_i} commits to the modular product of the value $c_{u_{i-1}}$ commits to and of $a^{2^i} \pmod{n}$ (which c_{v_i} commits to). Finally, Clause 14 proves (in a similar manner as the Clauses 12 and 13) that c_d commits to the result of the square & multiply algorithm and thus to $a^b \pmod{n}$.

Theorem 2. *Let c_a, c_b, c_d, and c_n be commitments to integers a, b, d, and n and let $c_{b_0}, \ldots, c_{b_{\ell-1}}, c_{v_1}, \ldots, c_{v_{\ell_b - 1}}, c_{u_0}, \ldots, c_{u_{\ell_b - 2}}$ be auxiliary commitments. Then, assuming computing discrete logarithms in G is infeasible, the protocol $S_\uparrow$ is a statistical zero-knowledge argument that the equation $a^b \equiv d \pmod{n}$ holds. The soundness error probability is 2^{-k}.*

Proof. The proof is straight forward from Theorem 1 and the explanations given above that $c_{b_0}, \ldots, c_{b_{\ell-1}}, c_{v_1}, \ldots, c_{v_{\ell_b}-1}, c_{u_0}, \ldots, c_{u_{\ell_b}-2}$, S implement the square & multiply algorithm step by step.   □

In the following, when denoting a protocol, we refer to the protocol $S_\uparrow$ by adding a clause like $(\alpha^\beta \equiv \gamma \pmod{\delta})$ to the statement that is proven and assume that the prover sends the verifier all necessary commitments; e.g.,

$$PK\Big\{(\alpha, \beta, \gamma, \delta, \tilde{\alpha}, \tilde{\beta}, \tilde{\gamma}, \tilde{\delta}) : c_a = g^\alpha h^{\tilde{\alpha}} \wedge c_b = g^\beta h^{\tilde{\beta}} \wedge c_d = g^\gamma h^{\tilde{\gamma}} \wedge$$
$$c_n = g^\delta h^{\tilde{\delta}} \wedge (\alpha^\beta \equiv \gamma \pmod{\delta})\Big\}.$$

3.3   Efficiency Analysis

We assume that G is chosen such that group elements can be represented with about $\log Q$ bits. For both S_+ and S_* the prover and the verifier both need to compute 5 multi-exponentiations per round. The communication per round is about $10 \log Q + 5\epsilon\ell$ bits in case of S_+ and S_*. In case of $S_\uparrow$, the prover and the verifier need to compute about $7\ell_b$ multi-exponentiations per round. Additionally, the prover needs to compute about $3\ell_b$ multi-exponentiations in advance of the protocol (these are the computations of the commitments to the intermediary results of the square & multiply algorithm). The communication cost per round is about $14\ell_b \log Q + 4\ell_b\epsilon\ell$ bits and an initial $3\ell_b$ group element which are the commitments to the intermediary results of the square & multiply algorithm.

3.4   Extension to a General Multivariate Polynomial

Let us outline how the correct computation of a general multivariate polynomial equation of form

$$f(x_1, \ldots, x_t, a_1, \ldots, a_l, b_{1,1}, \ldots, b_{l,t}, n) = \sum_{i=1}^{l} a_i \prod_{j=1}^{t} x_j^{b_{i,j}} \equiv 0 \pmod{n}$$

can be shown, where all integers $x_1, \ldots, x_t, a_1, \ldots, a_l, b_{1,1}, \ldots, b_{l,t}$, and n might only given as commitments: The prover commits to all the summands $s_1 := a_1 \prod_{j=1}^{t} x_j^{b_{1,j}} \pmod{n}, \ldots, s_l := a_l \prod_{j=1}^{t} x_j^{b_{l,j}} \pmod{n}$ and shows that the sum of these summands is indeed zero modulo n. Then, she commits to all the product terms $p_{1,1} := x_1^{b_{1,1}} \pmod{n}, \ldots, p_{t,l} := x_t^{b_{l,t}} \pmod{n}$ of the product and shows that $s_i \equiv a_i \prod_{j=1}^{t} p_{i,j} \pmod{n}$. Finally, she shows that $p_{i,j} \equiv x_j^{b_{i,j}} \pmod{n}$ (using the protocol $S_\uparrow$) and that for all i the same x_j is in $p_{i,j}$. This extends easily to several polynomial equations, where some variables appear in more than one equation.

4   A Proof that a Secret Number Is a Prime

In this section we describe how a prover and a verifier can carry out a primality test for an integer hidden in a commitment. Some primality tests reveal information about the structure of the prime and are hence not suited unless one is willing to expose this information. Examples of such tests are the Miller-Rabin test [30,33] or the one based on Pocklington's theorem. A test that does not reveal such information is due to Lehmann [27] and described in the next subsection.

4.1   Lehmann's Primality Test

Lehmann's test is variation of the Solovay-Strassen [36] primality test and based on the following theorem [26]:

Theorem 3. *An odd integer $n > 1$ is prime if and only if*

$$\forall a \in \mathbb{Z}_n^* : a^{(n-1)/2} \equiv \pm 1 \pmod{n} \text{ and } \exists a \in \mathbb{Z}_n^* : a^{(n-1)/2} \equiv -1 \pmod{n} .$$

This theorem suggest the following probabilistic primality test [27]:

- Choose k random bases $a_1, \ldots, a_k \in \mathbb{Z}_n^*$,
- check whether $a_i^{(n-1)/2} \equiv \pm 1 \pmod{n}$ holds for all i's, and
- check whether $a_i^{(n-1)/2} \equiv -1 \pmod{n}$ if true for at least one $i \in \{1, \ldots, k\}$.

The probability that a non-prime n passes this test is at most 2^{-k}, and that a prime n does not pass this test is at most 2^{-k} as well. Note that in case n and $(n-1)/2$ are both odd, the condition that $a_i^{(n-1)/2} \equiv -1 \pmod{n}$ holds for at least one i can be omitted. In this special case of Lehmann's test is equivalent to the Miller-Rabin test [33] and the failure probability is at most 4^{-k}.

4.2   Proving the Primality of a Committed Number

We now show how the prover and the verifier can apply Lehmann's primality test to a number committed to by the prover such that the verifier is convinced that the test was correctly done but does not learn any other information. The general idea is that the prover commits to t random bases a_i (of course, the verifier must be assured that the a_i's are chosen at random) and then proves that for these bases $a_i^{(n-1)/2} \equiv \pm 1 \pmod{n}$ holds. Furthermore, to conform with the second condition in Theorem 3, the prover must commit to a base, say $\tilde{a}$, such that $\tilde{a}^{(n-1)/2} \equiv -1 \pmod{n}$ holds.

Let ℓ be an integer such that $n < 2^\ell$ holds and let $\epsilon > 1$ be a security parameter. As in the previous section, a group G of prime order $Q > 2^{2\epsilon\ell+5}$ and two generators g and h are chosen, such that $\log_g h$ is not known. Let $c_n := g^n h^{r_n}$ with $r_n \in_R \mathbb{Z}_Q$ be the prover's commitment to the integer on which the primality test should be performed.

The following four steps constitute the protocol.

1. The prover picks random $\hat{a}_i \in_R \mathbb{Z}_n$ for $i = 1, \ldots, t$ and commits to them as $c_{\hat{a}_i} := g^{\hat{a}_i} h^{r_{\hat{a}_i}}$ with $r_{\hat{a}_i} \in_R \mathbb{Z}_Q$ for $i = 1, \ldots, t$. She sends $c_{\hat{a}_1}, \ldots, c_{\hat{a}_t}$ to the verifier.
2. The verifier picks random integers $-2^\ell < \breve{a}_i < 2^\ell$ for $i = 1, \ldots, t$ and sends them to the prover.
3. The prover computes $a_i := \hat{a}_i + \breve{a}_i \pmod{n}$, $c_{a_i} := g^{a_i} h^{r_{a_i}}$ with $r_{a_i} \in_R \mathbb{Z}_Q$, $d_i := a_i^{(n-1)/2} \pmod{n}$, and $c_{d_i} := g^{d_i} h^{r_{d_i}}$ with $r_{d_i} \in_R \mathbb{Z}_Q$ for all $i = 1, \ldots, t$. Moreover, the prover commits to $(n-1)/2$ by $c_b := g^{(n-1)/2} h^{r_b}$ with $r_b \in_R \mathbb{Z}_Q$. Then the prover searches a base $\tilde{a}$ such that $\tilde{a}^{(n-1)/2} \equiv -1 \pmod{n}$ holds and commits to $\tilde{a}$ by $c_{\tilde{a}} := g^{\tilde{a}} h^{r_{\tilde{a}}}$ with $r_{\tilde{a}} \in_R \mathbb{Z}_Q$.
4. The prover sends $c_b, c_{\tilde{a}}, c_{a_1}, \ldots, c_{a_t}, c_{d_1}, \ldots, c_{d_t}$ to the verifier and then they carry out the following (sub-)protocol k times.

$$S_p := PK\Big\{ (\alpha, \beta, \gamma, \nu, \xi, \varrho, \omega, (\delta_i, \varepsilon_i, \zeta_i, \eta_i, \vartheta_i, \pi_i, \varrho_i, \kappa_i, \mu_i, \psi_i)_{i=1}^t) :$$

$$c_b = g^\alpha h^\beta \ \wedge \ -2^{\tilde{\ell}} < \alpha < 2^{\tilde{\ell}} \ \wedge \tag{15}$$

$$c_n = g^\nu h^\xi \ \wedge \ -2^{\tilde{\ell}} < \nu < 2^{\tilde{\ell}} \ \wedge \tag{16}$$

$$c_b^2 g / c_n = h^\gamma \ \wedge \tag{17}$$

$$c_{\tilde{a}} = g^\varrho h^\omega \ \wedge \ (\varrho^\alpha \equiv -1 \pmod{\nu}) \ \wedge \tag{18}$$

$$c_{\hat{a}_1} = g^{\delta_1} h^{\varepsilon_1} \ \wedge \ldots \wedge \ c_{\hat{a}_t} = g^{\delta_t} h^{\varepsilon_t} \ \wedge \tag{19}$$

$$c_{a_1}/g^{\breve{a}_1} = g^{\delta_1} c_n^{\zeta_1} h^{\eta_1} \ \wedge \ldots \wedge \ c_{a_t}/g^{\breve{a}_t} = g^{\delta_t} c_n^{\zeta_t} h^{\eta_t} \ \wedge \tag{20}$$

$$-2^{\tilde{\ell}} < \delta_1 < 2^{\tilde{\ell}} \ \wedge \ldots \wedge \ -2^{\tilde{\ell}} < \delta_t < 2^{\tilde{\ell}} \ \wedge \tag{21}$$

$$-2^{\tilde{\ell}} < \zeta_1 < 2^{\tilde{\ell}} \ \wedge \ldots \wedge \ -2^{\tilde{\ell}} < \zeta_t < 2^{\tilde{\ell}} \ \wedge \tag{22}$$

$$c_{a_1} = g^{\varrho_1} h^{\kappa_1} \ \wedge \ldots \wedge \ c_{a_t} = g^{\varrho_t} h^{\kappa_t} \ \wedge \tag{23}$$

$$\big(c_{d_1}/g = h^{\vartheta_1} \vee c_{d_1} g = h^{\vartheta_1}\big) \wedge \ldots \wedge \big(c_{d_t}/g = h^{\vartheta_t} \vee c_{d_t} g = h^{\vartheta_t}\big) \wedge \tag{24}$$

$$c_{d_1} = g^{\mu_1} h^{\psi_1} \ \wedge \ldots \wedge \ c_{d_t} = g^{\mu_t} h^{\psi_t} \ \wedge \tag{25}$$

$$(\varrho_1^\alpha \equiv \mu_1 \pmod{\nu}) \ \wedge \ldots \wedge \ (\varrho_t^\alpha \equiv \mu_t \pmod{\nu}) \Big\} \tag{26}$$

Let us analyze the protocol. In Step 1 and 2 of the protocol, the prover and the verifier together choose the random bases $a_1, \ldots, a_t$ for the primality test. Each base is the sum (modulo n) of the random integer the verifier chose and the one the prover chose. Hence, both parties are ensured that the bases are random, although the verifier does not get any information about the bases finally used in the primality test. That the bases are indeed chosen according to this procedure is shown in the Clauses 19–23 of the sub-protocol S_p, where the correct generation of the random values a_i, committed in c_{a_i}, is proved. The Clauses 16–17 prove that indeed $(n-1)/2$ is committed in c_b and the Clause 18 shows that there exists a base $\tilde{a}$ such that $\tilde{a}^{(n-1)/2} \equiv -1 \pmod{n}$. In the Clause 24 it is shown that the values committed in c_{d_i} are either equal to -1 or to 1. Finally, in Clause 26 (together with the Clauses 15, 16, 23, and 25) it is proved that $a_i^{(n-1)/2} \equiv d_i \pmod{n}$, i.e., $a_i^{(n-1)/2} \pmod{n} \in \{-1, 1\}$ and thus the conditions that n is a prime with error-probability 2^{-t} are met.

Note that all modular exponentiations in Clause 26 have the same b and n and hence the proofs for these parts can be optimized. In particular, this is the case for the Clauses 3, 4, and 11–14 in $S_\uparrow$.

Theorem 4. *Assume computing discrete logarithms in G is infeasible. Then, the above protocol is a statistical zero-knowledge argument that the integer committed to by c_n is a prime. The soundness error probability is at most $2^{-k} + 2^{-t}$.*

Proof. The proof is straight forward from the Theorems 1, 2, and 3.   □

In the sequel, we abbreviate the above protocol by adding a clause such as $\alpha \in \mathbf{primes}(t)$ to the statement that is proven, where t denotes the number of bases used in the primality test.

Remark. If $(n-1)/2$ is odd and the prover is willing to reveal this, she can additionally prove that she knows χ and ψ such that $c_b/g = (g^2)^\chi h^\psi$ and $-2^{\tilde{\ell}} < \chi < 2^{\tilde{\ell}}$ holds and skip the Clause 18. This results in a statistical zero-knowledge proof that n of form $n = 2w + 1$ is prime and w is odd with error-probability at most 2^{-2t}.

4.3   Efficiency Analysis

Assume that the commitment to the prime n is given. Altogether $t+1$ protocols that a modular exponentiation holds are carried out where the exponents are about $\log n$ bits. Thus, prover and verifier need to compute about $7t \log n$ multi-exponentiations per round each. Additionally, the prover needs to compute about $2t \log n$ multi-exponentiations for the commitments to the intermediary results of the square & multiply algorithm. (Note that the exponents in Clause 26 is the same in all relations and hence the commitments to its bits need to be computed only once.) The communication cost per round is about $14t \log n \log Q + 4t \log n\epsilon\ell$ bits and an initial $2t \log n$ group elements which are the commitments to the intermediary results of the square & multiply algorithm and the commitments to the bases of the primality test.

5   Proving that an RSA Modulus Consists of Two Safe Primes

We finally present protocols for proving that an RSA modulus consists of two safe primes. First, we restrict ourselves to the case where not the modulus but only a commitment to it is not known to the verifier. Later, we will discuss improvements for cases when the RSA modulus is known to the verifier.

5.1   A Protocol for a Secret RSA Modulus

Let 2^ℓ be an upper-bound on the length of the largest factor of the modulus and let $\epsilon > 1$ be a security parameter. Furthermore, a group G of prime order

$Q > 2^{2\epsilon\ell+5}$ and two generators g and h are chosen, such that $\log_g h$ is not known and computing discrete logarithms is infeasible. Let $c_n := g^n h^{r_n}$ be the prover's commitment to an integer n, where she has chosen $r_n \in_R \mathbb{Z}_Q$, and let p and q denote the two prime factors of n. The following is a protocol that allows her to convince the verifier that c_n commits to the product of two safe primes.

1. The prover computes the commitments $c_p := g^p h^{r_p}$, $c_{\tilde{p}} := g^{(p-1)/2} h^{r_{\tilde{p}}}$, $c_q := g^q h^{r_b}$, and $c_{\tilde{q}} := g^{(q-1)/2} h^{r_{\tilde{p}}}$ with $r_p, r_{\tilde{p}}, r_q, r_{\tilde{q}} \in_R \mathbb{Z}_Q$ and sends all these commitments to the verifier.
2. The two parties sequentially carry out the following protocol k times.

$$S_{51} := PK\{(\alpha, \beta, \gamma, \delta, \varrho, \nu, \xi, \chi, \epsilon, \zeta, \eta) :$$

$$c_{\tilde{p}} = g^\alpha h^\beta \;\wedge\; c_{\tilde{q}} = g^\gamma h^\delta \;\wedge\; c_p = g^\varrho h^\nu \;\wedge\; c_q = g^\xi h^\chi \;\wedge \tag{27}$$

$$c_p/(c_{\tilde{p}}^2 g) = h^\epsilon \;\wedge\; c_q/(c_{\tilde{q}}^2 g) = h^\zeta \;\wedge\; c_n = c_p^\xi h^\eta \;\wedge \tag{28}$$

$$\alpha \in \mathbf{primes}(t) \;\wedge\; \gamma \in \mathbf{primes}(t) \;\wedge \tag{29}$$

$$\varrho \in \mathbf{primes}(t) \;\wedge\; \xi \in \mathbf{primes}(t)\}, \tag{30}$$

where t denotes the number of bases used in Lehmann's primality tests. (The length conditions on α, γ, ϱ, and ξ are shown in the $\mathbf{primes}(t)$-parts of the protocol.)

Theorem 5. *Assume computing discrete logarithms in G is infeasible. Then, the above protocol is a statistical zero-knowledge argument that the integer committed to by c_n is the product of product of two integers p and q and p, q, $(p-1)/2$ and $(q-1)/2$ are primes. The soundness error probability is at most $2^{-k} + 2^{-t}$.*

Proof. The proof is straight forward from the Theorems 1, 2, and 4. □

The computational and communication costs of this protocol are reigned by the primality-protocols and thus about four times as high as for a single primality-protocol (cf. Subsection 4.3).

5.2   A Protocol for a Publicly Known RSA Modulus

In cases the number n is publicly known and fulfills some side-conditions (see below), much less rounds of the Lehmann test will be sufficient if the prover and the verifier first run the protocol due to Gennaro et al. [22] (which includes the protocols proposed by Peralta & van de Graaf [37] and by Boyar et al. [2]). This protocol is a statistical zero-knowledge proof system that there exist two integers $a, b \geq 1$ such that n consists of two primes $p = 2\tilde{p}^a + 1$ and $q = 2\tilde{q}^b + 1$ with $p, q, \tilde{p}, \tilde{q} \not\equiv 1 \pmod 8$, $p \not\equiv q \pmod 8$, $\tilde{p} \not\equiv \tilde{q} \pmod 8$ and $\tilde{p}, \tilde{q}$ are primes. Given the fact that $(p-1)/2$ is a prime power, and assuming that it is not prime, the probability that it passes a single round of Lehmann's primality test for any $a > 1$ is at most $\tilde{p}^{1-a} \leq \sqrt{2/(p-1)}$ (for q the corresponding statement hold). Hence, if p and q are sufficiently large, a single round of Lehmann's primality test on $(p-1)/2$ and $(q-1)/2$ will be sufficient to prove their primality with overwhelming probability. Thus, the resulting protocol to prove that n is the product of two safe primes is as follows.

1. First the prover computes $c_p := g^p h^{r_p}$, $c_{\tilde{p}} := g^{(p-1)/2} h^{r_{\tilde{p}}}$, $c_q := g^q h^{r_b}$, and $c_{\tilde{q}} := g^{(q-1)/2} h^{r_{\tilde{p}}}$ with $r_p, r_{\tilde{p}}, r_q, r_{\tilde{q}} \in_R \mathbb{Z}_Q$ and sends these commitments together with n to the verifier.
2. The prover and the verifier carry out the protocol by Gennaro et al. [22]
3. and then k times the protocol denoted by

$$S_{52} := PK\{(\alpha, \beta, \gamma, \delta, \varrho, \epsilon, \xi, \chi, \varepsilon, \zeta, \eta) :$$

$$c_{\tilde{p}} = g^\alpha h^\beta \ \wedge \ c_{\tilde{q}} = g^\gamma h^\delta \ \wedge \ c_p = g^\varrho h^\epsilon \ \wedge \ c_q = g^\xi h^\chi \ \wedge \tag{31}$$

$$c_p/(c_{\tilde{p}}^2 g) = h^\varepsilon \ \wedge \ c_q/(c_{\tilde{q}}^2 g) = h^\zeta \ \wedge \ g^n = c_{\tilde{p}}^\xi h^\eta \ \wedge \tag{32}$$

$$\alpha \in \text{primes}(1) \ \wedge \ \gamma \in \text{primes}(1)\}, \tag{33}$$

where the length conditions on α and γ are hidden within in the sub-protocols primes(1).

Theorem 6. *Let n be the product of two primes p and q such that $p = 2\tilde{p}^a + 1$ and $q = 2\tilde{q}^b + 1$ with $p, q, \tilde{p}, \tilde{q} \not\equiv 1 \pmod 8$, $p \not\equiv q \pmod 8$, $\tilde{p} \not\equiv \tilde{q} \pmod 8$, $a, b \geq 1$ and $\tilde{p}, \tilde{q}$ are primes. Assume computing discrete logarithms in G is infeasible. Then, the protocol S_{52} is a statistical zero-knowledge argument that n is the product of two integers p and q and that p, q, $(p-1)/2$, and $(q-1)/2$ are primes. Assume $p > q$. Then the soundness error probability is at most $2^{-k} + \sqrt{2/(q-1)}$.*

The computational and communication costs of this protocol is dominated by the costs of a single round (i.e., $t = 1$) of the primality protocol described in the previous section and the costs of protocol of Gennaro et al. [22].

It is obvious how to apply our techniques to get a protocol for proving that n is the product of two *strong* primes [24] (i.e., $(p-1)/2$, $(q-1)/2$, $(p+1)/2$ and $(q+1)/2$ are primes or have a large prime factor) or, more general, two primes p and q such that $\Phi_k(p)$ and $\Phi_k(q)$ are not smooth, where Φ_k is the k-th cyclotomic polynomial. (Recall that smoothness of $\Phi_k(p)$ or $\Phi_k(q)$ for any integer $k > 0$, $k = O(\log n)$ allows to factor n efficiently [1]). Lower bounds on p, q, and on n might also be shown. Also, factors r other than 2 in $(p-1)/r$ could easily be incorporated.

6   Acknowledgments

The authors are grateful to Christian Cachin, Ivan Damgård, Markus Stadler, and Robert Zuccherato for valuable discussions and helpful comments. While the second author was with Ubilab, UBS, he was supported by the Swiss National Foundation (SNF/SPP project no. 5003-045293).

References

1. E. Bach and J. Shallit. Factoring with cyclotomic polynomials. In *26th FOCS*, IEEE, pp. 443–450, 1985.

2. J. Boyar, K. Friedl, and C. Lund. Practical zero-knowledge proofs: Giving hints and using deficiencies. *Journal of Cryptology*, 4(3):185–206, 1991.
3. S. Brands. Untraceable off-line cash in wallets with observers. In *Advances in Cryptology — CRYPTO '93*, volume 773 of *LNCS*, pp. 302–318, 1993.
4. S. Brands. Rapid demonstration of linear relations connected by boolean operators. In *Advances in Cryptology — EUROCRYPT '97*, volume 1233 of *LNCS*, pp. 318–333. Springer Verlag, 1997.
5. G. Brassard, D. Chaum, and C. Crépeau. Minimum disclosure proofs of knowledge. *Journal of Computer and System Sciences*, 37(2):156–189, Oct. 1988.
6. J. Camenisch and M. Michels. Proving in zero-knowledge that a number n is the product of two safe primes. Technical Report RS-98-29, BRICS, Departement of Computer Science, University of Aarhus, Nov. 1998.
7. J. Camenisch and M. Michels. A group signature scheme based on an RSA-variant. Tech. Rep. RS-98-27, BRICS, Departement of Computer Science, University of Aarhus, Nov. 1998. Preliminary version appeared in *Advances in Cryptology — ASIACRYPT '98*, volume 1514 of *LNCS*, pages 160–174. Springer Verlag, 1998.
8. J. Camenisch and M. Stadler. Efficient group signature schemes for large groups. In *Advances in Cryptology — CRYPTO '97*, volume 1296 of *LNCS*, pp. 410–424. Springer Verlag, 1997.
9. J. Camenisch and M. Stadler. Proof systems for general statements about discrete logarithms. Technical Report TR 260, Institute for Theoretical Computer Science, ETH Zürich, Mar. 1997.
10. J. L. Camenisch. *Group Signature Schemes and Payment Systems Based on the Discrete Logarithm Problem*. PhD thesis, ETH Zürich, 1998. Diss. ETH No. 12520.
11. A. Chan, Y. Frankel, and Y. Tsiounis. Easy come – easy go divisible cash. In *Advances in Cryptology — EUROCRYPT '98*, volume 1403 of *LNCS*, pp. 561–575. Springer Verlag, 1998. Revised version available as GTE Technical Report.
12. D. Chaum, J.-H. Evertse, and J. van de Graaf. An improved protocol for demonstrating possession of discrete logarithms and some generalizations. In *Advances in Cryptology — EUROCRYPT '87*, volume 304 of *LNCS*, pp. 127–141. Springer-Verlag, 1988.
13. D. Chaum, J.-H. Evertse, J. van de Graaf, and R. Peralta. Demonstrating possession of a discrete logarithm without revealing it. In *Advances in Cryptology — CRYPTO '86*, volume 263 of *LNCS*, pp. 200–212. Springer-Verlag, 1987.
14. D. Chaum and T. P. Pedersen. Wallet databases with observers. In *Advances in Cryptology — CRYPTO '92*, volume 740 of *LNCS*, pp. 89–105. Springer-Verlag, 1993.
15. H. Cohen. *A Course in Computational Algebraic Number Theory*. Number 138 in Graduate Texts in Mathematics. Springer-Verlag, Berlin, 1993.
16. R. Cramer and I. Damgård. Zero-knowledge proof for finite field arithmetic, or: Can zero-knowledge be for free? In *Advances in Cryptology — CRYPTO '98*, volume 1642 of *LNCS*, pp. 424–441, Berlin, 1998. Springer Verlag.
17. R. Cramer, I. Damgård, and B. Schoenmakers. Proofs of partial knowledge and simplified design of witness hiding protocols. In *Advances in Cryptology — CRYPTO '94*, volume 839 of *LNCS*, pp. 174–187. Springer Verlag, 1994.
18. E. Fujisaki and T. Okamoto. Statistical zero knowledge protocols to prove modular polynomial relations. In *Advances in Cryptology — CRYPTO '97*, volume 1294 of *LNCS*, pp. 16–30. Springer Verlag, 1997.
19. E. Fujisaki and T. Okamoto. A practical and provably secure scheme for publicly verifiable secret sharing and its applications. In *Advances in Cryptology — EUROCRYPT '98*, volume 1403 of *LNCS*, pp. 32–46. Springer Verlag, 1998.

20. R. Gennaro, S. Jarecki, H. Krawczyk, and T. Rabin. Robust and efficient sharing of RSA functions. In *Advances in Cryptology — CRYPTO '96*, volume 1109 of *LNCS*, pp. 157–172, Berlin, 1996. IACR, Springer Verlag.
21. R. Gennaro, H. Krawczyk, and T. Rabin. RSA-based undeniable signatures. In *Advances in Cryptology — CRYPTO '97*, volume 1296 of *LNCS*, pp. 132–149. Springer Verlag, 1997.
22. R. Gennaro, D. Micciancio, and T. Rabin. An efficient non-interactive statistical zero-knowledge proof system for quasi-safe prime products. In *5rd ACM Conference on Computer and Communicatons Security*, 1998.
23. O. Goldreich, S. Micali, and A. Wigderson. How to prove all NP statements in zero-knowledge and a methodology of cryptographic protocol design. In *Advances in Cryptology — CRYPTO '86*, volume 263 of *LNCS*, pp. 171–185. Springer-Verlag, 1987.
24. J. Gordon. Strong RSA keys. *Electronics Letters*, 20(12):514–516, 1984.
25. K. Koyama, U. Maurer, T. Okamoto, and S. Vanstone. New public-key schemes based on elliptic curves over the ring Z_n. In *Advances in Cryptology — CRYPTO '91*, volume 576 of *LNCS*, pp. 252–266. Springer-Verlag, 1992.
26. E. Kranakis. *Primality and Cryptography*. Wiley-Teubner Series in Computer Science, 1986.
27. D. J. Lehmann. On primality tests. *SIAM Journal of Computing*, 11(2):374–375, May 1982.
28. M. Liskov and B. Silverman. A Statisical limited-knowledge proof for secure RSA keys. manuscript, (1998).
29. W. Mao. Verifable Partial Sharing of Integer Factors. to appear in Proc. *SAC'98*, 1998.
30. G. L. Miller. Riemann's hypothesis and tests for primality. *Journal of Computer and System Sciences*, 13:300–317, 1976.
31. T. P. Pedersen. Non-interactive and information-theoretic secure verifiable secret sharing. In *Advances in Cryptology – CRYPTO '91*, volume 576 of *LNCS*, pp. 129–140. Springer Verlag, 1992.
32. J. M. Pollard. Theorems on factorization and primality testing. *Proc. Cambridge Philosophical Society*, 76:521–528, 1974.
33. M. O. Rabin. Probabilistic algorithm for testing primality. *Journal of Number Theory*, 12:128–138, 1980.
34. A. de Santis, L. di Crescenzo, G. Persiano, M. Yung. On Monotone Formula Closure of SZK. *35th FOCS*, IEEE, pp. 454–465, 1994.
35. C. P. Schnorr. Efficient signature generation for smart cards. *Journal of Cryptology*, 4(3):239–252, 1991.
36. R. Solovay and V. Strassen. A fast monte-carlo test for primality. *SIAM Journal on Computing*, 6(1):84–85, Mar. 1977.
37. J. van de Graaf and R. Peralta. A simple and secure way to show the validity of your public key. In *Advances in Cryptology — CRYPTO '87*, volume 293 of *LNCS*, pp. 128–134. Springer-Verlag, 1988.
38. H. C. Williams. A $p + 1$ method of factoring. *Mathematics of Computation*, 39(159):225–234, 1982.
39. X9.31 - 1998 Digital Signatures using reversible public key cryptography for the financial services industry (rDSA). *American National Standard, Working Draft*, 59 pages, 1998.

Secure Hash-and-Sign Signatures
Without the Random Oracle

Rosario Gennaro, Shai Halevi, and Tal Rabin

IBM T.J.Watson Research Center, PO Box 704, Yorktown Heights, NY 10598, USA
{rosario,shaih,talr}@watson.ibm.com

Abstract. We present a new signature scheme which is existentially unforgeable under chosen message attacks, assuming some variant of the RSA conjecture. This scheme is not based on "signature trees", and instead it uses the so called "hash-and-sign" paradigm. It is unique in that the assumptions made on the cryptographic hash function in use are well defined and reasonable (although non-standard). In particular, we *do not* model this function as a random oracle.

We construct our proof of security in steps. First we describe and prove a construction which operates in the random oracle model. Then we show that the random oracle in this construction can be replaced by a hash function which satisfies some strong (but well defined!) computational assumptions. Finally, we demonstrate that these assumptions are reasonable, by proving that a function satisfying them exists under standard intractability assumptions.

Keywords: Digital Signatures, RSA, Hash and Sign, Random Oracle, Smooth Numbers, Chameleon Hashing.

1 Introduction

Digital signatures are a central cryptographic primitive, hence the question of their (proven) security is of interest. In [12], Goldwasser, Micali and Rivest formally defined the strongest notion of security for digital signatures, namely "existential unforgeability under an adaptive chosen message attack". Since then, there have been many attempts to devise practical schemes which are secure even in the presence of such attacks.

Goldwasser, Micali and Rivest presented a scheme in [12] which provably meets this definition (under some standard computational assumption). Their scheme is based on *signature trees*, where the messages to be signed are associated with the leaves of a binary tree, and each node in the tree is authenticated with respect to its parent. Although this scheme is feasible, it is not very practical, since a signature on a message involves many such authentication steps (one for each level of the tree). This was improved by Dwork and Naor [9] and Cramer and Damgård [7], who use "flat trees" with high degree and small depth, resulting in schemes where (for a reasonable setting of the parameters) it only takes about four basic authentication steps to sign a message. Hence in these schemes the time for generating a signature and its verification (and the size of the signatures)

J. Stern (Ed.): EUROCRYPT'99, LNCS 1592, pp. 123–139, 1999.

is about four times larger than in the RSA signature scheme, for which no such proof of security exist. Besides efficiency concerns, another drawback of these schemes is their "stateful" nature, i.e. the signer has to store some information from previously signed messages.

Another line of research concentrates on *hash-and-sign* schemes, where the message to be signed is hashed using a so called "cryptographic hash function" and the result is signed using a "standard signature scheme" such as RSA or DSA. Although hash-and-sign schemes are very efficient, they only enjoy a heuristic level of security: the only known security proofs for hash-and-sign schemes are carried out in a model where the hash function is replaced by a random oracle. It is hoped that these schemes remain secure as long as the hash function used is "complicated enough" and "does not interact badly" with the rest of the signature scheme. This "random oracle paradigm" was introduced by Bellare and Rogaway in [2], where they show how it can be used to devise signature schemes from any trapdoor permutation. They later described concrete implementations for the RSA and Rabin functions (with some security improvements) in [3]. Also, Pointcheval and Stern proved similar results with respect to ElGamal-like schemes in [15].

Security proofs in the random oracle model, however, can only be considered a heuristic. A recent result by Canetti, Goldreich and Halevi [5] demonstrates that "behaving like a random oracle" is not a property that can be realized in general, and that security proofs in the random-oracle model do not always imply the security of the actual scheme in the "real world". Although this negative result does not mean that the schemes in [2,3,15] cannot be proven secure in the standard model, to this day nobody was able to formalize precisely the requirements on the cryptographic hash functions in these schemes, or to construct functions that can provably replace the random oracle in any of them.

Our result. We present a new construction of a hash-and-sign scheme (similar to the standard hash-and-sign RSA), for which we can prove security in a standard model, *without a random oracle*. Instead, the security proof is based on a stronger version of the RSA assumption and on some specific *constructible* properties that we require from the hash function. At the same time, our scheme enjoys the same level of efficiency of typical hash-and-sign schemes. Compared to tree-based schemes this new algorithm fares better in terms of efficiency (typically 2.5 times faster), size of keys and signatures and does not require the signer to keep state (other than the secret signature key).

1.1   The New Construction

Our scheme resembles the standard RSA signature algorithm, but with a novel and interesting twist. The main difference is that instead of encoding the message in the base of the exponent and keeping the public exponent fixed, we encode the message in the exponent while keeping a fixed public base.

Set up. The public key is an RSA modulus $n = pq$ and a random element $s \in \mathbb{Z}_n^*$.

Signing. To sign a message M with respect to the public key (n, s), the signer first applies a hash function to compute the value $e = H(M)$, and then uses it as an exponent, i.e. he finds the eth root of s mod n. Hence a signature on M is an integer σ such that $\sigma^{H(M)} = s$ mod n.

Assumptions and requirements. In our case, it is necessary to choose p, q as "safe" or "quasi-safe" primes (i.e., such that $(p - 1)/2, (q - 1)/2$ are either primes or prime powers.) In particular, this choice implies that $p - 1, q - 1$ do not have any small prime factors other than 2, and that finding an odd integer which is not co-prime with $\phi(n)$ is as hard as factoring n. This guarantees that extracting e^{th} roots when $e = H(M)$ is always possible (short of factoring n).

Intuitively, the reason that we can prove the security of our scheme without viewing H as a random oracle, is that in RSA the base must be random, but the exponent can be arbitrary. Indeed, it is widely believed that the RSA conjecture holds for any fixed exponent (greater than one). Moreover, if e_1, e_2 are two different exponents, then learning the e_1'th root of a random number s does not help in computing the e_2'th root of s, as long as e_2 does not divide e_1. Hence, it turns out that the property of H that is needed for this construction is that it is hard to find a sequence of messages $M_1, M_2, \ldots$ such that for some i, $H(M_i)$ divides the other $H(M_j)$'s. In the sequel, we call this property of the hash function *division intractability*.

In our scheme, an attacker who on input (n, s) can find both an exponent e and the eth root of s, may have the ability to forge messages. Thus our formal security proof is based on the assumption that such a task is computationally infeasible. This stronger variant of the RSA assumption has already appeared in the literature, in a recent work of Barić and Pfitzmann for constructing fail-stop signatures without trees [1].

The proof. We present our proof in three steps:

1. First, we prove the security of the scheme in the random oracle model. This step already presents some technical difficulties. One of the main technical problems for this part is to prove that a random oracle is division-intractable. We prove this using some facts about the density of smooth numbers.

2. Next, we show that the random oracle in the proof of Step 1 can be replaced by a hash function which satisfies some (well defined) computational assumptions. We believe that this part is the main conceptual contribution of this work.

   We introduce a new computational assumption which is quite common in complexity theory, yet we are unaware of use of this type of assumptions in cryptography. Instead of assuming that there is no efficient algorithm that solves some problem, we assume that *there is no efficient reduction* between two problems. We elaborate on this issue in Subsection 5.2.

3. As we have introduced these non-standard assumptions, we need to justify that they are "reasonable". (Surely, we should explain why they are more reasonable than assuming that a hash function "behaves like a random oracle").

We do this by showing how to construct functions that satisfy these assumptions from any collision-intractable hash function [8] and Chameleon commitment scheme [4]. It follows, for example, that such functions exist if factoring is hard. As we explained above, this is in sharp contrast to the hash functions that are needed in previous hash-and-sign schemes, for which no provable construction is known.

2 Preliminaries

Before discussing our scheme, let us briefly present some notations and definitions which are used throughout the paper. In the sequel we usually denote integers by lowercase English letters, and strings by uppercase English letters. We often identify integers with their binary representation. The set of positive integers is denoted by $\mathcal{N}$.

Families of hash functions. We usually consider hash functions which map strings of arbitrary length into strings of a fixed length. In some constructions we allow these functions to be randomized. Namely, we consider functions of the type $h : \$ \times \{0,1\}^* \rightarrow \{0,1\}^k$ for some set of coins $\$$ and some integer k. We write either $h(X) = Y$ or $h(R; X) = Y$, where $R \in \$, X \in \{0,1\}^*$, and Y is the output of h on the the input X (and the coins R, if they are specified).

A family of hash function is a sequence $\mathcal{H} = \{H_k\}_{k \in \mathcal{N}}$, where each H_k is a collection of functions as above, such that each function in H_k maps arbitrary-length strings into strings of length k. The properties of such hashing families that are of interest to us, are *collision-intractability* which was defined by Damgård in [8], and *division-intractability* (which we define below). For the latter, we view the output of the hash function as the binary representation of an integer. For our scheme we use hash functions with the special property that their output is *always an odd integer*. Such a function can be easily obtained from an arbitrary hash function by setting $h'(X) = h(X)|1$ (or just setting the lowest bit of $h(X)$ to one).

Definition 1 (Collision intractability [8]). *A hashing family $\mathcal{H}$ is collision intractable if it is infeasible to find two different inputs that map to the same output. Formally, for every probabilistic polynomial time algorithm A there exists a negligible function* negl() *such that*

$$\Pr_{h \in H_k} [A(h) = \langle X_1, X_2 \rangle \ s.t. \ X_1 \neq X_2 \ and \ h(X_1) = h(X_2)] = \mathsf{negl}(k)$$

If h is randomized, we let the adversary algorithm A choose both the input and the randomness. That is, A is given a randomly chosen function h from H_k, and it needs to find two pairs $(R_1, X_1), (R_2, X_2)$ such that $X_1 \neq X_2$ but $h(R_1; X_1) = h(R_2; X_2)$.

Definition 2 (Division intractability). *A hashing family $\mathcal{H}$ is division intractable if it is infeasible to find distinct inputs $X_1, \ldots, X_n, Y$ such that $h(Y)$*

divides the product of the $h(X_i)$'s. Formally, for every probabilistic polynomial time algorithm A there exists a negligible function $\mathsf{negl}()$ such that

$$\Pr_{h \in H_k} \left[\begin{array}{l} A(h) = \langle X_1, \ldots, X_n, Y \rangle \\ s.t.\ Y \neq X_i \ for\ i = 1 \ldots n, \\ and\ h(Y)\ divides\ the\ product\ \prod_{i=1}^{n} h(X_i) \end{array} \right] = \mathsf{negl}(k)$$

Again, if h is randomized then we let A choose the inputs and the randomness. It is easy to see that a division intractable hashing family must also be collision intractable, but the converse does not hold.

Signature schemes. Recall that a signature scheme consists of three algorithms: a randomized key generation algorithm Gen, and (possibly randomized) signature and verification algorithms, Sig and Ver. The algorithm Gen is used to generate a pair of public and secret keys, Sig takes as input a message, the public and secret key and produces a signature, and Ver checks if a signature on a given message is valid with respect to a given public key. To be of any use, it must be the case that signatures that are generated by the Sig algorithm are accepted by the Ver algorithm. The strongest notion of security for signature schemes was defined by Goldwasser, Micali and Rivest as follows:

Definition 3 (Secure signatures [12]). *A signature scheme $S = \langle$Gen, Sig, Ver$\rangle$ is existentially unforgeable under an adaptive chosen message attack if it is infeasible for a forger who only knows the public key to produce a valid (message, signature) pair, even after obtaining polynomially many signatures on messages of its choice from the signer.*

Formally, for every probabilistic polynomial time forger algorithm $\mathcal{F}$, there exists a negligible function $\mathsf{negl}()$ such that

$$\Pr \left[\begin{array}{l} \langle \mathsf{pk}, \mathsf{sk} \rangle \leftarrow \mathsf{Gen}(1^k); \\ for\ i = 1 \ldots n \\ \quad M_i \leftarrow F(\mathsf{pk}, M_1, \sigma_1, \ldots, M_{i-1}, \sigma_{i-1});\ \sigma_i \leftarrow \mathsf{Sig}(\mathsf{sk}, M_i); \\ \langle M, \sigma \rangle \leftarrow F(\mathsf{pk}, M_1, \sigma_1, \ldots, M_n, \sigma_n), \\ M \neq M_i \ for\ i = 1 \ldots n, \ and\ \mathsf{Ver}(\mathsf{pk}, M, \sigma) = accept \end{array} \right] = \mathsf{negl}(k)$$

3   The Construction

Key generation. The key-generation algorithm in our construction resembles that of standard RSA. First, two random primes p, q of the same length are chosen, and the RSA modulus is set to $n = p \cdot q$. In our case, we assume that p, q are chosen as "safe" or "quasi-safe" primes (i.e., that $(p-1)/2, (q-1)/2$ are either primes or prime-powers.) In particular, this choice implies that $p-1, q-1$ do not have any small prime factors other than 2, and that finding an odd integer which is not co-prime with $\phi(n)$ is as hard as factoring n. After the modulus n is set, an element $s \in \mathbb{Z}_n^*$ is chosen at random.

Finally, since we use a hash-and-sign scheme, a hash function has to be chosen from a hashing family. The properties that we need from the hashing family are discussed in the security proof (but recall that we use a hash function whose output is always an odd integer). Below we view the hash function h as part of the public key, but it may also be a system parameter, so the same h can be used by everyone. The public key consists of n, s, h. The secret key is the primes p and q.

Signature and verification. To sign a message M, the signer first hashes M to get an odd exponent $e = h(M)$. Then, using its knowledge of p, q, the signer computes the signature σ as the e'th root of the public base s modulo n. If the hash function h is randomized, then the signature consists also of the coins R which were used for computing $e = h(R; M)$.

To verify a signature σ (resp. $\langle \sigma, R \rangle$) on message M with respect to the hash function h, RSA modulus n and public base s, one needs to compute $e = h(M)$ (resp. $e = h(R; M)$) and check that indeed $\sigma^e = s \pmod{n}$.

3.1   A Few Comments

1. Note that with overwhelming probability, the exponent $e = h(M)$ will be co-prime with $\phi(n)$. This is since finding an odd number e which is not co-prime with $\phi(n)$ is as hard as factoring n, for the class of moduli used in this scheme.

2. The output length of the hash function is relevant for the efficiency of the scheme. If we let the output of the hash function be $|n|$-bit long then signature generation will take roughly twice as long as standard RSA (since the signer must first compute $e^{-1} \bmod \phi(n)$ and then a modular exponentiation to compute σ). Also signature verification takes a full exponentiation modulo n. The efficiency can be improved by shortening the output length for h. However (as it will become clear from the proof of security), in order for h to be division intractable, its output must be sufficiently long. Our current experimental results suggest that to get equivalent security to a 1024-bit RSA, the output size of the hash should be about 512 bits. For this choice of hash output length we have that computing a signature will be less than 1.5 times slower than for a standard RSA signature.

3. When a key for our scheme is certified, it is possible for the signer to prove that the modulus n has been chosen correctly (i.e. the product of two quasi-safe primes) by using a result from [11].

4   Security in the Random-Oracle Model

As we have stated, for the security of our scheme we must use the "strong RSA conjecture" which was introduced recently by Barić and Pfitzmann. The difference between this conjecture and the standard RSA conjecture is that here the adversary is given the freedom to choose the exponent e. Stated formally:

Conjecture 4 (Strong-RSA [1]) *Given a randomly chosen RSA modulus n, and a random element $s \in \mathbb{Z}_n^*$, it is infeasible to find a pair $\langle e, r \rangle$ with $e > 1$ such that $r^e = s \pmod{n}$.*

The meaning of the "randomly chosen RSA modulus" in this conjecture depends on the way this modulus is chosen in the key generation algorithm. In our case, this is a product of two randomly chosen "safe" (or "quasi-safe") primes of the same length.

We start by analyzing the security of this construction in a model where the hash function h is replaced by a random oracle.[1]

Theorem 5. *In the random oracle model, the above signature scheme is existentially unforgeable under an adaptive chosen message attack, assuming the strong-RSA conjecture.*

Proof. Let $\mathcal{F}$ be a forger algorithm. We assume w.l.o.g. that $\mathcal{F}$ always queries the oracle about a message M before it either asks the signer to sign this message, or outputs (M, σ) as a potential forgery. Also, let v be some polynomial upper bound on the number of queries that $\mathcal{F}$ makes to the random oracle.

Using the same method as in Shamir's pseudo-random generator [17], we now show an efficient algorithm $\mathcal{A}_1$ (which we call *the attacker*), that uses $\mathcal{F}$ as a subroutine, such that if $\mathcal{F}$ has probability ϵ of forging a signature, then $\mathcal{A}_1$ has probability $\epsilon' \approx \epsilon/v$ of breaking the strong RSA conjecture.

The random-oracle attacker. The attacker $\mathcal{A}_1$ is given an RSA modulus n (chosen as in the key generation algorithm) and a random element $t \in \mathbb{Z}_n^*$, and its goal is to find e, r (with $e > 1$) such that $r^e = t \pmod{n}$.

First, $\mathcal{A}_1$ prepares the answers for the oracle queries that $\mathcal{F}$ will ask. He does so by picking at random v odd k-bit integers $e_1 \ldots e_v$ and an index $j \in \{1 \ldots v\}$. Intuitively, $\mathcal{A}_1$ bets on the chance that $\mathcal{F}$ will use its j'th oracle query to generate the forgery.[2]

Next, $\mathcal{A}_1$ prepares the answers for signature queries that $\mathcal{F}$ will ask. $\mathcal{A}_1$ computes $E = (\prod_i e_i)/e_j$ (i.e., E is the product of all the e_i's except e_j). If e_j divides E, then $\mathcal{A}_1$ outputs "failure" and halts. Otherwise, it sets $s = t^E \pmod{n}$, and initializes the forger $\mathcal{F}$, giving it the public key (n, s). The attacker then runs the forger algorithm $\mathcal{F}$, answering:

1. the i'th oracle query with the odd integer e_i. Namely, if the forger makes oracle queries $M_1 \ldots M_v$, then $\mathcal{A}_1$ answers these queries by setting $h(M_i) = e_i$.

2. signature query for message M_i for $i \neq j$ with the answer $\sigma_i = t^{E/e_i} \pmod{n}$ (recall that E/e_i is an integer for all $i \neq j$).

If $\mathcal{F}$ asks signature query for message M_j, or halts with an output other than (M_j, σ) then $\mathcal{A}_1$ outputs "failure" and halts. If $\mathcal{F}$ does output (M_j, σ) for which $\sigma^{e_j} = s \pmod{n}$, then $\mathcal{A}_1$ proceeds as follows. Using the extended Euclidean gcd algorithm, it computes $g = GCD(e_j, E)$, and also two integers a, b such that

---

[1] Also here, we assume that the random oracle always return an odd integer as output. Namely, the answer of the oracle on every given query is a randomly chosen odd k-bit integer.

[2] This is where we get the $1/v$ factor in the success probability. Interestingly, this factor *only shows up in the random oracle model*, so we get a tighter reduction in the standard model.

$ae_j + bE = g$. Then $\mathcal{A}_1$ sets $e = e_j/g$ and $r = t^a \cdot \sigma^b \pmod{n}$, and outputs (e, r) as its solution to this instance of the strong-RSA problem.

Analysis of $\mathcal{A}_1$. If $\mathcal{A}_1$ does not output "failure", then it outputs a correct solution for the strong RSA instance at hand (except with a negligible probability): First, since e_j does not divide E, then $g < e_j$, which means that $e = e_j/g > 1$. Moreover, we have

$$r^e = \left(t^a \cdot \sigma^b\right)^{e_j/g} = t^{ae_j/g} \cdot \sigma^{be_j/g} \stackrel{*}{=} t^{ae_j/g} \cdot t^{bE/g} = t^{(ae_j+bE)/g} = t \pmod{n}$$

Equality $(*)$ holds because: (a) $\sigma^{e_j} = s = t^E \pmod{n}$, which implies that also $\sigma^{be_j} = t^{bE} \pmod{n}$; and (b) e_j is co-prime with $\phi(n)$ (except with negligible probability), which means that so is g. Therefore, there is a single element $x \in \mathbb{Z}_n^*$ satisfying $x^g = \sigma^{be_j} = t^{bE} \pmod{n}$.

It is left to show, therefore, that the event in which $\mathcal{A}_1$ does not output "failure" happens with probability $\epsilon' \approx \epsilon/v$. Denote by DIV the event in which e_j divides E. Conditioned on the complement of DIV, $\mathcal{F}$ sees the same transcript when interacting with $\mathcal{A}_1$ as when it interacts with the real signer, and so it outputs a valid forgery for M_j with probability ϵ/v (since j is chosen at random between 1 and v). It follows that the probability that $\mathcal{A}_1$ does not output "failure" is $\epsilon' \geq \epsilon/v - \Pr[DIV]$. In Lemma 6 we prove that when the output length of the random oracle is k, then $\Pr[DIV] \leq 2^{-\sqrt{k}}$, which completes the proof of Theorem 5.

Lemma 6. *Let $e_1 \ldots e_v$ be random odd k-bit integers, let j be any integer $j \in \{1 \ldots v\}$, and denote $E = (\prod_i e_i)/e_j$. Then, the probability that e_j divides E is less than $2^{-\sqrt{k}}$.*

Proof. As before, we denote the above event by DIV. To prove Lemma 6, we use some facts about the density of smooth numbers. Recall that when x, y are integers, $0 < y \leq x$, we say that x is y-*smooth* if all the prime factors of x are no larger than y, and let $\Psi(x, y)$ denote the number of integers in the interval $[0, x]$ which are y-smooth. The following fact can be found in several texts on number-theory (e.g., [14]).

Proposition 7. *Fix some real number $\epsilon > 0$, let x be an integer $x \geq 10$, let y be another integer such that $\log x > \log y \geq (\log x)^\epsilon$, and denote $\mu \stackrel{def}{=} \log x/\log y$ (namely, $y = x^{1/\mu}$). Then $\Psi(x, y)/x = \mu^{-\mu(1-f(x,\mu))}$, where $f(x, \mu) \to 0$ as $\mu \to \infty$, uniformly in x.*

Below we write somewhat informally $\Psi(x, x^{1/\mu}) = \mu^{-\mu(1-o(1))}$. Substituting 2^k for x and $\sqrt{k}/2$ for μ in the expression above, we get

$$\Psi(2^k, 2^{2\sqrt{k}})/2^k = \left(\sqrt{k}/2\right)^{-\sqrt{k}(1-o(1))/2} < 2^{-\sqrt{k}\log k/16} < 2^{-2\sqrt{k}}.$$

We comment that the same bound also holds when we talk about odd k-bit integers, (this can be shown using the fact that an even k-bit integer x is smooth if and only if the $(k-1)$-bit integer $x/2$ is also smooth). If we denote by $SMOOTH$

the event in which the integer e_j is $2^{2\sqrt{k}}$-smooth, then by the bound above, $\Pr[SMOOTH] \leq 2^{-2\sqrt{k}}$.

Assume, then, that the event $SMOOTH$ does not happen. Then e_j has at least one prime factor $p > 2^{2\sqrt{n}}$. In this case, the probability that e_j divides the product of the other e_i's is bounded by the probability that at least one of these e_i's is divisible by p. But since all the other e_i's are chosen at random, then the probability that any specific e_i is divisible by p is at most $1/p < 2^{-2\sqrt{k}}$, and the probability that there exists one which is divisible by k is at most $v \cdot 2^{-2\sqrt{k}}$. As v is polynomial in k, we get $v \cdot 2^{-2\sqrt{k}} < 2^{-1.5\sqrt{k}}$. Combining the two bounds, we get $\Pr[DIV] < \Pr[SMOOTH] + \Pr[DIV \mid \neg SMOOTH] < 2^{-2\sqrt{k}} + 2^{-1.5\sqrt{k}} < 2^{-\sqrt{k}}$.

4.1   The Value of k

The above bound on $\Pr[DIV]$ is very weak. For example, to get security level of 2^{-80}, this bound suggest a value of $k \approx 6000$. Although the equations above can be optimized, they still only give a very crude bound. One reason for this is that we only bound the probability that p, the largest prime factor of e_j, divides the product of the e_i's. If e_j is rather smooth, then e_j/p is still rather large, so even if p divides one of the e_i's, the probability that e_j/p divides the product of the e_i's is still rather small. We therefore performed some experiments to get a practical estimate for the value of k. Our experiments suggest that $\Pr[DIV]$ is in fact much smaller than the bound $2^{-\sqrt{k}}$. (In fact, for the values of k which we tested, we got $\Pr[DIV] \approx 2^{-k/8}$.) See more details in Appendix A.

5   Eliminating the Random Oracle

Below we show that the random oracle in the above proof can be replaced by a randomized hash function with certain properties. Clearly, this hash function should be division-intractable, since violating division intractability immediately yields an attack on the signature scheme. However, this property alone is not sufficient: even if we assume that the hash function is division intractable, we still face problems carrying out the above security proof in the standard model. Specifically, recall that in the previous proof, the attacker $\mathcal{A}_1$ had to simulate the signer for $\mathcal{F}$, and do it without knowing the prime factorization of the modulus.

$\mathcal{A}_1$ was able to carry out this task since it could choose the outputs of the oracle (the e_i's) before seeing the inputs, and so it was able to "tailor" the public base s to these specific e_i's. In a standard model this is no longer the case: Clearly, if h is deterministic, then the forger's choice of M_i's uniquely determines the e_i's, and the attacker has no room to play with these values. But even if h is randomized this does not help the attacker due to the fact that h is also division-intractable which implies that it is one-way. Thus, if the attacker first chooses e and then sees M, it cannot find randomness R for which $e = h(R; M)$ (even if such R exists).

As a first step towards overcoming this difficulty, we note that the hardness of finding such randomness R is in some sense "unrelated" to the hardness of solving the strong RSA problem. Namely, our intuition is that being able to find R should not help anyone solving strong RSA.[3] We formalize this intuition by replacing the strong RSA conjecture (which asserts that there is no efficient algorithm to solve strong RSA), with the "funny looking" conjecture which asserts that there is *no efficient reduction* between finding the randomness R and solving strong RSA. Technically, this is done by asserting that the strong RSA conjecture remains valid *even in a relativized world where there is an oracle that finds this randomness*.

5.1   The Hashing Family

To be able to carry the security proof in a standard world, we have to make the following assumptions on the hashing family $\mathcal{H}$ used in the scheme and its relation to the strong RSA conjecture.

We say that a hashing family $\mathcal{H}$ is *suitable* if

1. For any $h \in \mathcal{H}$, the outputs of h are always odd integers.
2. $\mathcal{H}$ is division-intractable.
3. For every $h \in \mathcal{H}$ and every two messages M_1, M_2, the distributions $h(R; M_1)$, $h(R; M_2)$, induced by the random choice of R, are statistically close.[4]
4. The strong RSA conjecture also holds in a model where there exists an oracle that on input h, M, e, returns a random $R \in \$$ subject to $h(R; M) = e$. [5]

We discuss these assumptions further in Section 5.2 below. But first let us prove that our signature scheme is secure when using a suitable hashing family $\mathcal{H}$. We stress that although one of our computational assumptions holds in a relativized world, we then prove the security of the scheme in the "real world".

Theorem 8. *If $\mathcal{H}$ is suitable, then the construction from Section 3 is existentially unforgeable under an adaptive chosen message attack.*

Proof. The proof proceeds similarly to the proof of Theorem 5, i.e. we construct an attacker $\mathcal{A}_2$ which will use the forger $\mathcal{F}$. The main difference is that the attacker A_2 operates in a relativized model, given in addition access to the oracle from Condition 4 of the suitable hash function. We show that if the forger $\mathcal{F}$ has probability ϵ of breaking the scheme (in the "real world"!) then the attacker has probability $\epsilon' \approx \epsilon$ of solving strong RSA in the relativized world. (Note that this reduction is tighter than the reduction in the random-oracle model.)

The oracle-assisted attacker. We again assume a bound of v on the number of signatures that the forger $\mathcal{F}$ asks to see before it outputs its forgery. As before, A_2 is given an RSA modulus n and a random element $t \in \mathbb{Z}_n^*$ (chosen as in the key generation algorithm), and its goal is to find e, r (with $e > 1$) such

---

[3] For example, if one thinks of h as SHA-1, then we have a very strong intuition that finding collisions in SHA-1 provides no help in violating the RSA conjecture.

[4] Together with the collision-intractability, this implies that $\mathcal{H}$ is a statistically hiding string-commitment scheme.

[5] Such R must exist because of Condition 3.

that $r^e = t \pmod{n}$. It starts by picking at random a hash function $h \in \mathcal{H}$ to be used for the forger. And v arbitrary values $e_1, ..., e_v$ in the range of the function. This can be done for example by picking v arbitrary "dummy messages" $M'_1 ... M'_v$ and computing $e_i = h(R'_i; M'_i)$ (for random R'_i's).

Then A_2 computes $E = \prod_i e_i$, sets $s = t^E \pmod{n}$ and initializes $\mathcal{F}$ with the public key (n, s, h). Whenever $\mathcal{F}$ asks for a signature on a message M_i, A_2 queries its randomness-finding oracle for a randomness R_i for which $h(R_i; M_i) = e_i$, and then computes the signature by setting $\sigma_i = t^{E/e_i} \pmod{n}$. A_2 returns the pair $\langle R_i, \sigma_i \rangle$ to $\mathcal{F}$.

It is important to note that because of Condition 3 on $\mathcal{H}$, the distribution that $\mathcal{F}$ sees in this simulation is statistically close to the distribution it sees when interacting with the real signer. In particular, since $\mathcal{H}$ is division intractable, then $\mathcal{F}$ has only a negligible probability of finding M', R' such that $e' = h(R'; M')$ divides the product of the e_i's.

It follows that with probability $\epsilon' \geq \epsilon - \mathsf{negl}$, $\mathcal{F}$ outputs a forgery M', R', σ such that $e' = h(R'; M')$ does not divide the product of the other e_i's, and yet $\sigma^{e'} = s \pmod{n}$. When this happens, the attacker A_2 uses the same gcd procedure as above to find (e, r) with $e > 1$ such that $r^e = t \pmod{n}$.

5.2   Discussion

The proof in the previous section eliminates the random oracle, but substitutes it with a non-standard assumption: the strong RSA assumption must still be true even in a relativized world where finding randomness for h is not hard. Is this a more reasonable assumption than just assuming that h "behaves like a random oracle"? We strongly believe it is. The assumption we use has a very concrete interpretation in the real world, meaning that there is no reduction from the problem of randomness-finding for h to the problem of solving the strong RSA problem. In other words the difficulty of the two problems are somewhat "independent". Moreover we show later that suitable families of hash functions are actually constructible. On the other hand the notion of "behaving as a random oracle" has no concrete counterpart in the real world, and there are no provable constructions of "good hash functions" for previously known schemes.

It is interesting to ask if our technique of substituting the random oracle in the security proof with a relativized assumption can be used in other proofs that employ random oracles (such as [2,3,15]). Unfortunately, it does not appear to be likely. The main reason our technique seems to fail in those proofs, is that their requirement from h is that the forger cannot find a message M for which he "knows" something about $h(M)$. In our scheme instead we were able to pin down the specific combinatorial property we require from h and flesh it out as a specific assumption.

In the next section we describe a construction of a suitable family of hash functions. The main purpose of this construction is to prove that the assumptions we make can be realized. However the construction requires the signer to search for a prime exponent in a large subset and thus it might require a significant amount of time. It is however plausible to conjecture that families built from widely used collision-resistant hash functions such as SHA-1 [10] can be suitable. The rationale is that such functions have been designed in a way that destroys any "structure" in the input-output relationship. In particular it is very unlikely (although we don't know how to prove it) that division intractability does not hold for such functions. A possible candidate would be to define h as following

$$h(R_1; R_2; R_3; R_4; M)$$
$$= 1 \,|\, SHA1(M|1|R_1) \,|\, SHA1(M|2|R_2) \,|\, SHA1(M|3|R_3) \,|\, SHA1(M|4|R_4) \,|\, 1$$

for a 642-bit exponent (this is the definition of a single h, a family could be constructed via any standard method of extending SHA-1 to a family, for example by keying the IV).

6   Implementing the Hashing Family $\mathcal{H}$

To argue that Conditions 1-4 are "reasonable" we at least need to show that they could be met. Namely, that there exists a function family $\mathcal{H}$ satisfying these conditions (under some standard assumptions). Below we show that such families exist, assuming that collision-intractable families and Chameleon commitment families exist. In particular, it follows that such families exist under the factoring conjecture (which is weaker than our "strong RSA" conjecture), or under the Discrete-log conjecture.[6]

We construct $\mathcal{H}$ in two steps: first we show how to transform any collision-intractable hashing family into a (randomized) division-intractable family, and then we show how to take any division-intractable hash function and transform it into one that satisfy Conditions 1 through 4.

6.1   From Collision-Intractable to Division-Intractable

The idea of this transformation is simply to force the output of the hash functions to be a prime number. Then, the function is division-intractable if and only if it is collision intractable. A heuristic for doing just that was suggested by Barić and Pfitzmann in [1]: If h is collision intractable with output length k, then define a randomized function $\tilde{h}$ with output length of (say) $2k$ bits, by setting for $r = 0, \ldots, 2^k - 1$, $\tilde{h}(r; X) = 2^k \cdot h(X) + r$, provided that $h(X) + r$ is an odd prime ($\tilde{h}(r; X)$ is undefined otherwise).

It is obvious that $\tilde{h}$ is still collision-intractable, and that it always outputs primes, so it is also division-intractable. However, to argue that $\tilde{h}$ is efficiently computable, we must assume that the density of primes in the interval $[2^k h(X),$ $2^k(h(X)+1)]$ is high enough (say, $1/poly(k)$ fraction). Hence, to use this heuristic,

---

[6] The way we set the definitions in this paper, Condition 4 on $\mathcal{H}$ implies the strong RSA conjecture, so formally there is no point in using any other conjecture. This technicality can be dealt with in some ways, but we chose to ignore it in this preliminary report.

one must rely on some number-theoretic conjecture about the density of primes in small intervals.

Below we show a simple technique that allows us to get rid of this extra conjecture: Just as in the above heuristic, we let the output size of $\tilde{h}$ be larger than that of h (letting $\tilde{h}$ output $3k$ bits is sufficient for our purposes), and partition the space of outputs in such a way that each output of the original h is identified with a different subset of the possible outputs of $\tilde{h}$. However, we choose the partition in a randomized manner, so we can prove that (with high probability) each one of the subsets is dense with primes.

The main tool that we use in this transformation is *universal hashing families* as defined by Carter and Wegman in [6]. Recall that a universal family of hash functions from a domain D to a range R is a collection U of such functions, such that for all $X_1 \neq X_2 \in D$ and all $Y_1, Y_2 \in R$, $\Pr_f[f(X_1) = Y_1$ and $f(X_2) = Y_2] = 1/|R|^2$ (the probability is taken over the uniformly random choice of $f \in U$). Several constructions of such universal families were described in [6].

In our case, we use universal hash functions which maps $3k$ bits to k bits, with the property that given a function $f \in U$ and a k-bit string Y, it is possible to efficiently sample uniformly from the space $\{X \in \{0,1\}^{3k} : f(X) = Y\}$. For any function $f : \{0,1\}^{3k} \to \{0,1\}^k$, we associate a partition of the set of outputs ($\{0,1\}^{3k}$) into 2^k subsets according to the values assigned by f. Each output value of the original h (which is a k-bit string Y) is then associated with the subset $f^{-1}(Y)$. The modified function $\tilde{h}$, on input X, outputs a random odd prime from the set $f^{-1}(h(X))$. Again, it is clear that $\tilde{h}$ is collision-intractable if h is, and that it only outputs primes, hence it is division-intractable. On the other hand, a standard hashing lemma shows that with high probability over the random choice of f, the subset $f^{-1}(h(X)) \subset \{0,1\}^{3k}$ is dense with primes (for all X). Thus, $\tilde{h}$ is also efficiently computable.

Lemma 9. *Let U be a universal family from $\{0,1\}^{3k}$ to $\{0,1\}^k$. Then, for all but a 2^{-k} fraction of the functions $f \in U$, for every $Y \in \{0,1\}^k$ a fraction of at least $1/ck$ of the elements in $f^{-1}(Y)$ are primes, for some small constant c.*
Proof omitted.

6.2   From Division-Intractable to Suitable

Finally, we show how to take any division-intractable hashing family (that always output odd integers) and transform it into a suitable one (i.e. one that satisfies Conditions 1 through 4 from Subsection 5.1). To this end, we use *Chameleon commitment schemes*, as defined and constructed by Brassard, Chaum and Crepeau [4]. In fact we use them as Chameleon Hashing exactly as defined and required in [18].

The Chameleon Hashing is a function $ch(\cdot; \cdot)$ which on input a random string R and a message M is easily computed. Furthermore, it is associated with a value known as the "trapdoor". It satisfies the following properties:

- Without knowledge of the trapdoor there is no efficient algorithm that can find pairs M_1, R_1 and M_2, R_2 such that $ch(M_1, R_1) = ch(M_2, R_2)$.

- There is an efficient algorithm that given the trapdoor, a pair M_1, R_1 and M_2 can compute R_2 such that $ch(M_1, R_1) = ch(M_2, R_2)$.
- For any pair of messages M_1, M_2 and for randomly chosen R the distribution $ch(M_1, R_1)$ and $ch(M_2, R_2)$ are statistically close.

To transform a division intractable hash function h into one that also satisfies Conditions 3 and 4 from Subsection 5.1, we simply apply it to the hash string $c = ch(R; M)$ instead of to the message M itself. A little more formally, we have the following construction.

Let $\mathcal{H}$ be a division-intractable family, and let CH be a Chameleon hashing scheme. We construct a randomized family $\tilde{\mathcal{H}}$ in which each function is associated with a function $h \in \mathcal{H}$ and an instance $ch \in CH$. We denote this by writing $\tilde{h}_{h,ch}$. This function is defined as $\tilde{h}_{h,ch}(R; M) = h(ch(R; M))$. (if h itself is randomized, then we have $\tilde{h}_{h,ch}(R_1, R_2; M) = h(R_2; ch(R_1; M))$). It is easy to see that $\tilde{\mathcal{H}}$ enjoys the following properties

1. $\tilde{\mathcal{H}}$ always outputs odd integers if $\mathcal{H}$ does.
2. $\tilde{\mathcal{H}}$ is collision intractable, since violating division-intractability requires either finding two different messages with the same hash string, or violating the division-intractability of $\mathcal{H}$.
3. $\tilde{\mathcal{H}}$ is a statistically hiding hashing scheme (since CH is, and $\mathcal{H}$ is collision intractable).

It is left to show that $\tilde{\mathcal{H}}$ also satisfies the last condition. This is shown in the following proposition:

Proposition 10. *If the Strong RSA conjecture holds, then it also holds in a relativized world where there is a randomness-finding oracle for $\tilde{\mathcal{H}}$.*

Proof. We need to show that an efficient algorithm for solving strong RSA in a relativized world where there is a randomness-finding oracle for $\tilde{\mathcal{H}}$ can be used to solve strong RSA also in the "real world". To do that, we use the trapdoor for the chameleon hashing scheme to implement the randomness-finding oracle in the real world.

A little more precisely, if there exists an efficient reduction algorithm A that solves strong RSA in the relativized world, then we construct an efficient algorithm that solves strong RSA (without the oracle) by picking a Chameleon hashing instance ch *together with its trapdoor*. Now, we execute the algorithm A, and whenever the forger asks a query concerning the hash, A turns to the randomness-finding oracle, which uses the randomness-finding algorithm of CH with the trapdoor to answer that query.

Since Chameleon hashing exists based on the factoring conjecture (which, in turn, is implied by the strong RSA conjecture) we have

Corollary 11. *Under the Strong RSA conjecture, suitable hashing families exist.*

7   Conclusions

We present a new signature scheme which has advantages in terms of both security and efficiency. In terms of efficiency, this scheme follows the "hash-and-sign" paradigm, i.e. the message is first hashed via a specific kind of hash function and then an RSA-like function is applied. Thus, in total the scheme requires a hashing operation and the only one modular exponentiation. These is no need to maintain trees and to rely on some stored information on the history of previous signatures.

The security of the scheme is based on two main assumptions. One is the "strong RSA" assumption: although this assumption has already appeared previously in the literature, it is still quite new and we think it needs to be studied carefully. The other assumption is the existence of division-intractable hash functions. We showed that such functions exist and that efficient implementations (like the one in Section 5.2) are possible based on conjectures which seem to be supported by experimental results and which we invite the research community to explore. In any case the proof of security is still based on *concrete* computational assumptions rather than on idealized models of computation (like the random oracle model).

References

1. N. Barić, and B. Pfitzmann. Collision-free accumulators and Fail-stop signature schemes without trees. In *Advances in Cryptology - Eurocrypt '97*, LNCS vol. 1233, Springer, 1997, pages 480-494.
2. M. Bellare and P. Rogaway. Random Oracles are Practical: a Paradigm for Designing Efficient Protocols. In *1st Conf. on Computer and Communications Security*, ACM, pages 62-73, 1993.
3. M. Bellare and P. Rogaway. The Exact Security of Digital Signatures: How to Sign with RSA and Rabin. In Advances in Cryptology – Eurocrypt '96, LNCS vol. 1070, Springer-Verlag, 1996, pages 399-416.
4. G. Brassard, D. Chaum, and C. Crépeau. Minimum disclosure proofs of knowledge. *JCSS*, 37(2):156–189, 1988.
5. R. Canetti, O. Goldreich and S. Halevi. The Random Oracle Methodology, Revisited. STOC '98, ACM, 1998, pages ???-???.
6. L. Carter and M. Wegman. Universal Hash Functions. J. of Computer and System Science 18, 1979, pp. 143-154.
7. R. Cramer and I. Damgård. New generation of secure and practical RSA-based signatures. In Advances in Cryptology – CRYPTO '96, LNCS vol. 1109, Springer-Verlag, 1996, pages 173-185.
8. I. Damgård. Collision free hash functions and public key signature schemes. In *Advances in Cryptology - Eurocrypt '87*, LNCS vol. 304, Springer, 1987, pages 203-216.
9. C. Dwork and M. Naor. An efficient existentially unforgeable signature scheme and its applications. In *J. of Cryptology*, 11(3), Summer 1998, pp. 187–208
10. National Institute for Standards and Technology. Secure Hash Standard, April 17 1995.

11. R. Gennaro, D. Micciancio, and T. Rabin. An Efficient Non-Interactive Statistical Zero-Knowledge Proof System for Quasi-Safe Prime Products. Proceedings of 1998 ACM Conference on Computers and Communication Security.
12. S. Goldwasser, S. Micali, and R. Rivest. A digital signature scheme secure against adaptive chosen-message attacks. *SIAM J. Computing*, 17(2):281–308, April 1988.
13. National Institute of Standards and Technology. Digital Signature Standard (DSS), Technical report 169, August 30, 1991.
14. A.K. Lenstra and H.W. Lenstra, Jr. Algorithms in number theory. In Handbook of theoretical computer science, Volume A (Algorithms and Complexity), J. Van Leeuwen (editor), MIT press/Elsevier, 1990. Pages 673-715.
15. D. Pointcheval and J. Stern. Security Proofs for Signature Schemes. In *Advances in Cryptology – Proceedings of EUROCRYPT'96*, LNCS vol. 1070, Springer-Verlag, pages 387–398.
16. R. Rivest, A. Shamir and L. Adelman. A Method for Obtaining Digital Signature and Public Key Cryptosystems. *Comm. of ACM*, 21 (1978), pp. 120–126
17. A. Shamir. On the generation of cryptographically strong pseudorandom sequences. ACM Trans. on Computer Systems, 1(1), 1983, pages 38-44.
18. H.Krawczyk and T.Rabin. Chameleon Hashing and Signatures. manuscript.

A   Experimental Results

Here we describe the results of some experiments which we performed to estimate the "true complexity" of the division property. We tried to measure how many random k-bit integers need to be chosen until we have a good chance of finding one that divides the product of all the others.

We carried out these experiments for bit-lengths 16 through 96 in increments of 8 (namely $k = 16, 24, \ldots, 88, 96$). For each bit length we performed 200 experiments in which we counted how many random integers of this length were chosen until one of them divides the product of the others. For each length, we took the second-smallest result (out the of the 200 experiments) as our estimate for the number of integers we need to choose to get a 1% chance of violating the division-intractability requirement.[7]

We repeated this experiment twice: in one experiment we chose random k-bit integers, and in the other we forced the least- and most-significant bits to '1'. The results are described in Figure 1. It can be seen that the number of integers seems to behave exponentially in the bit-length. Specifically for the bit-lengths $k = 16 \ldots 96$, it seems to behave more or less as $2^{k/8}$ (in fact even a little more). Forcing the low and high bits to '1' seems to increase the complexity slightly.

---

[7] Taking the 2nd-smallest of 200 experiments seems like a slightly better estimate than taking the smallest of 100 experiments, and our equipment didn't allow us to do more than 200 experiments for each bit-length.

Bit length	16	24	32	40	48	56	64	72	80	88	96
random	5	9	23	50	151	307	691	1067	2786	3054	8061
msb=lsb=1	8	17	39	63	160	293	710	1472	3198	4013	8124

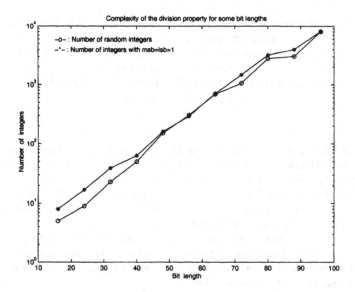

Fig. 1. Experimental results. The line –o– describes the number of random k-bit integers, and the line –*– describes the number of random k-bit integers with the first and last bits set to '1'.

A Note on the Limits of Collusion-Resistant Watermarks

Funda Ergun[1], Joe Kilian[2], and Ravi Kumar[3]

[1] Bell Laboratories, 700 Mountain Avenue, Murray Hill, NJ 07974
    fergun@research.bell-labs.com
[2] NEC Research Institute, 4 Independence Way, Princeton, NJ
    joe@research.nj.nec.com
[3] IBM Almaden Research Center/K53C San Jose , CA 95120-6099
    ravi@almaden.ibm.com

Abstract. In one proposed use of digital watermarks, the owner of a document D sells slightly different documents, $D^1, D^2, \ldots$ to each buyer; if a buyer posts his/her document D^i to the web, the owner can identify the source of the leak. More general attacks are however possible in which k buyers create some composite document D^*; the goal of the owner is to identify at least one of the conspirators.

We show, for a reasonable model of digital watermarks, fundamental limits on their efficacy against collusive attacks. In particular, if the effective document length is n, then at most $O(\sqrt{n/\ln n})$ adversaries can defeat any watermarking scheme.

Our attack is, in the theoretical model, oblivious to the watermarking scheme being used; in practice, it uses very little information about the watermarking scheme. Thus, using a proprietary system seems to give only a very weak defense.

Keywords: Watermarking, Intellectual Property Protection, Collusion Resistance.

1 Introduction

1.1 The General Problem

The very properties that have made digital media so attractive present difficult, not clearly surmountable, security problems. The ability to cheaply copy and transmit perfect copies of text, audio, and video opens up new avenues both for electronic commerce and for electronic piracy. The advent of ubiquitous high speed networks and network caching algorithms further amplifies this problem. Anyone will have the capability to cheaply distribute any movie, song, book, or picture (which we will generically call a *document*) in their possession to anyone else on the planet. The challenge is to maintain intellectual property in this environment.

There are a number of approaches to this problem; we concentrate on methods related to digital watermarking, also known as digital fingerprinting. In one

J. Stern (Ed.): EUROCRYPT'99, LNCS 1592, pp. 140–149, 1999.

general approach, the media to be distributed is altered so that it contains a hidden "do not copy" signal (the "watermark"). Most or all of the hardware for viewing, copying, or transmitting the media look for this signal and prevent illicit use. Two major problems with this approach are preventing the construction of illicit hardware that ignores the safeguards and preventing the erasure of the hidden signal. The latter problem is aggravated by the fact that one has to effectively distribute oracles (e.g., copying machines) that give feedback as to whether the signal can still be detected.

We know of no such watermarking scheme that has survived a serious attack. Indeed, with one commercially distributed scheme for watermarking images, the mark was so delicate that owners would accidentally destroy it themselves (such as by resizing the image prior to selling it).

A less ambitious use of watermarking is to identify pirates after the fact. That is, nothing prevents a pirate from anonymously posting ones intellectual property to the web, but one should be able to identify who did so. The general approach is to, given a document D, perturb it in an unobtrusive manner to generate documents $D^1, D^2, \ldots$, giving each buyer a distinct copy. If the i-th buyer posts D^i to the web, the document owner can identify him/her as the pirate.

Innumerable schemes have been proposed for both uses; we refer to [3] for a discussion of many of these schemes.

1.2 Modeling Collusion Attacks

Of course, a pirate may not be so cooperative as to simply post its document unchanged. It may attempt to alter it or, perhaps in concert with others, combine several documents to produce a document that cannot be linked with any of the "original" marked documents.

The first theoretical modeling and treatment of collusion of attacks was given by Boneh and Shaw [1]. We instead use a model suggested by Cox *et. al.* [3]. We refer to [3,5] for a more extensive introduction to this model, described briefly below.

First, we model a document D as a sequence of real numbers $\langle D_1, \ldots, D_n \rangle$. This should not be thought of as a literal and complete description of the document, but as an indication of the values of "critical values" that might be changed by the watermarking process. For example, they may be coefficients in a wavelet decomposition of an image or audio stream. In [3], it is posited that these should be orthogonal, independent attributes; [5] primarily analyzes the case where they are uniformly distributed. We do not make any such assumptions.

We model collusion attacks as follows. First, we model a watermarking scheme as a pair of functions Mark and Detect. Mark(D, m) defines a distribution on sequences $D^1, \ldots, D^m$, where m is the total number of documents produced; Mark may be viewed as a randomized procedure for producing $D^1, \ldots, D^m$.

A t-collusion attacker is modeled by a probabilistic polynomial time procedure Attack, and a distribution on distinct $i_1, \ldots, i_t$. In all of our discussions we assume that $i_1, \ldots, i_t$ is chosen uniformly from all t-element subsets. On input

$$(i_1, \ldots, i_t, D^{i_1}, \ldots, D^{i_t}),$$

Attack generates a distribution on its output, D^*.

On input $D, D^1, \ldots, D^m, D^*$, Detect returns a distribution on its output $i \in [1, m] \cup \emptyset$. Returning an index indicates an accusation; returning $\emptyset$ indicates that no one has been caught. For notational simplicity, we omit the $D, D^1, \ldots, D^m$ arguments when they are fixed or clear, writing simply $\mathsf{Detect}(D^*)$.

We now specify our requirements for Mark, Detect, and Attack. First, we consider the fidelity of the marked documents and the attacked documents. We require that $d(D^i, D) \leq \Delta/2$, where d denotes the Euclidean metric. We require a successful attack to achieve $d(D^*, D) \leq \Delta'/2$; the closer Δ' is to Δ, the better the attack. Intuitively, $\Delta/2$ indicates the degree to which the watermarking algorithm is willing to distort D, and $\Delta'/2$ indicates the amount of distortion past which the document is no longer worth stealing or protecting.

(We use $\Delta/2$ instead of Δ to simplify the analysis. By the triangle inequality, our condition enforces that $d(D^i, D^j) \leq \Delta$; this turns out to be the more natural condition to consider.)

Next, we consider the efficacy of the detection algorithm. Detect succeeds if it returns an $i \in \{i_1, \ldots, i_t\}$. Detect can fail in two ways: (i) The owner can fail to identify any of the pirates by returning $\emptyset$ (a false negative), or (ii) the owner can falsely conclude that an innocent person is a pirate (a false positive). A false negative is unfortunate; a false positive is catastrophic. If one fails to catch a pirate 90% of the time, the 10% may deter some (but not all), but if one misidentifies an innocent person 1% of the time one may not be able to ever credibly accuse anyone of piracy.

1.3 Our Result

We show a generic attack procedure Attack that defeats all watermarking schemes for the above model. It is oblivious to the Mark and Detect schemes. It has the following properties:

1. The attack uses $t = \frac{\alpha}{\epsilon}\sqrt{n/\ln n}$ documents, where α is a parameter (the larger the parameter, the more effective the attack), and ϵ controls the fidelity of the attack (we ignore integer rounding issues).

2. With high probability, it produces an attack document D^* such that

$$d(D^*, D) \leq (\Delta/2)(1 + 2\epsilon^2 + o(1)).$$

3. Suppose Detect succeeds with probability above, say $2/\sqrt{\ln n}$, then it must incur a false positive probability of $\Omega(n^{-c})$, for some c, where c depends on α. More general tradeoffs are implied by our analysis.

1.4 Related Work

Boneh and Shaw introduced the first formal model of collusion resistance. They consider a more abstract model in which one may insert a sequence of marks into a document; each mark has a value associated with it (most usually boolean) . They assume that if for all the documents available to the attacker, the i-th mark has the same value, then the attacker cannot remove this mark. If, however, two of the documents disagree on the value of the i-th mark, the attacker can change its value as it sees fit. In this model, they show upper and lower bounds for the collusion resistance as a function of the number of marks. Further improvements and additions to their basic scheme appear in [9,10,8].

It is impossible to directly compare this model and its models with that of Cox *et. al.* The model of [3] gives a more low-level model for watermarking. One pleasing aspect of our result is that it essentially matches to within a constant factor some lower bounds on collusion resistance proven by [5]. For the case where $m = n^{O(1)}$, they show that one can achieve collusion resistance of $\Omega(\sqrt{n/\ln n})$, given a very specialized assumption about the distribution of D (or given a very restricted class of attacks). Our bounds show that this is essentially the best one can hope for, regardless of the assumptions one makes about the distribution of the documents. In contrast, there is a substantial gap in the upper and lower bounds known for the Boneh-Shaw model.

Along a similar vein, Chor, Fiat, and Naor [2] introduce *traitor tracing* schemes. In their scenario, a large amount of data is broadcast, or made publicly available (say by DVD disks) in encrypted form; keys allowing the data to be decrypted are individually sold. Subsequent work in this area includes [6,7]; a further twist on key protection is given in [4]. In one respect, these models have a similar flavor to the scenario we consider, in that one wishes to identify those who publish or resell their keys. This work, however, is intended for the regime where the plaintext is so large that it is hard to (re)broadcast it. Watermarking hopes to protect much smaller data (hundreds of kilobytes).

1.5   Road Map

In Section 2 we describe our attack. In Section 3 we analyze its efficacy. In Section 4 we present conclusions and open problems.

2   The Attack

Our attack is parameterized by a collusion parameter t and a noise parameter σ. We will analyze the case where $t = (\alpha/\epsilon)\sqrt{n/\ln n}$, α is some (typically constant) parameter, and $\sigma = \epsilon\Delta/(2\sqrt{n})$, where n is the length of the attacked document; i.e., the dimension of D, and ϵ is a (typically small constant) parameter. Let $N(\mu, \sigma^2)$ be the Gaussian (normal) distribution with mean μ and standard deviation σ.

Described in words, the colluding attack is to average the t vectors and perturb with a random Gaussian noise at each component. σ is to be determined later.

$\mathsf{Attack}_{t,\sigma}(i_1, \ldots, i_t, D^{i_1}, \ldots, D^{i_t})$

1. First, compute $\overline{D^*} = \dfrac{1}{t} \sum\limits_{j=1}^{t} D^{i_j}$, where the sum is performed coordinate-wise.

   That is, each coordinate of $\overline{D^*}$ is set to be the average of the corresponding values of the sample documents.
2. Let n denote the length of $\overline{D^*}$. Choose $R = \langle r_1, \ldots, r_n \rangle$ by choosing r_j independently according to $N(0, \sigma^2)$, for $1 \le j \le n$. Compute $D^* = \overline{D^*} + R$.

Observe that in the abstract model, Attack uses no information about Mark, except for σ. We discuss more practical issues in Section 4.

There is a tension in our choice of t and σ. As we will see, the larger the values of t and σ, the more effective the attack. However, we would like to minimize the number t of adversaries (document copies) needed, and increasing σ weakens the fidelity of the attacked copy.

3   Analysis

We analyze the efficacy of Attack as a function of the parameters t and σ. First we analyze the fidelity of the attack, and then we show, for any choice of Detect, a tradeoff between the probability that it generates a false positive and the probability that it generates a false negative.

3.1   The Fidelity of the Attack

For the rest of our discussion, high probability mean with probability $1 - o(1)$ as n grows large.

Lemma 1. *Suppose that $\sigma = \epsilon \Delta / \sqrt{n}$. Then with high probability, $d(D, D^*) \le (\Delta/2)(1 + 2\epsilon^2 + o(1))$.*

Proof. (Sketch) Consider the triangle formed by D, D^*, and $\overline{D^*}$. Let $a = d(D, D^*)$, $b = d(D^*, \overline{D^*})$, and $c = d(D, \overline{D^*})$. Let θ be $\angle DD^*\overline{D^*}$. Then $c^2 = a^2 + b^2 - 2ab \cos\theta$. First, by the convexity of the Euclidean norm, it follows that $a \le \Delta/2$ (D^* is the centroid of points all within $\Delta/2$ of D). Now, $b^2 = \langle r_1^2, \ldots, r_n^2 \rangle$ (where r_i is as in Attack); hence, b^2 is a χ^2 distribution with mean $\sigma^2 n = \epsilon^2 \Delta^2$. Using simple bounds on the tail of χ^2 distributions, we have that with high probability, $b^2 \le (1 + o(1))\epsilon^2 \Delta^2$. It remains to bound the magnitude of $\cos\theta$. By the spherical symmetry of the distribution on R, θ has the same distribution as the angle between two random unit rays from the origin. For this case, it is well known that $|\cos\theta|$ is $O(\ln n / \sqrt{n})$ with high probability. Hence, with high probability,

$$c^2 \le (\Delta/2)^2 + (1 + o(1))\epsilon^2 \Delta^2 + O(\epsilon \Delta^2 \ln n / \sqrt{n}).$$

The lemma follows.

3.2   A Tradeoff between Errors

Attack ignores the values of $i_1, \ldots, i_t$. To simplify our notation, we assume without loss of generality that the attacking coalition is $1, \ldots, t$.

Suppose on D^* Detect outputs a valid value of $i \in \{1, \ldots, t\}$ with probability at least ρ, where the probability is taken over the randomness used by Attack and any randomness used by Detect. Assume without loss of generality that Player 1 is the player most often detected. Thus, $\text{Detect}(D, D^1, \ldots, D^m, D^*) = 1$, with probability $\geq \rho/t$. The idea is to produce another document D' with a slightly different colluding set that does not include Player 1. When t is sufficiently large, D^* and D' cannot be reliably distinguished by Detect (or any other distinguisher). Hence Detect will output $i = 1$, yielding a false positive, with an unacceptably large probability.

Consider the output of Attack on $D^2, \ldots, D^{t+1}$. We define $\overline{D'}$ and D' by

$$\overline{D'} = \frac{1}{t} \sum_{i=2}^{t+1} D^i, \text{ and}$$

$$D' = \overline{D'} + N(0, \sigma^2)^n.$$

That is, D' is distributed according to the output of Attack. Note that D^1 is not part of the set that produces D'.

Fixing, $D, D^1, \ldots, D^m$, we consider D' and D^* as defining probability measures on the document space. We claim that $\text{Detect}(D')$ still outputs 1 with unacceptably high probability if $\text{Detect}(D^*)$ outputs 1 with a reasonably large probability.

We now proceed to show that there is a tradeoff between the false positive and false negative probabilities. First we define a parameterized set of problematic documents for which the false positive probability is low.

Definition 2. *Given probability measure D' and D^*, and a parameter γ, we define the bad set B_γ by*

$$B_\gamma = \{x \mid \Pr_{x \leftarrow D'}[x] \leq \gamma \Pr_{x \leftarrow D^*}[x]\}.$$

This set is bad for the attacker, because Detect can safely output 1 without incurring too large a probability of producing a false positive. Lemma 3 bounds the probability that Detect makes a false positive as a function of γ.

Lemma 3. $\Pr_{x \leftarrow D'}[\text{Detect}(x) = 1] \geq \gamma \cdot \left(\frac{\rho}{t} - \Pr_{D^*}[B_\gamma] \right).$

Proof. We have

$$\Pr_{x \leftarrow D^*}[\text{Detect}(x) = 1 \land x \notin B_\gamma] \geq \Pr_{x \leftarrow D^*}[\text{Detect}(x) = 1] - \Pr_{x \leftarrow D^*}[x \in B_\gamma]$$

$$\geq \frac{\rho}{t} - \Pr_{x \leftarrow D^*}[x \in B_\gamma].$$

Thus,

$$\Pr_{x \leftarrow D'}[\text{Detect}(x) = 1] \geq \Pr_{x \leftarrow D'}[\text{Detect}(x) = 1 \wedge x \notin B_\gamma]$$
$$\geq \gamma \cdot \Pr_{x \leftarrow D^*}[\text{Detect}(x) = 1 \wedge x \notin B_\gamma]$$
$$\geq \gamma \cdot \left(\frac{\rho}{t} - \Pr_{x \leftarrow D^*}[x \in B_\gamma]\right). \square$$

We now obtain, for some reasonable setting of parameters, a lower bound on the false positive probability.

Lemma 4. *Let* $t \geq \frac{\alpha}{\epsilon}\sqrt{n/\ln n}$ *and* $\sigma = \epsilon\Delta/(2\sqrt{n})$. *If* $\Pr_{x \leftarrow D^*}[\text{Detect}(x) = 1] \geq \rho/t$, *for* $1/\rho = o(\ln n)$, *then*

$$\Pr_{x \leftarrow D'}[\text{Detect}(x) = 1] \geq \frac{\epsilon}{\alpha}\rho n^{-\beta - 1/2}\sqrt{\ln n},$$

where $\beta = (2/\alpha)(1 + 1/\alpha)$ *and* n *is sufficiently large.*

Proof. For the proof, we set some of the parameters in the expression given in Lemma 3 and use the lemma to lower bound the probability of a false positive. The value of $\gamma > 0$ must be chosen to balance between two competing considerations imposed by the γ term and the $\rho/t - \Pr_{x \leftarrow D^*}[x \in B_\gamma]$ term. Intuitively, when γ is close to 1, then x is often in B_γ, but this is not so advantageous for the Detect; when γ is small, it is indeed good for Detect to have $x \in B_\gamma$, but this hardly ever happens.

We will choose γ such that $\Pr_D[B_\gamma] \leq \rho/(2t)$; γ will be n^β for some constant β. Since $\Pr_{x \leftarrow D^*}[\text{Detect}(x) = 1] \geq \rho/t$, $\Pr_{x \leftarrow D^*}[\text{Detect}(x) = 1 \wedge x \notin B_\gamma] \geq \rho/(2t)$. Then the probability of a false positive for document instances from D', will be at least $\gamma\rho/(2t)$.

Although each point x we consider is an n-dimensional quantity, we can exploit the spherical symmetry of n-dimensional Gaussian distributions as follows. Given a point x, let $x_\parallel$ denote the projection of x onto the line L connecting $\overline{D^*}$ and $\overline{D'}$. We define $d_\parallel(x)$ to be $d(\overline{D^*}, x_\parallel)$ if $\overline{D^*}$ is between $x_\parallel$ and $\overline{D'}$, and $-d(\overline{D^*}, x_\parallel)$ otherwise. We define $d_\perp(x)$ to be the distance from x to L.

Now, by the spherical symmetry of Gaussian distributions, we have

$$\Pr_{x \leftarrow D^*}[x] = c_n \exp\left(\frac{-d^2(x, \overline{D^*})}{2\sigma^2}\right) \text{ and}$$
$$\Pr_{x \leftarrow D'}[x] = c_n \exp\left(\frac{-d^2(x, \overline{D'})}{2\sigma^2}\right).$$

where c_n is some normalization constant, depending on n. Let $\delta = d(\overline{D^*}, \overline{D'})$; note that $\delta \leq \Delta/t$. By the Pythagorean theorem and elementary geometry, we have

$$d^2(x, \overline{D^*}) = d_\parallel^2(x) + d_\perp^2(x) \text{ and}$$
$$d^2(x, \overline{D'}) = (d_\parallel(x) + \delta)^2 + d_\perp^2(x).$$

Hence,

$$\Pr_{x \leftarrow D^*}[x] = c_n \exp - \left(\frac{d_\parallel^2(x) + d_\perp^2(x)}{2\sigma^2} \right) \text{ and }$$

$$\Pr_{x \leftarrow D'}[x] = c_n \exp - \left(\frac{(d_\parallel(x) + \delta)^2 + d_\perp^2(x)}{2\sigma^2} \right).$$

Let $b(x) \stackrel{\text{def}}{=} \dfrac{\Pr_{x \leftarrow D'}[x]}{\Pr_{x \leftarrow D^*}[x]} = \exp \dfrac{-(2d(x_\parallel)\delta + \delta^2)}{2\sigma^2}$. By definition, $B_\gamma = \{x \mid b(x) \leq \gamma\}$. Let $\sigma = \epsilon\Delta/(2\sqrt{n})$, where ϵ is to be determined later. Thus, $\delta \leq 2\sigma\sqrt{n}/(\epsilon t)$, hence

$$b(x) \geq \exp - \left(\frac{2d(x_\parallel)\sqrt{n}}{t\epsilon\sigma} + \frac{2n}{\epsilon^2 t^2} \right).$$

If $b(x) \leq \gamma$, we get

$$d(x_\parallel) \geq \sigma \left(\frac{\epsilon t}{2\sqrt{n}} \ln \frac{1}{\gamma} - \frac{\sqrt{n}}{\epsilon t} \right).$$

Thus, we can bound $\Pr_{x \leftarrow D^*}[B_\gamma]$ as

$$\Pr_{x \leftarrow D^*}[B_\gamma] \leq \int_{d(x_\parallel) \geq \sigma \left(\frac{\epsilon t}{2\sqrt{n}} \ln \frac{1}{\gamma} - \frac{\sqrt{n}}{\epsilon t} \right)} \frac{1}{\sqrt{2\pi}\sigma} \exp \frac{-x^2}{2\sigma^2},$$

which is upper bounded by

$$\frac{1}{\sqrt{2\pi}} \cdot \left(\frac{1}{\frac{\epsilon t}{2\sqrt{n}} \ln \frac{1}{\gamma} - \frac{\sqrt{n}}{\epsilon t}} \right) \exp - \frac{1}{2} \left(\frac{\epsilon t}{2\sqrt{n}} \ln \frac{1}{\gamma} - \frac{\sqrt{n}}{\epsilon t} \right)^2,$$

when $\dfrac{\epsilon t}{2\sqrt{n}} \ln \dfrac{1}{\gamma} - \dfrac{\sqrt{n}}{\epsilon t} > 0$. Here, we are exploiting the spherical symmetry of our n-dimensional Gaussian distribution: projecting onto a line gives a 1-dimensional Gaussian distribution.

We are interested in the case when this is at most $\rho/(2t)$. Now, $t = \frac{\alpha}{\epsilon}\sqrt{n/\ln n}$. Set $\gamma = n^{-\beta}$. Then, the above bound is at most

$$\frac{1}{\sqrt{2\pi \ln n}} \cdot \frac{1}{\left(\frac{\alpha\beta}{2} - \frac{1}{\alpha} \right)} \cdot n^{-\frac{1}{2}\left(\frac{\alpha\beta}{2} - \frac{1}{\alpha} \right)^2}.$$

If we set $\beta = (2/\alpha)(1 + 1/\alpha)$, then for large enough n this is less than

$$\frac{1}{\sqrt{2\pi n \ln n}} < \frac{\rho}{2t},$$

for $\rho = 1/o(\ln n)$. We must also ensure that $\dfrac{\epsilon t}{2\sqrt{n}} \ln \dfrac{1}{\gamma} - \dfrac{\sqrt{n}}{\epsilon t} > 0$. When $t = \frac{\alpha}{\epsilon}\sqrt{n/\ln n}$, for a given α, our choice of $\beta = (2/\alpha)(1 + 1/\alpha) > 2/\alpha^2$ guarantees $\alpha\beta/2 - 1/\alpha > 0$.

3.3   The Final Calculation

Lemma 4 gives a criterion for when the output of

$$\mathsf{Attack}(i_1, \ldots, i_t, D^{i_1}, \ldots, D^{i_t})$$

will cause Detect to have a high false positive rate; however, these bad indices may be very uncommon, and almost never encountered by the Detect procedure, since we assume the adversary receives a uniformly chosen subset. It remains to bound how likely it is for such a bad $i_1, \ldots, i_t$ to be chosen. There are many ways of doing so; a very simple argument will make our point.

First, we show a high false positive rate under a different distribution of indices, defined by the following procedure.

1. Choose $I = i_1, \ldots, i_t$ uniformly,
2. Determine the j maximizing the probability that, after Attack produces D^*, Detect returns i_j.
3. Remove i_j from I and replace it with a new element, chosen at random (without replacement), giving I'.

The sets I, I' are completely analogous to $\{1, \ldots, t\}$ and $\{2, \ldots, t+1\}$ in the previous analysis. Lemma 2 implies that whenever Detect catches the colluders (correctly) on set I with probability ρ, for $\rho > 1/\sqrt{\ln n}$ (and n sufficiently large), it will falsely accuse someone with probability at least $\rho\phi$, where

$$\phi \stackrel{\mathrm{def}}{=} \frac{\epsilon}{\alpha} n^{-\beta - 1/2} \sqrt{\ln n}$$

when the attack is based on set I'. Note that the $1/\sqrt{\ln n}$ term can be replaced by any function $f(n)$ where $1/\ln n = o(f(n))$.

By a simple probability calculation, if Detect is successful with probability q (catches a correct colluder), when I is chosen uniformly (as in the procedure above), it will make a false accusation with probability $(q - 1/\sqrt{\ln n})\phi$ on sets chosen according to the distribution of I' in the procedure above.

We next observe that for any t-set I^*, $\Pr[I' = I^*] \le t\Pr[I = I^*]$. That is, the distribution on I' assigns at most t times the weight to some subset than would the uniform distribution. To see this, note that for any I^* there are at most $t(m - t)$ values of I in the above procedure such that I' could possibly be equal to I^* (there are that many ways of swapping an index out). For each of these possibilities, the probability that I' is indeed equal to I^* is either 0 (when the index that needed to be swapped out was not the maximally accused index) or exactly $1/(m - t)$ (the probability of swapping the right index in).

By the "flatness" property we have shown for I', it then follows that if the false positive rate is $(q - 1/\sqrt{\ln n})\phi$ when the sets are chosen according to I', then the false positive rate is at least $(q - 1/\sqrt{\ln n})\phi/t$ when the sets are chosen uniformly.

4   Conclusion and Open Problems

We have shown that in the framework of [3], $O(\sqrt{n/\ln n})$ adversaries suffice to break the watermarking scheme. Within this framework, the attack is essentially oblivious to the actual watermarking method. In practice, a real document consists of much more than the n-vector assumed for the theoretical model; the relationship between a document and its corresponding n-vector may be more obscure. As soon as this correspondence (and a way of computing inverses) is figured out, our attack is applicable.

An interesting open question is to generalize our result for a general class of metrics. One criticism of the Euclidean distance is that it is not always a good measure of fidelity; one would like to choose ones notion of fidelity.

A more important open question is to properly model the "do not copy" problem for watermarking. Whereas for the problem we consider, the question is what the right bound for the adversaries should be, for the other problem it is unclear whether there is a theoretically defensible solution at all.

Acknowledgements

Uri Feige provided us with crucial assistance. Based on our preliminary notes, Steven Mitchell, Bob Tarjan, and Francis Zane have written and shared with us an alternate exposition of our methods.

References

1. D. Boneh and J. Shaw. Collusion-secure fingerprinting for digital data. *Proc. Advances in Cryptology — CRYPTO*, Springer LNCS 963:452–465, 1995.
2. B. Chor, A. Fiat, and M. Naor. Tracing traitors. *Proc. Advances in Cryptology — CRYPTO*, Springer LNCS 839:257–270, 1994.
3. I. Cox, J. Kilian, T. Leighton, and T. Shamoon. A secure, robust watermark for multimedia. *IEEE Transaction on Image Processing*, 6(12):1673–1687, 1997.
4. C. Dwork, J. Lotspiech, and M. Naor. Digital signets: Self-enforcing protection of digital information (preliminary version). *Proc. 28th ACM Symposium on Theory of Computing*, pp. 489–498, 1996.
5. J. Kilian, T. Leighton, L. R. Matheson, T. G. Shamoon, R. E. Tarjan, and F. Zane. Resistance of digital watermarks to collusive attacks. *Technical Report TR-585-98*, Department of Computer Science, Princeton University, 1998.
6. M. Naor and B. Pinkas. Threshold Traitor Tracing. *Proc. Advances in Cryptology — CRYPTO*, Springer LNCS 1462:502–517, 1998.
7. B. Pfitzmann. Trials of traced traitors. *Proc. 1st International Workshop on Information Hiding*, Springer LNCS 1174:49–64, 1996.
8. B. Pfitzmann and M. Schunter. Asymmetric fingerprinting (extended abstract). *Proc. Advances in Cryptology — EUROCRYPT*, Springer LNCS 1070:84–95, 1996.
9. B. Pfitzmann and M. Waidner. Anonymous fingerprinting. *Proc. Advances in Cryptology — EUROCRYPT*, Springer LNCS 1233:88–102, 1997.
10. B. Pfitzmann and M. Waidner. Asymmetric fingerprinting for larger collusions. *Proc. 4th ACM Conference on Computer and Communications Security*, pp. 151–160, 1997.

Coin-Based Anonymous Fingerprinting

Birgit Pfitzmann and Ahmad-Reza Sadeghi

Universität des Saarlandes, Fachbereich Informatik,
D-66123 Saarbrücken, Germany
{pfitzmann, sadeghi}@cs.uni-sb.de

Abstract. Fingerprinting schemes are technical means to discourage
people from illegally redistributing the digital data they have legally
purchased. These schemes enable the original merchant to identify the
original buyer of the digital data. In so-called asymmetric fingerprinting
schemes the fingerprinted data item is only known to the buyer after a
sale and if the merchant finds an illegally redistributed copy, he obtains
a proof convincing a third party whom this copy belonged to. All these
fingerprinting schemes require the buyers to identify themselves just for
the purpose of fingerprinting and thus offer the buyers no privacy. Hence
anonymous asymmetric fingerprinting schemes were introduced, which
preserve the anonymity of the buyers as long as they do not redistribute
the data item.

In this paper a new anonymous fingerprinting scheme based on the principles of digital coins is introduced. The construction replaces the general
zero-knowledge techniques from the known certificate-based construction
by explicit protocols, thus bringing anonymous fingerprinting far nearer
to practicality.

1 Introduction

Fingerprinting schemes are cryptographic techniques supporting the copyright
protection of digital data. They do not require tamper-resistant hardware, i.e.,
they do not belong to the class of copyright protection methods which *prevent*
copying. It is rather assumed that buyers obtain data in digital form and can copy
them. Buyers who redistribute copies disregarding the copyright conditions are
called *traitors*. Fingerprinting schemes discourage traitors by enabling the original merchant to identify a traitor who originally purchased the data item. Every
sold copy is slightly different from the original data item and unique to its buyer.
Obviously the differences to the original represent the information embedded in
the data item, which must be imperceptible. As several traitors might collude
and compare their copies to find and eliminate the differences, cryptographic
methods were used to make fingerprinting schemes *collusion tolerant*. There are
different classes of fingerprinting schemes called *symmetric* [BMP86, BS95] and
asymmetric [PS96, PW97a, BM97]. In contrast to symmetric schemes, asymmetric schemes require the data item to be fingerprinted via an interactive protocol
between the buyer and the merchant where the buyer also inputs her own secret. At the end of this protocol only the buyer knows the fingerprinted data

J. Stern (Ed.): EUROCRYPT'99, LNCS 1592, pp. 150–164, 1999.

item. However, after finding a redistributed copy the merchant can extract information which not only enables him to identify a traitor but also provides him with a proof of treachery that convinces any third party. The main construction in [PS96] was based on general primitives; an *explicit* construction, i.e., without such primitives, was only given for the case without significant collusions. Explicit collusion-tolerant constructions were given in [PW97a, BM97]. A special variant of fingerprinting is traitor tracing [CFN94, NP98]; here the keys for broadcast encryption [FN94] are fingerprinted. Asymmetric traitor tracing was introduced in [Pfi96] with a construction based on general primitives. Explicit constructions for this case were also given in [PW97a]. Even more efficient construction were given in [KD98]; however, they are not asymmetric in the usual sense but "arbitrated", i.e., a certain number of predefined arbiters can be convinced by the merchant (similar to the difference between arbitrated authentication codes and asymmetric signature schemes).

As in "real-life" market places, it is desired that electronic market places offer privacy to the customers. It should be possible to buy different articles (pictures, books, etc.) anonymously, since buying items can reveal a lot of behavioristic information about an individual. To allow this also for buying fingerprinted items, *anonymous asymmetric fingerprinting schemes* were proposed [PW97b]. Note that in normal fingerprinting (symmetric and asymmetric) the buyer has to identify herself during each purchase. In anonymous fingerprinting the anonymity of the buyers is preserved as long as they do not redistribute copies of the data item.

In this paper, we introduce a new anonymous asymmetric fingerprinting scheme based on the principles of digital coins. Our protocols are explicit, in contrast to the scheme in [PW97b], where general theorems like "every NP-language has a zero-knowledge proof system" were used, and thus far more efficient. The anonymity is information-theoretic and security computational. Security of one party relies on a restrictiveness assumption about the underlying payment system, which we formulate precisely.

2   The Model of Anonymous Fingerprinting

The involved parties in the model are merchants $\mathcal{M}$, buyers $\mathcal{B}$, registration centers $\mathcal{RC}$ and arbiters $\mathcal{A}$. For the purpose of fingerprinting it is required in this model that buyers register themselves to a registration center $\mathcal{RC}$ (e.g., their bank). The required trust in $\mathcal{RC}$ should be minimum such that a cheating $\mathcal{RC}$ can only refuse a registration.[1] It is assumed that $\mathcal{B}$ can generate signatures (using an arbitrary signature scheme) under her "real" identity $ID_\mathcal{B}$ and that the corresponding public keys have already been distributed. Furthermore there

---

[1] In particular, even a collusion of $\mathcal{M}$ and $\mathcal{RC}$ should not be able to trace a buyer who did not redistribute a data item. Otherwise one can trivially use any known non-anonymous asymmetric scheme and simply let the initial key pair of the buyer be certified under a pseudonym whose owner is only known to the certification authority. This was overlooked in [Dom98].

is no special restriction on the arbiter $\mathcal{A}$. Any third party having access to the corresponding public keys should be convinced by the proof. The main subprotocols of the construction are registration, fingerprinting, identification, and trial. Identification includes a variant "enforced identification" for the case where $\mathcal{RC}$ refuses to cooperate. The main security properties are:

- Security for the merchant: As long as collusions do not exceed a certain size, the merchant will be able to identify a traitor for each illegally redistributed item and to convince any honest arbiter. (In case $\mathcal{RC}$ colludes with the traitors the identified traitor may be $\mathcal{RC}$.)
- Security for the buyer and $\mathcal{RC}$: Nobody is unduly identified as a traitor; at least no honest arbiter will believe it.
- Anonymity as sketched above; different purchases by one buyer should also be unlinkable.

For the detailed definitions of the subprotocols and security properties we refer the interested reader to [PW97b].

3   Overview of the Construction

To see more precisely what we achieve, note that [PW97b] contains a modular construction: The first part, called framework, is a construction based on certificates. At the end, the merchant holds a commitment *com* to a certain value *emb* and possibly other information. The buyer can open the commitment and may also hold other information. This framework guarantees that whenever the merchant later obtains *emb*, he can identify this buyer and win a trial against him. In the second part, the value *emb* is embedded into the data item in a way that does not release additional information about *emb*. The embedding procedure must guarantee that whenever a collusion of at most the maximum tolerated size redistributes a data item, the merchant will be able to reconstruct the value *emb* that was used by at least one traitor. For the second part, constructions were previously only known for the case of traitor tracing and for normal fingerprinting without collusion tolerance. The latter is explicit, and one can see quite easily that the former (based on [PW97a] Section 4) can also be made explicit by using the efficient key selection protocol from Section 2.3 of the same paper in the appropriate places. The main technical part in [PW97b] was to construct a suitable collusion-tolerant embedding procedure for normal fingerprinting.

In contrast, the framework part in [PW97b] is a comparatively simple construction where *emb*, the content of the commitment, is a signature with respect to a key that the merchant must not know, and the buyer proves with a general zero-knowledge technique that she knows such a key and a certificate by $\mathcal{RC}$ on it. It is this first part that we replace with the explicit and much more efficient coin-based construction. It can then be combined with the known second parts. This gives us explicit overall constructions for collusion-tolerant anonymous traitor tracing and for anonymous normal fingerprinting without collusion tolerance. For collusion-tolerant normal fingerprinting, the construction in [PW97b] is not explicit, but it is mentioned that all the steps in a secure 2-party computation look quite simple so that it should be possible to find simpler explicit realizations for them. This can in fact be done (each time exploiting either homomorphism – note that even the Reed-Solomon codes are a linear operation – or the efficient table-lookup from [PW97b], end of Section 2.3), but we do not attempt to include details of that here.

The basic idea for using digital cash systems with double-spender identification to construct an anonymous fingerprinting scheme is as follows: Registration will correspond to withdrawing a coin. (The "coins" serve only as a cryptographic primitive and have no monetary value.) The untraceability of the cash system will give us the unlinkability of the views of the registration center and the merchant. Redistribution of a data item should correspond to double-spending of the underlying coin, i.e., the value *emb* embedded in the data item will be similar to the response in a second payment with the coin. We could execute a complete first payment during fingerprinting, but actually our protocols can be simpler. (They are more like "zero-spendable" coins where each coin as such can be shown, but any response to a challenge leads to identification.) One new problem is that while in a payment the response is given to the merchant in clear, in our case it must be verifiably hidden in a commitment. Another problem is that double-spender identification is usually a binary decision. In our case, however, it must be decided reliably under what copyright conditions the redistributed item was bought – e.g., there may be items that can be redistributed after a while. In the formal security requirements this is a value *text* input in fingerprinting and also in a trial. Thus the identification information must be linked to such a text during fingerprinting in a way that even a collusion of a merchant and the registration center cannot forge, although such a collusion can sign additional coins that look like belonging to a specific buyer.

4 Construction

Our explicit construction employs the ideas from the digital cash scheme in [Bra94], or, as we do not have the same double-spender identification, at least from the underlying blind signature scheme [CP93]. We make the following conventions:

Algebraic Structure: All arithmetic operations are performed in a group G_q of order q for which efficient algorithms are known to multiply, invert, determine equality of elements, test membership and randomly select elements. Any group G_q satisfying these requirements and in which the computation of discrete logarithms is infeasible can be a candidate. For concrete constructions one can assume that G_q is the unique subgroup of prime order q of the multiplicative group $\mathbb{Z}_p^*$ where p is a prime such that $q|(p-1)$.

Hash Function: Hash functions are denoted by *hash*. We have to make the same assumptions as in [Bra94], which is behavior similar to a random oracle.

Commitment Scheme: We use two commitment schemes. The first one is based on discrete logarithms in G_q (see [BKK90] for the one-bit version and [CHP92, Ped92] based on [BCP88] for the general version). To commit to a value $b \in \mathbb{Z}_q$, the committer needs generators $g'', h'' \in_R G_q \setminus \{1\}$, which are typically randomly generated and sent by the recipient. Then the committer selects $x \in_R \mathbb{Z}_q$ and computes the commitment $y = BC^{DL}(x, b) = g''^b h''^x \bmod p$. To open y, the committer reveals (b, x). This scheme is information-theoretically hiding and it is binding under the discrete logarithm assumption in the corresponding group. Moreover due to the homomorphic property of this scheme one can commit to a number $r \in \mathbb{Z}_q^*$ using commitments to the bits of r: Let $r = \sum_{j=0}^{l-1} r_j 2^j$ be the binary representation of r and $BC^{DL}(x_j, r_j)$ be commitments to the bits. Then $\prod_{j=0}^{l-1} (BC^{DL}(x_j, r_j))^{2^j} = g''^r h''^x = BC^{DL}(x, r) \bmod p$ with $x = \sum_{j=0}^{l-1} x_j 2^j \bmod q$.

The second bit commitment scheme is based on quadratic residues (see [GM84, BCC88]). To commit to a bit b' the committer computes $y = BC^{QR}(x', b') = (-1)^{b'} x'^2 \bmod n$ where $x' \in_R \mathbb{Z}_n^*$ and n is a Blum integer chosen by the committer.

4.1   Key Distribution and Registration

Registration Center Key Distribution: $\mathcal{RC}$ randomly selects a group G_q and generators $g, g_1, g_2 \in_R G_q \setminus \{1\}$ and a number $x \in_R \mathbb{Z}_q^*$ as its secret key. $\mathcal{RC}$ also chooses a hash function $hash$. It publishes the group description, (g, g_1, g_2), its public key $h \equiv g^x \bmod p$ and $hash$.

The correctness of the group description (i.e., that p and q primes with $q|(p-1)$ and generated in a way believed to exclude trap doors) and whether the generators are elements of $G_q \setminus \{1\}$ should be verified by other parties when using them. However, other parties do not rely on the randomness of the generators.

Opening a one-time account: This phase is similar to [Bra94], but in order to bind the identification information to a specific purchase text, each "account" is used only once. To open an account, $\mathcal{B}$ chooses $i \in_R \mathbb{Z}_q^*$ randomly and secretly and computes $h_1 \equiv g^i \bmod p$ (with $h_1 g_2 \neq 1$). She gives h_1 to $\mathcal{RC}$ and proves that she knows i (using the zero-knowledge proof from [CEG88] or, more efficiently, Schnorr identification [Sch91]). First $\mathcal{RC}$ verifies that the account number h_1 has not been used before. Then it stores h_1 in its registration database together with the claimed normal identity $ID_\mathcal{B}$ of this buyer. $\mathcal{B}$ gives a signature sig_{coin} on h_1 (with a suitable explanation) under her normal identity $ID_\mathcal{B}$ and $\mathcal{RC}$ verifies it. This signature can be used in later trials to show that $\mathcal{B}$ is responsible for this "account number" h_1.

Withdrawal: The protocol is shown in Figure 1. Essentially, $\mathcal{RC}$ signs the common input $m \equiv h_1 g_2 \equiv g_1^i g_2 \bmod p$ using a restrictive blind signature as in [Bra94]. Thus $\mathcal{B}$ obtains a signature $\sigma' = (z', a', b', r')$ on $m' \equiv m^s \equiv g_1^{is} g_2^s \bmod p$ where $s \in_R \mathbb{Z}_q^*$ is chosen randomly and secretly by $\mathcal{B}$.

Our protocol is also similar to the withdrawal protocol in [Bra94] in that an additional value is included in the hashing to obtain the challenge c' and thus in the signing process. In our case it is the public key of the key pair (sk_{text}, pk_{text}) from an arbitrary signature scheme, here Schnorr's for concreteness [Sch91]. We call the triple (m', pk_{text}, σ') a coin.

$$
\begin{array}{lr}
\mathcal{B} & \mathcal{RC} \\
& z \leftarrow m^x \\
& w \in_R \mathbf{Z}_q \\
& a \leftarrow g^w \bmod p \\
& \xleftarrow{\quad z, a, b \quad} \quad b \leftarrow m^w \bmod p \\
s \in_R \mathbf{Z}_q^* & \\
m' \leftarrow m^s, \; z' \leftarrow z^s & \\
sk_{text} \in_R \mathbf{Z}_q^*, \; pk_{text} \leftarrow g_2^{sk_{text}} \bmod p & \\
u, v \in_R \mathbf{Z}_q & \\
a' \leftarrow a^u g^v, \; b' \leftarrow b^{su} m'^v & \\
c' \leftarrow hash(m', z', a', b', pk_{text}) \bmod q & \\
c \leftarrow c'/u \bmod q \quad\xrightarrow{\quad c \quad} & \\
g^r \stackrel{?}{\equiv} ah^c \bmod p, \; m^r \stackrel{?}{\equiv} bz^c \bmod p \; \xleftarrow{\quad r \quad} & r \leftarrow cx + w \bmod q \\
r' \leftarrow ru + v \bmod q &
\end{array}
$$

Fig. 1. The Withdrawal Part of the Registration Protocol

4.2 Fingerprinting

The fingerprinting subprotocol is executed between the (anonymous) buyer $\mathcal{B}$ and the merchant $\mathcal{M}$. This protocol differs (except for Step 1) from the payment protocol in [Bra94]. The common input is a text, *text*, describing the purchase item and licensing conditions.

Step 1: $\mathcal{B}$ selects an unused coin $coin' = (m', pk_{text}, \sigma')$. She uses the corresponding sk_{text} to make a Schnorr signature sig_{text} on *text* and sends $(coin', sig_{text})$ to $\mathcal{M}$. Now $\mathcal{M}$ first verifies the validity of $coin'$ by computing $c' \equiv hash(m', z', a', b', pk_{text}) \bmod q$ and checking whether $m' \neq 1$ and $g^{r'} \equiv a'h^{c'}$ and $m'^{r'} \equiv b'z'^{c'}$ mod p hold [Bra94]. We say that a coin is valid if and only if it passes these tests. He then verifies sig_{text} using pk_{text} from $coin'$.

Step 2: $\mathcal{B}$ takes the internal structure (is, s) of $m' \equiv g_1^{is} g_2^s$ as the value to be embedded in the data item. Hence $emb = (is, s)$, and let $r_1 = is$, $r_2 = s$.

Since $\mathcal{M}$ should not get any useful information on this value, $\mathcal{B}$ hides it in a commitment. While the certificate-based framework in [PW97b] could leave the type of commitment open, we have to provide an efficient link from the given representation of emb, i.e., m', to the type of commitments needed in the later embedding procedures. For normal fingerprinting with and without collusion tolerance, these are quadratic residue commitments to individual bits of *emb*.[2]

We start this by producing discrete logarithm commitments to a binary representation of r_1 and r_2: The merchant $\mathcal{M}$ sends generators g'' and h'' to $\mathcal{B}$, and $\mathcal{B}$ sends back commitments $com_{kj} = g''^{r_{kj}} h''^{x_{kj}}$ where $r_k = \sum_{j=0}^{l-1} r_{kj} 2^j$ for $k = 1, 2$ and $x_{kj} \in_R \mathbb{Z}_q$. $\mathcal{M}$ may choose the generators randomly once for all its buyers. $\mathcal{B}$ should verify that they are elements of $G_q \setminus \{1\}$. From these individual commitments, $\mathcal{M}$ computes the commitments to r_1 and r_2, i.e., $com_k = \prod_{j=0}^{l-1} (com_{kj})^{2^j}$ for $k = 1, 2$. Now $\mathcal{B}$ proves the following predicate P to $\mathcal{M}$: The content of the commitments com_k is a pair $(r_1, r_2) \in \mathbb{Z}_q^* \times \mathbb{Z}_q^*$ with $m' \equiv g_1^{r_1} g_2^{r_2} \bmod p$. This can be done in zero-knowledge as shown in Figure 2, similar to other proofs concerning knowledge of representations of numbers with respect to certain generators following [CEG88]. As usual, we could also use larger challenges c at the cost of the real zero-knowledge property.

Note that this protocol does not prove that the values r_{kj} are binary; such a proof will be a side effect of Step 3.

$\mathcal{B}$		$\mathcal{M}$
$k = 1, 2$ and $0 \le j \le l-1$		
$(r_1, r_2) \leftarrow emb = (is, s)$		
$r_k = \sum_{j=0}^{l-1} r_{kj} 2^j$		
$x_{kj} \in_R \mathbb{Z}_q$, $x_k \leftarrow \sum_{j=0}^{l-1} x_{kj} 2^j$		
$com_{kj} \leftarrow g''^{r_{kj}} h''^{x_{kj}}$	$\xrightarrow{com_{kj}}$	$com_k \leftarrow \prod_{j=0}^{l-1} com_{kj}^{2^j}$
REPEAT N times		
$e_k, y_k \in_R \mathbb{Z}_q^*$		
$V \leftarrow g_1^{e_1} g_2^{e_2} \bmod p$		
$rcom_k \leftarrow g''^{e_k} h''^{y_k}$	$\xrightarrow{rcom_k, V}$	
$res_k \leftarrow cr_k + e_k \bmod q$	$\xleftarrow{c}$	$c \in_R \mathbb{Z}_2$
$z_k \leftarrow cx_k + y_k \bmod q$	$\xrightarrow{res_k, z_k}$	$g_1^{res_1} g_2^{res_2} \stackrel{?}{\equiv} Vm'^c \bmod p$
		$g''^{res_k} h''^{z_k} \stackrel{?}{\equiv} rcom_k com_k^c \bmod p$
END REPEAT		

Fig. 2. Step 2 of the Fingerprinting Protocol

Step 3: Now $\mathcal{B}$ additionally computes quadratic residue commitments on the same values r_{kj}. As mentioned, these are needed as input to the embedding

[2] For traitor tracing, they are quadratic residue commitments to small blocks of *emb* represented in unary. Such a representation can also be derived efficiently from commitments to the bits.

procedures. We denote them by $com'_{kj} = BC^{QR}(x'_{kj}, r_{kj})$ for $x'_{kj} \in_R \mathbb{Z}_n^*$, where n is a Blum integer chosen by the buyer. $\mathcal{B}$ sends these commitments to $\mathcal{M}$ and proves in zero-knowledge that the contents in each pair (com_{kj}, com'_{kj}) are equal. Since one can only commit to bits when using BC^{QR}, the equality proof implies that the values r_{kj} in Step 2 were binary. An efficient proof can again be carried out by fairly standard techniques. For instance, one can see two pairs of commitments BC^{DL} and BC^{QR}, where both commitments in one pair contain "0" and in the other pair "1", as a cryptographic capsule and proceed similar to [Ben87]. This is a proof with one-bit challenges and thus the least efficient part of our protocol. However, even if one only compares it with the simplest embedding procedure that might follow, fingerprinting without collusion tolerance as in [PS96], one sees that quadratic residue commitments must be made on a portion of the data item significantly larger than the word emb to be embedded into it (so that the resulting changes are small). Thus the complexity of our last step is not larger than that of embedding.

4.3   Identification

After finding a redistributed copy of the data item, $\mathcal{M}$ tries to identify a traitor as follows:

Step 1: $\mathcal{M}$ extracts a value $emb = (r_1, r_2)$ from the redistributed data item using the extraction algorithm from the underlying embedding scheme. This pair is (is, s) with $s \neq 0$. $\mathcal{M}$ computes $m' \equiv g_1^{is} g_2^s \bmod p$ and retrieves $coin'$, $text$, and sig_{text} from the purchase record of the corresponding data item. If he does not find the coin identifier m', he gives up (the collusion tolerance of the underlying code may be exceeded). Otherwise he sends i to $\mathcal{RC}$.

Step 2: $\mathcal{RC}$ searches in its registration database for a buyer who is registered under the value $h_1 \equiv g_1^i$. It retrieves the values $(ID_\mathcal{B}, sig_{coin})$ and sends them to $\mathcal{M}$. Note that $\mathcal{M}$ can enforce $\mathcal{RC}$'s cooperation, see below.

Step 3: $\mathcal{M}$ verifies the signature sig_{coin} on h_1.

Enforced Identification This is a special case in identification if $\mathcal{RC}$ refuses to reveal the information requested by $\mathcal{M}$:

Step 1: $\mathcal{M}$ sends $proof_1 = (coin', (i, s))$ to an arbiter $\mathcal{A}$.

Step 2: $\mathcal{A}$ verifies the validity of $coin'$ using the algorithm from Step 1 of fingerprinting and that $m' \equiv g_1^{is} g_2^s \bmod p$. If this is wrong, $\mathcal{A}$ rejects $\mathcal{M}$'s claim. Otherwise she sends i to $\mathcal{RC}$ and requests the values $(ID_\mathcal{B}, sig_{coin})$. Then $\mathcal{A}$ verifies them as $\mathcal{M}$ does in Step 3 of the identification.

4.4   Trial

Now $\mathcal{M}$ tries to convince an arbiter $\mathcal{A}$ that $\mathcal{B}$ redistributed the data item bought under the conditions described in $text$. The values $ID_\mathcal{B}$ and $text$ are common inputs.

Step 1: $\mathcal{M}$ sends to $\mathcal{A}$ the proof string

$$proof = ((i, sig_{coin}), (coin', sig_{text}), s).$$

Step 2: $\mathcal{A}$ computes $h_1 \equiv g_1^i \bmod p$ and verifies that sig_{coin} is a valid signature on h_1 with respect to $ID_\mathcal{B}$. If yes, it means that i, the internal structure of an account number h_1 for which $\mathcal{B}$ was responsible, has been recovered by $\mathcal{M}$ and thus, as we will see, that $\mathcal{B}$ has redistributed some data item. Note that i alone is not enough evidence for $\mathcal{A}$ to find $\mathcal{B}$ guilty of redistributing a data item under the specific text, $text$.

Step 3: $\mathcal{A}$ verifies the validity of $coin'$, that $m' \equiv g_1^{is} g_2^s \bmod p$ holds, and the signature sig_{text} on the disputed text using the test key pk_{text} contained in $coin'$. These verifications imply that if the accused buyer owned this coin, she must have spent it in the disputed purchase on $text$. Now $\mathcal{A}$ must verify that this coin belongs to $\mathcal{B}$. It is not possible to do so by only showing the link between the coin and the withdrawal (which could be fixed by a signature from $\mathcal{B}$ under $\mathcal{RC}$'s view), because a collusion of $\mathcal{M}$ and $\mathcal{RC}$ could forge such a link. (Interested readers can find the attack in Appendix A.) Thus $\mathcal{A}$ performs the following last step where $\mathcal{B}$ is required to take part.

Step 4: $\mathcal{A}$ asks $\mathcal{B}$ whether she has withdrawn another valid coin, i.e., a tuple $coin^* = (m^*, pk_{text}^*, \sigma^*)$ with $pk_{text}^* \neq pk_{text}$ using the one-time account h_1. If yes, $\mathcal{B}$ has to show the representation of m^*, i.e., a value s^* such that $m^* \equiv g_1^{is^*} g_2^{s^*}$. If $\mathcal{B}$ can do that, then $\mathcal{A}$ decides that $\mathcal{RC}$ is guilty, otherwise $\mathcal{B}$.

5   Security of the Construction

We now present detailed proof sketches of our construction. We assume that all the underlying primitives are secure. The merchant's security only relies on the security of the underlying embedding scheme, the buyer's on standard cryptographic assumptions. The security for the registration center needs the restrictiveness of the blind signature scheme, and we will make a precise assumption for this. Anonymity is information-theoretic.

5.1   Security for the Merchant

Due to the properties of the underlying embedding scheme, we can assume that whenever the maximum tolerated size of a collusion is not exceeded, and the collusion redistributes a data item sufficiently similar to the original, then $\mathcal{M}$ can extract a value emb that belongs to a traitor with very high probability. More precisely emb is the value to which the traitor could open the final quadratic-residue commitments given in the corresponding purchase.

The zero-knowledge proofs in fingerprinting guarantee that this value emb is a pair (r_1, r_2) such that $g_1^{r_1} g_2^{r_2} \equiv m' \neq 1$, where m' is the identifier of the coin used in this purchase. Thus $\mathcal{M}$ can retrieve a valid $coin'$ and the pair $(i, s) \equiv (r_1/r_2, r_2)$. This enables him to ask $\mathcal{RC}$ for identification and, in the worst case, have it enforced by $\mathcal{A}$. Moreover, he can retrieve $text$ and a valid signature sig_{text}. Together with the values that $\mathcal{RC}$ must return, $\mathcal{M}$ therefore obtains a valid proof string $proof$ and passes the first three steps of the trial. This is sufficient because the Step 4 of the trial only concerns $\mathcal{A}$'s decision on whether $\mathcal{RC}$ or $\mathcal{B}$ is guilty.

$\mathcal{M}$ is also protected from making wrong accusations (and thus possibly damaging his reputation): Even if there are more than the tolerated number of traitors, $\mathcal{M}$'s verifications in identification guarantee that whenever he makes an accusation he will not lose in the trial.

5.2 Security for the Buyer

Consider an honest buyer $\mathcal{B}$ and a trial about the purchase on a specific text, $text$, for which $\mathcal{B}$ has not revealed the corresponding data item. She is secure if the attackers cannot convince an honest arbiter $\mathcal{A}$ in this trial, even if the other parties collude and obtain other data items that she bought (active attack). Such situations occur, e.g., if $\mathcal{B}$ is allowed to redistribute another item after a certain period of time.

Step 2 of the trial guarantees that $\mathcal{B}$ is only held responsible for one of her own one-time account numbers $h_1 \equiv g_1^i$, and that the attackers must know i.

First it is shown that the attackers cannot find i unless they obtain the result of a purchase where $\mathcal{B}$ has used a coin $coin^*$ withdrawn from the account h_1. The only knowledge the attackers can otherwise obtain about i in our protocol is: (1) h_1 itself, (2) the proof of knowledge of i in registration, and (3) the commitments on $emb = (is, s)$ and two zero-knowledge proofs in fingerprinting. Additionally, they might obtain information in embedding, but by definition of secure embedding this is not the case. If the proofs are actually zero-knowledge and the commitments semantically secure, computing i from all this information is as hard as computing it from h_1 alone, i.e., as computing a discrete logarithm. (If we use Schnorr identification and a similar proof in fingerprinting, security relies on the joint security of these identification protocols against retrieval of the secret key.)

Hence the only way for the attackers to find i is in fact to obtain the resulting data item in a purchase where $\mathcal{B}$ has used $coin^*$ based on h_1. Let $text^*$ be the text describing that purchase. By the precondition, we know that $text^* \neq text$. The secret key sk^*_{text} corresponding to pk^*_{text} in $coin^*$ is known only to $\mathcal{B}$, and $\mathcal{B}$ reveals no information about it during registration, fingerprinting, and redistribution except making one signature on $text^*$ with it. As we assume that the underlying signature scheme is secure against active attacks, this does not help the attackers to forge a valid signature with respect to pk^*_{text} on $text$.

Now, even if the attackers are a collusion of $\mathcal{M}$ and $\mathcal{RC}$ and succeed in constructing a wrong coin $coin_A$ with a self-made pk^A_{text} which passes all $\mathcal{A}$'s verifications so far, $\mathcal{B}$ wins by showing $coin^*$ with $pk^*_{text} \neq pk^A_{text}$.

5.3   Security for the Registration Center

An honest registration center $\mathcal{RC}$ should never be found guilty by an honest arbiter $\mathcal{A}$. This could happen in two cases:

1. In enforced identification if $\mathcal{M}$ can send a value $proof_1$ that convinces $\mathcal{A}$, but $\mathcal{RC}$ cannot reveal the required registration data under the identity i.
2. In a trial if the attackers (a collusion of $\mathcal{M}$ and $\mathcal{B}$) can generate two valid coins, $coin'$ and $coin^*$, withdrawn from the same account h_1, i.e., their representations correspond to the same h_1.

To exclude both cases, we need the restrictiveness of the underlying blind signature scheme. The corresponding assumption in [Bra94] is fairly informal, so we formalize it in a version that suffices for our purpose. Note that although this version seems closest to Brands' formulation, for his payment system a weaker assumption would be sufficient, and the corresponding definition in [FY93] is also of that weaker form: For the payment system one only needs that all the coins the attackers can construct correspond to some account of the attackers. For Case 2 of $\mathcal{RC}$'s security, we need that the attackers cannot even transform coins between their own accounts. (In [Bra94], Definition 5, this is somehow implicit because a one-to-one correspondence between constructed coins and withdrawal protocols is assumed a priori.) We only need the case with one-time accounts, but we formulate the general case, and also for any number of generators. We first introduce some notation:

- Let gen_{group} be an algorithm that generates a group from the given family (see Section 4) and a fixed number $n + 1$ (in our concrete case $n = 2$) of random generators g and $g_1, ..., g_n$. Its output $(p, q, g, g_1, ..., g_n)$ is denoted by $desc$.
- Let gen_{key} be the key generation algorithm that takes as input a value $desc$ and outputs a key pair (sk, pk), here (x, g^x).
- Let $valid(desc, pk, (m', par, \sigma'))$ be the predicate for validity of a coin as in Step 1 of fingerprinting (where par may be an arbitrary value in the place of pk_{text}).
- By the predicate $repr$ we denote that a vector $I = (i_1, ..., i_n) \in \mathbb{Z}_q^n$ is the representation of a message $m \in G_q$, i.e.,

$$repr(m, I) : \iff m = g_1^{i_1}...g_n^{i_n}.$$

- $blindsig$ is the protocol between a signer $\mathcal{S}$ and a (dishonest) recipient $\mathcal{R}$ shown in Figure 1. Signer and recipient are interactive probabilistic polynomial-time algorithms, and in our case $\mathcal{S}$ always follows the protocol and $\mathcal{R}$ need not. The signer inputs sk and a message m, and the recipient has an arbitrary auxiliary local variable aux as input; it models the attacker's memory between protocol executions. The signer has no final output, while the recipient outputs an update aux'' of its auxiliary variable. If the recipient wants to continue the attack, we also let him immediately output the message

m'' that the signer should sign next, and we also allow another output o''. We write one such interaction as

$$((aux'', m'', o''), -) \leftarrow blindsig(\mathcal{R}(aux), \mathcal{S}(sk, m)).$$

We now consider a dishonest recipient $\mathcal{R}$ who executes the withdrawal protocol with an honest Brands signer $\mathcal{S}$ polynomially many times. In addition, $\mathcal{R}$ has to show a representation I_j of each input message m_j. (Note that we make the assumption with actual outputs in places where the real protocols only have proofs of knowledge.) At the end, $\mathcal{R}$ outputs valid signatures, again with representations of the signed messages. The assumption (or, implicitly, the definition of strong restrictiveness) is that all the output representations are scalar multiples of input representations, and in a kind of one-to-one mapping. To express this easily, let $< \ldots >$ be a notation for a multiset (i.e., unordered like a set but possibly with repetition). We use $\subseteq$ to denote the subset relation for multisets, i.e., each element on the left must also occur on the right with at least the same multiplicity. Finally, for any vector (representation) $I \neq 0$ let $\overline{I}$ denote the line it generates, i.e., the set of its scalar multiples.

Assumption 1 (General Strong Restrictiveness) *Let $\mathcal{R}$ denote a probabilistic polynomial-time interactive algorithm, Q a polynomial and l a security parameter. Then for all Q and for all $\mathcal{R}$ that interact with the Brands signer for $k = Q(l)$ times:*

$P(\neg(< \overline{I'_1}, \ldots, \overline{I'_{k'}} > \subseteq < \overline{I_1}, \ldots, \overline{I_k} >)$
$\wedge \forall j = 1, \ldots, k : repr(m_j, I_j)$
$\wedge \forall j = 1, \ldots, k' : (repr(m'_j, I'_j) \wedge valid(desc, pk, (m'_j, par_j, \sigma'_j)))$
$\wedge$ *all pairs (m'_j, par_j) are different*
$::\quad desc \leftarrow gen_{group}(l); \ k := Q(l); (sk, pk) \leftarrow gen_{key}(desc);$
$((aux_1, m_1, I_1), -) \leftarrow \mathcal{R}(l, desc, pk);$
$((aux_2, m_2, I_2), -) \leftarrow blindsig(\mathcal{R}(aux_1), \mathcal{S}(sk, m_1));$

$\qquad \vdots$

$((aux_k, m_k, I_k), -) \leftarrow blindsig(\mathcal{R}(aux_{k-1}), \mathcal{S}(sk, m_{k-1}));$
$(aux_{k+1}, -) \leftarrow blindsig(\mathcal{R}(aux_k), \mathcal{S}(sk, m_k));$
$(k', ((m'_1, par_1, \sigma'_1), \ldots, (m'_{k'}, par_{k'}, \sigma'_{k'})), (I'_1, \ldots, I'_{k'})) \leftarrow \mathcal{R}(aux_{k+1}))$
$\leq 1/poly(l).$

Next we return to the security of fingerprinting and consider a successful attacker $\mathcal{R}^*$ against $\mathcal{RC}$'s security. As $\mathcal{RC}$ never uses its secret key except in the blind signature protocol within registration, $\mathcal{R}^*$ can be seen as an attacker against the blind signature protocol with $n = 2$. The only difference to the scenario in Assumption 1 up to the choice of aux_{k+1} is that $\mathcal{R}^*$ need not output the values I_j, but instead gives a proof of knowledge of i_j in the specific representation

$m_j = g_1^{i_j} g_2$. We combine $\mathcal{R}^*$ with the extractor for these proofs of knowledge to obtain another attacker $\mathcal{R}$ that also outputs I_j, so that Assumption 1 applies.

We now consider the two cases of how $\mathcal{RC}$ could be found guilty by an honest arbiter with the new attacker $\mathcal{R}$.

In Case 1, the attackers need a valid coin $coin' = (m', pk_{text}, \sigma')$ and a representation $I = (is, s)$ of m' such that no withdrawal with respect to $h_1 = g_1^i$ was performed. (If $\mathcal{RC}$ does find such a withdrawal it is clear that the retrieved values pass $\mathcal{A}$'s tests.) However, by Assumption 1, I is a scalar multiple of one of the original representations, say I_j. As I_j is of the form $(i_j, 1)$, this means that $i = i_j$ and thus a withdrawal with $h_1 = g_1^i$ was performed.

In Case 2, the attackers must show two valid coins $coin' = (m', pk_{text}, \sigma')$ and $coin^* = (m^*, pk_{text}^*, \sigma^*)$ with $pk_{text}^* \neq pk_{text}$ and representations (is, s) of m' and (is^*, s^*) of m^* with the same i. These output representations lie on the same line $\overline{I}$ with $I = (i, 1)$. By Assumption 1, this $\overline{I}$ and thus $h_1 = g_1^i$ also have to occur twice among the input representations. However, $\mathcal{RC}$ ensures that each account is only used once. This finishes the proof that such a successful attacker cannot exist.

5.4 Anonymity

Due to the properties of the underlying cash system the views of $\mathcal{RC}$ and $\mathcal{M}$ concerning the withdrawal of a coin and its verification are unlinkable.

We now consider the additional information in registration and fingerprinting: Obviously, sig_{coin} in registration is no problem because the only additional information it is based on ($\mathcal{B}$'s secret key) is not used in fingerprinting. Similarly, sig_{text} in fingerprinting is no problem: The only additional information it is based on is sk_{text}. However, an information-theoretic attacker can compute sk_{text} from pk_{text}, which is already transmitted in the coin, and we know that the coin-related data in the underlying cash system are not linkable.

The value emb in fingerprinting, which is based on the same i that is also used in fingerprinting, is hidden in bit commitments. The first commitment scheme used, based on discrete logarithms, is information-theoretically hiding and the second one, based on quadratic residues, is hiding under the Quadratic Residuosity Assumption. Moreover the zero-knowledge protocols used in fingerprinting do not leak information about emb and the embedding operation is assumed not to leak such information either.

Furthermore using one-time accounts implies that $\mathcal{B}$'s different purchases are unlinkable, even if a purchased data item has been redistributed and the information contained in it recovered.

6 Conclusion

We have presented an anonymous fingerprinting scheme where all protocols are explicit and fairly efficient. The complexity is lower than that of any known embedding procedure that might follow, so that the anonymity is currently not the bottleneck for implementation. A disadvantage of this scheme in comparison with the previous one based on general zero-knowledge techniques is that the buyer is needed to carry out a fair trial; we hope to find a scheme that combines explicitness and 2-party trials in the future.

Of course, embedding for any asymmetric scheme is still rather an expensive procedure because commitments on a significant fraction of the data are needed, and the data must be long because otherwise one could not hide enough bits in them, at least if one desires collusion tolerance.

Acknowledgments: We thank Matthias Schunter and Michael Waidner for generous and fruitful discussions.

References

[BCC88]   Gilles Brassard, David Chaum, Claude Crépeau: Minimum Disclosure Proofs of Knowledge; Journal of Computer and System Sciences 37 (1988) 156-189.

[BCP88]   Jurjen Bos, David Chaum, George Purdy: A Voting Scheme; unpublished manuscript, presented at the rump session of Crypto'88.

[Ben87]   Josh Cohen Benaloh: Cryptographic Capsules: A Disjunctive Primitive for Interactive Protocols; Crypto'86, LNCS 263, Springer-Verlag, Berlin 1987, 213-222.

[BKK90]   Joan F. Boyar, Stuart A. Kurtz, Mark W. Krentel: A Discrete Logarithm Implementation of Perfect Zero-Knowledge Blobs; Journal of Cryptology 2/2 (1990) 63-76.

[BM97]   Ingrid Biehl, Bernd Meyer: Protocols for Collusion-Secure Asymmetric Fingerprinting; STACS 97, LNCS 1200, Springer-Verlag, Berlin 1997, 399-412.

[BMP86]   G. R. Blakley, Catherine Meadows, George B. Purdy: Fingerprinting Long Forgiving Messages; Crypto'85, LNCS 218, Springer-Verlag, Berlin 1986, 180-189.

[Bra94]   Stefan Brands: Untraceable Off-line Cash in Wallet with Observers; Crypto'93, LNCS 773, Springer-Verlag, Berlin 1994, 302-318.

[BS95]   Dan Boneh, James Shaw: Collusion-Secure Fingerprinting for Digital Data; Crypto'95, LNCS 963, Springer-Verlag, Berlin 1995, 452-465.

[CEG88]   David Chaum, Jan-Hendrik Evertse, Jeroen van de Graaf: An Improved Protocol for Demonstrating Possession of Discrete Logarithms and some Generalizations; Eurocrypt'87, LNCS 304, Springer-Verlag, Berlin 1988, 127-141.

[CFN94]   Benny Chor, Amos Fiat, Moni Naor: Tracing Traitors; Crypto'94, LNCS 839, Springer-Verlag, Berlin 1994, 257-270.

[CHP92]   David Chaum, Eugène van Heijst, Birgit Pfitzmann: Cryptographically Strong Undeniable Signatures, Unconditionally Secure for the Signer; Crypto'91, LNCS 576, Springer-Verlag, Berlin 1992, 470-484.

[CP93]   David Chaum, Torben Pryds Pedersen: Wallet Databases with Observers; Crypto'92, LNCS 740, Springer-Verlag, Berlin 1993, 89-105.

[Dom98]   Josep Domingo-Ferrer: Anonymous Fingerprinting of Electronic Information with Automatic Identification of Redistributors; Electronics Letters 34/13 (1998) 1303-1304.

[FN94]   Amos Fiat, Moni Naor: Broadcast Encryption; Crypto'93, LNCS 773, Springer-Verlag, Berlin 1994, 480-491.

[FTY96]   Yair Frankel, Yiannis Tsiounis, Moti Yung: "Indirect Discourse Proofs": Achieving Efficient Fair Off-Line E-cash; Asiacrypt'96, LNCS 1163, Springer-Verlag, Berlin 1997, 287-300.

[FY93]   Matthew Franklin, Moti Yung: Secure and Efficient Off-Line Digital Money; 20th International Colloquium on Automata, Languages and Programming (ICALP), LNCS 700, Springer-Verlag, Berlin 1993, 265-276.

[GM84]   Shafi Goldwasser, Silvio Micali: Probabilistic Encryption; Journal of Computer and System Sciences 28 (1984) 270-299.

[KD98]   Kaoru Kurosawa, Yvo Desmedt: Optimum Traitor Tracing and Asymmetric Schemes; Eurocrypt'98, LNCS 1403, Springer-Verlag, Berlin 1998, 145-157.

[NP98]   Moni Naor, Benny Pinkas: Threshold Traitor Tracing; Crypto'98, LNCS 1462, Springer-Verlag, Berlin 1998, 502-517.

[Ped92]   Torben Pryds Pedersen: Non-Interactive and Information-Theoretic Secure Verifiable Secret Sharing; Crypto'91, LNCS 576, Springer-Verlag, Berlin 1992, 129-140.

[Pfi96]   Birgit Pfitzmann: Trials of Traced Traitors; Information Hiding, LNCS 1174, Springer-Verlag, Berlin 1996, 49-64.

[PS96]   Birgit Pfitzmann, Matthias Schunter: Asymmetric Fingerprinting; Eurocrypt'96, LNCS 1070, Springer-Verlag, Berlin 1996, 84-95.

[PW97a]   Birgit Pfitzmann, Michael Waidner: Asymmetric Fingerprinting for Larger Collusions; 4th ACM Conference on Computer and Communications Security, acm press, New York 1997, 151-160.

[PW97b]   Birgit Pfitzmann, Michael Waidner: Anonymous Fingerprinting; Eurocrypt'97, LNCS 1233, Springer-Verlag, Berlin 1997, 88-102.

[Sch91]   Claus-Peter Schnorr: Efficient Signature Generation by Smart Cards; Journal of Cryptology 4/3 (1991) 161-174.

A   Linking a Self-Made Coin to a Correct Withdrawal

In this appendix, we show the attack that motivates the last verification step in the trial, as mentioned in Section 4.4.

Let h_1 be an account number of $\mathcal{B}$ and $coin^* = (m', pk^*_{text}, \sigma')$ (with $\sigma' = (z', a', b', r')$) be the coin that $\mathcal{B}$ withdrew from this account and used in a purchase with the text $text^*$. Note that $a' \equiv a^u g^v = g^{uw+v} \bmod p$ and $b' \equiv b^{su} m'^v = m'^{wsu} m'^v = m'^{wu+v} \bmod p$.

Assume now that the attackers are a collusion of $\mathcal{M}$ and $\mathcal{RC}$ and have access to the data item that $\mathcal{B}$ has bought in this purchase. Then they know $emb = (is, s)$, w, x, $view_{withdraw} = (a, b, c)$, r, and the additional signature sig^*_{coin} that $\mathcal{B}$ would give to fix $view_{withdraw}$ before receiving r.

As sig^*_{coin} fixes the values (a, b, c) that can be used together with the given h_1, the attackers want to link a self-made coin $coin'_A$, in particular with a self-made pk^A_{text} for which the attackers know sk^A_{text}, to these values. For this, they randomly select $w_A \in_R \mathbb{Z}_q$ and compute $a'_A \equiv g^{w_A} \bmod p$ and $b'_A \equiv m'^{w_A} \bmod p$. Then they compute $c'_A = hash(m', z', a'_A, b'_A, pk^A_{text})$. Since c is fixed they first compute $u_A \equiv c'_A/c \bmod q$ and then $v_A \equiv w_A - w u_A$ and $r'_A \equiv r u_A + v_A \bmod q$. In this manner they obtain values which pass all verifications by $\mathcal{A}$.

Finally, the attackers can use sk^A_{text} to sign any $text$ they like with respect to this coin.

On the Performance of Hyperelliptic Cryptosystems

Nigel P. Smart

Hewlett-Packard Laboratories, Filton Road, Stoke Gifford, Bristol, BS12 6QZ, U.K.
nsma@hplb.hpl.hp.com

Abstract. In this paper we discuss various aspects of cryptosystems based on hyperelliptic curves. In particular we cover the implementation of the group law on such curves and how to generate suitable curves for use in cryptography. This paper presents a practical comparison between the performance of elliptic curve based digital signature schemes and schemes based on hyperelliptic curves. We conclude that, at present, hyperelliptic curves offer no performance advantage over elliptic curves.

Elliptic curve cryptosystems are now being deployed in the real world and there has been much work in recent years on their implementation. A natural generalization of such schemes was given by Koblitz [12], who described how the group law on a Jacobian of a hyperelliptic curve can be used to define a cryptographic system. Almost all of the standard discrete logarithm based protocols such as DSA and ElGamal have elliptic and hyperelliptic variants. This is because such protocols only require the presence of a finite abelian group, with a large prime order subgroup, within which the basic group operation is easy whilst the associated discrete logarithm problem is hard. We shall not discuss these protocols in this paper since everything that can be said for elliptic curve based protocols can usually be said for hyperelliptic curve based protocols. Instead we shall concentrate more on the underlying group: In particular how one performs the group operation and how one produces groups of the required type.

The Jacobian of a genus g hyperelliptic curve will have roughly q^g points on it, where q denotes the number of elements in the field of definition of the Jacobian. By choosing hyperelliptic curves of genus greater than one we can achieve the same order of magnitude of the group order with a smaller value for q when compared with elliptic curve based systems which have $g = 1$. This has led some people to suggest that hyperelliptic curves may offer some advantages over elliptic curves in some special situations. For example if we wanted to only perform arithmetic using single words on a 32-bit computer we could choose $g = 5$ or 6 to obtain group orders of around 160 to 192 bits.

One has to be a little careful as to how large one makes g, since for large genus there is a sub-exponential method to solve the discrete logarithm problem [1]. However this does not appear to affect the security of curves of genus less than 10 over field sizes of around 32 bits.

J. Stern (Ed.): EUROCRYPT'99, LNCS 1592, pp. 165–175, 1999.
© Springer-Verlag Berlin Heidelberg 1999

In this paper we give an overview of the group law on a curve of genus g in arbitrary characteristic. We shall give a more efficient reduction method than the standard method of Cantor [3]. This is an immediate extension of the method of Tenner reduction from [19]. We shall then describe various techniques for generating hyperelliptic curves for use in cryptography.

Finally we report on an actual implementation of a hyperelliptic digital signature algorithm. We will conclude that hyperelliptic systems, with current algorithms, are more efficient in characteristic two but appear to offer no practical advantage over elliptic curve systems.

1   Arithmetic

In this section we summarize the details and leave the reader to consult [12] for a fuller explanation. A hyperelliptic curve, C, of genus g will be given in the form

$$C : Y^2 + H(X)Y = F(X)$$

where $F(X)$ is a monic polynomial of degree $2g + 1$ and $H(X)$ is a polynomial of degree at most g. Both $H(X)$ and $F(X)$ have coefficients in $\mathbb{F}_q$. Such a curve is non-singular if for no point on $C(\bar{\mathbb{F}}_q)$ does there exist a point for which the two partial derivatives,

$$2Y + H(X) \text{ and } H'(X)Y - F'(X),$$

simultaneously vanish. We shall always assume that the curve C is non-singular.

In odd characteristic fields we will always assume that $H(X) = 0$, whilst in even characteristic fields we will assume that $H(X) = 1$, for reasons which will become clear later. Notice that if $H(X) = 1$ then in characteristic two any choice for the polynomial $F(X)$ will give rise to a non-singular curve.

The above representation gives rise to a so called 'imaginary' quadratic function field. It is given this name since there are no units of infinite order and the arithmetic in the Jacobian closely mirrors the arithmetic one uses for the class group of an imaginary quadratic number field.

We can also define a hyperelliptic curve of genus g to be given by an equation, like that above but, with $\deg F = 2g + 2$. This gives rise to a 'real' quadratic function field. It is easy to see that, unlike the number field situation, an imaginary quadratic function field can be viewed as a real quadratic function field after making a change of variables. However, just as in the case of the class group of real quadratic number fields, the arithmetic in the Jacobians of real quadratic hyperelliptic curves is more involved and requires the use of 'infrastructure'. The reader should consult [18] for an explanation of the algorithms required and [19] for a complexity analysis of the two situations. For the rest of this article we will concentrate on the imaginary quadratic representation, which is more suited to efficient implementations in practice.

Following Cantor and Koblitz, an element of the Jacobian of C will be given by two polynomials $a, b \in \mathbb{F}_q[x]$ which satisfy

i) $\deg b < \deg a \leq g$.
ii) b is a solution of the equation $b^2 + Hb - F \pmod{a}$.

Addition in the Jacobian is accomplished by two procedures: Composition and Reduction. Given (a_1, b_1) and (a_2, b_2) the composition of these two elements in the group of divisors is given by (a_3, b_3) using the following algorithm due to Cantor and Koblitz:

Composition

1. **Perform two extended gcd computations to compute**
$$d = \gcd(a_1, a_2, b_1 + b_2 + H) = s_1 a_1 + s_2 a_2 + s_3(b_1 + b_2 + H).$$
2. **Set** $a_3 = a_1 a_2 / d^2$ **and**
3. $b_3 = (s_1 a_1 b_2 + s_2 a_2 b_1 + s_3(b_1 b_2 + F))/d \pmod{a_3}$.

Note that a_3 will have degree at most $2g$ and hence (a_3, b_3) will most probably need to be reduced. We shall return to this later. Notice, however, that for cryptography the most important composition step is doubling, where $a_1 = a_2$ and $b_1 = b_2$. This is because in discrete logarithm based systems we wish to perform a multiplication operation on the Jacobian. Using window techniques this involves mainly the doubling of elements rather than a general composition. Hence it is important that doubling an element can be accomplished efficiently.

With our above choice of curves in odd and even characteristic we find:

Doubling in Odd Characteristic Fields

Since we have chosen $H(X) = 0$ the doubling operation simplifies to: Put $d = \gcd(a_1, 2b_1) = s_1 a_1 + s_3(2b_1)$ then $a_3 = (a_1/d)^2$ and $b_3 = (2s_1 a_1 b_1 + s_3(b_1^2 + F))/d$.

Doubling in Even Characteristic Fields

Now since we have $H(X) = 1$ the doubling operation simplifies to: Put $a_3 = a_1^2$ and $b_3 = b_1^2 + F \pmod{a_3}$. This is much simpler than the odd characteristic step and contributes to much faster times for the verifying of messages using curves over even characteristic fields, see below for details.

We shall now describe the reduction step, which given the result (a_3, b_3) of a composition will return an element, (a, b), of the Jacobian with $\deg a \leq g$. The element (a_3, b_3) represents an element in the group of divisors. Since we are in an imaginary quadratic situation every divisor class (and so every element in the Jacobian) can be represented by a unique, so called reduced, divisor. The reduction step takes the divisor represented by (a_3, b_3) and returns the unique reduced divisor (a, b) in the same divisor class as (a_3, b_3). As mentioned above we use a variant of Tenner reduction which is more efficient than the method given by Cantor and Koblitz.

Reduction

---

1. $a = (b_3^2 + b_3 H - F)/a_3$.
2. $(u, b) = \texttt{quo/rem}(-b_3 - H, a)$.
3. While $\deg a > g$
4. $a^* = a_3 + u(b_3 - b)$.
5. $a_3 = a, a = a^*, b_3 = b$.
6. $(u, b) = \texttt{quo/rem}(-b_3 - H, a)$.

This is exactly the same as the standard method except for Step 4. In this step we have replaced the division $a^* = (b^2 + Hb - F)/a$ with simpler operations, on noticing that u in general will have small degree whilst $\deg a$ in Step 4 could be at most $2g - 2$. To see that Step 4 is equivalent to the standard method we notice that $u = (-b_3 - H - b)/a$ and so

$$a^* = a_3 + (b_3 - b)\left(\frac{-b_3 - H - b}{a}\right)$$
$$= (b^2 + Hb - F)/a.$$

In [6] the extended Euclidean algorithm is analyzed in the context of hyperelliptic cryptosystems. As we have already pointed out for even characteristic fields for the most important operation, point doubling, no extended Euclidean algorithm is required. Most of the effort in performing a sign or verify operation is in the reduction step. Hence analyzing the reduction step is far more important, luckily this has already been done in [19], where it is shown that the above reduction step takes $12g^2 + O(g)$ field operations, in [18] the standard method is stated to take $3g^3 + O(g^2)$ field operations. However, a complexity analysis can often be inappropriate since complexity only deals with the assymptotics of an algorithm. In real life the relative performance of algorithms in small ranges can depend on factors such as cache size and processor type.

2 Curve Generation

There are many ways, in theory, that one could proceed if one wanted to produce curves suitable for use in cryptography. Many of the methods are analogues of those used in the elliptic curve case. The order of $|J(\mathbb{F}_q)|$ can be computed in polynomial time using methods due to Adleman, Huang and Pila, see [2] and [20], which are themselves generalizations of the method of Schoof [25] used in the elliptic case. There is no implementation of this method for genus greater than one at the present time. This is probably because the algorithm, although easy to understand, appears very hard to implement. Another reason is that there is no known analogue of the improvements made by Atkins and Elkies to the original Schoof algorithm. Hence only the 'naive' Schoof algorithm is available in genus greater than one. Such an algorithm appears hopeless as a method, since the 'naive' Schoof algorithm is far too inefficient even for elliptic curves.

The fact that it seems unlikely that anyone can compute the order of $J(\mathbb{F}_q)$ for a general curve of genus 5 or 6 could lead one to propose that one should not worry. For example, if I do not believe that someone can compute the order of $J(\mathbb{F}_q)$ then I do not need to worry about many of the attacks on such systems, since most attacks such as Pohlig-Hellman require knowledge of the group order. This of course also means that our protocols need to be changed so that they do not require knowledge of the group order. Although this is a possible approach, it is to be rejected as it is assumes that someone will not make a known polynomial time algorithm run efficiently. Our security is therefore not built on the difficulty of some underlying mathematical problem but on the difficulty of programming a known algorithm efficiently.

Just as for elliptic curves one can compute hyperelliptic curves using the theory of Complex Multiplication (CM). This has been worked out in detail for the case of $g = 2$ in [30] and uses the class groups of complex quadratic extensions of real quadratic number fields, which are the quartic CM fields. Clearly the class numbers of any such field used should be small, and hence the curves which are produced will in some sense be 'special'. In the CM method for hyperelliptic curves multi-variable analogues of the Hilbert polynomial are constructed, the roots of which modulo p gives the j-invariants of the curve. The curve is then recovered from its j-invariants.

This method is only currently effective in genus two since the j-invariants of a hyperelliptic curve have only been worked out for genus less than three. The invariants used are the Igusa Invariants [11] which are linked to the classical 19th Century invariants of quintic and sextic polynomials. After the demise of classical invariant theory at the end of the 19th Century the drive to compute invariants of the higher order quantics, as they were then called, died out. Even today with the advent of computer algebra systems this seems a daunting task. One way around this problem, which still uses CM, is to use reductions of hyperelliptic curves defined over $\mathbb{Q}$ which have global complex multiplication, see [4]. However, here one is restricting to an even more special type of hyperelliptic curve than the general CM method above.

Another technique is to use the theory of the modular curves, $X_0(N)$, see [8] and [15]. Such curves are well studied and much is known about them. This enables us to compute the orders of the Jacobians of such curves in a much easier way than other general curves. However, paranoid readers should beware since they are well understood curves with special properties they may be susceptible to some new attack which makes use of the fact that they are modular.

Koblitz, in [13], suggests using curves of the form

$$v^2 + v = u^n,$$

over some finite prime field $\mathbb{F}_p$. Given such curves he then gives a procedure to determine the group order by evaluating a Jacobi sum of a certain character. We refer the reader to Koblitz's book for details. However once again we are restricting to a very special type of curve which may be susceptible to some, as yet unknown, attack.

In characteristic two one can use curves defined over subfields [12] just as one can do for elliptic curves. For example a simple search found the curves in Table 1, which all have subgroups of their Jacobians of 'large' prime order; We could also use such a technique to generate curves over $\mathbb{F}_p$, where p is a small odd prime and look at the Jacobian over $\mathbb{F}_{p^n}$.

Table 1. Curves of the form $Y^2 + Y = F(X)$

$\mathbb{F}_q$	$F(X)$	$\log_2 p$ where $p \mid \#J(\mathbb{F}_q)$
$\mathbb{F}_{2^{31}}$	$X^{11} + X^5 + 1$	150
$\mathbb{F}_{2^{29}}$	$X^{13} + X^{11} + X^3 + X$	157
$\mathbb{F}_{2^{29}}$	$X^{13} + X^{11} + X^7 + X + 1$	153
$\mathbb{F}_{2^{29}}$	$X^{13} + X^{11} + X^7 + X^3 + 1$	169
$\mathbb{F}_{2^{29}}$	$X^{13} + X^{11} + X^9 + X^5 + 1$	170
$\mathbb{F}_{2^{29}}$	$X^{13} + X^{11} + X^9 + X^7 + X^3 + X + 1$	152
$\mathbb{F}_{2^{31}}$	$X^{13} + X^{11} + X^7 + X^3 + X$	162
$\mathbb{F}_{2^{31}}$	$X^{13} + X^{11} + X^9 + X + 1$	154
$\mathbb{F}_{2^{31}}$	$X^{13} + X^{11} + X^9 + X^5$	158
$\mathbb{F}_{2^{31}}$	$X^{13} + X^{11} + X^9 + X^7$	178
$\mathbb{F}_{2^{31}}$	$X^{13} + X^{11} + X^9 + X^7 + X^3 + X + 1$	181
$\mathbb{F}_{2^{31}}$	$X^{15} + X$	207
$\mathbb{F}_{2^{31}}$	$X^{15} + X^5 + X^3 + X$	200

Apart from the, currently unimplemented, method of Schoof, Pila et al the above methods do not seem very pleasing. It is a good general principle never to choose a curve with 'special structure', and all of the above schemes use 'special' properties of the curves to make the group order computation easier.

To see why one should avoid special curves one only has to look at the history of elliptic curve cryptography. In the past various authors proposed using supersingular or anomalous curves as they offered some advantages over other more general curves. However, both types of curves are now known to be weak, see [14], [24], [26] and [27]. Hence it is probably worth adopting the principle of always avoiding special curves of any shape or form. In the current authors opinion this is the major open problem with using hyperelliptic curves for cryptographic purposes: How to choose a suitable curve efficiently ?

3   The Discrete Logarithm Problem in Hyperelliptic Jacobians

The security of hyperelliptic cryptosystems is based upon the difficulty of solving the discrete logarithm problem in the Jacobian of the curve. We summarize the main characteristics of the possible attacks on the hyperelliptic discrete logarithm problem below. The reader should note that in all but one case they closely mirror analogues for the elliptic curve discrete logarithm problem.

Apart from the generic discrete logarithm algorithms such as the baby-step / giant-step and the rho/kangaroo method there are three known methods which are specific to hyperelliptic curves. Two of these give rise to two weak classes of hyperelliptic curve cryptosystems:

1. Curves of order n over $\mathbb{F}_q$ such that $q^l \equiv 1 \pmod{n}$ for some small value of n. This is due to a generalization of the method of Menezes et al [14] for supersingular elliptic curves due to Frey and Rück [9].
2. Anomalous curves over $\mathbb{F}_p$ and in general curves which have a large subgroup of order p in a field of characteristic p. This attack uses a generalization due to Rück [21] of the anomalous curve attack for elliptic curves due to Semaev, Satoh, Araki and Smart, see [24], [26] and [27].

However, such cases are easy to check for and only eliminate a small fraction of all possible curves.

For hyperelliptic curves the most interesting case, from a theoretical standpoint, is when the genus is large in comparison to the size of the field of definition of the Jacobian. In this case there are conjectured subexponential methods. The first of these was due to Adleman, De Marrais and Huang which is based on the number field sieve factoring method.

Paulus [17] and Flassenberg and Paulus [7] have implemented such a method for solving discrete logarithms in Jacobians of hyperelliptic curves. Flassenberg and Paulus did not, however, use the method of Adleman, De Marrais and Huang directly. Instead they made use of the fact that our hyperelliptic curves correspond to real quadratic function field extensions. Using the analogy between quadratic function fields and quadratic number fields, Flassenberg and Paulus adapt the class group method of Hafner and McCurley [10] (see also [5]). Then combining this with a sieving method they obtain a working method which can be applied to hyperelliptic curves of relatively small genus. It should be pointed out that although Flassenberg and Paulus do not actually solve discrete logarithm problems their methods are such that they can be easily extended so that they do.

Flassenberg and Paulus compared their algorithm to the baby-step / giant-step approach. Over finite prime fields, $\mathbb{F}_p$, their implementation of the Hafner-McCurley method beat the baby-step / giant-step method, as soon as $3g > \log p$. However, this is only given a very small sample size. But it would appear, for theoretical reasons as well, to be a good rule of thumb to avoid curves for which $2g > \log q$. Hence if $q \approx \mathbb{F}_{2^{31}}$ then we should avoid curves whose genus is larger than eleven.

4   Implementation

In [22] the number of bit operations for implementing a hyperelliptic cryptosystem is studied and compared with both ECC and RSA systems which offer roughly the same level of security. It is concluded that hyperelliptic cryptosystem could be efficient enough in practice to use in real life situations. Following on from this work in [23] an implementation of such a system is described. However this implementation makes no use of Tenner reduction and generally uses field sizes which require more than a single word to represent each field element.

We decided to implement the group law in the Jacobian for curves of arbitrary genus over $\mathbb{F}_{2^n}$ and $\mathbb{F}_p$, where p is a prime. We decided to choose values of p and n such that p and 2^n are less than 2^{32}. This choice was to make sure that our basic arithmetic could all be fitted into single words on our computer. Such curves and fields have attracted some interest in the community in recent years since they may offer some implementation advantages. In even characteristic we used a trinomial basis while in odd characteristic we used a small in-lined machine code subroutine to perform the modular multiplication. Field inversion in both cases was carried out using a modification of the binary method.

The general multiplication algorithm on the Jacobian for curves defined over odd characteristic fields ended up being around twice as slow as that for even characteristic fields, of an equivalent size, in genus two. In genus five the odd characteristic fields were nearly three times slower. This fact led us to only implement a full digital signature scheme in characteristic two.

For the signing operation the multiplication performed is on the fixed group generator. Hence this can be efficiently accomplished using a precomputed table of powers of the generator. The verification step requires two multiplications, one of the generator and one of a general point. Hence for verification we cannot use precomputed tables and the difficulty of doubling an element will dominate the computation. For the general multiplication, used in the verification step, we used a signed window method, since negation in the Jacobian of a hyperelliptic curve comes virtually for free.

Our timings, in milliseconds, for a hyperelliptic variant of the DSA method (HCDSA) are given in Table 2. These timings were obtained on a Pentium Pro 334MHz, running Windows NT, using the Microsoft Visual C++ compiler. We also give an estimate of the timings for an elliptic curve (ECDSA) system with approximately the same group order.

The elliptic curve implementation made no use of special field representations, such as using the subfield structure. The even characteristic field representation for the elliptic curve system was a standard polynomial basis. The odd characteristic field (of size approximately 2^{161}) used for the elliptic curve system used a Montgomery representation.

So we see that even though the finite field elements fit into a single word the extra cost of the polynomial arithmetic needed for operations in the Jacobian makes the time needed to perform the complete set of hyperelliptic curve operations over four times slower than in the elliptic curve case. If more efficient

Table 2. HCDSA and ECDSA Timings in Milliseconds

Curve	Field	Sign	Verify
HCDSA $g = 5$	$\mathbb{F}_{2^{31}}$	18	71
HCDSA $g = 6$	$\mathbb{F}_{2^{31}}$	26	98
HCDSA $g = 7$	$\mathbb{F}_{2^{31}}$	40	156
ECDSA	$\mathbb{F}_{2^{161}}$	4	19
ECDSA	$\mathbb{F}_p$	3	17

elliptic curve techniques were used then the relative performance of the HCDSA algorithm would degrade even more.

Given the relative difficulty of finding hyperelliptic curves for use in cryptography which do not possess some addition structure and the relatively poor performance of the HCDSA algorithm when compared to ECDSA there seems no benefit in using hyperelliptic curves.

Of course further work could result in significant speed improvements for hyperelliptic systems. For example at present there appears to be no notion akin to the projective representation in elliptic curves. Another possible avenue for improvement is to use Frobenius expansions. Not as much work has been carried out in the hyperelliptic case to the study of Frobenius expansions compared to the elliptic curve case. These are useful for curves defined over small subfields, such as those used above. The only cases having been considered in the hyperelliptic case are in [12]. However, for elliptic curves Frobenius expansions techniques can be made very fast in all characteristics, see [16], [28] and [29].

References

1. L. Adleman, J. De Marrais, and M.-D. Huang. A subexponential algorithm for discrete logarithms over the rational subgroup of the Jacobians of large genus hyperelliptic curves over finite fields. In *ANTS-1 : Algorithmic Number Theory*, Editors L.M. Adleman and M-D. Huang, Springer-Verlag, LNCS 877, pp 28–40, 1994.
2. L. Adleman and M.-D. Huang. Counting rational points on curves and abelian varieties over finite fields. In *ANTS-2 : Algorithmic Number Theory*, Editor H. Cohen, Springer-Verlag, LNCS 1122, pp 1–16, 1996.
3. D.G. Cantor. Computing in the Jacobian of a hyper-elliptic curve. *Math. Comp.*, **48**, 95–101, 1987.
4. J. Chao, N. Matsuda and S. Tsujii. Efficient construction of secure hyperelliptic discrete logarithms. In *Information and Communications Security*, Editors Y. Han, T. Okamoto and S. Quing, Springer-Verlag, LNCS 1334, pp 292–301, 1997.
5. H. Cohen. A Course In Computational Algebraic Number Theory. Springer-Verlag, GTM 138, 1993.
6. A. Enge. The extended Euclidean algorithm on polynomials, and the efficiency of hyperelliptic cryptosystems. Preprint, 1998.
7. R. Flassenberg and S. Paulus. Sieving in function fields. *Preprint*, 1997.

8. G. Frey and M. Müller. Arithmetic of modular curves and its applications. Preprint, 1998.

9. G. Frey and H.-G. Rück. A remark concerning m-divisibility and the discrete logarithm problem in the divisor class group of curves. *Math. Comp.*, **62**, 865–874, 1994.

10. J.L. Hafner and K.S. McCurley. A rigorous subexponential algorithm for computation of class groups. *J. AMS*, **2**, 837–850, 1989.

11. J.I. Igusa. Arithmetic variety of moduli for genus two. *Ann. Math.*, **72**, 612–649, 1960.

12. N. Koblitz. Hyperelliptic cryptosystems. *J. of Crypto.*, **1**, 139–150, 1989.

13. N. Koblitz, Algebraic aspects of cryptography. *Vol. 3, Algorithms and Computation in Mathematics*, Springer-Verlag, Berlin, 1998.

14. A. Menezes, T. Okamoto and S. Vanstone. Reducing elliptic curve logarithms to a finite field. *IEEE Trans. on Inform. Theory*, **39**, 1639–1646, 1993.

15. M. Müller. Algorithmische Konstruktion hyperelliptischer Kurven mit kryptographischer Relevanz und einem Endomorphismenring echt grösser als $\mathbb{Z}$. *Phd Thesis*, Universität Essen, 1996.

16. V. Müller. Fast multiplication on elliptic curves over small fields of characteristic two. *J. Crypto.*, **11**, 219–234, 1998.

17. S. Paulus. An algorithm of sub-exponential type computing the class group of quadratic orders over principal ideal domains. In *ANTS-2 : Algorithmic Number Theory*. Editor H. Cohen, Springer-Verlag, LNCS 1122, pp 243–257, 1996.

18. S. Paulus and H.-G. Rück. Real and imaginary quadratic representation of hyperelliptic function fields. To appear *Math. Comp.*.

19. S. Paulus and A. Stein. Comparing real and imaginary arithmetics for divisor class groups of hyperelliptic curves. In *ANTS-3 : Algorithmic Number Theory*, Editor J. Buhler, Springer-Verlag, LNCS 1423, pp 576–591, 1998.

20. J. Pila. Frobenius maps of abelian varieties and finding roots of unity in finite fields. *Math. Comp.*, **55** , 745–763, 1996.

21. H.-G. Rück. On the discrete logarithm problem in the divisor class group of curves. Preprint 1997.

22. Y. Sakai, K. Sakurai and H. Ishizuka. Secure hyperelliptic cryptosystems and their performance. In *Public Key Cryptography*, Editors H. Imai and Y. Zheng, Springer-Verlag, LNCS 1431, pp 164–181, 1998.

23. Y. Sakai and K. Sakurai. Design of hyperelliptic cryptosystems in small characteristic and a software implementation over $\mathbb{F}_{2^n}$. In *Advances in Cryptology, ASIACRYPT 98*, Editors K. Ohta and D. Pei, Springer-Verlag, LNCS 1514, pp 80–94, 1998.

24. T. Satoh and K. Araki. Fermat quotients and the polynomial time discrete log algorithm for anomalous elliptic curves. *Comm. Math. Univ. Sancti Pauli*, **47**, 81–92, 1998.

25. R. Schoof. Elliptic curves over finite fields and the computation of square roots mod p. *Math. Comp.*, **44**, 483–494, 1985.

26. I.A. Semaev. Evaluation of discrete logarithms on some elliptic curves. *Math. Comp.*, **67**, 353–356, 1998.

27. N.P. Smart. The discrete logarithm problem on elliptic curves of trace one. To appear *J. Crypto.*, 1999.

28. N.P. Smart. Elliptic curves over small fields of odd characteristic. To appear *J. Crypto.*, 1999.

29. J.A. Solinas. An improved algorithm for arithmetic on a family of elliptic curves. In *Advances in Cryptology, CRYPTO 97*, Editor B. Kaliski, Springer Verlag, LNCS 1294, pp 357-371, 1997.

30. A-M. Spallek. Kurven vom Geschlecht 2 und ihre Anwendung in Public-Key-Kryptosytemen. *Phd Thesis*, Universität Essen, 1994.

Fast Elliptic Curve Algorithm Combining Frobenius Map and Table Reference to Adapt to Higher Characteristic

Tetsutaro Kobayashi, Hikaru Morita, Kunio Kobayashi, and Fumitaka Hoshino

NTT Laboratories
Nippon Telegraph and Telephone Corporation
1-1 Hikari-no-oka, Yokosuka-shi, Kanagawa-ken, 239-0847 Japan
`kotetsu@isl.ntt.co.jp`

Abstract. A new elliptic curve scalar multiplication algorithm is proposed. The algorithm offers about twice the troughput of some conventional OEF-base algorithms because it combines the Frobenius map with the table reference method based on base-ϕ expansion. Furthermore, since this algorithm suits conventional computational units such as 16, 32 and 64 bits, its base field $\mathbf{F}_{p^m}$ is expected to enhance elliptic curve operation efficiency more than $\mathbf{F}_q$ (q is a prime) or $\mathbf{F}_{2^n}$.

Keywords: Elliptic curve cryptosystem, Scalar multiplication, OEF, Finite field, Frobenius map, Table reference method.

1 Introduction

While speeding up modular exponentiation has been a prime approach to speeding up the RSA scheme, scalar multiplication of an elliptic curve point can speed up elliptic curve schemes such as EC-DSA and EC-ElGamal. In particular, elliptic curves over $\mathbf{F}_q$ (q is a prime) or $\mathbf{F}_{2^n}$ have been implemented by many companies and standardized by several organizations such as IEEE P1363 and ISO/IEC JTC1/SC27.

For the $\mathbf{F}_{2^n}$ type, many efficient computational algorithms have been proposed. Koblitz introduced a base-ϕ expansion method that uses a Frobenius map to multiply $\mathbf{F}_{2^n}$-rational points over the elliptic curve defined over $\mathbf{F}_2, \mathbf{F}_4, \mathbf{F}_8$ or $\mathbf{F}_{16}$ in [2]. Müller [7] and Cheon et. al. [5] extended the base-ϕ expansion method to elliptic curves defined over $\mathbf{F}_{2^r}$, where r is a small integer. Koblitz also expanded the base-ϕ expansion method to $\mathbf{F}_3, \mathbf{F}_7$ in [3].

However, since the calculation over small characteristic fields does not offer adequate speed on general purpose machines, very high-capacity tables or special-purpose machines are needed. If you select $\lceil \log_2 p \rceil$ (the bit size of a prime number p) to match the operation unit of an individual computer, the scalar multiplication of $\mathbf{F}_{p^m}$ could be calculated faster than that of $\mathbf{F}_q$ or $\mathbf{F}_{2^n}$ where $\lceil \log_2 p^m \rceil$ should be close to $\lceil \log_2 q \rceil$ or $\lceil \log_2 2^n \rceil (= n)$ under the condition of the same security level. Bailey and Paar newly proposed an elliptic curve scheme on OEF (Optimal Extension Fields), or an $\mathbf{F}_{p^m}$ type, at Crypto'98 [1].

J. Stern (Ed.): EUROCRYPT'99, LNCS 1592, pp. 176–189, 1999.

Their method represents the elliptic curve points using a polynomial basis. They showed that multiplication as well as addition and subtraction can be efficiently computed by introducing a binomial as a minimal polynomial.

Though the original OEF method simply indicated how to compute addition and multiplication on $\mathbf{F}_{p^m}$, efficient computational techniques similar to those developed for the $\mathbf{F}_{2^n}$ type have not been introduced to the OEF world.

This paper extends the base-ϕ extension method from $\mathbf{F}_{2^n}$ to the general finite field $\mathbf{F}_{p^m}$ by using a table reference method. Several table reference methods have been developed for schemes using fixed primitive points – base points – such as the DSA scheme [6]. Ours is the first to combine the Frobenius map and the table reference method and so does not need any pre-computation. It can be applied to any higher-characteristic elliptic curve as well as the small-characteristic. When p equals two, this method is reduced to Koblitz's method. Different from Cheon's method, this method isn't limited to an elliptic curve defined on $\mathbf{F}_{2^r}$. The method works over OEF-type elliptic curves because the table reference method is effective even if p is large. If you select p close to 2^{16}, 2^{32} or 2^{64}, that are suitable operation units for computers, our method is about twice as fast as ordinary OEF methods.

Section 2 describes the idea of our proposed method. Its procedure is given in Sect. 3. Section 4 shows how to construct the proposed OEF parameters. Its efficiency and further techniques are given in Sects. 5 and 6. Section 7 concludes this paper.

2 Approach

2.1 Frobenius Map

In this section, we define the Frobenius map. Let $E/\mathbf{F}_p$ denote a non-supersingular elliptic curve defined over a finite field $\mathbf{F}_p$ where p is a prime or any power of a prime. $P = (x, y)$ is an $\mathbf{F}_{p^m}$-rational point of elliptic curve E defined over $\mathbf{F}_p$. The Frobenius map ϕ is defined as

$$\phi : (x, y) \rightarrow (x^p, y^p).$$

The Frobenius map is an endomorphism over $E(\mathbf{F}_{p^m})$. It satisfies the equation

$$\phi^2 - t\phi + p = 0, \qquad -2\sqrt{p} \le t \le 2\sqrt{p}. \tag{1}$$

Since E is non-supersingular, the endomorphism ring of E is an order of the imaginary quadratic field $\mathbf{Q}(\sqrt{t^2 - 4p})$ [8]. The ring $\mathbf{Z}[\phi]$ is a subring of the endomorphism ring.

To compute the Frobenius map ϕ takes negligible time, provided that element $a \in \mathbf{F}_{p^m}$ is represented using a normal basis of $\mathbf{F}_{p^m}$ over $\mathbf{F}_p$.

2.2   Normal Basis and Polynomial Basis

The elements of the field $\mathbf{F}_{p^m}$ can be represented in several different ways; for example, "polynomial basis" and "normal basis." In polynomial basis, element $a \in \mathbf{F}_{p^m}$ is represented as

$$a = a_{m-1}\alpha^{m-1} + \cdots + a_1\alpha + a_0. \tag{2}$$

where $a_i \in \mathbf{F}_p$ and α is a defining element of $\mathbf{F}_{p^m}$ over $\mathbf{F}_p$.

In normal basis, $a \in \mathbf{F}_{p^m}$ is represented as

$$a = a_{m-1}\alpha^{p^{m-1}} + \cdots + a_1\alpha^p + a_0\alpha \tag{3}$$

where $a_i \in \mathbf{F}_p$ and α is a generator of normal basis.

Addition and subtraction in $\mathbf{F}_{p^m}$ are quite fast in both representation forms. When you choose polynomial basis, multiplication and squaring can be done with reasonable speed.

When you choose normal basis, the p-th power operation, which is equal to the Frobenius map, is quite fast. Though multiplication isn't fast in general normal basis, there are several techniques for fast multiplication in $\mathbf{F}_{2^m}$ such as the optimal normal basis [9]. Thus, fast algorithms for scalar multiplication using the Frobenius map [2] have been developed using $\mathbf{F}_2$ or its extension field represented by normal basis.

On the other hand, we developed a fast Frobenius map algorithm for OEF [1] which has a special polynomial basis.

2.3   Frobenius Map for OEF

Let OEF be the finite field $\mathbf{F}_{p^m}$ that satisfies the following:

 - p is a prime less than but close to the word size of the processor,
 - $p = 2^n \pm c$, where $\log_2 c \leq n/2$ and
 - An irreducible binomial $f(x) = x^m - \omega$ exists.

Although the paper [1] showed that OEF has an efficient algorithm for multiplication and squaring, there was no discussion of the Frobenius map. In this section, we present a new algorithm to compute the Frobenius map in OEF.

We consider the following polynomial basis representation of an element $a \in \mathbf{F}_{p^m}$:

$$a = a_{m-1}\alpha^{m-1} + \cdots + a_1\alpha + a_0$$

where $a_i \in \mathbf{F}_p$, $\alpha \in \mathbf{F}_{p^m}$ is a root of $f(x)$. Since we choose $\lceil \log_2 p \rceil$ to be less than the processor's word size, we can represent a using m registers.

The Frobenius map moves a to a^p;

$$\phi(a) = a^p = a_{m-1}\alpha^{(m-1)p} + \cdots + a_1\alpha^p + a_0. \tag{4}$$

Since α is a root of $f(x) = 0$, $\alpha^m = \omega$,

$$\alpha^{ip} = \alpha^{(ip \bmod m)}\omega^{\lfloor ip/m \rfloor}$$

where $\lfloor x \rfloor$ is the maximum integer not exceeding x.

Assuming $\gcd(m, p) = 1$, $(i_1 p \bmod m) = (i_2 p \bmod m)$ is equivalent to $i_1 = i_2$. Thus, the map $\pi(i) \overset{\triangle}{=} ip \bmod m$ is bijective.

We rewrite Equation (4) using $\pi(i)$ as follows:

$$a^p = a'_{m-1}\alpha^{m-1} + \cdots + a'_1\alpha + a'_0,$$

$$\text{where } a'_{\pi(i)} \overset{\triangle}{=} a_i\omega^{\lfloor \frac{ip}{m} \rfloor}.$$

Since p, m and ω are independent of an element a, we can pre-compute $\omega_i = \omega^{\lfloor \frac{ip}{m} \rfloor}$ before computing the Frobenius map. Accordingly, the complete procedure to compute the Frobenius map to an element on OEF is as follows;

[Frobenius Map Procedure for OEF]

> **Input:** $[a_0, \dots, a_{m-1}]$ $(= a)$
> **Output:** $[a'_0, \dots, a'_{m-1}]$ $(= \phi(a))$
> **Step 1:** compute $b_i = a_i\omega_i$,   for $i = 1$ to $m - 1$.
> **Step 2:** compute $a'_{\pi(i)} = b_i$,   for $i = 1$ to $m - 1$.
> **Step 3:** $a'_0 = a_0$.

This procedure needs only $m - 1$ multiplications on $\mathbf{F}_p$. This takes negligible time compared to multiplication on $\mathbf{F}_{p^m}$, which needs m^2 multiplications [1] on $\mathbf{F}_p$.

2.4   Base-ϕ Scalar Multiplication Method

This section describes the basic idea of base-ϕ scalar multiplication given by Koblitz[2].

Consider scalar multiplication, kP where k and P represent a scalar multiplier and an elliptic curve point P, respectively. Consider $k = 15$ as an example. By using the binary method, $15P$ is calculated as $2(2(2P + P) + P) + P$ by three elliptic curve doublings and three elliptic curve additions. If you use the signed-binary method, $15P$ is calculated as $2(2(2(2P))) - P$ by four elliptic curve doublings and one elliptic curve subtraction. General computational times are given by Table 1 where $n = \lceil \log_2 p^m \rceil$.

Base-ϕ expansion is generally calculated as follows: If an intermediate multiplier k_i is represented by $k_i = x_i + y_i\phi$ where x_i and y_i are integers, $x_0 = k$ and $y_0 = 0$, the equation is modified to $k_i = u_i + k_{i+1}\phi$, where u_i is defined as an integer such that $u_i = x_i \pmod{p}$ and $-\frac{p}{2} < u_i \leq \frac{p}{2}$.

---

[1]   This is the straightforward method.

Table 1. Computational Times for Binary Method

		EC Doubling	EC Addition	Total
Binary	(maximum)	n	n	$2n$
	(avarage)	n	$\dfrac{n}{2}$	$\dfrac{3n}{2}$
Signed Binary	(maximum)	n	$\dfrac{n}{2}$	$\dfrac{3n}{2}$
	(avarage)	n	$\dfrac{3n}{8}$	$\dfrac{11n}{8}$

$k_{i+1} = x_{i+1} + y_{i+1}\phi$, $x_{i+1} = y_i + t\dfrac{x_i - u_i}{p}$ and $y_{i+1} = -\dfrac{x_i - u_i}{p}$ by using $\phi^2 - t\phi + p = 0$. Iterating this operation, k is expanded to

$$k = \sum_{i=0}^{l} u_i \phi^i, \qquad \text{where} \quad -\frac{p}{2} \leq u_i \leq \frac{p}{2}. \tag{5}$$

l is an integer and is discussed in Sect. 3.

In the case of $k = 15, p = 2$, the elliptic curve is $E/\mathbf{F}_2 : y^2 + xy = x^3 + 1$ (trace t is 1 as an example), $15 = -1 + \phi^4 - \phi^8$. Accordingly, $15P$ is calculated by two elliptic curve additions, which is much faster than the signed or unsigned binary method.

Koblitz[2] presented the scalar multiplication algorithm for $\mathbf{F}_{2^m}$-rational points over $E/\mathbf{F}_2$. Solinas[4] improved it. In those papers $u_i \in \{-1, 0, 1\}$. Thus it needs at most l elliptic curve additions and computation of the Frobenius map to calculate kP.

On the other hand, we must limit the elliptic curve defined over $E/\mathbf{F}_p$ for $\mathbf{F}_{p^m}$-rational points to utilize a base-ϕ scalar multiplication method. Since there is only an exponential time attack such as [10], it is not obstacle to use the elliptic curves for elliptic curve cryptosystems.

2.5   Generalized Base-ϕ Scalar Multiplication

We consider the fact that the cost of $\phi^i P$ can be reduced very much for OEF as shown at Section 2.3. Though traditional base-ϕ scalar multiplication has been applied to finite fields with small characteristics, it can be applied to more general cases such as OEF.

This, however, makes each coefficient u_i in Equation (6) large, because $0 \leq |u_i| \leq p/2$. When u_i is large, the delay time to calculate $u_i \phi^i P$ from $\phi^i P$ becomes a bottle neck. Thus, the traditional base-ϕ method is not always faster than the binary method. This is one reason why the base-ϕ scalar multiplication method was applied only to fields with small characteristics.

To solve this problem, we introduce the idea of the table reference scalar multiplication method. After each value of $\phi^i P$ is stored in a memory table, we should perform addition on the elliptic curve. There are two different ways to look up the table and add.

One method uses only addition. If $15P + 13\phi^2 P + 2\phi^3 P$ is to be calculated, $X \leftarrow P$ and $Y \leftarrow O$ then $Y \leftarrow Y + X$ is computed twice. $X \leftarrow X + \phi^2 P$ then $Y \leftarrow Y + X$ is computed eleven times. $X \leftarrow X + \phi^3 P$ then $Y \leftarrow Y + X$ is computed twice. Generally speaking, when you compute

$$k = \sum_{i=0}^{l} u_i \phi^i \tag{6}$$

where $-\dfrac{p}{2} \le u_i \le \dfrac{p}{2}$, the elliptic curve addition should be computed roughly as $l + \dfrac{p}{2}$. This idea of the original table reference method was created for the non-elliptic curve scheme by Brickell et al. [6]. By introducing the base-ϕ method, this method can be enhanced to handle any primitive. Thus, our method supports not only signature generation by EC-DSA but also verification which involves multiplication of unpredictable elliptic curve points.

The second method uses both doubling and addition. If $15P + 13\phi^2 P + 2\phi^3 P = (1111)_2 P + (1101)_2 \phi^2 P + (0010)_2 \phi^3 P$ is calculated, $X \leftarrow P + \phi^2 P$ is computed then doubled. $X \leftarrow X + P + \phi^2 P$ is computed then doubled. $X \leftarrow X + P + \phi^3 P$ is computed then doubled. Finally, $X \leftarrow X + P + \phi^2 P$ is computed. In general, when you compute $k = \sum_{i=0}^{l} u_i \phi^i$ where $-\dfrac{p}{2} \le u_i \le \dfrac{p}{2}$, the elliptic curve addition and doubling should be computed roughly $(l+1)(\lceil \log_2 p \rceil - 1)/2$ and $\lceil \log_2 p \rceil - 2$ times, respectively. If the reference table contains not only P and $\phi^i P$ but also their combinations such as $P + \phi^2 P$, $P + \phi^3 P$ and so on, the addition and doubling times could be reduced by at least $\lceil \log_2 p \rceil - 2$.

We found that there are some trade offs. Which of the two methods is better? How many values should the memory table store? It depends on the case.

3   Procedure

3.1   Base-ϕ Scalar Multiplication

In this section, we describe the base-ϕ scalar multiplication method. The following procedure computes $Q = kP$ for inputs P and k, where $0 < k < N_m$ and P is an $\mathbf{F}_{p^m}$-rational point and is not an $\mathbf{F}_p$-rational point on E. We use $w_H(x)$ to represent the Hamming weight of x expressed in signed binary digit, N_m denotes the number of $\mathbf{F}_{p^m}$-points on E, and t denotes a trace of E.

[Base-ϕ Scalar Multiplication Procedure]

Input: k, P, E, t, p
Output: $Q \quad (= kP)$
Step 1: Base-ϕ Expansion of k
  Step 1-1: $i \leftarrow 0, \quad x \leftarrow k, \quad y \leftarrow 0, \quad u_j \leftarrow 0 \quad$ for $\forall j$.
  Step 1-2: if $(x = 0$ and $y = 0)$ then go to **Step 2:**.
  Step 1-3: $u_i \leftarrow x \bmod p$.
  Step 1-4: $v \leftarrow (x - u_i)/p, \quad x \leftarrow tv + y, \quad y \leftarrow -v, \quad i \leftarrow i + 1$.
  Step 1-5: go to **Step 1-2:**.
Step 2: Optimization of Base-ϕ Expansion
  Step 2-1: $d_i \leftarrow u_i + u_{i+m} + u_{i+2m} \quad$ for $0 \le i < m$.
  Step 2-2: $c_i \leftarrow d_i - z \quad$ for $0 \le i \le m - 1$,
    where z is an integer that minimizes $\sum_i w_H(c_i)$.
Step 3: Table Reference Multiplication
  Step 3-1: $P_i \leftarrow \phi^i P \quad$ for $0 \le i < m$.
  Step 3-2: $Q \leftarrow \mathcal{O}, \quad j \leftarrow \lceil \log_2 p \rceil + 1$.
  Step 3-3: $Q \leftarrow 2Q$.
  Step 3-4: for $(i = 0$ to $m - 1)$ {
        if $(c_{ij} = 1)$ then $Q \leftarrow Q + P_i$.
      }
  Step 3-5: $j \leftarrow j - 1$.
  Step 3-6: if $(j \ge 0)$ then go to **Step 3-3**.

First, the procedure finds u_i such that $k = \sum_{i=0}^{l} u_i \phi^i$ in **Step 1** by using $\phi^2 - t\phi + p = 0$. This part of the procedure is nearly equal to the procedure in [7] and integer l is nearly equal to $2m + 3$. This is discussed in Sect. 3.2.

Next, it reduces the series of base-ϕ expansion $\{u_0, \ldots, u_l\}$ into $\{c_0, \ldots, c_{m-1}\}$ in **Step 2**. Detailed explanation is given in Sect. 3.3.

Finally, it calculates kP using $\{c_0, \ldots, c_{m-1}\}$ in **Step 3**. **Step 3** requires $\lceil \log_2 p \rceil$ elliptic curve doublings and $\dfrac{m\lceil \log_2 p \rceil}{2}$ elliptic curve additions at most. On the other hand, we can use the following **Step 3'** in stead of **Step 3** to compute kP using the Frobenius map. **Step 3'** requires $p + m + 2$ elliptic curve additions at most. We can choose the method which has lower computation cost.

[Another Table Reference Multiplication Procedure]

Step 3': Table Reference Multiplication

  Step 3'-1: $Q \leftarrow O, \quad S \leftarrow O, \quad d \leftarrow \max_i \{c_i\}$.
  Step 3'-2: for $(i = 0$ to $m - 1)$ {
        if $d = c_i$ then $S = S + \phi^i P$.
      }
  Step 3'-3: $Q \leftarrow Q + S, \quad d \leftarrow d - 1$.
  Step 3'-4: if $d \ne 0$ then go to **Step 3'-2**.

3.2 Loop Number of Step 1

In this section, we discuss l in **Step 1**.

Theorem 1. [7] *Let $p \geq 4$ and let $k \in \mathbf{Z}[\phi]$. If we set $l = \lceil 2 \log_p \|k\| \rceil + 3$, then there exist rational integers $-p/2 \leq u_i \leq p/2$, $0 \leq i \leq l$, such that*

$$k = \sum_{i=0}^{l} u_i \phi^i \qquad (7)$$

where $\|k\| := \sqrt{k\bar{k}}$ and $\bar{k}$ is the complex conjugate of k and $\lceil x \rceil$ is the minimum integer greater than or equal to x.

Since the proof of Theorem 1 in [7] does not assume p to be a small power of two, the loop in **Step 1** ends at most in $i \leq \lceil 2 \log_p \|k\| \rceil + 3$ for general p.

3.3 Optimization of Base-ϕ Expansion

This section explains the background of the procedure in **Step 2**.

Step 2-1 If k is randomly chosen from $0 < k < N_m$, we can assume $k \simeq p^m$ and $l = \lceil 2 \log_p k \rceil + 3 \simeq 2m + 3$. However, the series of base-ϕ expansion $\{u_0, \ldots, u_{2m+3}\}$ can be easily reduced to $\{d_0, \ldots, d_{m-1}\}$ by using the following equation;

$$\phi^m = 1 \quad \text{in End}_E. \qquad (8)$$

This is because $x^{p^m} = x$ for $\forall x \in \mathbf{F}_{p^m}$. Thus,

$$\sum_{i=0}^{\lceil 2 \log_p k \rceil + 3} u_i \phi^i = \sum_{i=0}^{m-1} (u_i + u_{i+m} + u_{i+2m}) \phi^i$$
$$= \sum_{i=0}^{m-1} d_i \phi^i.$$

Step 2-2 We can accelerate **Step 3** by decreasing the density of '1's in the bit expression of d_i by using Equation (9).

$$\sum_{i=0}^{m-1} \phi^i = 0 \qquad (9)$$

Since P is not an $\mathbf{F}_p$-point over E, $\phi P \neq P$. Equation (9) is derived from Equation (8) and $\phi \neq 1$.

The theoretical required time for scalar multiplication for the case of $m = 7$ and $A = \lceil \log_2 p \rceil$ is shown in Table 2. "Type I expansion" denotes the proposed procedure using d_i instead of c_i at **Step 3** and "Type II expansion" denotes the full proposed procedure.

Table 2. Required Time for Scalar Multiplication ($m = 7$)

Algorithm	EC Addition	EC Doubling	Total
binary	$\frac{7}{2}A$	$7A$	$\frac{21}{2}A$ $(\simeq 10.5A)$
signed binary	$\frac{21}{8}A$	$7A$	$\frac{77}{8}A$ $(\simeq 9.6A)$
Type I expansion	$\frac{7}{2}A$	A	$\frac{9}{2}A$ $(\simeq 4.5A)$
Type II expansion	$\frac{77}{32}A$	A	$\frac{109}{32}A$ $(\simeq 3.4A)$

4   Elliptic Curve Generation

In this section, we discuss how to generate elliptic curves for the base-ϕ expansion method.

Let p be a prime, where $p > 3$ and let E be the elliptic curve

$$Y^2 = X^3 + aX + b \qquad (10)$$

over $\mathbf{F}_p (a, b \in \mathbf{F}_p)$. We should define elliptic curve E over $\mathbf{F}_p$ to use base-ϕ expansion. In such a case, we can easily compute N_m by using Theorem 2.

Theorem 2 (Weil Conjecture [8, pp.132-137]). *Suppose E is an elliptic curve over $\mathbf{F}_p$ and $t := p + 1 - N_1$. The number of $\mathbf{F}_{p^m}$-points on E is*

$$N_m = p^m + 1 - (\alpha^m + \beta^m),$$

where α, β are the roots of $x^2 - tx + p$.

From the view point of cryptography, E is a "good" elliptic curve if N_m has a large prime factor. Since $N_n \simeq p^n$ and N_n divides N_m if n divides m, we have the best chance of getting a large prime factor of N_m when m is a prime. We can generate elliptic curve $E/\mathbf{F}_p$ with N_m that has a large prime factor by using the following procedure.

[**Elliptic-Curve Generation Procedure for Base-ϕ Expansion**]

> **Input:** p, m
> **Output:** $E/\mathbf{F}_p, N_m$
> **Step 1:** Generate $E/\mathbf{F}_p$ randomly and find its order $N_1 = p + 1 - t$.
> **Step 2:** Find N_m using the Weil conjecture.
> **Step 3:** If N_m doesn't have a large enough prime factor, go to **Step 1**.

For example, let $p = 2^{31} - 1$, $m = 7$ and $M := \dfrac{N_m}{N_1}$ $(N_1 \simeq 2^{31}, N_m \simeq 2^{186})$.
We can find some parameters such that M becomes prime. One example is
$a = -3, b = -212, \lceil \log_2 M \rceil = 186$.

5   Further Speed Up Techniques

5.1   Affine Coordinates

Points on an elliptic curve can be represented by some different coordinate systems: for example, affine coordinates or Jacobian coordinates, as shown in [11] and the Appendix. The number of operations on a finite field differ with the system. If you choose Jacobian coordinates, no inversion on the finite field is needed, but "elliptic curve addition" needs ten or more multiplications on the field. On the other hand, if you choose affine coordinates, "elliptic curve addition" needs one inversion and two multiplications. Thus, if inversion is faster than 8 multiplications, affine coordinates are faster than Jacobian coordinates. The implementation in [1] used the Jacobian coordinates because no efficient inversion algorithm for OEF has been proposed. Therefore, if we have a fast enough inversion algorithm, affine coordinates can accelerate elliptic curve operation.

In this section, we present a fast inversion algorithm for OEF. We consider the polynomial basis representation of a field element $a \in \mathbf{F}_{p^m}$:

$$a = a_{m-1}\alpha^{m-1} + \cdots + a_1\alpha + a_0$$

where $a_i \in \mathbf{F}_p$, $\alpha \in \mathbf{F}_{p^m}$ is a primitive root of $x^m - \omega$.

The inversion c of a is defined as

$$ac = (a_{m-1}\alpha^{m-1} + \cdots + a_1\alpha + a_0)(c_{m-1}\alpha^{m-1} + \cdots + c_1\alpha + c_0) = 1.$$

Since α is a root of $x^m - \omega$,

$$ac = \left(\left(\sum_{0 \le u \le m-1} (a_u c_{m-1-u})\right)\alpha^{m-1}\right) + \sum_{0 \le t \le m-2}\left(\left(\sum_{0 \le u \le t} (a_u c_{t-u}) + \left(\sum_{t+1 \le u \le m-1} (a_u c_{t+m-u})\right)\omega\right)\alpha^t\right) = 1. \tag{11}$$

We introduce c_i from Equation(11).

$$\begin{pmatrix} c_0 \\ c-1 \\ \vdots \\ c_{m-2} \\ c_{m-1} \end{pmatrix} = \begin{pmatrix} a_0 & a_{m-1}\omega & a_{m-2}\omega & \cdots & a_2\omega & a_1\omega \\ a_1 & a_0 & a_{m-1}\omega & \cdots & \cdots & a_2\omega \\ a_2 & a_1 & a_0 & \ddots & & \vdots \\ \vdots & & & \ddots & a_0 & a_{m-1}\omega \\ a_{m-1} & a_{m-2} & \cdots & \cdots & a_1 & a_0 \end{pmatrix}^{-1} \begin{pmatrix} 1 \\ 0 \\ \vdots \\ 0 \\ 0 \end{pmatrix}. \tag{12}$$

Table 3. Calculation Cost for Each Coordinate System over OEF($m = 3$)

Coordinates	EC Doubling	EC Addition
Affine	57M	51M
Chudnovsky Jacobian	81M	117M
Modified Jacobian	72M	153M

For example, if $m = 3$ then

$$\begin{pmatrix} c_0 \\ c_1 \\ c_2 \end{pmatrix} = (a_0^3 - 3a_0a_1a_2\omega + a_1^3\omega + a_2^3\omega^2)^{-1} \begin{pmatrix} a_0^2 - a_1a_2\omega \\ a_2^2\omega - a_0a_1 \\ a_1^2 - a_0a_2 \end{pmatrix}.$$

[Inverse Procedure for OEF, $m = 3$]

Input: $[a_0, a_1, a_2]$ $(= a)$
Output: $[c_0, c_1, c_2]$ $(= c = a^{-1})$
Step 1: Compute $b_0 \leftarrow a_0^2$, $b_1 \leftarrow a_1^2$, $b_2 \leftarrow a_2^2\omega$,
     $e_0 \leftarrow a_0a_1$, $e_1 \leftarrow a_1a_2\omega$, $e_2 \leftarrow a_0a_2$,
     $e_3 \leftarrow a_0b_0$, $e_4 \leftarrow a_1b_1\omega$, $e_5 \leftarrow a_2(b_2 - 3e_0)\omega$.
Step 2: Compute $d \leftarrow (e_3 + e_4 + e_5)^{-1}$.
Step 3: Compute $c_0 \leftarrow d(b_0 - e_1)$, $c_1 \leftarrow d(b_2 - e_0)$, $c_2 \leftarrow d(b_1 - e_2)$

Since we can normally use ω as a small integer such as 2 or 3, we ignore multiplication by ω and 3 to count computing cost. Using this procedure, inversion over $\mathbf{F}_{p^3}$ needs 12 multiplications and one inversion over $\mathbf{F}_p$.

Let M denote the cost of multiplication over $\mathbf{F}_p$ and let the cost of inversion over $\mathbf{F}_p$ be $15M$. Then, the costs of multiplication, squaring, and inversion over $\mathbf{F}_{p^m}$ are $9M$, $6M$, and $27M$, respectively. In this case, the elliptic curve operation costs in each coodinate are as shown in Table 3. The operations over affine coordinates are about twice as fast as those over Jacobian coordinates.

Though the proposed inversion algorithm needs $O(m^3)$ computing cost, it is efficient enough for small m.

6   Total Efficiency

We show the total efficiency of the base-ϕ scalar multiplication method.

Table 4 shows the current results of our elliptic curve implementation. We implemented our algorithms on a 500 MHz DEC Alpha workstation which has a 64-bit architecture and a 400 MHz Intel Pentium II PC which has a 32-bit architecture.

We executed the elliptic curve generation algorithm shown in the procedure described in Sect. 4 for the word sizes of 16 and 32.

Table 4. Scalar Multiplication Speed

Base-ϕ Expansion Method

Platform	Order (bit)	Size of Base Field	EC-Add (μsec)	EC-Double (μsec)	Scalar Mult. (msec)	
P II 400	186	$2^{31} - 1$	19.7	13.2	**1.95**	**Base-ϕ**
P II 400	186	$2^{31} - 1$	19.7	13.2	3.89	Signed Binary
P II 400	156	$2^{13} - 1$	32.1	22.3	**2.66**	**Base-ϕ**
P II 400	156	$2^{13} - 1$	32.1	22.3	5.50	Signed Binary

"P II 400" denotes 400 MHz Pentium II PC.

Affine Coordinates

Platform	Order (bit)	Size of Base Field	EC-Add (μsec)	EC-Double (μsec)	Scalar Mult. (msec)	
Alpha 500	183	$2^{61} - 1$	4.64	5.25	**0.994**	**Affine**
Alpha 500	183	$2^{61} - 1$	7.8	6.24	1.58	Jacobian(Bailey[1])

"Alpha 500" denotes 500 MHz DEC Alpha workstation.

Speed in Previous Works

Platform	Order (bit)	Size of Base Field	EC-Add (μsec)	EC-Double (μsec)	Scalar Mult. (msec)	
Sparc4	180	2^5	*[a]	*[a]	59.2	Muller[7]
P 133	177	2	306	309	72	De Win[12]
Alpha 500	160	$2^{32} - 5$	20	16.2	3.62	Bailey[1]
Alpha 500	160	$2^{16} - 165$	207	166	37.1	Bailey[1]

"Sparc4" denotes SparcStation4.
"P 133" denotes 133 MHz Pentium PC.

---

[a] No information in [7].

The parameters used in the implementation are as follows:

[**64-bit OEF**]
$p = 2^{61} - 1$, $m = 3$, $f(x) = x^3 - 37$,
$E : y^2 = x^3 - ax - a$,
    where $a = 17986158219035990087\alpha^2 + 2579024427385917720\alpha$
         $+ 13732791713385999842$,
        α is a root of $f(x)$,

[**32-bit OEF**]
$p = 2^{31} - 1$, $m = 7$, $f(x) = x^7 - 3$,
$E : y^2 = x^3 - 3x - 212$,

[**16-bit OEF**]
$p = 2^{13} - 1$, $m = 13$, $f(x) = x^{13} - 2$,
$E : y^2 = x^3 - 3x + 30$,

where $f(x)$ is a minimal polynomial.

We implemented 16-bit and 32-bit cases on the Pentium II ("P II" in Table 4) to examine the effectiveness of the base-ϕ scalar multiplication method. We implemented the 64-bit case on the DEC Alpha ("Alpha" in Table 4) to examine effectiveness of affine coordinates. "Speed in Previous Works" in Table 4 is shown as reference.

The results clarify that the proposed base-ϕ expansion method speeds up scalar multiplication by a factor of two over the traditional signed binary method in the 16-bit and 32-bit OEF cases. In the case of 64-bit OEF, the new inversion algorithm is about 60% faster for scalar multiplication.

7   Conclusions

This paper proposed a new algorithm that computes the Frobenius map and inversion over OEF-type finite field $\mathbf{F}_{p^m}$. We need only $m-1$ multiplications over $\mathbf{F}_p$ to compute the Frobenius map. The inversion algorithm needs one inversion and $O(m^3)$ multiplications over $\mathbf{F}_p$, and it is quite efficient for small m.

Consequently, we expanded the base-ϕ scalar multiplication method to suit finite fields with higher characteristic (such as OEF) by introducing the table reference method. When the proposed algorithm is applied to OEF-type elliptic curves, the algorithm is about twice as fast as some conventional OEF-base algorithms.

We proved the total efficiency of the proposed algorithm by implementation. In the case of 16-bit and 32-bit OEF, the base-ϕ expansion method is twice as fast as traditional techniques. In the case of 64-bit OEF, the calculation time is 1.6 times shorter due to use of the new inversion algorithm.

Acknowledgments

We are very grateful to Kazumaro Aoki of NTT for implementing OEF primitives on the Pentium II architecture, and Eisaku Teranishi of NTT Advanced Technology for implementing our base-ϕ method.

References

1. D. V. Bailey and C. Paar, "Optimal Extension Fields for Fast Arithmetic in Public-Key Algorithms," Advances in Cryptology – CRYPTO '98, Lecture Notes in Computer Science 1462, pp.472-485, Springer, 1998.
2. N. Koblitz, "CM-Curves with Good Cryptographic Properties," Advances in Cryptology – CRYPTO'91, Lecture Notes in Computer Science 576, pp.279-287, Springer-Verlag, 1992.
3. N. Koblitz, "An Elliptic Curve Implementation of the Finite Field Digital Signature Algorithm," Advances in Cryptology – CRYPTO'98, Lecture Notes in Computer Science 1462, pp.327-337, Springer-Verlag, 1998.
4. J. A. Solinas "An Improved Algorithm for Arithmetic on a Family of Elliptic Curves," Advances in Cryptology – CRYPTO'97, Lecture Notes in Computer Science 1294, pp.357-371, Springer, 1997.

5. J. H. Cheon, S. Park and S. Park, D. Kim, "Two Efficient Algorithms for Arithmetic of Elliptic Curves Using Frobenius Map," Public Key Cryptography: Proceedings of the First international workshop, PKC '98, Lecture Notes in Computer Science 1431, pp.195-202, Springer, 1998.
6. E. F. Brickell, D. M. Gordon, K. S. McCurley and D. B. Wilson, "Fast Exponentiation with Precomputation," Advances in Cryptology – EUROCRYPT'92, Lecture Notes in Computer Science 658, pp.200-207, Springer, 1993.
7. V. Müller, "Fast Multiplication on Elliptic Curves over Small Fields of Characteristic Two," Journal of Cryptology(1998) 11, pp.219-234, 1998.
8. J. H. Silverman, *The Arithmetic of Elliptic Curves*, Springer-Verlag, New York, 1986.
9. R. Mullin, I. Onyszchuk, S. Vanstone, and R. Wilson, "Optimal Normal Basis in $GF(p^n)$," Discrete Applied Mathematics, 22:149-161, 1988.
10. M. J. Wiener and R. J. Zuccherato, "Faster Attacks on Elliptic Curve Cryptosystems," Fifth Annual Workshop on Selected Areas in Cryptography – SAC'98 pp.196-207, Workshop Record, 1998.
11. H. Cohen, A. Miyaji and T. Ono, "Efficient Elliptic Curve Exponentiation Using Mixed Coordinates," Advances in Cryptology – ASIACRYPT'98, Lecture Notes in Computer Science 1514, pp.51-65, Springer-Verlag, 1998.
12. E. De Win, A. Bosselaers and S. Vandenberghe, "A Fast Software Implementation for Arithmetic Operations in GF(2^n)," Advances in Cryptology – ASIACRYPT'96, Lecture Notes in Computer Science 1163, pp.65-76, Springer-Verlag, 1996.

Appendix: Coordinates

Let

$$E : y^2 = x^3 + ax + b \qquad (a, b \in \mathbf{F}_p, 4a^3 + 27b^2 \neq 0)$$

be the equation of an elliptic curve E over $\mathbf{F}_p$.

For Jacobian coordinates, with $x = X/Z^2$ and $y = Y/Z^3$, a point on elliptic curve P is represented as $P = (X, Y, Z)$. In order to make addition faster, the Chudnovsky Jacobian coordinates represents a Jacobian point as the quintuple (X, Y, Z, Z^2, Z^3). On the other hand, in order to make doubling faster, the modified Jacobian coordinates represents a Jacobian point as the quadruple (X, Y, Z, aZ^4).

The number of operations needed to compute elliptic curve doubling and addition is shown in Table 5.

Table 5. Operations for Each Coordinate

Coordinates	Elliptic Curve Doubling	Elliptic Curve Addition
Affine	2 M + 2 S + 1 I	2 M + 1 S + 1 I
Chudnovsky Jacobian	5 M + 6 S	11 M + 3 S
Modified Jacobian	4 M + 4 S	13 M + 6 S

M: Multiplication,   S: Squaring,   I: Inversion.

Comparing the MOV and FR Reductions
in Elliptic Curve Cryptography

Ryuichi Harasawa[1], Junji Shikata[1], Joe Suzuki[1], and Hideki Imai[2]

[1] Department of Mathematics, Graduate School of Science, Osaka University,
1-1 Machikaneyama, Toyonaka, Osaka 560-0043, Japan
{harasawa, shikata, suzuki}@math.sci.osaka-u.ac.jp
[2] Institute of Industrial Science, University of Tokyo, 7-22-1 Roppongi,
Minatoku, Tokyo 106-8558, Japan
imai@iis.u-tokyo.ac.jp

Abstract. This paper addresses the discrete logarithm problem in elliptic curve cryptography. In particular, we generalize the Menezes, Okamoto, and Vanstone (MOV) reduction so that it can be applied to some non-supersingular elliptic curves (ECs); decrypt Frey and Rück (FR)'s idea to describe the detail of the FR reduction and to implement it for actual elliptic curves with finite fields on a practical scale; and based on them compare the (extended) MOV and FR reductions from an algorithmic point of view. (This paper has primarily an expository role.)

1 Introduction

This paper addresses the discrete logarithm problem (DLP) in elliptic curve (EC) cryptography. ECs have been intensively studied in algebraic geometry and number theory. In recent years, they have been used in devising efficient algorithms for factoring integers [11] and primality proving [3], and in the construction of public key cryptosystems [15,9]. In particular, EC cryptography whose security is based on the intractability of the DLP in ECs (ECDLP) has drawn considerable public attention in recent years.

Let E/F_q be an EC given by the Weierstrass equation:

$$y^2 + a_1 xy + a_3 y = x^3 + a_2 x^2 + a_4 x + a_6, \ a_1, a_2, a_3, a_4, a_6 \in \mathbb{F}_q, \qquad (1)$$

where $\mathbb{F}_q$ is a finite field with $q = p^m$ elements (p: prime, and $m \geq 1$). The ECDLP in E/F_q is defined to find $0 \leq l \leq n - 1$ such that $R = lP := \underbrace{P + P + \cdots + P}_{l}$ given $P \in E(\mathbb{F}_q)$ and $R \in\ <P>$, where n is the order of the finite cyclic group $<P>$. Through the paper, we denote for $E(\mathbb{K}) := \{(x, y) \in \mathbb{K} \times \mathbb{K} | (x, y) \text{ satisfies } Eq.(1)\} \cup \{O\}$, the addition is defined in such a way that $E := E(\bar{\mathbb{K}})$ makes an abelian group, where $\bar{\mathbb{K}}$ is the algebraic closure of $\mathbb{K}$, and O is the identity element of the group [22].

The main reason why EC cryptosystems are getting more accepted compared to the conventional schemes is that it is believed that the ECDLP in E/F_q

J. Stern (Ed.): EUROCRYPT'99, LNCS 1592, pp. 190–205, 1999.
© Springer-Verlag Berlin Heidelberg 1999

generally requires an exponential time in $\log q$ to solve it (V. Miller [15], and J. Silverman and J. Suzuki [23]) while the DLP in $\mathbb{F}_q^*$ can be solved at most within a subexponential time.

In other words, if EC cryptosystems provide equivalent security as the existing schemes, then the key lengths will be shorter. Having short key lengths means smaller bandwidth and memory requirements and can be a crucial factor in some applications, for example the design of smart card systems.

However, it has been reported that for specific cases the ECDLP is no more difficult than the DLP by considering injective homomorphisms that map in a polynomial time from $< P >$ to $\mathbb{F}_q$ or $\mathbb{F}_{q^k}^*$, where $\mathbb{F}_{q^k}^*$ is a suitable extension field of $\mathbb{F}_q$. (For attacks against hyper-EC cryptography, L. Adleman, J. DeMarrais, and M. Huang gave a heuristic argument that under certain assumptions, the DLP in the group of rational points on the Jacobian of a genus g hyper-EC over $\mathbb{F}_p$ is solved in a subexponential time for sufficiently large g and odd p with $\log p \leq (2g+1)^{0.98}$. For the detail, see [1].)

For the reduction to $\mathbb{F}_q$, recently only the case of anomalous ECs, i.e. the case of $q = p$ and $\#E(\mathbb{F}_p) = p$, and its simple generalization have been solved [21,24,18].

On the other hand, for the reduction to $\mathbb{F}_{q^k}^*$, A. Menezes, T. Okamoto, and S. Vanstone [13] proposed the so-called MOV reduction that makes it possible to solve the case of supersingular ECs, i.e. the case of $p|t$ with $t := q + 1 - \#E(\mathbb{F}_q)$. In other words, for supersingular ECs the ECDLP in $E/\mathbb{F}_q$ is reduced to the DLP in $\mathbb{F}_{q^k}^*$ for some k that is solved in a subexponential time. The DLP obtained in that way is defined in $\mathbb{F}_{q^k}^*$, so that the input size is multiplied by k. In actual, the value of k is the minimum positive integer such that $E[n] \subseteq E(\mathbb{F}_{q^k})$, where $E[n] := \{T \in E|nT = O\}$. Menezes, Okamoto, and Vanstone found in [13] that if $E/\mathbb{F}_q$ is supersingular, such a k is at most six, and constructed a probabilistic polynomial time algorithm to find $Q \in E[n]$ such that the Weil pairing $e_n(P,Q)$ [22] has order n in $\mathbb{F}_{q^k}^*$.

Concerning the reduction to $\mathbb{F}_{q^k}^*$, after the MOV reduction appeared, G. Frey and H. Rück [7] proposed another injective homomorphism based on the Tate pairing (FR reduction). The FR reduction is applied when $n|q - 1$. Also, by extending the definition field from $\mathbb{F}_q$ to $\mathbb{F}_{q^k}$, the reduction is possible even for the case of $n|q^k - 1$. In this case, k is the minimum positive integer such that $n|q^k - 1$. Then, as in the MOV reduction, the input size of the DLP is multiplied by k. But the Ref. [7] dealt with only the conceptual aspect.

At this point, we should be aware that there is a gap between the conditions to which the MOV and FR reductions are applied. In fact, according to R. Schoof [19], if $p \nmid n$, $E[n] \subseteq E(\mathbb{F}_{q^k})$ is equivalent to $n|q^k - 1$ and other two conditions.

In this paper, we generalize the MOV reduction so that it can be applied to some non-supersingular ECs satisfying $E[n] \subseteq E(\mathbb{F}_{q^k})$ for some k (Section 2). This extension is never straightforward since no algorithm has been proposed to efficiently find for non-supersingular ECs some $Q \in E[n]$ such that $e_n(P,Q)$ is a primitive n^{th} root of unity. We construct a polynomial time algorithm to realize it although those ECs do not cover all the ones satisfying $E[n] \subseteq E(\mathbb{F}_{q^k})$.

Moreover, we prove that it is possible to immediately find such a $Q \in E[n]$ for the MOV reduction unless $c_2 n | c_1$ when we express the group structure as[1]

$$E(\mathbb{F}_q) \cong \mathbb{Z}_{n_1} \oplus \mathbb{Z}_{n_2}, \quad E[n] \subseteq E(\mathbb{F}_{q^k}), \quad \text{and} \quad E(\mathbb{F}_{q^k}) \cong \mathbb{Z}_{c_1 n_1} \oplus \mathbb{Z}_{c_2 n_1}$$

with $n_2 | n_1$ and $c_2 | c_1$ (See [13,14]).

On the other hand, quite recently, R. Balasubramanian and N. Koblitz [5] showed that if n is a prime, $n \nmid q$, and $n \nmid q - 1$, then $E[n] \subseteq E(\mathbb{F}_{q^k})$ is equivalent to $n | q^k - 1$.

In this sense, if n is a prime, the following are the cases that the (extended) MOV reduction cannot deal with but the FR can:

1. $n | q - 1$; and
2. $E[n] \subseteq E(\mathbb{F}_{q^k})$, $c_2 n | c_1$.

Next, we describe the detail algorithm for the FR reduction, and analyze the computational property (in Section 3). We actually implement the FR reduction for many cases. In addition, we compare it with the extended MOV reduction except for those two cases (in Section 4). Consequently, we should suggest that the FR is better than the MOV in any situation.

Through the paper, for brevity, we assume

1. the order n of $< P >$ is a prime.

If the given $n = \prod_i p_i^{e_i}$ is not a prime. the problem is reduced to finding for each i, $l \bmod p_i$ such that $R = lP$. Then, we can obtain the values of $l \bmod p_i^{e_i}$ for all i using the Pohlig-Hellman's algorithm [17] to determine $l \bmod n$ using the Chinese Remainder Theorem. Further, without loss of generality, we can further assume the following two conditions:

2. $p \nmid t$ (non-supersingularity), and
3. $p \nmid n$ (non-anomalousness) i.e. $p \neq n$

because for those cases, the ECDLP has been already solved in subexponential and polynomial times, respectively.

This paper has primarily an expository role.

2   Extending the MOV Reduction

The framework of the MOV reduction can be described as follows ([13], page 71 in [14]). The idea is to extend the definition field from $\mathbb{F}_q$ to $\mathbb{F}_{q^k}$ for some k so that $E[n] \subseteq E(\mathbb{F}_{q^k})$.

Algorithm 1
Input: *an element $P \in E(\mathbb{F}_q)$ of order n, and $R \in < P >$.*
Output: *an integer l such that $R = lP$*

---

[1] Through the paper, $\mathbb{Z}_n$ denotes $\mathbb{Z}/n\mathbb{Z}$.

Step 1: *determine the smallest integer k such that $E[n] \subseteq E(\mathbb{F}_{q^k})$.*
Step 2: *find $Q \in E[n]$ such that $\alpha = e_n(P, Q)$ has order n.*
Step 3: *compute $\beta = e_n(R, Q)$.*
Step 4: *compute l, the discrete logarithm of β to the base α in $\mathbb{F}_{q^k}^*$.*

Let μ_n be the group of nth roots of unity, $e_n \colon E[n] \times E[n] \to \mu_n$ the Weil pairing [22], and $Q \in E[n]$ such that $e_n(P, Q)$ is a primitive n^{th} root of unity. Then, from the property of the Weil pairing, $\mu_n \subseteq \mathbb{F}_{q^k}^*$ holds. Thus, the group isomorphism $< P > \to \mu_n$ defined by $S \mapsto e_n(S, Q)$ gives an injective homomorphism $< P > \to \mathbb{F}_{q^k}^*$ [13].

It is known that for any $E/\mathbb{F}_q$ there is a pair (n_1, n_2) such that $E(\mathbb{F}_q) \cong \mathbb{Z}_{n_1} \oplus \mathbb{Z}_{n_2}$ with $n_2 | n_1$ [14]. Ref. [13] proved that if $E/\mathbb{F}_q$ is supersingular,

1. k is at most 6, and
2. if put $E(\mathbb{F}_{q^k}) \cong \mathbb{Z}_{c_1 n_1} \oplus \mathbb{Z}_{c_2 n_1}$ for appropriate c_1 and c_2 with $c_2 | c_1$, then $c_1 = c_2$.

In general, the values of c_1 and c_2 can be obtained by the following:

1. Count $\#E(\mathbb{F}_q)$, using Schoof's method [20] or its variant [6,2].
2. For each k,
   (a) compute $\#E(\mathbb{F}_{q^k})$ from $\#E(\mathbb{F}_q)$, using the Weil Theorem [14];
   (b) factor $\#E(\mathbb{F}_{q^k})$; and
   (c) find n_1' and n_2' such that $E(\mathbb{F}_{q^k}) \cong \mathbb{Z}_{n_1'} \oplus \mathbb{Z}_{n_2'}$, using Miller's algorithm [16] ($c_1 = n_1'/n_1$ and $c_2 = n_2'/n_1$).

However, it would be time-consuming to follow these steps: the first two steps take polynomial times, the third takes a subexponential time, and the last takes a probabilistic polynomial time, provided k is small enough compared to q. However, in Ref. [13], Algorithm 1', which will be mentioned later, is constructed concretely based on the following facts concerning supersingular ECs:

1. there are six classes of supersingular ECs;
2. the values of k and c ($= c_1 = c_2$) are uniquely determined by the class; and
3. the class is uniquely determined by the value of $t = q + 1 - \#E(\mathbb{F}_q)$, where t is the trace of qth-power Frobenius endomorphism.

That is, for supersingular ECs, the following algorithm was proposed in [13] [14].
Algorithm 1'

Input: an element $P \in E(\mathbb{F}_q)$ of order n, and $R \in < P >$.
Output: an integer l such that $R = lP$

Step 1': determine the smallest integer k such that $E[n] \subseteq E(\mathbb{F}_{q^k})$.
Step 2': pick $Q' \in E(\mathbb{F}_{q^k})$ randomly, and compute $Q = [cn_1/n]Q'$.
Step 3': compute $\alpha = e_n(P, Q)$ and $\beta = e_n(R, Q)$.
Step 4': compute l' by solving the discrete logarithm of β to the base α in $\mathbb{F}_{q^k}^*$.
Step 5': check if $l'P = R$ holds. If it does, set $l = l'$. Otherwise, go to Step 2'.

It can be easily seen that Algorithms 1 and 1' are essentially the same although they take different step. At this point, we pay attention to how to determine an element $Q \in E[n]$. The correct l is obtained with probability $1 - 1/n$ ($\phi(n)/n$ if n is not a prime) after Steps 1'-5' of Algorithm 1'.

Since n is large, the expected number of trials is close to one.

Since we consider non-supersingular ECs, we cannot use the above three facts. Let (e, r) be such that $c_1/c_2 = n^e r$ with $e \geq 0$ and $(n, r) = 1$. We propose the details of Step 2 in Algorithm 1 for non-supersingular ECs as follows:

Step 2-1: pick $Q' \in E(\mathbb{F}_{q^k})$ randomly.
Step 2-2: set $Q = [c_1 n_1 / n^{e+1}] Q' \in E[n^{e+1}] \cap E(\mathbb{F}_{q^k})$.
Step 2-3: if $Q \notin E[n]$, i.e. if $nQ \neq O$, go to Step 2-1.
Step 2-4: compute $\alpha = e_n(P, Q)$. If $\alpha = 1$, go to Step 2-1.

We should note here that the above modification provides a generalization of the MOV reduction: previously, the MOV can be applied if the EC is supersingular, i.e. $e = 0$ and $r = 1$. If $e = 0$, Step 2-3 can be omitted. The following theorem suggests from a computational point of view that the extension of the MOV reduction in this paper is useful if and only if $e = 0$.

Theorem 1 *The probability that $Q \in E[n^{e+1}] \cap E(\mathbb{F}_{q^k})$ obtained in Step 2-2 satisfies both $Q \in E[n]$ and $e_n(P, Q) \neq 1$ is $\dfrac{1}{n^e}(1 - \dfrac{1}{n})$.*

Proof: Consider the map:

$$f : E(\mathbb{F}_{q^k}) \to E(\mathbb{F}_{q^k}) , \quad f(Q) = [c_1 n_1 / n^{e+1}] Q .$$

Then, since $E(\mathbb{F}_{q^k}) \cong \mathbb{Z}_{c_1 n_1} \oplus \mathbb{Z}_{c_2 n_1}$, the image of f is isomorphic to $\mathbb{Z}_{n^{e+1}} \oplus \mathbb{Z}_n$. Let Ω be the set of Q such that $Q \in E[n]$ and $e_n(P, Q) \neq 1$. From the property of the Weil pairing [22], $e_n(P, Q) = 1$ with $P \neq O$ if and only if $Q \in \langle P \rangle$. Thus, $\#\Omega = n^2 - n$. If $Q' \in E(\mathbb{F}_{q^k})$ is randomly selected in Step 2-1, the probability of success in Step 2-4 is obtained as:

$$\frac{\#\mathrm{Ker}f \times \#\Omega}{\#E(\mathbb{F}_{q^k})} = \frac{\dfrac{c_1 n_1 \times c_2 n_1}{n^{e+1} \times n} \times (n^2 - n)}{c_1 n_1 \times c_2 n_1} = \frac{1}{n^e}(1 - \frac{1}{n})$$

$\square$

Corollary 1 *In Steps 2-1 through 2-4 of Algorithm 1, the expected number of iterations is $n^{e+1}/(n - 1) \approx n^e$.*

Proof: From Kac's lemma [8], the expected time is the reciprocal number of the probability $(1 - 1/n)/n^e$ that has been obtained in Theorem 1, i.e.

$$1/[\frac{1}{n^e}(1 - \frac{1}{n})] = \frac{n^{e+1}}{n - 1} .$$

$\square$

Recall $n = O(q)$, which means Step 2-3 requires an exponential time on average if $e \geq 1$.

If we have $c_2 n | c_1$ during the field extension when we apply the MOV reduction, we must give up the reduction process. Such a probability may be small, and we might in the future come up with an alternative method that can deal with even such a case. However, we should keep in mind that there is much additional computation to realize the MOV reduction for nonsupersingular ECs: counting $\#E(\mathbb{F}_q)$, factoring $\#E(\mathbb{F}_{q^k})$, finding the pair (c_1, c_2) for the group structure $E(\mathbb{F}_{q^k})$ (more precisely, the value of $c_1 n_1 / n^{e+1}$ in Step 2-2), etc., even when $E[n] \subseteq E(\mathbb{F}_{q^k})$ and $c_2 n \nmid c_1$.

3 Implementing the FR Reduction

In this section, assuming $\mathbb{K} := \mathbb{F}_{q^k}$ for some k. We consider the realization of the FR reduction.

In the original paper by Frey and Rück [7], only the conceptual aspect was stated, and it seems that no realization on the FR reduction has been published because the FR reduction appears to be less familiar to the cryptography community than the MOV reduction. We first describe an algorithm for realizing Frey and Rück's idea, where we assume that k is the minimum integer such that $n | q^k - 1$.

Algorithm 2
Input: *an element $P \in E(\mathbb{F}_q)$ of order n, and $R \in< P >$.*
Output: *an integer l such that $R = lP$.*

Step 1: *determine the smallest integer k such that $n | q^k - 1$, and set $\mathbb{K} := \mathbb{F}_{q^k}$.*
Step 2: *pick $S, T \in E(\mathbb{K})$ randomly.*
Step 3: *compute the element $f \in \mathbb{K}(E)^*$ such that $div(f) = n((P) - (O))$, and compute $\alpha = f(S)/f(T)$*
Step 4: *compute $\gamma = \alpha^{\frac{q^k-1}{n}}$. If $\gamma = 1$, then go to Step 2.*
Step 5: *compute the element $g \in \mathbb{K}(E)^*$ such that $div(g) = n((R) - (O))$, and compute $\beta = g(S)/g(T)$, and $\delta = \beta^{\frac{q^k-1}{n}}$.*
Step 6: *solve the DLP $\delta = \gamma^l$ in $\mathbb{K}^*$, i.e. the logarithm of δ to the base γ in $\mathbb{K}^*$.*

3.1 Frey and Rück's Idea

Let $Div(E)$ be the divisor group of E and $supp(D) := \{P \in E(\bar{\mathbb{K}}) : n_P \neq 0\}$ for $D = \sum_{P \in E} n_P(P) \in Div(E)$. Then, since E is defined over $\mathbb{K}$, the Galois group $G_{\bar{\mathbb{K}}/\mathbb{K}}$ acts on $Div(E)$ as $D^\sigma = \sum_{P \in E} n_P(P^\sigma)$ for $D = \sum_{P \in E} n_P(P) \in Div(E)$ and $\sigma \in G_{\bar{\mathbb{K}}/\mathbb{K}}$. We say that $D \in Div(E)$ is defined over $\mathbb{K}$ if $D^\sigma = D$ for all $\sigma \in G_{\bar{\mathbb{K}}/\mathbb{K}}$, and denote by $Div_\mathbb{K}(E)$ the subset of $Div(E)$ whose elements are defined over $\mathbb{K}$.

For $f \in \bar{\mathbb{K}}(E)^*$, the divisor $div(f)$ is defined by $div(f) := \sum_{P \in E} ord_P(f)(P)$, where $ord_P(f)$ is the multiplicity of zeros (if positive) or poles (if negative) at $P \in E$ with respect to $f \in \bar{\mathbb{K}}(E)^*$, and we refer to such a divisor as the principal divisor.

Let $Div^0(E) := \{D \in Div(E) | deg(D) = 0\}$, where $deg(D) := \sum n_P$, and $Prin(E)$ the subset of $Div^0(E)$ whose elements are principal divisors. Then, we can define the following surjective map:

$$Div^0(E) \to Pic^0(E) := Div^0(E)/Prin(E) , \quad D \mapsto \bar{D}$$

and denote $D_1 \sim D_2$ if two divisors D_1 and D_2 have the same image, i.e. $\bar{D}_1 = \bar{D}_2$ in $Pic^0(E)$. We further define $Pic_{\mathbb{K}}^0(E)$ to be the set of all divisor classes in $Pic^0(E)$ that have a representative element defined over $\mathbb{K}$, which is a subgroup of $Pic^0(E)$. Moreover, $Pic_{\mathbb{K}}^0(E)_n := \{\bar{D} \in Pic_{\mathbb{K}}^0(E) | n\bar{D} = 0\}$.

It is known that by the isomorphism

$$E(\mathbb{K}) \to Pic_{\mathbb{K}}^0(E) , \quad Q \mapsto \overline{(Q) - (O)} ,$$

we can identify $E(\mathbb{K})$ with $Pic_{\mathbb{K}}^0(E)$ [22], and denote $\overline{(Q) - (O)}$ by $\bar{Q}$.

Let A be a divisor such that $\bar{A} \in Pic_{\mathbb{K}}^0(E)_n$ and B another divisor $\sum_i a_i(Q_i) \in Div_{\mathbb{K}}^0(E)$ such that $supp(A) \cap supp(B) = \phi$. Since $nA \sim 0$, there exists an element f_A in the function field $\bar{\mathbb{K}}(E)$ such that $div(f_A) = nA$ [22], so that we can put $f_A(B) := \prod_i f_A(Q_i)^{a_i}$.

Then, Frey and Rück [7] proved the following:

Proposition 1 ([7]) *If $n | q - 1$, $\{\bar{A}, \bar{B}\}_{0,n} := f_A(B)$ defines a nondegenerate bilinear pairing:*

$$\{,\}_{0,n} : \ E(\mathbb{K})[n] \times E(\mathbb{K})/nE(\mathbb{K}) \to \mathbb{K}^*/(\mathbb{K}^*)^n$$

where $E(\mathbb{K})[n] := E[n] \cap E(\mathbb{K})$.

Then the mapping $\mathbb{K}^* \to \mathbb{K}^*$ defined by $\alpha \mapsto \alpha^{\frac{q^k-1}{n}}$ gives $\mathbb{K}^*/(\mathbb{K}^*)^n \cong \mu_n \subseteq \mathbb{K}^*$, where μ_n is the group of nth roots of unity. From the nondegeneracy of the pairing $\{,\}_{0,n}$, there exists $Q \in E(\mathbb{K})/nE(\mathbb{K})$ such that $\{\bar{P}, \bar{Q}\}_{0,n}^{\frac{q^k-1}{n}}$ is a primitive n^{th} root of unity. Thus, the group isomorphism $< P > \to \mu_n$ defined by $S \mapsto \{\bar{S}, \bar{Q}\}_{0,n}^{\frac{q^k-1}{n}}$ gives an injective homomorphism $< P > \to \mathbb{F}_{q^k}^*$.

The pairing $\{,\}_{0,n}$ can be said to be a variant of the Tate pairing [25].

3.2   Theoretical Analysis

In [7], the computation of Steps 2-5 is supposed to be within a probabilistic polynomial time, now we actually evaluate the computation for each step in Algorithm 2. We assume that the usual multiplication algorithms are used, so that multiplying two elements of length N takes time $O(N^2)$.

For Step 2, we first pick an element $x = a$ in $\mathbb{K}$ to substitute it to Eq. (1). Then, we check if the quadratic equation with respect to y has a solution in $\mathbb{K}$, i.e. if the discriminant is a quadratic residue in $\mathbb{K}$. The probability of the success is approximately a half. If it is successful, it suffices to solve the quadratic equation in a usual manner. The computation to solve the quadratic equation dominants one to compute quadratic roots in $\mathbb{K}$. This takes expected running time $O((\log q^k)^3) = O(k^3 (\log q)^3)$ (for the detail, see [4], [10]). We do this process twice to obtain $S, T \in E(\mathbb{K})$.

For Step 3, there is a standard procedure to compute the function $f \in \bar{\mathbb{K}}(E)$ from a principal divisor $div(f) \in Prin(E)$ (see for example pages 63-64 in [14]). Basically, this can be done by the following:

1. Write $div(f) = \sum_i a_i((P_i) - (O))$.
2. For each i, compute $P_i' \in E$ and $f_i \in \bar{\mathbb{K}}(E)$ such that

$$a_i((P_i) - (O)) = (P_i') - (O) + div(f_i) \ .$$

3. Add the divisors $(P_i') - (O) + div(f_i)$, for all i.

Then, we can add two divisors as follows: if two divisor D, D' are expressed by

$$D = (P) - (O) + div(f) \ , \qquad D' = (P') - (O) + div(f')$$

with $P, P' \in E$ and $f, f' \in \bar{\mathbb{K}}(E)^*$, then

$$D + D' = (P + P') - (O) + div(f f' g)$$

where $g = l/v$ with l and v are the lines through P and P' and through $P + P'$ and O (in particular, $P' = -P$ implies $v \equiv 1$). We can obtain the value of $\alpha = f(S)/f(T)$ by substituting S, T to the aforementioned f, f', g and multiplying them. Hence, Step 3 takes $O((\log q^k)^2) \times O(\log n) = O(k^2 (\log q)^3)$.

For Step 4, the computation of $\gamma = \alpha^{\frac{q^k - 1}{n}}$ takes $O(\log(\frac{q^k - 1}{n})) \times O((\log q^k)^2) = O(k^3 (\log q)^3)$. Moreover, we should evaluate the probability of going back to Step 2 so that we can measure how long it takes to compute the whole steps. The crucial point here is that we should efficiently find $\bar{Q} \in Pic_{\mathbb{K}}^0(E)/nPic_{\mathbb{K}}^0(E)$ such that $\{\bar{P}, \bar{Q}\}_{0,n}$ can be a generator of $\mathbb{K}^*/(\mathbb{K}^*)^n$. We prove the following theorem.

Theorem 2 *Let k be the smallest positive integer such that $n | q^k - 1$ (in this case, $\mathbb{K} = \mathbb{F}_{q^k}$). Then the probability of going back from Step 4 to Step 2 is $1/n$.*

Proof: Note that $E(\mathbb{K}) \cong \mathbb{Z}_{n_1} \oplus \mathbb{Z}_{n_2}$, $n_2 | n_1$, and $E[n] \cong \mathbb{Z}_n \oplus \mathbb{Z}_n$. Thus,

$$E(\mathbb{K})[n] \cong \begin{cases} \mathbb{Z}_n & (n \nmid n_2) \\ \mathbb{Z}_n \oplus \mathbb{Z}_n & (n | n_2). \end{cases}$$

Also, from the nondegeneracy of the FR reduction,

$$E(\mathbb{K})/nE(\mathbb{K}) \cong \begin{cases} \mathbb{Z}_n & (n \nmid n_2) \\ \mathbb{Z}_n \oplus \mathbb{Z}_n & (n | n_2). \end{cases}$$

We consider the two cases separately.

1. $E[n] \not\subseteq E(\mathbb{K})$, i.e. $n \nmid n_2$: if we pick $Q \in E(\mathbb{K})$ randomly, the probability of $\{\bar{P}, \bar{Q}\}_{0,n} \notin (\mathbb{K}^*)^n$ is

$$\frac{\#E(\mathbb{K}) - \#nE(\mathbb{K})}{\#E(\mathbb{K})} = \frac{n_1 n_2 - n_1 n_2/n}{n_1 n_2} = 1 - 1/n.$$

2. $E[n] \subseteq E(\mathbb{K})$, i.e. $n|n_2$: let $T := \{Q \in E(\mathbb{K})/nE(\mathbb{K}) \mid \{\bar{P}, \bar{Q}\}_{0,n} \notin (\mathbb{K}^*)^n\}$. Then, $\#T = n^2 - n$. Since the map $\varphi : E(\mathbb{K}) \to E(\mathbb{K})/nE(\mathbb{K})$ is a module homomorphism, the probability of $\{\bar{P}, \bar{Q}\}_{0,n} \notin (\mathbb{K}^*)^n$ is

$$\frac{\#\varphi^{-1}(T)}{\#E(\mathbb{K})} = \frac{\#\mathrm{Ker}(\varphi) \times \#T}{\#E(\mathbb{K})} = \frac{(n_1 n_2/n^2)(n^2 - n)}{n_1 n_2} = 1 - 1/n.$$

□

The probability of going back from Step 4 to Step 2 is almost close to zero since we assume that n is considerably large.

For Step 5, we can estimate the computation as $O(k^3 (\log q)^3)$.

From the above insight, if k can be assumed to be small enough compared to q, the expected running time of the FR reduction (from Step 2 to Step 5 in Algorithm 2) is $O((\log q)^3)$.

3.3   Implementation

We made several experiments including the following four cases. The CPU is Pentium 75MHz (SONY Quarter L, QL-50NX, the second cache capacity: 256kB)

In Examples 1 and 2, the FR reduction was applied to ECs with trace 2.

Example 1 (EC with trace 2, i.e. $\#E(\mathbb{F}_p) = p - 1$) *Suppose that the curve $E/\mathbb{F}_p : y^2 = x^3 + ax + b$, the base point $P = (x_0, y_0) \in E(\mathbb{F}_p)$, the order n of P, and a point $R = [l]P = (x_1, y_1)$ are given as follows:*
$p = 23305425500899$ *(binary 45-bits, $p - 1 = 2 \times 3^2 \times 1137869^2$),*
$a = 13079575536215$, $b = 951241857177$,
$n = 1137869$,
$x_0 = 17662927853004$, $y_0 = 1766549410280$,
$x_1 = 2072411881257$, $y_1 = 5560421985272$.
Then, we find that $l = 709658$.

Example 2 (EC with trace 2, i.e. $\#E(\mathbb{F}_p) = p - 1$) *Suppose that the curve $E/\mathbb{F}_p : y^2 = x^3 + ax + b$, the base point $P = (x_0, y_0) \in E(\mathbb{F}_p)$, the order n of P, and a point $R = [l]P = (x_1, y_1)$ are given as follows:*
$p = 93340306032025588917032364977153$
(binary 107-bits, $p - 1 = 2^{10} \times 7^2 \times 163 \times 847321^2 \times 3986987^2$),
$a = 71235469403697021051902688366816$, $b = 47490312935798014034601792244544$,
$n = 3986987$,
$x_0 = 103624099299650416143178356924 63$, $y_0 = 79529049191468905652172306035573$,
$x_1 = 15411349585423321468944221089888$, $y_1 = 94160529078832780887 82335830033$.

For Example 2, the reduction process was implemented as follows:

1) Choose random points $S, T \in E(\mathbb{F}_p)$:
   $R = (x_2, y_2)$,
   $x_2 = 7818312665362296556444425568155546$, $y_2 = 78588945135854560800493672181265$,
   $S = (x_3, y_3)$,
   $x_3 = 58714658884321859706339658012314$, $y_3 = 29352359294307548304481400079114$.
   The time of computation : 177 sec.

2) Compute the FR pairing:
   Set $div(f) := n((P) - (O))$, $div(g) := n((R) - (O))$ and $D := (S) - (T)$,
   then
   $\{\bar{P}, \bar{D}\}_{0,n} = \frac{f(S)}{f(T)} = 28089673702084922579189210362050$,
   $\left(\frac{f(S)}{f(T)}\right)^{\frac{p-1}{n}} = 86048548119736537511939909279595$,
   $\{\bar{Q}, \bar{D}\}_{0,n} = \frac{g(S)}{g(T)} = 54538105615281807032380914744128$,
   $\left(\frac{g(S)}{g(T)}\right)^{\frac{p-1}{n}} = 44179423723975173427344893182175$.
   The time of computation:

   computation of $f(S)$: 982 sec, computation of $f(T)$: 996 sec,
   computation of $g(S)$: 971 sec, computation of $g(T)$: 968 sec,
   computation of $\left(\frac{f(S)}{f(T)}\right)^{\frac{p-1}{n}}$: 5 sec, computation of $\left(\frac{g(S)}{g(T)}\right)^{\frac{p-1}{n}}$: 6 sec.

3) Solve the DLP: $(86048548119736537511939909279595)^l$
   $= 44179423723975173427344893182175 \bmod p$,
   find that $l = 764009$.

Next, in Examples 3 and 4, the FR and MOV reductions were applied to supersingular-ECs, and experimental data in the both reductions were analyzed and compared.

Example 3 (Supersingular-EC) *Suppose that the curve $E/\mathbb{F}_p$: $y^2 = x^3 + ax + b$, the base point $P = (x_0, y_0) \in E(\mathbb{F}_p)$, the order n of P, and a point $R = [l]P = (x_1, y_1)$ are given as follows:*
$p = 23305425500899$ *(binary 45-bits,* $p + 1 = 2^2 \times 5^2 \times 29 \times 1217 \times 6603413$*),*
$a = 1, b = 0$,
$n = 6603413$,
$x_0 = 18414716422748$, $y_0 = 9607997424906$,
$x_1 = 22829488331658$, $y_1 = 15463570264423$.

Since $E(\mathbb{F}_p) \cong \mathbb{Z}_{p+1}$, $E(\mathbb{F}_{p^2}) \cong \mathbb{Z}_{p+1} \oplus \mathbb{Z}_{p+1}$ [13], the definition field $\mathbb{F}_p$ is extended to $\mathbb{F}_{p^2}$ to apply the FR and MOV reductions. Then, we find that $l = 4500974$.

Example 4 (Supersingular-EC) *Suppose that the curve $E/\mathbb{F}_p$: $y^2 = x^3 + ax + b$, the base point $P = (x_0, y_0) \in E(\mathbb{F}_p)$, the order n of P, and a point $R = [l]P = (x_1, y_1)$ are given as follows:*
$p = 1020213065766829380028651032779469420609306683196983$

(binary 163-bits, $p+1 = 2^3 \times 3^3 \times 59 \times 113$
$\times 708445873377740404845389902584519528254847$),
$a = 1, b = 0$,
$n = 708445873377740404845389902584519528254847$,
$x_0 = 636140843166014501847273496446991894972799363117$,
$y_0 = 222428572612516351526464210931959631877226149291$,
$x_1 = 179140020238388209409497264852379835824276605 0148$,
$y_1 = 666228287982545247994555402829685728224357263 5001$.

Since $E(\mathbb{F}_p) \cong \mathbb{Z}_{\frac{p+1}{2}} \oplus \mathbb{Z}_2$, $E(\mathbb{F}_{p^2}) \cong \mathbb{Z}_{p+1} \oplus \mathbb{Z}_{p+1}$ *[13]*, *the definition field*
$\mathbb{F}_p$ *is extended to* $\mathbb{F}_{p^2}$ *to apply the FR and MOV reductions. Set* $g(\alpha) := \alpha^2 + 1$.
Then $\mathbb{F}_{p^2} \cong \mathbb{F}_p[\alpha]/g(\alpha)$.

For Example 4, the FR and MOV reductions process were implemented as follows:

(FR reduction):

1) Choose random points $S, T \in E(\mathbb{F}_p)$:
   $S = (x_2, y_2)$,
   $x_2 = 5$,
   $y_2 = 278527964102001851794759488558715840137459875224 9\alpha$
   $T = (x_3, y_3)$,
   $x_3 = 338530611385145171186893854505822118617259793 7436$,
   $y_3 = 498677065440695353174518618475802696104861959 8992$.
   The time of computation : 2245 sec;

2) Compute the FR pairing:
   Set $div(f) := n((P) - (O))$, $div(g) := n((R) - (O))$ and $D := (S) - (T)$,
   then
   $\{\bar{P}, \bar{D}\}_{0,n} = \frac{f(S)}{f(T)}$
   $= 353316662547946563279907394908121139779745626897 4\alpha$
   $+400149665628249304288065611973616699622145275 1615$,
   $(\frac{f(S)}{f(T)})^{\frac{p^2-1}{n}} = 501035026731987279504884889683664624292006059759 2\alpha$
   $+684597904528238743074511834101748764895625936 7889$,
   $\{\bar{R}, \bar{D}\}_{0,n} = \frac{g(S)}{g(T)}$
   $= 761805382122428568738346617472025239650166349941 6\alpha$
   $+591026751695345226866965976208822232514317607 4230$,
   $(\frac{g(S)}{g(T)})^{\frac{p^2-1}{n}} = 135433531518182121168248536585921809875527887737 8\alpha$
   $+865441031838417931745198119632221028739343235 4847$.
   The time of computation :

   computation of $f(S)$: 39667 sec, computation of $f(T)$: 40023 sec,
   computation of $g(S)$: 39634 sec, computation of $g(T)$: 39646 sec,
   computation of $(\frac{f(S)}{f(T)})^{\frac{p^2-1}{n}}$: 116 sec, computation of $(\frac{g(S)}{g(T)})^{\frac{p^2-1}{n}}$: 136 sec.

3) Solve the DLP
   $(501035026731987279504884889683664624292006059759 2\alpha$

$+6845979045282387430745118341017487648956259367889)^l$
$= 135433531518182121168248536585921809875527887737 8\alpha$
$+86544103183841793174519811963222102873934323548 47$ in $\mathbb{F}_{p^2}$,
find $l = 388267735689900037826187381399337 8$.

(MOV reduction): Let R, S be as in FR reduction.

1) Compute $Q = (x_4, y_4) = [\frac{p+1}{n}]S$ with order n.
   $x_4 = 26860739989985619529342332046329044964185363851 38$,
   $y_4 = 769368303013534155401573490515765808450022343 9095\alpha$.
   The time of computation Q : 1203 sec.

2) Compute the Weil pairing:
   Set $div(f) = n((P + S) - (S))$, $div(g) = n((R + S) - (S))$ and
   $div(h) = (Q + T) - (T)$,
   then
   $e_n(P, Q) = \frac{f(Q+T)}{f(T)} \times \frac{h(S)}{h(P+S)}$
   $= 51917803903484210078162543811102958180106225993 91\alpha$
   $+6845979045282387430745118341017487648956259367889$,
   $e_n(R, Q) = \frac{g(Q+T)}{g(T)} \times \frac{h(S)}{h(R+S)}$
   $= 88477953424864725911826179120877239621754043196 05\alpha$
   $+86544103183841793174519811963222102873934323548 47$.
   The time of computation:

   computation of $f(Q + T)$: 39972 sec, computation of $f(T)$: 39720 sec,
   computation of $h(S)$: 39626 sec, computation of $h(P + S)$: 39850 sec,
   computation of $g(Q + T)$: 39992 sec, computation of $g(T)$: 39956 sec,
   computation of $h(R + S)$: 39862 sec.

3) Solve the DLP:
   $(51917803903484210078162543811102958180106225993 91\alpha$
   $+6845979045282387430745118341017487648956259367889)^l$
   $= 88477953424864725911826179120877239621754043196 05\alpha$
   $+86544103183841793174519811963222102873934323548 47$ in $\mathbb{F}_{p^2}$,
   find $l = 388267735689900037826187381399337 8$.

When we implement the FR and MOV reductions, two random points are
needed. The numbers of function values needed to compute the pairings for
the FR and MOV reductions are four and seven, respectively. In the both reductions, the computation of function values dominates the whole computation
time (Table 1).

From the implementation data and the above consideration, the computation
of function values needed to implement the FR and MOV reductions may be a
heavy load. For each reduction, the computation of pairings actually dominates
the whole computation time while other steps theoretically take $O((\log q)^3)$ as
well. We find that the running time of the FR reduction is almost $4/7$ times as
much as that of the MOV reduction.

Table 1. The time of computation in Examples 1-4

Type	$\log q$	k	Running time(sec)	
Example 1	46	1	FR reduction	419
Example 2	108	1	FR reduction	4105
Example 3	46	2	FR reduction	999
			MOV reduction	1872
Example 4	164	2	FR reduction	161467
			MOV reduction	282426

$\log q$ and k are the binary size of the definition field and the necessary minimum extension degree, respectively.

4   Comparing the (Extended) MOV and FR Reductions

We extended the MOV reduction so that it can be applied to some non-supersingular ECs, and implemented the FR reduction to understand the whole process. Now time to compare the two reductions.

4.1   On the Extension Degrees

Bad news for the MOV reduction is the following fact on group structures, which is due to R. Schoof [19]

Proposition 2 ([20]) *The following two conditions are equivalent:*

1. $E[n] \subseteq E(\mathbb{F}_{q^k})$;

2. $n|q^k - 1$, $n^2|\#E(\mathbb{F}_{q^k})$, *and either* $\phi \in Z$ *or* $\mathcal{O}(\dfrac{t_k^2 - 4q^k}{n^2}) \subseteq \mathrm{End}_{\mathbb{F}_{q^k}}(E)$,

where ϕ *and* t_k *denote the* q^k-*Frobenius endomorphism of* $E/\mathbb{F}_{q^k}$ *and its trace, respectively, and* $\mathcal{O}(\frac{t_k^2 - 4q^k}{n^2})$ *and* $\mathrm{End}_{\mathbb{F}_{q^k}}(E)$ *are the order of discriminant* $\frac{t_k^2 - 4q^k}{n^2}$ *and the endomorphism ring of* $E/\mathbb{F}_{q^k}$ *in which the isogenies are defined over* $\mathbb{F}_{q^k}$, *respectively.*

In this sense, the condition under which the FR can be applied generally includes the one under which the MOV can be applied.

On the contrary, here's good news for the MOV reduction: the difference is not so large between the two conditions for extension degree k under which the MOV and FR reductions can be applied. In fact, R. Balasubramanian and N. Koblitz [5] proved the following:

Proposition 3 ([5]) *Suppose* $n|\#E(\mathbb{F}_q)$, *and that* n *is a prime with* $p \neq n$, $n \nmid q - 1$. *Then,*

$$E[n] \subseteq E(\mathbb{F}_{q^k}) \Longleftrightarrow n|q^k - 1$$

Based on the proof of Proposition 3 [5], we show the following result that provides us with important information for comparing the extension degrees for the MOV and FR reductions although it may be clear from Ref. [5].

Remark 1 *Suppose $E[n] \not\subseteq E(\mathbb{F}_q)$, and that n is a prime. If $n|q-1$*

$$E[n] \subseteq E(\mathbb{F}_{q^k}) \iff k = nj \text{ with } j \geq 1$$

Proof: We pick the basis $\{P, T\}$ of $E[n]$ so that the matrix expression on $E[n]$ of the q-Frobenius endomorphism ϕ is given by

$$M_\phi = \begin{pmatrix} 1 & a \\ 0 & q \end{pmatrix} = \begin{pmatrix} 1 & a \\ 0 & 1 \end{pmatrix} \in GL_2(\mathbb{Z}_n) .$$

(Recall $q \equiv 1 \mod n$.) Then, the matrix that ϕ^k expresses is $M_{\phi^k} = \begin{pmatrix} 1 & ka \\ 0 & 1 \end{pmatrix}$.
Thus,

$$\phi^k(T) = T \iff ka \equiv 0 \mod n \iff k \equiv 0 \mod n ,$$

where we have used $a \not\equiv 0 \mod n$ since $E[n] \not\subseteq E(\mathbb{F}_{q^k})$. Thus, $k = nj$ with $j \geq 1$.

□

If $E[n] \not\subseteq E(\mathbb{F}_q)$ and $n|q-1$, Remark 2 implies that the extension degree k is no less than n, which further means that an exponential number of extensions are needed in the MOV reduction. Hence, then, we will have to give up applying the MOV reduction.

4.2 On the Efficiency of the Reductions

In the following, assuming $n \nmid q-1$, we compare the efficiency of the MOV and FR reductions.

We exclude the following computation in the pre-processing:

1. counting $\#E(\mathbb{F}_q)$, say by Schoof's algorithm [20,6,2], and
2. factoring $\#E(\mathbb{F}_q)$.

Moreover, suppose that the DLP that is obtained by the both reductions from the ECDLP essentially has the same difficulty. Then, all we should compare is the main part of the reductions, i.e. Steps 2-3 in Algorithms 1 and Step 2-5 in Algorithm 2.

However, as considered in Section 2, compared to the FR reduction, additional computation is needed to find the group structure for $E(\mathbb{F}_{q^k})$ for the proposed MOV reduction, although it is computed in a subexponential time.

Moreover, as for application of the MOV reduction, we must give up the application if $e \geq 1$ in Theorem 1. Besides, we should notice that computing the Weil pairing requires almost twice time that the pairing in the FR reduction takes.

4.3   The Actual Difference of the Conditions Between the Two Reductions

At present, we find that there are still two conditions under which the FR can be applied but the MOV cannot:

1. $n|q-1$; and
2. $E[n] \subseteq E(\mathbb{F}_{q^k})$, $c_2 n | c_1$.

Besides, the factorization of $\#E(\mathbb{F}_{q^k})$ is needed to apply Miller's algorithm. (This might be solved immediately because Miller's algorithm sometimes does not require complete factorization.)

Even if the second condition is cleared in the future, the FR reduction is superior to the MOV reduction for the computation of the main part, i.e. for computing the pairings, the MOV requires almost twice time that the FR takes.

In this regard, we must conclude that practically, in any situation the FR reduction is better than the MOV reduction from an algorithmic point of view.

Acknowledgement

Authors would like to thank the program commitee for their comments.

References

1. L. Adleman, J. DeMarrais, and M. Huang, *A subexponential algorithm for discrete logarithms over the rational subgroup of the Jacobians of large genus hyperelliptic curves over finite fields*, Algorithmic Number Theory, Lecture Notes in Computer Science, volume 877, Springer-Verlag, 1994, pp. 28-40.
2. A. O. Atkin, *The number of points on an elliptic curve modulo a prime*, preprint, 1988.
3. A. Atkin and F. Morain, *Elliptic curves and primality proving*, Mathematics of Computation **61** (1993), 29-68.
4. E.Bach,*Algorithmic Number Theory,Volume I; Efficient Algorithms*, MIT Press, Cambridge, Massachusetts, 1996.
5. R. Balasubramanian and N. Koblitz, *The improbability that an elliptic curve has subexponential discrete log problem under the Menezes-Okamoto-Vanstone algorithm*, Journal of Cryptography **11** (1998), 141-145.
6. N. Elkies, *Explicit isogenies*, preprint, 1991.
7. G. Frey and H. Rück, *A remark concerning m-divisibility and the discrete logarithm in the divisor class group of curves*, Mathematics of Computation **62** (1994), 865-874.
8. M. Kac, *On the notion of recurrence in discrete stochastic processes*, Ann. of Math. Statist., **53** (1947), 1002-1010.
9. N. Koblitz, *Elliptic curve cryptosystems*, Mathematics of Computation **48** (1987), 203-209.
10. N. Koblitz, *A Course in Number Theory and Cryptography*, Springer-Verlag, New York, 2nd edition, 1996.

11. H. W. Lenstra, *Factoring integers with elliptic curves*. Annals of Mathematics, 126 (1987), 649-673.

12. R. Lercier and F. Morain, *Counting the number of points on elliptic curves over finite fields: strategy and performance*, Advances in Cryptology, EUROCRYPT'95, Lecture Notes in Computer Science 921 (1995), 79-94.

13. A. Menezes, T. Okamoto, and S. Vanstone, *Reducing elliptic curve logarithms to logarithms in a finite field*, IEEE Transactions on Information Theory **39** (1993), 1639-1646.

14. A. Menezes, *Elliptic curve public key cryptosystems*, Kluwer Academic Publishes (1994).

15. V. S. Miller, *Use of elliptic curves in cryptography*, Advances in Cryptography CRYPTO '85 (Lecture Notes in Computer Science, vol 218), Springer-Verlag, 1986, pp. 417-426.

16. V. Miller, *Short program for functions on curves*, unpublished manuscript, 1986.

17. S.C.Pohlig and M.E.Hellman, *An improved algorithm for computing logarithms over GF(p) and its cryptographic significance*, IEEE Transactions on Information Theory, **24** (1978),pp 106-110.

18. T. Satoh and K. Araki, *Fermat quotients and the polynomial time discrete log algorithm for anomalous elliptic curves*, Commentarii Math,Univ,Saniti Pauli, **47** ,1,pp 88 - 92 (1998).

19. R. Schoof, *Nonsingular plane cubic curves over finite fields*, Journal of Combination Theory, Vol. A. 46 (1987), 183-211.

20. R. Schoof, Elliptic curves over finite fields and the computation of square roots modulo p Math. Comp. **44** (1985), 483-494.

21. I. Semaev, *Evaluation of discrete logarithms in a group of p-torsion points of an elliptic curve in characteristic p*, Mathematics of Computation **67** (1998), 353-356.

22. J. H. Silverman, *The Arithmetic of Elliptic Curves*, Graduate Texts in Math., vol. 106, Springer-Verlag, Berlin and New York, 1994.

23. J. H. Silverman, J. Suzuki, *Elliptic curve discrete logarithms and index calculus*, ASIACRYPT'98 (Lecture Notes in Computer Science), to appear.

24. N. Smart, *The discrete logarithm problem on elliptic curves of trace one*, preprint, 1997.

25. J. Tate, *WC-groups over p-adic fields*, ann Sci. Ecole, Norm. Sup. **2** (1969), 521-560

Unbalanced Oil and Vinegar Signature Schemes

Aviad Kipnis[1], Jacques Patarin[2], and Louis Goubin[2]

[1] NDS Technologies, 5 Hamarpe St. Har Hotzvim, Jerusalem - Israel,
`akipnis@ndsisrael.com`
[2] Bull SmartCards and Terminals, 68 route de Versailles - BP45,
78431 Louveciennes Cedex - France,
`{J.Patarin,L.Goubin}@frlv.bull.fr`

Abstract. In [16], J. Patarin designed a new scheme, called "Oil and Vinegar", for computing asymmetric signatures. It is very simple, can be computed very fast (both in secret and public key) and requires very little RAM in smartcard implementations. The idea consists in hiding quadratic equations in n unknowns called "oil" and $v = n$ unknowns called "vinegar" over a finite field K, with linear secret functions. This original scheme was broken in [10] by A. Kipnis and A. Shamir. In this paper, we study some very simple variations of the original scheme where $v > n$ (instead of $v = n$). These schemes are called "Unbalanced Oil and Vinegar" (UOV), since we have more "vinegar" unknowns than "oil" unknowns. We show that, when $v \simeq n$, the attack of [10] can be extended, but when $v \geq 2n$ for example, the security of the scheme is still an open problem. Moreover, when $v \simeq \frac{n^2}{2}$, the security of the scheme is exactly equivalent (if we accept a very natural but not proved property) to the problem of solving a random set of n quadratic equations in $\frac{n^2}{2}$ unknowns (with no trapdoor). However, we show that (in characteristic 2) when $v \geq n^2$, finding a solution is generally easy. Then we will see that it is very easy to combine the Oil and Vinegar idea and the HFE schemes of [14]. The resulting scheme, called HFEV, looks at the present also very interesting both from a practical and theoretical point of view. The length of a UOV signature can be as short as 192 bits and for HFEV it can be as short as 80 bits.

Note: An extended version of this paper can be obtained from the authors.

1 Introduction

Since 1985, various authors (see [7], [9], [12], [14], [16], [17], [18], [21] for example) have suggested some public key schemes where the public key is given as a set of multivariate quadratic (or higher degree) equations over a small finite field K.

The general problem of solving such a set of equations is NP-hard (cf [8]) (even in the quadratic case). Moreover, when the number of unknowns is, say, $n \geq 16$, the best known algorithms are often not significantly better than exhaustive search (when n is very small, Gröbner bases algorithms are more efficient, cf [6]).

J. Stern (Ed.): EUROCRYPT'99, LNCS 1592, pp. 206–222, 1999.
© Springer-Verlag Berlin Heidelberg 1999

The schemes are often very efficient in terms of speed or RAM required in a smartcard implementation. (However, the length of the public key is generally ≥ 1 Kbyte. Nevertheless, it is sometimes useful to notice that secret key computations can be performed without the public key). The most serious problem is that, in order to introduce a trapdoor (to allow the computation of signatures or to allow the decryption of messages when a secret is known), the generated set of public equations generally becomes a small subset of all the possible equations and, in many cases, the algorithms have been broken. For example [7] was broken by their authors, and [12], [16], [21] were broken. However, many schemes are still not broken (for example [14], [17], [18], [20]), and also in many cases, some very simple variations have been suggested in order to repair the schemes. Therefore, at the present, we do not know whether this idea of designing public key algorithms with multivariate polynomials over small finite fields is a very powerful idea (where only some too simple schemes are insecure) or not.

In this paper, we will present two new schemes: UOV and HFEV. UOV is a very simple scheme: the original Oil and Vinegar signature scheme (of [16]) was broken (see [10]), but if we have significantly more "vinegar" unknowns than "oil" unknowns (a definition of the "oil" and "vinegar" unknowns can be found in section 2), then the attack of [10] does not work and the security of this more general scheme (called UOV) is still an open problem. We will also study Oil and Vinegar schemes of degree three (instead of two). Then, we will present another scheme, called HFEV. HFEV combines the ideas of HFE (of [14]) and of vinegar variables. HFEV looks more efficient than the original HFE scheme. Finally, in section 13, we present what we know about the main schemes in this area of multivariate polynomials.

2   The (Original and Unbalanced) Oil and Vinegar of Degree Two

Let $K = \mathbf{F}_q$ be a small finite field (for example $K = \mathbf{F}_2$). Let n and v be two integers. The message to be signed (or its hash) is represented as an element of K^n, denoted by $y = (y_1, ..., y_n)$. Typically, $q^n \simeq 2^{128}$ (in section 8, we will see that $q^n \simeq 2^{64}$ is also possible). The signature x is represented as an element of K^{n+v} denoted by $x = (x_1, ..., x_{n+v})$.

Secret Key

The secret key is made of two parts:

1. A bijective and affine function $s : K^{n+v} \rightarrow K^{n+v}$. By "affine", we mean that each component of the output can be written as a polynomial of degree one in the $n + v$ input unknowns, and with coefficients in K.
2. A set $(\mathcal{S})$ of n equations of the following type:

$$\forall i,\ 1 \leq i \leq n,\ y_i = \sum \gamma_{ijk} a_j a'_k + \sum \lambda_{ijk} a'_j a'_k + \sum \xi_{ij} a_j + \sum \xi'_{ij} a'_j + \delta_i \quad (\mathcal{S}).$$

The coefficients γ_{ijk}, λ_{ijk}, ξ_{ij}, ξ'_{ij} and δ_i are the secret coefficients of these n equations. The values a_1, ..., a_n (the "oil" unknowns) and a'_1, ..., a'_v (the "vinegar" unknowns) lie in K. Note that these equations (S) contain no terms in $a_i a_j$.

Public Key

Let A be the element of K^{n+v} defined by $A = (a_1, ..., a_n, a'_1, ..., a'_v)$. A is transformed into $x = s^{-1}(A)$, where s is the secret, bijective and affine function from K^{n+v} to K^{n+v}. Each value y_i, $1 \le i \le n$, can be written as a polynomial P_i of total degree two in the x_j unknowns, $1 \le j \le n + v$. We denote by $(\mathcal{P})$ the set of the following n equations:

$$\forall i, \ 1 \le i \le n, \ y_i = P_i(x_1, ..., x_{n+v}) \qquad (\mathcal{P}).$$

These n quadratic equations $(\mathcal{P})$ (in the $n + v$ unknowns x_j) are the public key.

Computation of a Signature (with the Secret Key)

The computation of a signature x of y is performed as follows:

Step 1: We find n unknowns a_1, ..., a_n of K and v unknowns a'_1, ..., a'_v of K such that the n equations (S) are satisfied. This can be done as follows: we randomly choose the v vinegar unknowns a'_i, and then we compute the a_i unknowns from (S) by Gaussian reductions (because – since there are no $a_i a_j$ terms – the (S) equations are affine in the a_i unknowns when the a'_i are fixed).

Remark: If we find no solution, then we simply try again with new random vinegar unknowns. After very few tries, the probability of obtaining at least one solution is very high, because the probability for a $n \times n$ matrix over $\mathbf{F}_q$ to be invertible is not negligible. (It is exactly $\left(1 - \frac{1}{q}\right)\left(1 - \frac{1}{q^2}\right)...\left(1 - \frac{1}{q^{n-1}}\right)$. For $q = 2$, this gives approximately 30 %, and for $q > 2$, this probability is even larger.)

Step 2: We compute $x = s^{-1}(A)$, where $A = (a_1, .., a_n, a'_1, ..., a'_v)$. x is a signature of y.

Public Verification of a Signature

A signature x of y is valid if and only if all the $(\mathcal{P})$ are satisfied. As a result, no secret is needed to check whether a signature is valid: this is an asymmetric signature scheme.

Note: The name "Oil and Vinegar" comes from the fact that – in the equations (S) – the "oil unknowns" a_i and the "vinegar unknowns" a'_j are not all mixed together: there are no $a_i a_j$ products. However, in $(\mathcal{P})$, this property is hidden by the "mixing" of the unknowns by the s transformation. Is this property "hidden enough" ? In fact, this question exactly means: "is the scheme secure ?". When

$v = n$, we call the scheme "Original Oil and Vinegar", since this case was first presented in [16]. This case was broken in [10]. It is very easy to see that the cryptanalysis of [10] also works, exactly in the same way, when $v < n$. However, the cases $v > n$ are, as we will see, much more difficult. When $v > n$, we call the scheme "Unbalanced Oil and Vinegar".

3   Cryptanalysis of the Case $v = n$ (from [10])

The idea of the attack of [10] is essentially the following: In order to separate the oil variables and the vinegar variables, we look at the quadratic forms of the n public equations of $(\mathcal{P})$, we omit for a while the linear terms. Let G_i for $1 \le i \le n$ be the respective matrix of the quadratic form of P_i of the public equations $(\mathcal{P})$. The quadratic part of the equations in the set $(\mathcal{S})$ is represented as a quadratic form with a corresponding $2n \times 2n$ matrix of the form : $\begin{pmatrix} 0 & A \\ B & C \end{pmatrix}$, the upper left $n \times n$ zero submatrix is due to the fact that an oil variable is not multiplied by an oil variable. After hiding the internal variables with the linear function s, we get a representation for the matrices $G_i = S \begin{pmatrix} 0 & A_i \\ B_i & C_i \end{pmatrix} S^t$, where S is an invertible $2n \times 2n$ matrix.

Definition 3.1:   We define the oil subspace to be the linear subspace of all vectors in K^{2n} whose second half contains only zeros.

Definition 3.2:   We define the vinegar subspace as the linear subspace of all vectors in K^{2n} whose first half contains only zeros.

Lemma 1. *Let E and F be a $2n \times 2n$ matrices with an upper left zero $n \times n$ submatrix. If F is invertible then the oil subspace is an invariant subspace of EF^{-1}.*

Proof: see [10].   □

Definition 3.4:   For an invertible matrix G_j, define $G_{ij} = G_i G_j^{-1}$.

Definition 3.5:   Let O be the image of the oil subspace by S^{-1}.
     In order to find the oil subspace, we use the following theorem:

Theorem 3.1. *O is a common invariant subspace of all the matrices G_{ij}.*

Proof:

$$G_{ij} = S \begin{pmatrix} 0 & A_i \\ B_i & C_i \end{pmatrix} S^t (S^t)^{-1} \begin{pmatrix} 0 & A_j \\ B_j & C_j \end{pmatrix}^{-1} S^{-1} = S \begin{pmatrix} 0 & A_i \\ B_i & C_i \end{pmatrix} \begin{pmatrix} 0 & A_j \\ B_j & C_j \end{pmatrix}^{-1} S^{-1}$$

The two inner matrices have the form of E and F in lemma 1. Therefore, the oil subspace is an invariant subspace of the inner term and O is an invariant subspace of $G_i G_j^{-1}$. The problem of finding common invariant subspace of set of matrices is studied in [10]. Applying the algorithms in [10] gives us O. We then pick V to be an arbitrary subspace of dimension n such that $V + O = K^{2n}$, and they give an equivalent oil and vinegar separation. Once we have such a separation, we bring back the linear terms that were omitted, we pick random values for the vinegar variables and left with a set of n linear equations with n oil variables.   □

Note: Lemma 1 is not true any more when $v > n$. The oil subspace is still mapped by E and F into the vinegar subspace. However F^{-1} does not necessary maps the image by E of the oil subspace back into the oil subspace and this is why the cryptanalysis of the original oil and vinegar is not valid for the unbalanced case.

4   Cryptanalysis when $v > n$ and $v \simeq n$

In this section, we will describe a modification of the above attack, that is applicable as long as $v - n$ is small (more precisely the expected complexity of the attack is approximately $q^{(v-n)-1} \cdot n^4$).

Definition 4.1:   We define in this section the oil subspace to be the linear subspace of all vectors in K^{n+v} whose last v coordinates are only zeros.

Definition 4.2:   We define in this section the vinegar subspace to be the linear subspace of all vectors in K^{n+v} whose first n coordinates are only zeros.

Here in this section, we start with the homogeneous quadratic terms of the equations: we omit the linear terms for a while. The matrices G_i have the representation

$$G_i = S \begin{pmatrix} 0 & A_i \\ B_i & C_i \end{pmatrix} S^t$$

where the upper left matrix is the $n \times n$ zero matrix, A_i is a $n \times v$ matrix, B_i is a $v \times n$ matrix, C_i is a $v \times v$ matrix and S is a $(n+v) \times (n+v)$ invertible linear matrix.

Definition 4.3:   Define E_i to be $\begin{pmatrix} 0 & A_i \\ B_i & C_i \end{pmatrix}$.

Lemma 2. *For any matrix E that has the form $\begin{pmatrix} 0 & A \\ B & C \end{pmatrix}$, the following holds:*

a) *E transforms the oil subspace into the vinegar subspace.*
b) *If the matrix E^{-1} exists, then the image of the vinegar subspace by E^{-1} is a subspace of dimension v which contains the n-dimensional oil subspace in it.*

Proof: a) follows directly from the definition of the oil and vinegar subspaces. When a) is given then b) is immediate. □

The algorithm we propose is probabilistic. It looks for an invariant subspace of the oil subspace after it is transformed by S. The probability for the algorithm to succeed on the first try is small. Therefore we need to repeat it with different inputs. We use the following property: any linear combination of the matrices $E_1, ..., E_n$ is also of the form $\begin{pmatrix} 0 & A \\ B & C \end{pmatrix}$. The following theorem explains why an invariant subspace may exist with a certain probability.

Theorem 4.1. *Let F be an invertible linear combination of the matrices $E_1, ..., E_n$. Then for any k such that E_k^{-1} exists, the matrix FE_k^{-1} has a non trivial invariant subspace which is also a subspace of the oil subspace, with probability not less than $\frac{q-1}{q^{2d}-1}$ for $d = v - n$.*

Proof: See the extended version of this paper. □

Note: It is possible to get a better result for the expected number of eigenvectors and with much less effort: I_1 is a subspace with dimension not less than $n-d$ and is mapped by FE_k^{-1} into a subspace with dimension n. The probability for a non zero vector to be mapped to a non zero multiple of itself is $\frac{q-1}{q^n-1}$. To get the expected value, we multiply it by the number of non zero vectors in I_1. It gives a value which is not less than $\frac{(q-1)(q^{n-d}-1)}{q^n-1}$. Since every eigenvector is counted $q - 1$ times, then the expected number of invariant subspcaes of dimension 1 is not less than $\frac{q^{n-d}-1}{q^n-1} \sim q^{-d}$.

We define O as in section 3 and we get the following result for O:

Theorem 4.2. *Let F be an invertible linear combination of the matrices $G_1, ..., G_n$. Then for any k such that G_k^{-1} exists, the matrix FG_k^{-1} has a non trivial invariant subspace, which is also a subspace of O with probability not less than $\frac{q-1}{q^{2d}-1}$ for $d = v - n$.*

Proof:
$$FG_k^{-1} = (\alpha_1 G_1 + ... + \alpha_n G_n)G_k^{-1}$$

$$= S(\alpha_1 E_1 + ... + \alpha_n E_n)S^t(S^t)^{-1}E_k^{-1}S^{-1} = S(\alpha_1 E_1 + ... + \alpha_n E_n)E_k^{-1}S^{-1}.$$

The inner term is an invariant subspace of the oil subspace with the required probability. Therefore, the same will hold for FG_k^{-1}, but instead of a subspace of the oil subspace, we get a subspace of O. □

How to find O ?

We take a random linear combination of $G_1, ..., G_n$ and multiply it by an inverse of one of the G_k matrices. Then we calculate all the minimal invariant subspaces of this matrix (a minimal invariant subspace of a matrix A contains no non trivial invariant subspaces of the matrix A – these subspaces corresponds to irreducible factors of the characteristic polynomial of A). This can be done in probabilistic polynomial time using standard linear algebra techniques. This matrix may have an invariant subspace wich is a subspace of O.

The following lemma enables us to distinguish between subspaces that are contained in O and random subspaces.

Lemma 3. *If H is a linear subspace and $H \subset O$, then for every x, y in H and every i, $G_i(x, y) = 0$ (here we regard G_i as a bilinear form).*

Proof: There are x' and y' in the oil subspace such that $x' = xS$ and $y' = yS$.

$$G_i(x, y) = xS \begin{pmatrix} 0 & A_i \\ B_i & C_i \end{pmatrix} S^t y^t = x' \begin{pmatrix} 0 & A_i \\ B_i & C_i \end{pmatrix} (y')^t = 0.$$

The last term is zero because x' and y' are in the oil subspace. $\square$

Lemma 3 gives a polynomial test to distinguish between subspaces of O and random subspaces. If the matrix we used has no minimal subspace which is also a subspace of O, then we pick another linear combination of G_1, ..., G_n, multiply it by an inverse of one of the G_k matrices and try again. After repeating this process approximately q^{d-1} times, we find with good probability at least one zero vector of O. We continue the process until we get n independent vectors of O. These vectors span O. The expected complexity of the process is proportional to $q^{d-1} \cdot n^4$. We use here the expected number of tries until we find a non trivial invariant subspace and the term n^4 covers the computational linear algebra operations we need to perform for evey try.

5   The Cases $v \simeq \frac{n^2}{2}$ (or $v \geq \frac{n^2}{2}$)

Property

Let $(\mathcal{A})$ be a random set of n quadratic equations in $(n+v)$ variables x_1, ..., x_{n+v}. (By "random" we mean that the coefficients of these equations are uniformly and randomly chosen). When $v \simeq \frac{n^2}{2}$ (and more generally when $v \geq \frac{n^2}{2}$), there is probably – for most of such $(\mathcal{A})$ – a linear change of variables $(x_1, ..., x_{n+v}) \mapsto (x'_1, ..., x'_{n+v})$ such that the set $(\mathcal{A}')$ of $(\mathcal{A})$ equations written in $(x'_1, ..., x'_{n+v})$ is an "Oil and Vinegar" system (i.e. there are no terms in $x'_i \cdot x'_j$ with $i \leq n$ and $j \leq n$).

An Argument to Justify the Property

Let

$$\begin{cases} x_1 & = \alpha_{1,1}x'_1 + \alpha_{1,2}x'_2 + \cdots + \alpha_{1,n+v}x'_{n+v} \\ \quad \vdots \\ x_{n+v} & = \alpha_{n+v,1}x'_1 + \alpha_{n+v,2}x'_2 + \cdots + \alpha_{n+v,n+v}x'_{n+v} \end{cases}$$

By writing that the coefficient in all the n equations of $(\mathcal{A})$ of all the $x_i' \cdot x_j'$ ($i \leq n$ and $j \leq n$) is zero, we obtain a system of $n \cdot n \cdot \frac{n+1}{2}$ quadratic equations in the $(n+v) \cdot n$ variables $\alpha_{i,j}$ ($1 \leq i \leq n+v$, $1 \leq j \leq n$). Therefore, when $v \geq$ approximately $\frac{n^2}{2}$, we may expect to have a solution for this system of equations for most of $(\mathcal{A})$.

Remarks:

1. This argument is very natural, but this is not a complete mathematical proof.
2. The system may have a solution, but finding the solution might be a difficult problem. This is why an Unbalanced Oil and Vinegar scheme might be secure (for well chosen parameters): there is always a linear change of variables that makes the problem easy to solve, but finding such a change of variables might be difficult.
3. In section 7, we will see that, despite the result of this section, it is not recommended to choose $v \geq n^2$ (at least in characteristic 2).

6   Solving a Set of n Quadratic Equations in k Unknowns, $k > n$, Is NP-hard

(See the extended version of this paper.)

7   A Generally (but Not Always) Efficient Algorithm for Solving a Random Set of n Quadratic Equations in n^2 (or More) Unknowns

In this section, we describe an algorithm that solves a system of n randomly chosen quadratic equations in $n+v$ variables, when $v \geq n^2$.

Let $(\mathcal{S})$ be the following system:

$$(\mathcal{S}) \quad \begin{cases} \displaystyle\sum_{1 \leq i \leq j \leq n+v} a_{ij1} x_i x_j + \sum_{1 \leq i \leq n+v} b_{i1} x_i + \delta_1 = 0 \\ \qquad\qquad\qquad \vdots \\ \displaystyle\sum_{1 \leq i \leq j \leq n+v} a_{ijn} x_i x_j + \sum_{1 \leq i \leq n+v} b_{in} x_i + \delta_n = 0 \end{cases}$$

The main idea of the algorithm consists in using a change of variables such as:

$$\begin{cases} x_1 = \alpha_{1,1} y_1 + \alpha_{2,1} y_2 + \ldots + \alpha_{n+v,1} y_{n+v} \\ \qquad \vdots \\ x_{n+v} = \alpha_{1,n+v} y_1 + \alpha_{2,n+v} y_2 + \ldots + \alpha_{n+v,n+v} y_{n+v} \end{cases}$$

whose $\alpha_{i,j}$ coefficients (for $1 \leq i \leq n$, $1 \leq j \leq n+v$) are found step by step, in order that the resulting system $(\mathcal{S}')$ (written with respect to these new variables $y_1, \ldots, y_{n+v}$) is easy to solve.

- We begin by choosing randomly $\alpha_{1,1}, ..., \alpha_{1,n+v}$.
- We then compute $\alpha_{2,1}, ..., \alpha_{2,n+v}$ such that $(\mathcal{S}')$ contains no $y_1 y_2$ terms. This condition leads to a system of n linear equations on the $(n + v)$ unknowns $\alpha_{2,j}$ $(1 \leq j \leq n + v)$:

$$\sum_{1 \leq i \leq j \leq n+v} a_{ijk} \alpha_{1,i} \alpha_{2,j} = 0 \qquad (1 \leq k \leq n).$$

- We then compute $\alpha_{3,1}, ..., \alpha_{3,n+v}$ such that $(\mathcal{S}')$ contains neither $y_1 y_3$ terms, nor $y_2 y_3$ terms. This condition is equivalent to the following system of $2n$ linear equations on the $(n + v)$ unknowns $\alpha_{3,j}$ $(1 \leq j \leq n + v)$:

$$\begin{cases} \displaystyle\sum_{1 \leq i \leq j \leq n+v} a_{ijk} \alpha_{1,i} \alpha_{3,j} = 0 & (1 \leq k \leq n) \\ \displaystyle\sum_{1 \leq i \leq j \leq n+v} a_{ijk} \alpha_{2,i} \alpha_{3,j} = 0 & (1 \leq k \leq n) \end{cases}$$

- ...
- Finally, we compute $\alpha_{n,1}, ..., \alpha_{n,n+v}$ such that $(\mathcal{S}')$ contains neither $y_1 y_n$ terms, nor $y_2 y_n$ terms, ..., nor $y_{n-1} y_n$ terms. This condition gives the following system of $(n-1)n$ linear equations on the $(n + v)$ unknowns $\alpha_{n,j}$ $(1 \leq j \leq n + v)$:

$$\begin{cases} \displaystyle\sum_{1 \leq i \leq j \leq n+v} a_{ijk} \alpha_{1,i} \alpha_{n,j} = 0 & (1 \leq k \leq n) \\ \qquad\qquad \vdots & \\ \displaystyle\sum_{1 \leq i \leq j \leq n+v} a_{ijk} \alpha_{n-1,i} \alpha_{n,j} = 0 & (1 \leq k \leq n) \end{cases}$$

In general, all these linear equations provide at least one solution (found by Gaussian reductions). In particular, the last system of $n(n-1)$ equations and $(n+v)$ unknowns generally gives a solution, as soon as $n + v > n(n-1)$, i.e. $v > n(n-2)$, which is true by hypothesis.

Moreover, the n vectors $\begin{pmatrix} \alpha_{1,1} \\ \vdots \\ \alpha_{1,n+v} \end{pmatrix}, ..., \begin{pmatrix} \alpha_{n,1} \\ \vdots \\ \alpha_{n,n+v} \end{pmatrix}$ are very likely to be linearly independent for a random quadratic system $(\mathcal{S})$.

The remaining $\alpha_{i,j}$ constants (i.e. those with $n + 1 \leq i \leq n + v$ and $1 \leq j \leq n + 1$) are randomly chosen, so as to obtain a bijective change of variables.

By rewriting the system $(\mathcal{S})$ with respect to these new variables y_i, we are led to the following system:

$$(\mathcal{S}') \begin{cases} \displaystyle\sum_{i=1}^{n} \beta_{i,1} y_i^2 + \sum_{i=1}^{n} y_i L_{i,1}(y_{n+1}, ..., y_{n+v}) + Q_1(y_{n+1}, ..., y_{n+v}) = 0 \\ \qquad\qquad\qquad\qquad\qquad \vdots \\ \displaystyle\sum_{i=1}^{n} \beta_{i,n} y_i^2 + \sum_{i=1}^{n} y_i L_{i,n}(y_{n+1}, ..., y_{n+v}) + Q_n(y_{n+1}, ..., y_{n+v}) = 0 \end{cases}$$

where each $L_{i,j}$ is an affine function and each Q_i is a quadratic function.

We then compute $y_{n+1}, ..., y_{n+v}$ such that:

$$\forall i, \; 1 \leq i \leq n, \; \forall j, \; 1 \leq j \leq n+v, \; L_{i,j}(y_{n+1}, ..., y_{n+v}) = 0.$$

This is possible because we have to solve a linear system of n^2 equations and v unknowns, which generally provides at least one solution, as long as $v \geq n^2$. We pick one of these solutions. In general, this gives the y_i^2 by Gaussian reduction.

Then, in characteristic 2, since $x \mapsto x^2$ is a bijection, we will then find easily a solution for the y_i from this expression of the y_i^2. In characteristic $\neq 2$, it will also succeed when 2^n is not too large (i.e. when $n \leq 40$ for example). When n is large, there is also a method to find a solution, based on the general theory of quadratic forms. Due to the lack of space, this method will be found in the extended version of this paper.

8 A Variation with Twice Smaller Signatures

In the UOV described in section 2, the public key is a set of n quadratic equations $y_i = P_i(x_1, ..., x_{n+v})$, for $1 \leq i \leq n$, where $y = (y_1, ..., y_n)$ is the hash value of the message to be signed. If we use a collision-free hash function, the hash value must at least be 128 bits long. Therefore, q^n must be at least 2^{128}, so that the typical length of the signature, if $v = 2n$, is at least $3 \times 128 = 384$ bits.

As we see now, it is possible to make a small variation in the signature design in order to obtain twice smaller signatures. The idea is to keep the same polynomial P_i (with the same associated secret key), but now the public equations that we check are:

$$\forall i, \; P_i(x_1, ..., x_{n+v}) + L_i(y_1, ..., y_n, x_1, ..., x_{n+v}) = 0,$$

where L_i is a linear function in $(x_1, ..., x_{n+v})$ and where the coefficients of L_i are generated by a hash function in $(y_1, ..., y_n)$.

For example $L_i(y_1, ..., y_n, x_1, ..., x_{n+v}) = \alpha_1 x_1 + \alpha_2 x_2 + ... + \alpha_{n+v} x_{n+v}$, where $(\alpha_1, \alpha_2, ..., \alpha_{n+v}) = \text{Hash}(y_1, ..., y_n \| i)$. Now, n can be chosen such that $q^n \geq 2^{64}$ (instead $q^n \geq 2^{128}$). (Note: q^n must be $\geq 2^{64}$ in order to avoid exhaustive search on a solution x). If $v = 2n$ and $q^n \simeq 2^{64}$, the length of the signature will be $3 \times 64 = 192$ bits.

9 Oil and Vinegar of Degree Three

The Scheme

The quadratic Oil and Vinegar schemes described in section 2 can easily be extended to any higher degree. In the case of degree three, the set (S) of hidden equations are of the following type: for all $i \leq n$,

$$y_i = \sum \gamma_{ijk\ell} a_j a'_k a'_\ell + \sum \mu_{ijk\ell} a'_j a'_k a'_\ell + \sum \lambda_{ijk} a'_j a'_k$$
$$+ \sum \nu_{ijk} a'_j a'_k + \sum \xi_{ij} a_j + \sum \xi'_{ij} a'_j + \delta_i \qquad (S).$$

The coefficients γ_{ijk}, $\mu_{ijk\ell}$, λ_{ijk}, ν_{ijk}, ξ_{ij}, ξ'_{ij} and δ_i are the secret coefficients of these n equations. Note that these equations (S) contain no terms in $a_j a_k a_\ell$ or in $a_j a_k$: the equations are affine in the a_j unknowns when the a'_k unknowns are fixed.

The computation of the public key, the computation of a signature and the verification of a signature are done as before.

First Cryptanalysis of Oil and Vinegar of Degree Three when $v \leq n$

We can look at the quadratic part of the public key and attack it exactly as for an Oil and Vinegar of degree two. This is expected to work when $v \leq n$.

Note: If there is no quadratic part (*i.e.* is the public key is homogeneous of degree three), or if this attack does not work, then it is always possible to apply a random affine change of variables and to try again.

Cryptanalysis of Oil and Vinegar of Degree Three when $v \leq (1 + \sqrt{3})n$ and K Is of Characteristic $\neq 2$ (from an Idea of D. Coppersmith, cf [4])

The key idea is to detect a "linearity" in some directions. We search the set V of the values $d = (d_1, ..., d_{n+v})$ such that:

$$\forall x, \forall i, \ 1 \leq i \leq n, \ P_i(x + d) + P_i(x - d) = 2P_i(x) \qquad (\#).$$

By writing that each x_k indeterminate has a zero coefficient, we obtain $n \cdot (n+v)$ quadratic equations in the $(n + v)$ unknowns d_j.

(Each monomial $x_i x_j x_k$ gives $(x_j + d_j)(x_k + d_k)(x_\ell + d_\ell) + (x_j - d_j)(x_k - d_k)(x_\ell - d_\ell) - 2x_j x_k x_\ell$, i.e. $2(x_j d_k d_\ell + x_k d_j d_\ell + x_\ell d_j d_k).$)

Furthermore, the cryptanalyst can specify about $n - 1$ of the coordinates d_k of d, since the vectorial space of the correct d is of dimension n. It remains thus to solve $n \cdot (n + v)$ quadratic equations in $(v + 1)$ unknowns d_j. When v is not too large (typically when $\frac{(v+1)^2}{2} \leq n(n + v)$, i.e. when $v \leq (1 + \sqrt{3})n$), this is expected to be easy. As a result when $v \leq$ approximately $(1 + \sqrt{3})n$ and $|K|$ is odd, this gives a simple way to break the scheme.

Note 1: When v is sensibly greater than $(1 + \sqrt{3})n$ (this is a more unbalanced limit than what we had in the quadratic case), we do not know at the present how to break the scheme.

Note 2: Strangely enough, this cryptanalysis of degre three Oil and Vinegar schemes does not work on degree two Oil and Vinegar schemes. The reason is that – in degree two –writing

$$\forall x, \forall i, \ 1 \leq i \leq n, \ P_i(x + d) + P_i(x - d) = 2P_i(x)$$

only gives n equations of degree two on the $(n + v)$ d_j unknowns (that we do not know how to solve). (Each monomial $x_j x_k$ gives $(x_j + d_j)(x_k + d_k) + (x_j - d_j)(x_k - d_k) - 2x_j x_k$, i.e. $2d_j d_k.$)

Note 3: In degree two, we have seen that Unbalanced Oil and Vinegar public keys are expected to cover almost all the set of n quadratic equations when $v \simeq \frac{n^2}{2}$. In degree three, we have a similar property: the public keys are expected to cover almost all the set of n cubic equations when $v \simeq \frac{n^3}{6}$ (the proof is similar).

10 Another Scheme: HFEV

In the "most simple" HFE scheme (we use the notations of [14]), we have $b = f(a)$, where:

$$ f(a) = \sum_{i,j} \beta_{ij} a^{q^{\theta_{ij}} + q^{\varphi_{ij}}} + \sum_i \alpha_i a^{q^{\xi_i}} + \mu_0, \qquad (1) $$

where β_{ij}, α_i and μ_0 are elements of the field $\mathbf{F}_{q^n}$. Let v be an integer (v will be the number of extra x_i variables, or the number of "vinegar" variables that we will add in the scheme). Let $a' = (a'_1, ..., a'_v)$ be a v-uple of variables of K. Let now each α_i of (1) be an element of $\mathbf{F}_{q^n}$ such that each of the n components of α_i in a basis is a secret random linear function of the vinegar variables a'_1, ..., a'_v. And in (1), let now μ_0 be an element of $\mathbf{F}_{q^n}$ such that each one of the n components of μ_0 in a basis is a secret random quadratic function of the variables a'_1, ..., a'_v. Then, the $n + v$ variables a_1, ..., a_n, a'_1, ..., a'_v will be mixed in the secret affine bijection s in order to obtain the variables x_1, ..., x_{n+v}. And, as before, $t(b_1, ..., b_n) = (y_1, ..., y_n)$, where t is a secret affine bijection. Then the public key is given as the n equations $y_i = P_i(x_1, ..., x_{n+v})$. To compute a signature, the vinegar values a'_1, ..., a'_v will simply be chosen at random. Then, the values μ_0 and α_i will be computed. Then, the monovariate equations (1) will be solved (in a) in $\mathbf{F}_{q^n}$.

Example: Let $K = \mathbf{F}_2$. In HFEV, let for example the hidden polynomial be:

$$ f(a) = a^{17} + \beta_{16}a^{16} + a^{12} + a^{10} + a^9 + \beta_8 a^8 + a^6 + a^5 + \beta_4 a^4 + a^3 + \beta_2 a^2 + \beta_1 a + \beta_0, $$

where $a = (a_1, ..., a_n)$ (a_1, ..., a_n are the "oil" variables), β_1, β_2, β_4, β_8 and β_{16} are given by n secret linear functions on the v vinegar variables and β_0 is given by n secret quadratic functions on the v vinegar variables. In this example, we compute a signature as follows: the vinegar variables are chosen at random and the resulting equation of degree 17 is solved in a.

Note: Unlike UOV, in HFEV we have terms in oil×oil (such as a^{17}, a^{12}, a^{10}, etc), oil×vinegar (such as $\beta_{16}a^{16}$, $\beta_8 a^8$, etc) and vinegar×vinegar (in β_0).

Simulations

Nicolas Courtois did some simulations on HFEV and, in all his simulations, when the number of vinegar variables is ≥ 3, there is no affine multiple equations of small degree (which is very nice). See the extended version of this paper for more details.

11   Concrete Examples of Parameters for UOV

At the present, it seems possible to choose for example $n = 64$, $v = 128$ (or $v = 192$) and $K = \mathbf{F}_2$. The signature scheme is the one of section 8, and the length of a signature is only 192 bits (or 256 bits) in this case. More examples of possible parameters are given in the extended version of this paper.

Note: If we choose $K = \mathbf{F}_2$ then the public key is often large. So it is often more practical to choose a larger K and a smaller n: then the length of the public key can be reduced a lot. However, even when K and n are fixed, it is always feasible to make some easy transformations on a public key in order to obtain the public key in a canonical way such that this canonical expression is slightly shorter than the original expression. See the extended version of this paper for details.

12   Concrete Example of Parameters for HFEV

At the present, it seems possible to choose a small value for v (for example $v = 3$) and a small value for d (for example $n = 77$, $v = 3$, $d = 33$ and $K = \mathbf{F}_2$). The signature scheme is described in the extended version of this paper (to avoid the birthday paradox). Here the length of a signature is only 80 bits ! More examples of possible parameters are given in the extended version of this paper.

13   State of the Art (in May 1999) on Public-Key Schemes with Multivariate Polynomials over a Small Finite Field

Recently, many new ideas have been introduced to design better schemes, such as UOV or HFEV described in this paper. Another idea is to fix some variables to hide some algebraic properties, and another idea is to introduce a few really random quadratic equations and to mix them with the original equations: see the extended version of this paper. However, many new ideas have also been introduced to design better attacks on previous schemes, such as the – not yet published – papers [1], [2], [3], [5]. So the field is fast moving and it can look a bit confusing at first. Moreover, some authors use the word "cryptanalysis" for "breaking" and some authors use this word with the meaning "an analysis about the security" that does not necessary mean "breaking". In this section, we describe what we know at the present about the main schemes.

In the large families of the public key based on multivariate polynomials over a small finite field, we can distinguish between five main families characterized by the way the trapdoor is introduced or by the difficult problem on which the security relies. In the first family are the schemes "with a Hidden Monomial", *i.e.* the key idea is to compute an exponentiation $x \mapsto x^d$ in a finite field for secret key computation. In the second family are the schemes where a polynomial function (with more than one monomial) is hidden. In the third family, the security relies on an isomorphism problem. In the fourth family, the security relies on the difficulty of finding the decomposition of two multivariate quadratic polynomials from all or part of their composition. Finally, in the fifth family, the secret key computations are based on Gaussian computations. The main schemes in these families are described in the figure below. What may be the most interesting scheme in each family is in a rectangle.

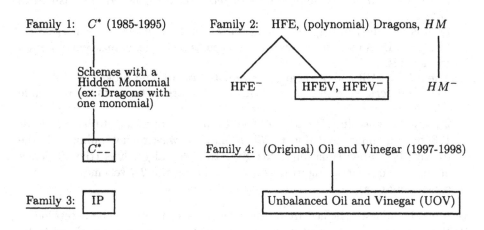

- C^* was the first scheme of all, and it can be seen as the ancestor of all these schemes. It was designed in [12] and broken in [13].
- Schemes with a Hidden Monomial (such as some Dragon schemes) were studied in [15], where it is shown that most of them are insecure. However, C^{*--} (studied in [20]) is (at the present) the most efficient signature scheme (in time and RAM) in a smartcard. The scheme is not broken (but it may seem too simple or too close to C^* to have a large confidence in its security ...).
- HFE was designed in [14]. The most recent results about its security are in [1] and [2]. In these papers, very clever attacks are described. However, at the present, it seems that the scheme is not broken since for well chosen and still reasonable parameters the computations required to break it are still too large. For example, the first challenge of US $500 given in the extended version of [14] has not been claimed yet (it is a pure HFE with $n = 80$ and $d = 96$ over $\mathbf{F}_2$).

- HFE⁻ is just an HFE where some of the public equations are not published. Due to [1] and [2], it may be recommended to do this (despite the fact that original HFE may be secure without it). In the extended version of [14] a second challenge of US $500 is described on a HFE⁻.
- HFEV is described in this paper. HFEV and HFEV⁻ look very hard to break. Moreover, HFEV is more efficient than the original HFE and it can give public key signatures of only 80 bits !
- HM and HM^- were designed in [20]. Very few analysis have been done in these schemes (but maybe we can recommend to use HM^- instead of HM ?).
- IP was designed in [14]. IP schemes have the best proofs of security so far (see [19]). IP is very simple and can be seen as a nice generalization of Graph Isomorphism.
- The original Oil and Vinegar was presented in [16] and broken in [10].
- UOV is described in this paper. With IP, they are certainly the most simple schemes.
- $2R$ was designed in [17] and [18]. Due to [3], it is necessary to have at least 128 bits in input, and due to [5], it may be wise to not publish all the (originally) public equations: this gives the $2R^-$ algorithms (the efficiency of the decomposition algorithms given in [5] on the $2R$ schemes is not yet completely clear).

Remark 1: These schemes are of theoretical interest but (at the exception of IP) their security is not directly relied to a clearly defined and considered to be difficult problem. So is it reasonable to implement them in real products ? We think indeed that it is a bit risky to rely all the security of sensitive applications on such schemes. However, at the present, most of the smartcard applications use secret key algorithms (for example Triple-DES) because RSA smartcards are more expensive. So it can be reasonable to put in a low-cost smartcard one of the previous public key schemes in <u>addition</u> to (not instead of) the existing secret key scheme. Then the security can only be increased and the price of the smartcard would still be low (no coprocessor needed). The security would then rely on a master secret key for the secret key algorithm (with the risk of depending on a master secret key) and on a new low-cost public-key scheme (with the risk that the scheme has no proof! ! ! of security). It can also be noticed that when extremely short signature length (or short block encryption) are required, there is no real choice: at the present only multivariate schemes can have length between 64 and 256 bits.

Remark 2: When a new scheme is found with multivariate polynomials, we do not necessary have to explain how the trapdoor has been introduced. Then we will obtain a kind of "Secret-Public Key scheme" ! The scheme is clearly a Public Key scheme since anybody can verify a signature from the public key (or can encrypt from the public key) and the scheme is secret since the way to compute the secret key computations (*i.e.* the way the trapdoor has been introduced) has not been revealed and cannot be guessed from the public key. For example, we could have done this for HFEV (instead of publishing it).

14   Conclusion

In this paper, we have presented two new public key schemes with "vinegar variables": UOV and HFEV. The study of such schemes has led us to analyze very general properties about the solutions of systems of general quadratic forms. Moreover, from the general view presented in section 13, we see that these two schemes are at the present among the most interesting schemes in two of the five main families of schemes based on multivariate polynomials over a small finite field. Will this still be true in a few years ?

References

1. Anonymous, *Cryptanalysis of the HFE Public Key Cryptosystem*, not yet published.
2. Anonymous, *Practical cryptanalysis of Hidden Field Equations (HFE)*, not yet published.
3. Anonymous, *Cryptanalysis of Patarin's 2-Round Public Key System with S Boxes*, not yet published.
4. D. Coppersmith, *personal communication*, e-mail.
5. Z. Dai, D. Ye, K.-Y. Lam, *Factoring-attacks on Asymmetric Cryptography Based on Mapping-compositions*, not yet published.
6. J.-C. Faugere, *personal communication*.
7. H. Fell, W. Diffie, *Analysis of a public key approach based on polynomial substitutions*, Proceedings of CRYPTO'85, Springer-Verlag, vol. 218, pp. 340-349
8. M. Garey, D. Johnson, *Computers and Intractability, a Guide to the Theory of NP-Completeness*, Freeman, p. 251.
9. H. Imai, T. Matsumoto, *Algebraic Methods for Constructing Asymmetric Cryptosystems*, Algebraic Algorithms and Error Correcting Codes (AAECC-3), Grenoble, 1985, Springer-Verlag, LNCS n°229.
10. A. Kipnis, A. Shamir, *Cryptanalysis of the Oil and Vinegar Signature Scheme*, Proceedings of CRYPTO'98, Springer, LNCS n°1462, pp. 257-266.
11. R. Lidl, H. Niederreiter, *Finite Fields*, Encyclopedia of Mathematics and its applications, volume 20, Cambridge University Press.
12. T. Matsumoto, H. Imai, *Public Quadratic Polynomial-tuples for efficient signature-verification and message-encryption*, Proceedings of EUROCRYPT'88, Springer-Verlag, pp. 419-453.
13. Jacques Patarin, *Cryptanalysis of the Matsumoto and Imai public Key Scheme of Eurocrypt'88*, Proceedings of CRYPTO'95, Springer-Verlag, pp. 248-261.
14. J. Patarin, *Hidden Fields Equations (HFE) and Isomorphisms of Polynomials (IP) : Two New Families of Asymmetric Algorithms*, Proceedings of EUROCRYPT'96, Springer, pp. 33-48.
15. Jacques Patarin, *Asymmetric Cryptography with a Hidden Monomial*, Proceedings of CRYPTO'96, Springer, pp. 45-60.
16. J. Patarin, *The Oil and Vinegar Signature Scheme*, presented at the Dagstuhl Workshop on Cryptography, september 1997 (transparencies).
17. J. Patarin, L. Goubin, *Trapdoor One-way Permutations and Multivariate Polynomials*, Proceedings of ICICS'97, Springer, LNCS n°1334, pp. 356-368.
18. J. Patarin, L. Goubin, *Asymmetric Cryptography with S-Boxes*, Proceedings of ICICS'97, Springer, LNCS n°1334, pp. 369-380.

19. J. Patarin, L. Goubin, N. Courtois, *Improved Algorithms for Isomorphisms of Polynomials*, Proceedings of EUROCRYPT'98, Springer, pp. 184-200.
20. J. Patarin, L. Goubin, N. Courtois, C^*_{-+} *and HM: Variations Around Two Schemes of T. Matsumoto and H. Imai*, Proceedings of ASIACRYPT'98, Springer, pp. 35-49.
21. A. Shamir, *A simple scheme for encryption and its cryptanalysis found by D. Coppersmith and J. Stern*, presented at the Luminy workshop on cryptography, september 1995.

Public-Key Cryptosystems Based on Composite Degree Residuosity Classes

Pascal Paillier[1,2]

[1] GEMPLUS
Cryptography Department
34 Rue Guynemer, 92447 Issy-Les-Moulineaux
paillier@gemplus.com
[2] ENST
Computer Science Department
46, rue Barrault, 75634 Paris Cedex 13
paillier@inf.enst.fr

Abstract. This paper investigates a novel computational problem, namely the Composite Residuosity Class Problem, and its applications to public-key cryptography. We propose a new trapdoor mechanism and derive from this technique three encryption schemes : a trapdoor permutation and two homomorphic probabilistic encryption schemes computationally comparable to RSA. Our cryptosystems, based on usual modular arithmetics, are provably secure under appropriate assumptions in the standard model.

1 Background

Since the discovery of public-key cryptography by Diffie and Hellman [5], very few convincingly secure asymetric schemes have been discovered despite considerable research efforts.

We refer the reader to [26] for a thorough survey of existing public-key cryptosystems. Basically, two major species of trapdoor techniques are in use today. The first points to RSA [25] and related variants such as Rabin-Williams [24,30], LUC, Dickson's scheme or elliptic curve versions of RSA like KMOV [10]. The technique conjugates the polynomial-time extraction of roots of polynomials over a finite field with the intractability of factoring large numbers. It is worthwhile pointing out that among cryptosystems belonging to this family, only Rabin-Williams has been proven equivalent to the factoring problem so far.

Another famous technique, related to Diffie-Hellman-type schemes (El Gamal [7], DSA, McCurley [14], etc.) combines the homomorphic properties of the modular exponentiation and the intractability of extracting discrete logarithms over finite groups. Again, equivalence with the primitive computational problem remains open in general, unless particular circumstances are reached as described in [12].

Other proposed mechanisms generally suffer from inefficiency, inherent security weaknesses or insufficient public scrutiny : McEliece's cryptosystem [15]

J. Stern (Ed.): EUROCRYPT'99, LNCS 1592, pp. 223–238, 1999.
© Springer-Verlag Berlin Heidelberg 1999

based on error correcting codes, Ajtai-Dwork's scheme based on lattice problems (cryptanalyzed by Nguyen and Stern in [18]), additive and multiplicative knapsack-type systems including Merkle-Hellman [13], Chor-Rivest (broken by Vaudenay in [29]) and Naccache-Stern [17] ; finally, Matsumoto-Imai and Goubin-Patarin cryptosystems, based on multivariate polynomials, were successively cryptanalyzed in [11] and [21].

We believe, however, that the cryptographic research had unnoticeably witnessed the progressive emergence of a third class of trapdoor techniques : firstly identified as *trapdoors in the discrete log*, they actually arise from the common algebraic setting of high degree residuosity classes. After Goldwasser-Micali's scheme [9] based on quadratic residuosity, Benaloh's homomorphic encryption function, originally designed for electronic voting and relying on prime residuosity, prefigured the first attempt to exploit the plain resources of this theory. Later, Naccache and Stern [16], and independently Okamoto and Uchiyama [19] significantly extended the encryption rate by investigating two different approaches : residuosity of smooth degree in $\mathbb{Z}_{pq}^*$ and residuosity of prime degree p in $\mathbb{Z}_{p^2q}^*$ respectively. In the meantime, other schemes like Vanstone-Zuccherato [28] on elliptic curves or Park-Won [20] explored the use of high degree residues in other settings.

In this paper, we propose a new trapdoor mechanism belonging to this family. By contrast to prime residuosity, our technique is based on *composite* residuosity classes *i.e.* of degree set to a hard-to-factor number $n = pq$ where p and q are two large prime numbers. Easy to understand, we believe that our trapdoor provides a new cryptographic building-block for conceiving public-key cryptosystems.

In sections 2 and 3, we introduce our number-theoretic framework and investigate in this context a new computational problem (the Composite Residuosity Class Problem), which intractability will be our main assumption. Further, we derive three homomorphic encryption schemes based on this problem, including a new trapdoor permutation. Probabilistic schemes will be proven semantically secure under appropriate intractability assumptions. All our polynomial reductions are simple and stand in the standard model.

Notations. We set $n = pq$ where p and q are large primes : as usual, we will denote by $\phi(n)$ Euler's totient function and by $\lambda(n)$ Carmichael's function[1] taken on n, i.e. $\phi(n) = (p-1)(q-1)$ and $\lambda(n) = \text{lcm}(p-1, q-1)$ in the present case. Recall that $|\mathbb{Z}_{n^2}^*| = \phi(n^2) = n\phi(n)$ and that for any $w \in \mathbb{Z}_{n^2}^*$,

$$\begin{cases} w^\lambda = 1 \bmod n \\ w^{n\lambda} = 1 \bmod n^2 \, , \end{cases}$$

which are due to Carmichael's theorem. We denote by RSA $[n, e]$ the (conventionally thought intractable) problem of extracting e-th roots modulo n where $n = pq$ is of unknown factorisation. The relation $P_1 \Leftarrow P_2$ (resp. $P_1 \equiv P_2$) will denote that the problem P_1 is polynomially reducible (resp. equivalent) to the problem P_2.

---

[1] we will adopt λ instead of $\lambda(n)$ for visual comfort.

2   Deciding Composite Residuosity

We begin by briefly introducing composite degree residues as a natural instance of higher degree residues, and give some basic related facts. The originality of our setting resides in using of a square number as modulus. As said before, $n = pq$ is the product of two large primes.

Definition 1. *A number z is said to be a n-th residue modulo n^2 if there exists a number $y \in \mathbb{Z}_{n^2}^*$ such that*

$$z = y^n \bmod n^2 .$$

The set of n-th residues is a multiplicative subgroup of $\mathbb{Z}_{n^2}^*$ of order $\phi(n)$. Each n-th residue z has exactly n roots of degree n, among which exactly one is strictly smaller than n (namely $\sqrt[n]{z} \bmod n$). The n-th roots of unity are the numbers of the form $(1 + n)^x = 1 + xn \bmod n^2$.

The problem of deciding n-th residuosity, *i.e.* distinguishing n-th residues from non n-th residues will be denoted by CR $[n]$. Observe that like the problems of deciding quadratic or higher degree residuosity, CR $[n]$ is a random-self-reducible problem that is, all of its instances are polynomially equivalent. Each case is thus an average case and the problem is either uniformly intractable or uniformly polynomial. We refer to [1,8] for detailed references on random-self-reducibility and the cryptographic significance of this feature.

As for prime residuosity (*cf.* [3,16]), deciding n-th residuosity is believed to be computationally hard. Accordingly, we will assume that :

Conjecture 2. There exists no polynomial time distinguisher for n-th residues modulo n^2, *i.e.* CR $[n]$ is intractable.

This intractability hypothesis will be refered to as the *Decisional Composite Residuosity Assumption* (DCRA) throughout this paper. Recall that due to the random-self-reducibility, the validity of the DCRA only depends on the choice of n.

3   Computing Composite Residuosity Classes

We now proceed to describe the number-theoretic framework underlying the cryptosystems introduced in sections 4, 5 and 6. Let g be some element of $\mathbb{Z}_{n^2}^*$ and denote by $\mathcal{E}_g$ the integer-valued function defined by

$$\mathbb{Z}_n \times \mathbb{Z}_n^* \longmapsto \mathbb{Z}_{n^2}^*$$
$$(x, y) \longmapsto g^x \cdot y^n \bmod n^2$$

Depending on g, $\mathcal{E}_g$ may feature some interesting properties. More specifically,

Lemma 3. *If the order of g is a nonzero multiple of n then $\mathcal{E}_g$ is bijective.*

We denote by $\mathcal{B}_\alpha \subset \mathbb{Z}_{n^2}^*$ the set of elements of order $n\alpha$ and by $\mathcal{B}$ their disjoint union for $\alpha = 1, \cdots, \lambda$.

Proof. Since the two groups $\mathbb{Z}_n \times \mathbb{Z}_n^*$ and $\mathbb{Z}_{n^2}^*$ have the same number of elements $n\phi(n)$, we just have to prove that $\mathcal{E}_g$ is injective. Suppose that $g^{x_1} y_1^n = g^{x_2} y_2^n \bmod n^2$. It comes $g^{x_2 - x_1} \cdot (y_2/y_1)^n = 1 \bmod n^2$, which implies $g^{\lambda(x_2 - x_1)} = 1 \bmod n^2$. Thus $\lambda(x_2 - x_1)$ is a multiple of g's order, and then a multiple of n. Since $\gcd(\lambda, n) = 1$, $x_2 - x_1$ is necessarily a multiple of n. Consequently, $x_2 - x_1 = 0 \bmod n$ and $(y_2/y_1)^n = 1 \bmod n^2$, which leads to the unique solution $y_2/y_1 = 1$ over $\mathbb{Z}_n^*$. This means that $x_2 = x_1$ and $y_2 = y_1$. Hence, $\mathcal{E}_g$ is bijective. $\qquad\square$

Definition 4. *Assume that $g \in \mathcal{B}$. For $w \in \mathbb{Z}_{n^2}^*$, we call n-th residuosity class of w with respect to g the unique integer $x \in \mathbb{Z}_n$ for which there exists $y \in \mathbb{Z}_n^*$ such that*

$$\mathcal{E}_g(x, y) = w .$$

Adopting Benaloh's notations [3], the class of w is denoted $[\![w]\!]_g$. It is worthwhile noticing the following property :

Lemma 5. $[\![w]\!]_g = 0$ *if and only if w is a n-th residue modulo n^2. Furthermore,*

$$\forall w_1, w_2 \in \mathbb{Z}_{n^2}^* \quad [\![w_1 w_2]\!]_g = [\![w_1]\!]_g + [\![w_2]\!]_g \bmod n$$

that is, the class function $w \mapsto [\![w]\!]_g$ is a homomorphism from $(\mathbb{Z}_{n^2}^*, \times)$ to $(\mathbb{Z}_n, +)$ for any $g \in \mathcal{B}$.

The n-th Residuosity Class Problem of base g, denoted Class $[n, g]$, is defined as the problem of computing the class function in base g : for a given $w \in \mathbb{Z}_{n^2}^*$, compute $[\![w]\!]_g$ from w. Before investigating further Class $[n, g]$'s complexity, we begin by stating the following useful observations :

Lemma 6. *Class $[n, g]$ is random-self-reducible over $w \in \mathbb{Z}_{n^2}^*$.*

Proof. Indeed, we can easily transform any $w \in \mathbb{Z}_{n^2}^*$ into a random instance $w' \in \mathbb{Z}_{n^2}^*$ with uniform distribution, by posing $w' = w g^\alpha \beta^n \bmod n^2$ where α and β are taken uniformly at random over $\mathbb{Z}_n$ (the event $\beta \notin \mathbb{Z}_n^*$ occurs with negligibly small probability). After $[\![w']\!]_g$ has been computed, one has simply to return $[\![w]\!]_g = [\![w']\!]_g - \alpha \bmod n$. $\qquad\square$

Lemma 7. *Class $[n, g]$ is random-self-reducible over $g \in \mathcal{B}$, i.e.*

$$\forall g_1, g_2 \in \mathcal{B} \quad Class\,[n, g_1] \quad \equiv \quad Class\,[n, g_2] .$$

Proof. It can easily be shown that, for any $w \in \mathbb{Z}_{n^2}^*$ and $g_1, g_2 \in \mathcal{B}$, we have

$$[\![w]\!]_{g_1} = [\![w]\!]_{g_2} [\![g_2]\!]_{g_1} \bmod n , \tag{1}$$

which yields $[\![g_1]\!]_{g_2} = [\![g_2]\!]_{g_1}^{-1}$ mod n and thus $[\![g_2]\!]_{g_1}$ is invertible modulo n. Suppose that we are given an oracle for Class $[n, g_1]$. Feeding g_2 and w into the oracle respectively gives $[\![g_2]\!]_{g_1}$ and $[\![w]\!]_{g_1}$, and by straightforward deduction :

$$[\![w]\!]_{g_2} = [\![w]\!]_{g_1} [\![g_2]\!]_{g_1}^{-1} \text{ mod } n .$$

□

Lemma 7 essentially means that the complexity of Class $[n, g]$ is independant from g. This enables us to look upon it as a computational problem which purely relies on n. Formally,

Definition 8. *We call Composite Residuosity Class Problem the computational problem Class $[n]$ defined as follows : given $w \in \mathbb{Z}_{n^2}^*$ and $g \in \mathcal{B}$, compute $[\![w]\!]_g$.*

We now proceed to find out which connections exist between the Composite Residuosity Class Problem and standard number-theoretic problems. We state first :

Theorem 9. *Class $[n]$ $\Leftarrow$ Fact $[n]$.*

Before proving the theorem, observe that the set

$$\mathcal{S}_n = \left\{ u < n^2 \mid u = 1 \text{ mod } n \right\}$$

is a multiplicative subgroup of integers modulo n^2 over which the function L such that

$$\forall u \in \mathcal{S}_n \quad \mathrm{L}(u) = \frac{u-1}{n}$$

is clearly well-defined.

Lemma 10. *For any $w \in \mathbb{Z}_{n^2}^*$, $\mathrm{L}(w^\lambda \bmod n^2) = \lambda [\![w]\!]_{1+n}$ mod n.*

Proof (of Lemma 10). Since $1 + n \in \mathcal{B}$, there exists a unique pair (a, b) in the set $\mathbb{Z}_n \times \mathbb{Z}_n^*$ such that $w = (1+n)^a b^n \bmod n^2$. By definition, $a = [\![w]\!]_{1+n}$. Then

$$w^\lambda = (1+n)^{a\lambda} b^{n\lambda} = (1+n)^{a\lambda} = 1 + a\lambda n \text{ mod } n^2,$$

which yields the announced result.

Proof (of Theorem 9). Since $[\![g]\!]_{1+n} = [\![1 + n]\!]_g^{-1}$ mod n is invertible, a consequence of Lemma 10 is that $\mathrm{L}(g^\lambda \bmod n^2)$ is invertible modulo n. Now, factoring n obviously leads to the knowledge of λ. Therefore, for any $g \in \mathcal{B}$ and $w \in \mathbb{Z}_{n^2}^*$, we can compute

$$\frac{\mathrm{L}(w^\lambda \bmod n^2)}{\mathrm{L}(g^\lambda \bmod n^2)} = \frac{\lambda [\![w]\!]_{1+n}}{\lambda [\![g]\!]_{1+n}} = \frac{[\![w]\!]_{1+n}}{[\![g]\!]_{1+n}} = [\![w]\!]_g \text{ mod } n , \tag{2}$$

by *virtue* of Equation 1.

□

Theorem 11. *Class* $[n]$ $\Leftarrow$ *RSA* $[n, n]$.

Proof. Since all the instances of Class $[n, g]$ are computationally equivalent for $g \in \mathcal{B}$, and since $1 + n \in \mathcal{B}$, it suffices to show that

$$\text{Class}\,[n, 1 + n] \quad \Leftarrow \quad \text{RSA}\,[n, n] \ .$$

Let us be given an oracle for RSA $[n, n]$. We know that $w = (1 + n)^x \cdot y^n \bmod n^2$ for some $x \in \mathbb{Z}_n$ and $y \in \mathbb{Z}_n^*$. Therefore, we have $w = y^n \bmod n$ and we get y by giving $w \bmod n$ to the oracle. From now,

$$\frac{w}{y^n} = (1 + n)^x = 1 + xn \bmod n^2 \ ,$$

which discloses $x = [\![w]\!]_{1+n}$ as announced.   □

Theorem 12. *Let D-Class* $[n]$ *be the decisional problem associated to Class* $[n]$ *i.e. given* $w \in \mathbb{Z}_{n^2}^*$, $g \in \mathcal{B}$ *and* $x \in \mathbb{Z}_n$, *decide whether* $x = [\![w]\!]_g$ *or not. Then*

$$CR\,[n] \quad \equiv \quad \text{D-Class}\,[n] \quad \Leftarrow \quad \text{Class}\,[n] \ .$$

Proof. The hierarchy D-Class $[n]$ $\Leftarrow$ Class $[n]$ comes from the general fact that it is easier to verify a solution than to compute it. Let us prove the left-side equivalence. ($\Rightarrow$) Submit $wg^{-x} \bmod n^2$ to the oracle solving CR $[n]$. In case of n-th residuosity detection, the equality $[\![wg^{-x}]\!]_g = 0$ implies $[\![w]\!]_g = x$ by Lemma 5 and then answer "Yes". Otherwise answer "No" or "Failure" according to the oracle's response. ($\Leftarrow$) Choose an arbitrary $g \in \mathcal{B}$ ($1 + n$ will do) and submit the triple $(g, w, x = 0)$ to the oracle solving D-Class $[n]$. Return the oracle's answer without change.   □

To conclude, the computational hierarchy we have been looking for was

$$CR\,[n] \equiv \text{D-Class}\,[n] \Leftarrow \text{Class}\,[n] \Leftarrow \text{RSA}\,[n, n] \Leftarrow \text{Fact}\,[n] \ , \qquad (3)$$

with serious doubts concerning a potential equivalence, excepted possibly between D-Class $[n]$ and Class $[n]$. Our second intractability hypothesis will be to assume the hardness of the Composite Residuosity Class Problem by making the following conjecture :

Conjecture 13. There exists no probabilistic polynomial time algorithm that solves the Composite Residuosity Class Problem, *i.e.* Class $[n]$ is intractable.

By contrast to the Decisional Composite Residuosity Assumption, this conjecture will be refered to as the *Computational Composite Residuosity Assumption* (CCRA). Here again, random-self-reducibility implies that the validity of the CCRA is only conditioned by the choice of n. Obviously, if the DCRA is true then the CCRA is true as well. The converse, however, still remains a challenging open question.

4   A New Probabilistic Encryption Scheme

We now proceed to describe a public-key encryption scheme based on the Composite Residuosity Class Problem. Our methodology is quite natural : employing $\mathcal{E}_g$ for encryption and the polynomial reduction of Theorem 9 for decryption, using the factorisation as a trapdoor.

Set $n = pq$ and randomly select a base $g \in \mathcal{B}$: as shown before, this can be done efficiently by checking whether

$$\gcd\left(L(g^\lambda \bmod n^2), n\right) = 1 . \tag{4}$$

Now, consider (n, g) as public parameters whilst the pair (p, q) (or equivalently λ) remains private. The cryptosystem is depicted below.

Encryption :

plaintext $m < n$

select a random $r < n$

ciphertext $c = g^m \cdot r^n \bmod n^2$

Decryption :

ciphertext $c < n^2$

plaintext $m = \dfrac{L(c^\lambda \bmod n^2)}{L(g^\lambda \bmod n^2)} \bmod n$

Scheme 1. Probabilistic Encryption Scheme Based on Composite Residuosity.

The correctness of the scheme is easily verified from Equation 2, and it is straightforward that the encryption function is a trapdoor function with λ (that is, the knowledge of the factors of n) as the trapdoor secret. One-wayness is based on the computational problem discussed in the previous section.

Theorem 14. *Scheme 1 is one-way if and only if the Computational Composite Residuosity Assumption holds.*

Proof. Inverting our scheme is by definition the Composite Residuosity Class Problem. ☐

Theorem 15. *Scheme 1 is semantically secure if and only if the Decisional Composite Residuosity Assumption holds.*

Proof. Assume that m_0 and m_1 are two known messages and c the ciphertext of either m_0 or m_1. Due to Lemma 5, c is the ciphertext of m_0 if and only if $cg^{-m_0} \bmod n^2$ is a n-th residue. Therefore, a successfull chosen-plaintext attacker could decide composite residuosity, and *vice-versa*. ☐

5   A New One-Way Trapdoor Permutation

One-way trapdoor permutations are very rare cryptographic objects : we refer the reader to [22] for an exhaustive documentation on these. In this section, we show how to use the trapdoor technique introduced in the previous section to derive a permutation over $\mathbb{Z}_{n^2}^*$.

As before, n stands for the product of two large primes and g is chosen as in Equation 4.

Encryption :

        plaintext $m < n^2$

        split m into m_1, m_2 such that $m = m_1 + nm_2$

        ciphertext $c = g^{m_1} m_2^{\ n} \bmod n^2$

Decryption :

        ciphertext $c < n^2$

Step 1. $m_1 = \dfrac{L(c^\lambda \bmod n^2)}{L(g^\lambda \bmod n^2)} \bmod n$

Step 2. $c' = cg^{-m_1} \bmod n$

Step 3. $m_2 = c'^{n^{-1} \bmod \lambda} \bmod n$

        plaintext $m = m_1 + nm_2$

Scheme 2. A Trapdoor Permutation Based on Composite Residuosity.

We first show the scheme's correctness. Clearly, Step 1 correctly retrieves $m_1 = m \bmod n$ as in Scheme 1. Step 2 is actually an unblinding phase which is necessary to recover $m_2^n \bmod n$. Step 3 is an RSA decryption with a public exponent $e = n$. The final step recombines[2] the original message m. The fact that Scheme 2 is a permutation comes from the bijectivity of $\mathcal{E}_g$. Again, trapdoorness is based on the factorisation of n. Regarding one-wayness, we state :

Theorem 16. *Scheme 2 is one-way if and only if* $RSA\,[n,n]$ *is hard.*

Proof. a) Since Class $[n]$ $\Leftarrow$ RSA $[n,n]$ (Theorem 11), extracting n-th roots modulo n is sufficient to compute m_1 from $\mathcal{E}_g(m_1, m_2)$. Retrieving m_2 then requires one more additionnal extraction. Thus, inverting Scheme 2 cannot be harder than extracting n-th roots modulo n. b) Conversely, an oracle which inverts Scheme 2 allows root extraction : first query the oracle to get the two

---

[2] note that every public bijection $m \leftrightarrow (m_1, m_2)$ fits the scheme's structure, but euclidean division appears to be the most natural one.

numbers a and b such that $1 + n = g^a b^n \bmod n^2$. Now if $w = y_0^n \bmod n$, query the oracle again to obtain x and y such that $w = g^x y^n \bmod n^2$. Since $1 + n \in \mathcal{B}$, we know there exists an x_0 such that $w = (1+n)^{x_0} y_0^n \bmod n^2$, wherefrom

$$w = (g^a b^n)^{x_0} y_0^n = g^{a x_0 \bmod n} \left(g^{a x_0 \text{ div } n} b^{x_0} y_0 \right)^n \bmod n^2 .$$

By identification with $w = g^x y^n \bmod n^2$, we get $x_0 = x a^{-1} \bmod n$ and finally $y_0 = y g^{-(a x_0 \text{ div } n)} b^{-x_0} \bmod n$ which is the wanted value. $\qquad\square$

Remark 17. Note that by definition of $\mathcal{E}_g$, the cryptosystem requires that $m_2 \in \mathbb{Z}_n^*$, just like in the RSA setting. The case $m_2 \notin \mathbb{Z}_n^*$ either allows to factor n or leads to the ciphertext zero for all possible values of m_1. A consequence of this fact is that our trapdoor permutation cannot be employed *ad hoc* to encrypt short messages *i.e.* messages smaller than n.

Digital Signatures. Finally, denoting by $h : \mathbb{N} \mapsto \{0,1\}^k \subset \mathbb{Z}_{n^2}^*$ a hash function see as a random oracle [2], we obtain a digital signature scheme as follows. For a given message m, the signer computes the signature (s_1, s_2) where

$$
\begin{cases}
s_1 = \dfrac{L(h(m)^\lambda \bmod n^2)}{L(g^\lambda \bmod n^2)} \bmod n \\[2mm]
s_2 = \left(h(m) g^{-s_1} \right)^{1/n \bmod \lambda} \bmod n
\end{cases}
$$

and the verifier checks that

$$h(m) \stackrel{?}{=} g^{s_1} s_2^n \bmod n^2 .$$

Corollary 18 (of Theorem 16). *In the random oracle model, an existential forgery of our signature scheme under an adaptive chosen message attack has a negligible success probability provided that RSA $[n, n]$ is intractable.*

Although we feel that the above trapdoor permutation remains of moderate interest due to its equivalence with RSA, the rarity of such objects is such that we find it useful to mention its existence. Moreover, the homomorphic properties of this scheme, discussed in section 8, could be of a certain utility regarding some (still unresolved) cryptographic problems.

6   Reaching Almost-Quadratic Decryption Complexity

Most popular public-key cryptosystems present a cubic decryption complexity, and this is the case for Scheme 1 as well. The fact that no faster (and still appropriately secure) designs have been proposed so far strongly motivates the search for novel trapdoor functions allowing increased decryption performances. This section introduces a slightly modified version of our main scheme (Scheme 1) which features an $\mathcal{O}\left(|n|^{2+\epsilon}\right)$ decryption complexity.

Here, the idea consists in restricting the ciphertext space $\mathbb{Z}_{n^2}^*$ to the subgroup $<g>$ of smaller order by taking advantage of the following extension of Equation 2. Assume that $g \in \mathcal{B}_\alpha$ for some $1 \leq \alpha \leq \lambda$. Then for any $w \in <g>$,

$$[\![w]\!]_g = \frac{L(w^\alpha \bmod n^2)}{L(g^\alpha \bmod n^2)} \bmod n \ . \tag{5}$$

This motivates the cryptosystem depicted below.

Encryption :
>   plaintext $m < n$
>   randomly select $r < n$
>   ciphertext $c = g^{m+nr} \bmod n^2$

Decryption :
>   ciphertext $c < n^2$
>   plaintext $m = \dfrac{L(c^\alpha \bmod n^2)}{L(g^\alpha \bmod n^2)} \bmod n$

Scheme 3. Variant with fast decryption.

Note that this time, the encryption function's trapdoorness relies on the knowledge of α (instead of λ) as secret key. The most computationally expensive operation involved in decryption is the modular exponentiation $c \to c^\alpha \bmod n^2$ which runs in complexity $\mathcal{O}\left(|n|^2|\alpha|\right)$ (to be compared to $\mathcal{O}\left(|n|^3\right)$ in Scheme 1). If g is chosen in such a way that $|\alpha| = \Omega\left(|n|^\epsilon\right)$ for some $\epsilon > 0$, then decryption will only take $\mathcal{O}\left(|n|^{2+\epsilon}\right)$ bit operations. To the best of our knowledge, Scheme 3 is the only public-key cryptosystem based on modular arithmetics whose decryption function features such a property.

Clearly, inverting the encryption function does not rely on the composite residuosity class problem, since this time the ciphertext is known to be an element of $<g>$, but on a weaker instance. More formally,

Theorem 19. *We call Partial Discrete Logarithm Problem the computational problem PDL $[n, g]$ defined as follows : given $w \in <g>$, compute $[\![w]\!]_g$. Then Scheme 3 is one-way if and only if PDL $[n, g]$ is hard.*

Theorem 20. *We call Decisional Partial Discrete Logarithm Problem the decisional problem D-PDL $[n, g]$ defined as follows : given $w \in <g>$ and $x \in \mathbb{Z}_n$, decide whether $[\![w]\!]_g = x$. Then Scheme 3 is semantically secure if and only if D-PDL $[n, g]$ is hard.*

The proofs are similar to those given in section 4. By opposition to the original class problems, these ones are not random-self-reducible over $g \in \mathcal{B}$ but over cyclic subgroups of $\mathcal{B}$, and present other interesting characteristics that we do not discuss here due to the lack of space. Obviously,

$$\text{PDL}\,[n,g] \quad \Leftarrow \quad \text{Class}\,[n] \quad \text{and} \quad \text{D-PDL}\,[n,g] \quad \Leftarrow \quad \text{CR}\,[n]$$

but equivalence can be reached when g is of maximal order $n\lambda$ and n the product of two safe primes. When $g \in \mathcal{B}_\alpha$ for some $\alpha < \lambda$ such that $|\alpha| = \Omega\,(|n|^\epsilon)$ for $\epsilon > 0$, we conjecture that both PDL $[n,g]$ and D-PDL $[n,g]$ are intractable.

In order to thwart Baby-Step Giant-Step attacks, we recommend the use of 160-bit prime numbers for αs in practical use. This can be managed by an appropriate key generation. In this setting, the computational load of Scheme 3 is smaller than a RSA decryption with Chinese Remaindering for $|n| \geq 1280$. Next section provides tight evaluations and performance comparisons for all the encryption schemes presented in this paper.

7   Efficiency and Implementation Aspects

In this section, we briefly analyse the main practical aspects of computations required by our cryptosystems and provide various implementation strategies for increased performance.

Key Generation. The prime factors p and q must be generated according to the usual recommandations in order to make n as hard to factor as possible. The fast variant (Scheme 3) requires additionally $\lambda = \text{lcm}(p-1, q-1)$ to be a multiple of a 160-bit prime integer, which can be managed by usual DSA-prime generation or other similar techniques. The base g can be chosen randomly among elements of order divisible by n, but note that the fast variant will require a specific treatment (typically raise an element of maximal order to the power λ/α). The whole generation may be made easier by carrying out computations separately mod p^2 and mod q^2 and Chinese-remaindering $g \bmod p^2$ and $g \bmod q^2$ at the very end.

Encryption. Encryption requires a modular exponentiation of base g. The computation may be significantly accelerated by a judicious choice of g. As an illustrative example, taking $g = 2$ or small numbers allows an immediate speed-up factor of $1/3$, provided the chosen value fulfills the requirement $g \in \mathcal{B}$ imposed by the setting. Optionally, g could even be fixed to a constant value if the key generation process includes a specific adjustment. At the same time, pre-processing techniques for exponentiating a constant base can dramatically reduce the computational cost. The second computation r^n or $g^{nr} \bmod n^2$ can also be computed in advance.

Decryption. Computing $L(u)$ for $u \in \mathcal{S}_n$ may be achieved at a very low cost (only one multiplication modulo $2^{|n|}$) by precomputing $n^{-1} \bmod 2^{|n|}$. The constant parameter

$$L(g^\lambda \bmod n^2)^{-1} \bmod n \quad \text{or} \quad L(g^\alpha \bmod n^2)^{-1} \bmod n$$

can also be precomputed once for all.

Decryption using Chinese-remaindering. The Chinese Remainder Theorem [6] can be used to efficiently reduce the decryption workload of the three cryptosystems. To see this, one has to employ the functions L_p and L_q defined over

$$S_p = \{x < p^2 \mid x = 1 \bmod p\} \quad \text{and} \quad S_q = \{x < q^2 \mid x = 1 \bmod q\}$$

by

$$L_p(x) = \frac{x-1}{p} \quad \text{and} \quad L_q(x) = \frac{x-1}{q} .$$

Decryption can therefore be made faster by separately computing the message mod p and mod q and recombining modular residues afterwards :

$$m_p = L_p(c^{p-1} \bmod p^2)\, h_p \bmod p$$
$$m_q = L_q(c^{q-1} \bmod q^2)\, h_q \bmod q$$
$$m = \text{CRT}(m_p, m_q) \bmod pq$$

with precomputations

$$h_p = L_p(g^{p-1} \bmod p^2)^{-1} \bmod p \quad \text{and}$$
$$h_q = L_q(g^{q-1} \bmod q^2)^{-1} \bmod q .$$

where $p - 1$ and $q - 1$ have to be replaced by α in the fast variant.

Performance evaluations. For each $|n| = 512, \cdots, 2048$, the modular multiplication of bitsize $|n|$ is taken as the unitary operation, we assume that the execution time of a modular multiplication is quadratic in the operand size and that modular squares are computed by the same routine. Chinese remaindering, as well as random number generation for probabilistic schemes, is considered to be negligible. The RSA public exponent is taken equal to $F_4 = 2^{16} + 1$. The parameter g is set to 2 in our main scheme, as well as in the trapdoor permutation. Other parameters, secret exponents or messages are assumed to contain about the same number of ones and zeroes in their binary representation.

Schemes	Main Scheme	Permutation	Fast Variant	RSA	ElGamal
One-wayness	Class $[n]$	RSA $[n, n]$	PDL $[n, g]$	RSA $[n, F_4]$	DH $[p]$
Semantic Sec.	CR $[n]$	none	D-PDL $[n, g]$	none	D-DH $[p]$
Plaintext size	$|n|$	$2|n|$	$|n|$	$|n|$	$|p|$
Ciphertext size	$2|n|$	$2|n|$	$2|n|$	$|n|$	$2|p|$

Encryption					
$|n|, |p| = 512$	5120	5120	4032	**17**	1536
$|n|, |p| = 768$	7680	7680	5568	**17**	2304
$|n|, |p| = 1024$	10240	10240	7104	**17**	3072
$|n|, |p| = 1536$	15360	1536	10176	**17**	4608
$|n|, |p| = 2048$	20480	20480	13248	**17**	6144

Decryption					
$|n|, |p| = 512$	768	1088	480	**192**	768
$|n|, |p| = 768$	1152	1632	480	**288**	1152
$|n|, |p| = 1024$	1536	2176	480	**384**	1536
$|n|, |p| = 1536$	2304	3264	**480**	576	2304
$|n|, |p| = 2048$	3072	4352	**480**	768	3072

These estimates are purely indicative, and do not result from an actual implementation. We did not include the potential pre-processing stages. Chinese remaindering is taken into account in cryptosystems that allow it *i.e.* all of them excepted ElGamal.

8   Properties

Before concluding, we would like to stress again the algebraic characteristics of our cryptosystems, especially those of Schemes 1 and 3.

Random-Self-Reducibility. This property actually concerns the underlying number-theoretic problems CR $[n]$ and Class $[n]$ and, to some extent, their weaker versions D-PDL $[n, g]$ and PDL $[n, g]$. Essentially, random-self-reducible problems are as hard on average as they are in the worst case : both RSA and the Discrete Log problems have this feature. Problems of that type are believed to yield good candidates for one-way functions [1].

Additive Homomorphic Properties. As already seen, the two encryption functions $m \mapsto g^m r^n \bmod n^2$ and $m \mapsto g^{m+nr} \bmod n^2$ are additively homomorphic on $\mathbb{Z}_n$. Practically, this leads to the following identities :

$$\forall m_1, m_2 \in \mathbb{Z}_n \quad \text{and} \quad k \in \mathbb{N}$$

$$\mathrm{D}(\mathrm{E}(m_1)\,\mathrm{E}(m_2) \bmod n^2) = m_1 + m_2 \bmod n$$

$$\mathrm{D}(\mathrm{E}(m)^k \bmod n^2) = km \bmod n$$

$$\mathrm{D}(\mathrm{E}(m_1)\,g^{m_2} \bmod n^2) = m_1 + m_2 \bmod n$$

$$\left.\begin{array}{c} \mathrm{D}(\mathrm{E}(m_1)^{m_2} \bmod n^2) \\ \mathrm{D}(\mathrm{E}(m_2)^{m_1} \bmod n^2) \end{array}\right\} = m_1 m_2 \bmod n \ .$$

These properties are known to be particularly appreciated in the design of voting protocols, threshold cryptosystems, watermarking and secret sharing schemes, to quote a few. Server-aided polynomial evaluation (see [27]) is another potential field of application.

Self-Blinding. Any ciphertext can be publicly changed into another one without affecting the plaintext :

$$\forall m \in \mathbb{Z}_n \quad \text{and} \quad r \in \mathbb{N}$$

$$\mathrm{D}(\mathrm{E}(m)\,r^n \bmod n^2) = m \qquad \text{or} \qquad \mathrm{D}(\mathrm{E}(m)\,g^{nr} \bmod n^2) = m \ ,$$

depending on which cryptosystem is considered. Such a property has potential applications in a wide range of cryptographic settings.

9   Further Research

In this paper, we introduced a new number-theoretic problem and a related trapdoor mechanism based on the use of composite degree residues. We derived three new cryptosystems based on our technique, all of which are provably secure under adequate intractability assumptions.

Although we do not provide any proof of security against chosen ciphertext attacks, we believe that one could bring slight modifications to Schemes 1 and 3 to render them resistant against such attacks, at least in the random oracle model.

Another research topic resides in exploiting the homomorphic properties of our systems to design distributed cryptographic protocols (multi-signature, secret sharing, threshold cryptography, and so forth) or other cryptographically useful objects.

10   Acknowledgments

The author is especially grateful to David Pointcheval for his precious comments and contributions to this work. We also thank Jacques Stern and an anonymous referee for having (independently) proved that Class $[n] \Leftarrow \text{RSA} [n, n]$. Finally, Dan Boneh, Jean-Sébastien Coron, Helena Handschuh and David Naccache are acknowledged for their helpful discussions and comments during the completion of this work.

References

1. D. Angluin and D. Lichtenstein, *Provable Security of Cryptosystems: A Survey*, Computer Science Department, Yale University, TR-288, 1983.

2. M. Bellare and P. Rogaway, *Random Oracles are Practical : a Paradigm for Designing Efficient Protocols*, In Proceedings of the First CCS, ACM Press, pp. 62–73, 1993.

3. J. C. Benaloh, *Verifiable Secret-Ballot Elections*, PhD Thesis, Yale University, 1988.

4. R. Cramer, R. Gennaro and B. Schoenmakers, *A Secure And Optimally Efficient Multi-Authority Election Scheme*, LNCS 1233, Proceedings of Eurocrypt'97, Springer-Verlag, pp. 103-118, 1997.

5. W. Diffie and M. Hellman, *New Directions in Cryptography*, IEEE Transaction on Information Theory, IT-22,6, pp. 644–654, 1995.

6. C. Ding, D. Pei and A. Salomaa, *Chinese Remainder Theorem - Applications in Computing, Coding, Cryptography*, World Scientific Publishing, 1996.

7. T. ElGamal, *A Public-Key Cryptosystem an a Signature Scheme Based on Discrete Logarithms*, IEEE Trans. on Information Theory, IT-31, pp. 469–472, 1985.

8. J. Feigenbaum, *Locally Random Reductions in Interactive Complexity Theory*, in Advances in Computational Complexity Theory, DIMACS Series on Discrete Mathematics and Theoretical Computer Science, vol. 13, American Mathematical Society, Providence, pp. 73–98, 1993.

9. S. Goldwasser and S. Micali, *Probabilistic Encryption*, JCSS Vol. 28 No 2, pp. 270–299, 1984.

10. K. Koyama, U. Maurer, T. Okamoto and S. Vanstone, *New Public-Key Schemes based on Elliptic Curves over the ring Zn*, LNCS 576, Proceedings of Crypto'91, Springer-Verlag, pp. 252–266, 1992.

11. T. Matsumoto and H. Imai, *Public Quadratic Polynomial-Tuples for Efficient Signature-Verification and Message-Encryption*, LNCS 330, Proceedings of Eurocrypt'88, Springer-Verlag, pp. 419–453, 1988.

12. U. Maurer and S. Wolf, *On the Complexity of Breaking the Diffie-Hellman Protocol.*

13. R. Merkle and M. Hellman, *Hiding Information and Signatures in Trapdoor Knapsacks*, IEEE Trans. on Information Theory, Vol. 24, pp. 525–530, 1978.

14. K. McCurley, *A Key Distribution System Equivalent to Factoring*, Journal of Cryptology, Vol. 1, pp. 95–105, 1988.

15. R. McEliece, *A Public-Key Cryptosystem Based on Algebraic Coding Theory*, DSN Progress Report 42-44, Jet Propulsion Laboratories, Pasadena, 1978.

16. D. Naccache and J. Stern, *A New Public-Key Cryptosystem Based on Higher Residues*, LNCS 1403, Advances in Cryptology, Proceedings of Eurocrypt'98, Springer-Verlag, pp. 308–318, 1998.

17. D. Naccache and J. Stern, *A New Public-Key Cryptosystem*, LNCS 1233, Advances in Cryptology, Proceedings of Eurocrypt'97, Springer-Verlag, pp. 27–36, 1997.

18. P. Nguyen and J. Stern, *Cryptanalysis of the Ajtai-Dwork Cryptosystem*, LNCS 1462, Proceedings of Crypto'98, Springer-Verlag, pp. 223–242, 1998.

19. T. Okamoto and S. Uchiyama, *A New Public-Key Cryptosystem as secure as Factoring*, LNCS 1403, Advances in Cryptology, Proceedings of Eurocrypt'98, Springer-Verlag, pp. 308–318, 1998.

20. S. Park and D. Won, *A Generalization of Public-Key Residue Cryptosystem*, In Proceedings of 1993 Korean-Japan Joint Workshop on Information Security and Cryptology, pp. 202–206, 1993.

21. J. Patarin, *The Oil and Vinegar Algorithm for Signatures*, presented at the Dagstuhl Workshop on Cryptography, 1997.

22. J. Patarin and L. Goubin, *Trapdoor One-Way Permutations and Multivariate Polynomials*, LNCS 1334, Proceedings of ICICS'97, Springer-Verlag, pp. 356–368, 1997.

23. R. Peralta and E. Okamoto, *Faster Factoring of Integers of a Special Form*, IEICE, Trans. Fundamentals, E79-A, Vol. 4, pp. 489–493, 1996.

24. M. Rabin, *Digital Signatures and Public-Key Encryptions as Intractable as Factorization*, MIT Technical Report No 212, 1979.

25. R. Rivest, A. Shamir and L. Adleman, *A Method for Obtaining Digital Signatures and Public-Key Cryptosystems*, Communications of the ACM, Vol. 21, No 2, pp. 120–126, 1978.

26. A. Salomaa, *Public-Key Cryptography*, Springer-Verlag, 1990.

27. T. Sander and F. Tschudin, *On Software Protection Via Function Hiding*, Proceedings of Information Hiding Workshop'98, 1998.

28. S. Vanstone and R. Zuccherato, *Elliptic Curve Cryptosystem Using Curves of Smooth Order Over the Ring Z_n*, IEEE Trans. Inf. Theory, Vol. 43, No 4, July 1997.

29. S. Vaudenay, *Cryptanalysis of the Chor-Rivest Cryptosystem*, LNCS 1462, Proceedings of Crypto'98, Springer-Verlag, pp. 243–256, 1998.

30. H. Williams, *Some Public-Key Crypto-Functions as Intractable as Factorization*, LNCS 196, Proceedings of Crypto'84, Springer-Verlag, pp. 66–70, 1985.

New Public Key Cryptosystems
Based on the Dependent–RSA Problems

David Pointcheval

LIENS – CNRS, École Normale Supérieure,
45 rue d'Ulm, 75230 Paris Cedex 05, France.
David.Pointcheval@ens.fr
http://www.dmi.ens.fr/~pointche

Abstract. Since the Diffie-Hellman paper, asymmetric encryption has been a very important topic, and furthermore ever well studied. However, between the efficiency of RSA and the security of some less efficient schemes, no trade-off has ever been provided.
In this paper, we propose better than a trade-off: indeed, we first present a new problem, derived from the RSA assumption, the "Dependent–RSA Problem". A careful study of its difficulty is performed and some variants are proposed, namely the "Decisional Dependent–RSA Problem".
They are next used to provide new encryption schemes which are both secure and efficient. More precisely, the main scheme is proven semantically secure in the standard model. Then, two variants are derived with improved security properties, namely against adaptive chosen-ciphertext attacks, in the random oracle model. Furthermore, all those schemes are more or less as efficient as the original RSA encryption scheme and reach semantic security.

Keywords: Public-Key Encryption, Semantic Security, Chosen-Ciphertext Attacks, the Dependent–RSA Problem

Introduction

Since the seminal Diffie-Hellman paper [9], which presented the foundations of the asymmetric cryptography, public-key cryptosystems have been an important goal for many people. In 1978, the RSA cryptosystem [20] was the first application and remains the most popular scheme. However, it does not satisfy any security criterion (*e.g.*, the RSA encryption standard PKCS #1 v1.5 has even been recently broken [4]) and was subject to numerous attacks (broadcast [13], related messages [7], etc).

Notions of Security. In 1984, Goldwasser and Micali [12] defined some security notions that an encryption scheme should satisfy, namely *indistinguishability of encryptions* (a.k.a. *polynomial security* or *semantic security*). This notion means that a ciphertext does not leak any useful information about the plaintext, but its length, to a polynomial time attacker. For example, if an attacker knows that the plaintext is either "sell" or "buy", the ciphertext does not help him.

J. Stern (Ed.): EUROCRYPT'99, LNCS 1592, pp. 239–254, 1999.
© Springer-Verlag Berlin Heidelberg 1999

By the meantime, El Gamal [11] proposed a probabilistic encryption scheme based on the Diffie-Hellman problem [9]. Its semantic security, relative to the Decisional Diffie-Hellman problem, was formally proven just last year [23], even if the result was informally well known. However this scheme never got very popular because of its computational load.

During the last ten years, beyond semantic security, a new security notion has been defined: the *non-malleability* [10]. Moreover, some stronger scenarios of attacks have been considered: the *(adaptive) chosen-ciphertext attacks* [16,19]. More precisely, the non-malleability property means that any attacker cannot modify a ciphertext while keeping any control over the relation between the resulting plaintext and the original one. On the other hand, the stronger scenarios give partial or total access to a decryption oracle to the attacker (against the semantic security or the non-malleability). Another kind of property for encryption schemes has also been defined, called *Plaintext-Awareness* [3], which means that no one can produce a valid ciphertext without knowing the corresponding plaintext. At last Crypto, Bellare *et al.* [1] provided a precise analysis of all these security notions. The main practical result is the equivalence between non-malleability and semantic security in adaptive chosen-ciphertext scenarios.

New Encryption Schemes. Besides all these strong notions of security, very few new schemes have been proposed. In 1994, Bellare and Rogaway [3] presented some variants of RSA semantically secure even in the strong sense (*i.e.* against adaptive chosen-ciphertext attacks) in the random oracle model [2]. But we had to wait 1998 to see other practical schemes with proofs of semantic security: Okamoto–Uchiyama [17], Naccache–Stern [15] and Paillier [18] all based on higher residues; Cramer–Shoup [8] based on the Decisional Diffie-Hellman problem. Nevertheless, they remain rather inefficient. Indeed, all of them are in a discrete logarithm setting and require many full-size exponentiations for the encryption process. Therefore, they are not more efficient than the El Gamal encryption scheme.

The random oracle model. The best security argument for a cryptographic protocol is a proof in the standard model relative to a well-studied difficult problem, such as RSA, the factorization or the discrete logarithm. But no really efficient cryptosystem can aspire to such an argument. Indeed, the best encryption scheme that achieves chosen-ciphertext security in the standard model was published last year [8], and still requires more than four exponentiations for an encryption.

In 1993, Bellare and Rogaway [2] defined a model, the so-called "Random Oracle Model", where some objects are idealized, namely hash functions which are assumed perfectly random. This helped them to design later OAEP [3], the most efficient encryption scheme known until now. In spite of a recent paper [6] making people to be careful with the random oracle model, the security of OAEP has been widely agreed. Indeed, this scheme is incorporated in SET, the Secure Electronic Transaction system [14] proposed by VISA and MasterCard, and will become the new RSA encryption standard PKCS #1 v2.0 [21].

Furthermore, an important feature of the random oracle model is to provide efficient reductions between a well-studied mathematical problem and an attack. Therefore, the reduction validates protocols together with practical parameters. Whereas huge-polynomial reductions, which can hardly be avoided in the standard model, only prove asymptotic security, for large parameters.

As a conclusion, it is better to get an efficient reduction in the random oracle model than a complex reduction in the standard model, since this latter does not prove anything for practical sizes!

Aim of our work. Because of all these inefficient or insecure schemes, it is clear that, from now, the main goal is to design a cryptosystem that combines both efficiency and security. In other words, we would like a *semantically secure scheme as efficient as RSA*.

Outline of the paper. Our feeling was that such a goal required new algebraic problems. In this paper, we first present the *Computational Dependent–RSA problem*, a problem derived from the RSA assumption. We also propose a decisional variant, the *Decisional Dependent–RSA problem*. Then, we give some arguments to validate the cryptographic purpose of those problems, with a careful study of their difficulty and their relations with RSA. Namely, the Computational Dependent–RSA problem is, in a way, equivalent to RSA.

Next, we apply them successfully to the asymmetric encryption setting, and we present a very efficient encryption scheme with the proof of its *semantic security* relative to the *Decisional Dependent–RSA problem* in the standard model. Thereafter, we present two techniques to make this scheme semantically secure both *against adaptive chosen-ciphertext attacks* and relative to the *Computational Dependent–RSA problem* in the random oracle model. Both techniques improve the security level at a very low cost.

1   The Dependent–RSA Problems

As claimed above, the only way to provide new interesting encryption schemes seems to find new algebraic problems. In this section, we focus on new problems with a careful study of both their difficulty and their relations.

1.1   Definitions

For all the problems presented below, we are given a large composite RSA modulus N and an exponent e relatively prime to $\varphi(N)$, the totient function of the modulus N. Let us define a first new problem called the *Computational Dependent–RSA Problem* (C–DRSA).

Definition 1 (The Computational Dependent–RSA: $C\text{-}DRSA(N, e)$).

Given: $\alpha \in \mathbb{Z}_N^\star$;
Find: $(a+1)^e \mod N$, *where* $\alpha = a^e \mod N$.

Notation: *We denote by* $\mathsf{Succ}(\mathcal{A})$ *the success probability of an adversary* $\mathcal{A}$:

$$\mathsf{Succ}(\mathcal{A}) = \Pr\left[\mathcal{A}(a^e \bmod N) = (a+1)^e \bmod N \,\middle|\, a \stackrel{R}{\leftarrow} \mathbb{Z}_N^\star\right].$$

As it has already been done with the Diffie-Hellman problem [9], we can define a decisional version of this problem, therefore called the *Decisional Dependent–RSA Problem* (D-DRSA): Given a candidate to the Computational Dependent–RSA problem, is it the right solution? This decisional variant will then lead to a semantically secure encryption scheme.

Definition 2 (The Decisional Dependent–RSA: $D\text{–}DRSA(N,e)$**).**

Problem: *Distinguish the two distributions*

$$\mathcal{R}and = \left\{(\alpha,\gamma) = (a^e \bmod N, c^e \bmod N) \,\middle|\, a,c \stackrel{R}{\leftarrow} \mathbb{Z}_N^\star\right\},$$

$$\mathcal{D}\mathcal{R}\mathcal{S}\mathcal{A} = \left\{(\alpha,\gamma) = (a^e \bmod N, (a+1)^e \bmod N) \,\middle|\, a \stackrel{R}{\leftarrow} \mathbb{Z}_N^\star\right\}.$$

Notation: *We denote by* $\mathsf{Adv}(\mathcal{A})$ *the advantage of a distinguisher* $\mathcal{A}$:

$$\mathsf{Adv}(\mathcal{A}) = \left|\Pr_{\mathcal{R}and}[\mathcal{A}(\alpha,\gamma)=1] - \Pr_{\mathcal{D}\mathcal{R}\mathcal{S}\mathcal{A}}[\mathcal{A}(\alpha,\gamma)=1]\right|.$$

1.2 The Dependent–RSA Problems and RSA

In order to study those Dependent–RSA problems, we define a new one, we call the *Extraction Dependent–RSA Problem* (E-DRSA):

Given: $\alpha = a^e \in \mathbb{Z}_N^\star$ and $\gamma = (a+1)^e \in \mathbb{Z}_N^\star$;
Find: $a \bmod N$.

One can then prove that extraction of e-th roots is easier to solve than the Computational Dependent–RSA problem and the Extraction Dependent–RSA problem together.

Theorem 3. $\mathbf{RSA(N,e)} \Longleftrightarrow \mathbf{E\text{-}DRSA(N,e)} + \mathbf{C\text{-}DRSA(N,e)}$.

Proof. Let $\mathcal{A}$ be an E-DRSA adversary and $\mathcal{B}$ a C-DRSA adversary. For a given $c = a^e \bmod N$, an element of $\mathbb{Z}_N^\star$, whose e-th root is wanted, one uses $\mathcal{B}$ to obtain $(a+1)^e \bmod N$ and gets a from $\mathcal{A}(a^e \bmod N, (a+1)^e \bmod N)$.

The opposite direction is trivial, since extraction of e-th roots helps to solve all the given problems. $\qquad\square$

Furthermore, it is clear that any decisional problem is easier to solve than its related computational version, and trying to extract a, it is easy to decide whether the given γ is the right one. Finally, for any (N,e), the global picture is

$$\mathbf{C\text{-}DRSA + E\text{-}DRSA} \Longleftrightarrow \mathbf{RSA} \Longrightarrow \mathbf{C\text{-}DRSA, E\text{-}DRSA} \Longrightarrow \mathbf{D\text{-}DRSA},$$

where $A \Longrightarrow B$ means that an oracle that breaks A can be used to break B within a time polynomial in the size of N.

2   How To Solve the Dependent–RSA Problems?

In order to use these problems in cryptography, we need to know their practical difficulty, for reasonable sizes. Hopefully, some of them have already been studied in the past. Indeed, they are related to many properties of the RSA cryptosystem, namely its malleability, its security against related-message attacks [7] and in the multicast setting [13].

Concerning the Extraction Dependent–RSA problem, some methods have been proposed by Coppersmith *et al.* [7], trying to solve the related-message system:

$$\begin{cases} \alpha = m^e \bmod N \\ \beta = (m+1)^e \bmod N \end{cases}$$

2.1   A First Method: Successive Eliminations

Let us assume that $e = 3$, then it is possible to successively eliminate the powers of m and express m from α and β:

$$\begin{cases} \alpha = m^3 \bmod N \\ \beta = (m+1)^3 = m^3 + 3m^2 + 3m + 1 \bmod N \\ \quad = \alpha + 3m^2 + 3m + 1 \bmod N \end{cases}$$

$$\begin{cases} m \times (\beta - \alpha) - 3\alpha = 3m^2 + m \bmod N \\ \beta - \alpha = (3m^2 + m) + 2m + 1 \bmod N \\ \quad = m \times (\beta - \alpha + 2) - 3\alpha + 1 \bmod N \end{cases}$$

$$\text{Then, } m = \frac{2\alpha + \beta - 1}{\beta - \alpha + 2} \bmod N.$$

First, Coppersmith *et al.* [7] claimed that for each e, there exist polynomials P and Q such that each can be expressed as rational polynomials in X^e and $(X+1)^e$, and such that $Q(X) = XP(X)$. Then $m = Q(m)/P(m)$. However, the explicit expression of m as a ratio of two polynomials in α and β requires $\Theta(e^2)$ coefficients, furthermore it is not obvious how to calculate them efficiently.

Consequently, this first method fails as soon as e is greater than, say 2^{40}.

2.2   A Second Method: Greatest Common Divisor

A second method comes from the remark that m is a root for both the polynomials P and Q over the ring $\mathbb{Z}_N$, where.

$$P(X) = X^e - \alpha \text{ and } Q(X) = (X+1)^e - \beta.$$

Then $X - m$ is a divisor of the gcd of P and Q. Furthermore, one can see that with high probability, it is exactly the gcd. A straightforward implementation of Euclid's algorithm takes $\mathcal{O}(e^2)$ operations in the ring $\mathbb{Z}_N$. More sophisticated techniques can be used to compute the gcd in $\mathcal{O}(e \log^2 e)$ time [22]. Then, this second method fails as soon as e is greater than 2^{60}.

2.3   Consequences on the Computational Dependent–RSA Problem

Since the RSA cryptosystem appeared [20], many people have attempted to find weaknesses. Concerning the malleability of the encryption, the multiplicative property is well-known. In other words, it is easy to derive the encryption of $m \times m'$ from the encryption of m, for any m', without knowing the message m itself. However, from the encryption of an unknown message m, nothing has been found to derive the encryption of $m + 1$ whatever the exponent e may be.

Concerning the Extraction Dependent–RSA problem, one can then state the following theorem:

Theorem 4. *There exist algorithms that solve the problem $E–DRSA(N, e)$ in $\mathcal{O}(|N|^2, e \times |e|^2)$ time.*

In conjunction with the Theorem 3, we can therefore claim that

Theorem 5. *There exists a reduction from the RSA problem to the Computational Dependent–RSA problem in $\mathcal{O}(|N|^2, e \times |e|^2)$ time.*

Then, for any fixed exponent e, $RSA(N, e)$ is reducible to $C–DRSA(N, e)$ polynomially in the size of N, since the Extraction Dependent–RSA problem is "easy" to solve, using the gcd technique (see the previous version).

Anyway, computation of e-th roots seems always required to solve the Computational Dependent–RSA problem, which is intractable for any exponent e, according to the RSA assumption.

Conjecture 6. The Computational Dependent–RSA problem is intractable for large enough RSA moduli.

Remark 7. Because of the Theorem 5, this conjecture holds for small exponents, since then C–DRSA is as hard as RSA.

2.4   About the Decisional Dependent–RSA Intractability

The gcd technique seems to be the best known attack against the Decisional Dependent–RSA problem and is impractical as soon as the exponent e is greater than 2^{60}. Which leads to the following conjecture:

Conjecture 8. The Decisional Dependent–RSA problem is intractable as soon as the exponent e is greater than 2^{60}, for large enough RSA moduli.

3   Security Notions for Encryption Schemes

For the formal definitions of all the kinds of attacks and of security notions, we refer the reader to the last Crypto paper [1]. However, let us briefly recall the main security notion, the *semantic security* (a.k.a. *indistinguishability of encryptions*) defined by Goldwasser and Micali [12]. For this notion, an attacker is seen as a two-stage ("find-and-guess") Turing machine which first chooses two

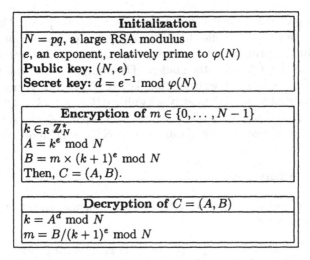

Fig. 1. The DRSA Encryption Scheme

messages, during the "find"-stage. In the second stage, the "guess"-stage, she receives a challenge, which is the encryption of one of both chosen messages, and has to guess which one is the corresponding plaintext.

In the public-key setting, any attacker can play a *chosen-plaintext attack*, since she can encrypt any message she wants. However, stronger attacks has been defined. First, Naor and Yung [16] defined the *chosen-ciphertext attack* (a.k.a. *lunchtime attack*) where the attacker has access to a decryption oracle during the "find"-stage, to choose the two plaintexts. Then, Rackoff and Simon [19] improved this notion, giving the decryption oracle access to the attacker in both stages (with the trivial restriction not to ask the challenge ciphertext). This attack is known as *adaptive chosen-ciphertext attack* and is the strongest that an attacker can play, in the classical model.

The aim of this paper is to provide a new efficient scheme, semantically secure against adaptive chosen-ciphertext attacks.

4   The DRSA Encryption Scheme

The Dependent–RSA problem can be used, like the Diffie-Hellman problem [9], to provide encryption schemes. An RSA version of the El Gamal encryption [11] is then proposed with some security properties, namely semantic security against chosen-plaintext attacks. In the next section, we propose two variants with very interesting improved security properties together with high efficiency.

4.1   Description

The scheme works as described in figure 1. We are in the RSA setting: each user publishes an RSA modulus N while keeping secret the prime factors p and q. He also chooses a public exponent e and its inverse d modulo $\varphi(N)$. The public key consists in the pair (N, e), while the secret key is the private exponent d (it can also consists in the prime factors p and q to improve the decryption algorithm efficiency, using the Chinese Remainders Theorem). To encrypt the message $m \in \{0, \ldots, N-1\}$ to Alice whose public key is (N, e), Bob chooses a random $k \in \mathbb{Z}_N^\star$ and computes $A = k^e \bmod N$ as well as $B = m \times (k+1)^e \bmod N$. He sends the pair (A, B) to Alice. When she receives a pair (A, B), Alice computes $k = A^d \bmod N$ and recovers the plaintext $m = B/(k+1)^e \bmod N$.

4.2   Security Properties

The same way as for the El Gamal encryption scheme, one can prove the semantic security of this scheme.

Theorem 9. *The DRSA encryption scheme is semantically secure against chosen-plaintext attacks relative to the Decisional Dependent-RSA problem.*

Proof. Let us consider an attacker $\mathcal{A} = (A_1, A_2)$ who can break the semantic security of this scheme within a time t and with an advantage, in the "guess"-stage, greater than ε.

In the figure beside, we construct a D–DRSA adversary, $\mathcal{B}$, who is able to break the Decisional Dependent–RSA problem for the given public key (N, e) with an advantage greater than $\varepsilon/2$ and a similar running time. The equivalence between the semantic security and the Decisional Dependent–RSA problem will follow, since the opposite direction is straightforward.

$\mathcal{B}(\alpha, \gamma)$:
  Run $A_1(pk)$
    Get m_0, m_1, s
  Randomly choose $b \in \{0, 1\}$
  $A = \alpha$, $B = m_b \cdot \gamma \bmod N$
  Run $A_2(s, m_0, m_1, (A, B))$
    Get c
  if $c = b$ Return 1
  else Return 0

On one hand, we have to study the probability for A_2 to answer $c = b$ when the pair (α, γ) comes from the random distribution. But in this case, one can see that the pair $(A, B) \in \{(r^e, m_b s^e) \mid r, s \in \mathbb{Z}_N^\star\}$ is uniformly distributed in the product space $\mathbb{Z}_N^\star \times \mathbb{Z}_N^\star$, hence independently of b. Then

$$\Pr_{\mathcal{R}and}[\mathcal{B}(\alpha, \gamma) = 1] = \Pr_{\mathcal{R}and}[c = b] = \frac{1}{2}.$$

On the other hand, when the pair (α, γ) comes from the $\mathcal{DRSA}$ distribution, one can remark that (A, B) is a valid ciphertext of m_b, following a uniform distribution among the possible ciphertexts. Then

$$\Pr_{\mathcal{DRSA}}[\mathcal{B}(\alpha, \gamma) = 1] = \Pr_{\mathcal{DRSA}}[c = b] = \Pr_b[A_2(s, m_0, m_1, \mathcal{E}(m_b)) = b] \stackrel{def}{=} \frac{1}{2} \pm \frac{\mathrm{Adv}^{\mathcal{A}}}{2}.$$

The advantage of $\mathcal{B}$ in distinguishing the $\mathcal{DRSA}$ and the $\mathcal{R}and$ distributions is $\mathrm{Adv}(\mathcal{B}) = \mathrm{Adv}^{\mathcal{A}}/2$, and therefore greater than $\varepsilon/2$. $\qquad\square$

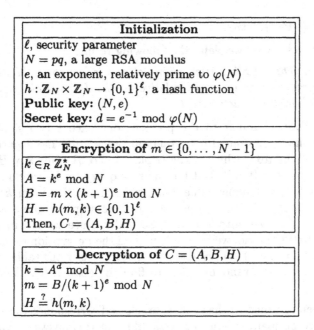

Fig. 2. First Variant: The DRSA-1 Encryption Scheme

5 Some Variants

As it has already been remarked, attackers can be in a stronger scenario than the chosen-plaintext one. Now, we improve the security level, making the scheme resistant to adaptive chosen-ciphertext attacks, in the random oracle model. In a second step, we weaken the algorithmic assumption: an attacker against the semantic security of the second variant, in an adaptive chosen-ciphertext scenario, can be used to efficiently break the Computational Dependent–RSA problem, and not only the Decisional Dependent–RSA problem.

Furthermore, it is important to remark that both improvements are very low-cost on both a computational point of view and the size of the ciphertexts.

5.1 Description of the First Variant: DRSA-1

The scheme works as described in figure 2, where h is a hash function, seen like a random oracle which outputs ℓ-bit numbers. The initialization is unchanged. To encrypt a message $m \in \{0, \ldots, N-1\}$ to Alice whose public key is (N, e), Bob chooses a random $k \in \mathbb{Z}_N^*$ and computes $A = k^e \bmod N$ as well as $B = m \times (k+1)^e \bmod N$ and the control padding $H = h(m, k)$. He sends the triple (A, B, H) to Alice. When she receives a triple (A, B, H), Alice first computes the random value $k = A^d \bmod N$ and recovers the probable plaintext $m = B/(k+1)^e \bmod N$. She then checks whether they both satisfy the control padding $H = h(m, k)$.

5.2   Security Properties

Concerning this scheme, we claim the following result:

Theorem 10. *The DRSA-1 encryption scheme is semantically secure against adaptive chosen-ciphertext attacks relative to the Decisional Dependent-RSA problem in the random oracle model.*

Proof. This proof is similar to the previous one except two simulations. Indeed, we first have to simulate the random oracle, and more particularly for the challenge ciphertext, which is the triple $(A = \alpha, B = m_b \times \gamma, H)$, where H is randomly chosen in $\{0, 1\}^\ell$. But for any new query to the random oracle, one simply returns a new random value. Furthermore, any query (m, k) to the random oracle is filtered: if $k^e = \alpha \bmod N$, then we stop the game, and whether $\gamma = (k + 1)^e \bmod N$ we output 1 or 0. Secondly, since we are in an adaptive chosen-ciphertext scenario, we have to simulate the decryption oracle: when the adversary asks a query (A', B', H'), the simulator looks in the table of the queries previously made to the random oracle to find the answer H'. Then, two cases may appear:

- H' has been returned by the random oracle and corresponds to a query (m, k) (there may be many queries corresponding to this answer). The simulator checks whether $A' = k^e \bmod N$ and $B' = m \times (k + 1)^e \bmod N$. Then it returns m as the decryption of the triple (A', B', H'). Otherwise, the simulator considers that it is an invalid ciphertext and returns the reject symbol "*".
- Otherwise, the simulator returns the reject symbol "*".

The bias is the same as above when all the simulations are correctly made. Concerning the simulation of the random oracle, it is perfectly made, because of the randomness of the answers. However, some decryptions may be incorrect, but only refusing a valid ciphertext: a ciphertext is refused if the query (m, k) has not been asked to the random oracle h. However, the attacker might have guessed the right value for $h(m, k)$ without having asked for it, but only with probability $1/2^\ell$.

Then, if the pair (α, γ) comes from the $\mathcal{DRSA}$ distribution, since the probability of success can be improved if the adversary guesses the e-th root of α, which had led to stop the game with an answer 1,

$$\Pr_{\mathcal{DRSA}}[\mathcal{B}(\alpha, \gamma) = 1] \geq \frac{1}{2} + \frac{\mathsf{Adv}^{\mathcal{A}}}{2} - \frac{q_d}{2^\ell},$$

where the adversary asks at most q_d queries to the decryption oracle. However, if the pair (α, γ) comes from the random distribution, for the same reason as in the previous proof, the adversary cannot gain any advantage, except the case where she had guessed the e-th root of α, but then, $\mathcal{B}$ likely outputs 0:

$$\Pr_{\mathcal{R}and}[\mathcal{B}(\alpha, \gamma) = 1] \leq \frac{1}{2} - \Pr[\alpha^d \text{ guessed}] \leq \frac{1}{2}.$$

Therefore, $\mathsf{Adv}(\mathcal{B}) \geq \dfrac{\mathsf{Adv}^{\mathcal{A}}}{2} - \dfrac{q_d}{2^\ell}.$ $\square$

5.3 Description of the Second Variant: DRSA-2

We can furthermore weaken the algorithmic assumption, making the scheme equivalent to the computational problem rather than to the decisional one. The variant works as described in figure 3, where h_1 and h_2 are two hash functions,

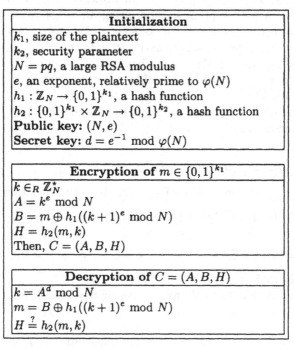

Fig. 3. Second Variant: The DRSA-2 Encryption Scheme

seen like random oracles which output k_1-bit numbers and k_2-bit numbers respectively. The initialization is unchanged. To encrypt a message $m \in \{0, 1\}^{k_1}$ to Alice whose public key is (N, e), Bob chooses a random $k \in \mathbb{Z}_N^*$ and computes $A = k^e \bmod N$. He can then mask the message in $B = m \oplus h_1((k + 1)^e \bmod N)$, a k_1-bit long string and compute the control padding $H = h_2(m, k) \in \{0, 1\}^{k_2}$. He sends the triple (A, B, H) to Alice. When she receives a ciphertext (A, B, H), Alice first computes the random value $k = A^d \bmod N$. She can therefore recover the probable plaintext $m = B \oplus h_1((k + 1)^e \bmod N)$. Then, she checks whether they both satisfy the control padding, $H = h_2(m, k)$.

Theorem 11. *The DRSA-2 encryption scheme is semantically secure against adaptive chosen-ciphertext attacks relative to the Dependent–RSA problem in the random oracle model.*

Proof. The result comes from the fact that any attacker cannot gain any advantage in distinguishing the original plaintext (in an information theoretical sense) if she has not asked for any $(\star, k)$ to h_2 (which is called "event 1" and denoted by E_1) or for $(k+1)^e \bmod N$ to h_1 (which is called "event 2" and denoted by E_2). Then, for a given $\alpha = a^e \bmod N$, either we learn the e-th root of α, or $(a+1)^e \bmod N$ is in the list of the queries asked to h_1. Both cases lead to the computation of $(a+1)^e \bmod N$.

More precisely, let $\mathcal{A} = (A_1, A_2)$ be an attacker against the semantic security of the DRSA-2 encryption scheme, using an adaptive chosen-ciphertext attacker. Within a time bound t, she asks q_d queries to the decryption oracle and q_h queries to the random oracles and distinguishes the right plaintext with an advantage greater than ε. We can use her to provide an algorithm that solves the Computational Dependent–RSA problem, simply filtering the queries asked to the random oracles.

Actually, because of the randomness of the random oracle h_1, if no critical queries have been asked,

$$
\begin{aligned}
\Pr_b[A_2(s, m_0, m_1, \mathcal{E}(m_b)) = b] &= \frac{1}{2} \pm \frac{\mathsf{Adv}^{\mathcal{A}}}{2} \\
&= \Pr_b[A_2 = b \wedge \neg(E_1 \vee E_2)] + \Pr_b[A_2 = b \wedge (E_1 \vee E_2)] \\
&= \Pr[\neg(E_1 \vee E_2)] \times 1/2 + \Pr_b[A_2 = b \wedge (E_1 \vee E_2)].
\end{aligned}
$$

Then, $\pm\mathsf{Adv}^{\mathcal{A}} = \Pr[E_1 \vee E_2] - 2 \times \Pr_b[A_2(s, m_0, m_1, \mathcal{E}(m_b)) = b \wedge (E_1 \vee E_2)]$, and both cases imply $\Pr[E_1 \vee E_2] \geq \mathsf{Adv}^{\mathcal{A}}$.

Using our simulations, namely for the decryption oracle, we obtain, as previously seen,

$$
\Pr[(E_1 \vee E_2) \wedge \text{ no incorrect decryption}] \geq \mathsf{Adv}^{\mathcal{A}} - q_d \times 2^{-k_2}.
$$

For the reduction, one just has to randomly choose the query which should correspond to $(a+1)^e \bmod N$. With probability greater than $1/q_h$, it is a good choice (or maybe, event 2 happens, but we assume the worst case). Then, with probability greater than $(\mathsf{Adv}^{\mathcal{A}} - q_d/2^{k_2})/q_h$, within roughly the same running time as the adversary $\mathcal{A}$, one obtains the right value for $(a+1)^e \bmod N$ corresponding to the given $\alpha = a^e \bmod N$. $\qquad\square$

6   Efficiency

Now that we know that these schemes are provably secure, let us compare them with other well-known cryptosystems from a computational point of view. And first, let us briefly recall the three other schemes we will consider:

El Gamal. An authority chooses and publishes two large prime numbers p and q such that q is a large prime factor of $p-1$, together with an element g of $\mathbb{Z}_p^*$ of order q. Each user chooses a secret key x in $\mathbb{Z}_q^*$ and publishes $y = g^x \bmod p$. To encrypt a message m, one has to choose a random element k in $\mathbb{Z}_q^*$ and sends the pair $(r = g^k \bmod p, s = m \times y^k \bmod p)$ as the ciphertext. The recipient can recover the message from a pair (r, s) since $m = s/r^x \bmod p$, where x is his secret key. To reach semantic security [23], this scheme requires m to be in the subgroup generated by g. To be practical, one can choose $p = 2q + 1$, a strong prime, which consequently increases the number of multiplications to be made for an encryption. We do not consider any variant of El Gamal, since all are much heavier to implement.

RSA. Each user chooses a large RSA modulus $N = pq$ of size n together with an exponent e. He publishes both and keeps secret the private exponent $d = e^{-1} \bmod \varphi(N)$. To encrypt a message m, one just has to send the string $c = m^e \bmod N$. To recover the plaintext, the recipient computes $c^d = m \bmod N$.

Optimal Asymmetric Encryption Padding. The RSA variant, OAEP, was the most efficient scheme, from our knowledge: An authority chooses and publishes two hash functions g and h which both output $n/2$-bit strings. Each user chooses as above a public key (N, e), where N is a n-bit long RSA modulus, and keeps secret the exponent d. To encrypt a message m, one has to choose a random element r, computes $A = (m\|0^{k_1}) \oplus g(r)$ and $B = r \oplus h(A)$ and finally sends $C = (A\|B)^e \bmod N$. The recipient can recover the message from C first computing $A\|B = C^d \bmod N$, then $r = B \oplus h(A)$ and $M = A \oplus g(r)$. If M ends with k_1 zero bits, then m is the beginning of M.

Both encryption schemes (the original RSA and OAEP) essentially require one exponentiation to the power e per encryption. And as one can remark, they depend on the message, and then has to be done online.

Precomputations. In the same vein as a last Eurocrypt paper [5], our scheme allows precomputations. Indeed, a user can precompute many pairs for a given recipient, *i.e.*, $(a^e \bmod N, (a + 1)^e \bmod N)$. Then an encryption only requires one multiplication, or even a XOR. However, to be fair, in the following, we won't consider this feature.

Efficiency Comparison. One can see, on figure 4, a brief comparison table involving our schemes together with the El Gamal encryption scheme (with a 512-bit long prime $p = 2q + 1$), the RSA cryptosystem and its OAEP version. Because of the new 140-digit record for factorization, for a similar security level between factorization-based schemes and discrete logarithm-based ones, we consider 1024-bit RSA-moduli: $n = |N| = 1024$, $e = 65537 = 2^{16} + 1$, and furthermore $k_1 = 64$ for OAEP. Concerning our DRSA encryption schemes, we also use a 1024-bit long modulus N. However, whereas we can use $e = 65537$ (even smaller, such as $e = 3$, since related-message attacks seem to not be applicable) in schemes based on the Computational Dependent–RSA problem (such as the DRSA-2 scheme), we need to use a larger exponent with the Decisional Dependent–RSA-based schemes, to avoid attacks presented above against the semantic security. Then, we use $e = 2^{67} + 3$, which is a prime integer, in the DRSA and in the DRSA-1 schemes.

Schemes	RSA 1024	OAEP 1024	El Gamal 512	DRSA 1024	DRSA-1 1024	DRSA-2 1024	
Security							
Inversion	RSA	RSA	DH	C-DRSA	C-DRSA	C-DRSA	
CPA-IND	–	RSA*	D-DH	D-DRSA	D-DRSA*	C-DRSA*	
CCA2-IND	–	RSA*	–	–	D-DRSA*	**C-DRSA***	
Size (in bits)							
Plaintext	1024	448	511	1024	1024	1024	2048
Ciphertext	1024	1024	1024	2048	2208	2208	3232
Expansion	1	2.3	2	2	2.2	2.2	1.6
Encryption							
Workload	17	17	384	139	139	35	35
Workload/kB	136	311	6144	1112	1112	280	**140**
Decryption							
Workload	384	384	192	523	523	419	419
Workload/kB	3072	7022	3072	4184	4184	3352	**1676**

* in the random oracle model

Fig. 4. Efficiency of Encryptions and Decryptions

Remark 12. In this table, the basic operation is the modular multiplication with a 1024-bit long modulus. We assume that the modular multiplication algorithm is quadratic in the modulus size and that modular squares are computed with the same algorithm. Furthermore, in the decryption phase, we use the CRT when it is possible.

CPA-IND and CCA2-IND both follow the notations of the Bellare *et al.* paper [1] and mean the indistinguishability of encryptions (a.k.a. *semantic security*) against chosen-plaintext attacks and adaptive chosen-ciphertext attacks respectively.

One can remark that our new scheme, in its basic version (DRSA–1024 bits), can encrypt **6 times faster** than El Gamal–512 bits and decrypt in essentially the same time. Therefore, the DRSA encryption schemes becomes the most efficient scheme provably semantically secure against chosen-plaintext attacks in the standard model.

If we consider the security in the random oracle model, the DRSA-1 scheme reaches the security against adaptive chosen-ciphertext attacks with an unchanged efficiency.

However, the most interesting scheme is the DRSA-2 cryptosystem that reaches semantic security both against adaptive chosen-ciphertext attacks and relative to the Computational Dependent–RSA problem, in a situation where it is practically equivalent to the RSA problem. Indeed, a smaller exponent, such as $e = 65537$ (or even 3), can be used, hence an improved efficiency is obtained: with $k_1 = |N| = 1024$, this scheme is already faster than OAEP, for both encryption and decryption. Furthermore, with larger k_1 (*e.g.* $k_1 = 2048$, such as in the last column), this scheme can reach higher rates, and even get **much faster than the original RSA encryption scheme**.

Conclusion

Therefore, we have presented three new schemes with security proofs and record efficiency. Indeed, the DRSA cryptosystem is semantically secure against chosen-plaintext attacks in the standard model, relative to a new difficult problem (the inversion problem is equivalent to RSA in many cases), with an encryption rate 6 times faster than El Gamal (with similar security levels: RSA-1024 bits vs. El Gamal-512 bits).

Next, we have presented two variants semantically secure against adaptive chosen-ciphertext attacks in the random oracle model (they can even be proven plaintext-aware [3,1]). Furthermore, the DRSA-2 scheme is more efficient than RSA, and therefore much more efficient than OAEP, with an equivalent security, since for those parameters, the Computational Dependent–RSA problem is practically equivalent to the RSA problem.

Acknowledgments

I would like to thank the anonymous Eurocrypt '99 referees for their valuable comments and suggestions, as well as Jacques Stern for fruitful discussions.

References

1. M. Bellare, A. Desai, D. Pointcheval, and P. Rogaway. Relations Among Notions of Security for Public-Key Encryption Schemes. In *Crypto '98*, LNCS 1462, pages 26–45. Springer-Verlag, 1998.
2. M. Bellare and P. Rogaway. Random Oracles are Practical: a Paradigm for Designing Efficient Protocols. In *Proc. of the 1st CCCS*, pages 62–73. ACM press, 1993.
3. M. Bellare and P. Rogaway. Optimal Asymmetric Encryption – How to Encrypt with RSA. In *Eurocrypt '94*, LNCS 950, pages 92–111. Springer-Verlag, 1995.
4. D. Bleichenbacher. A Chosen Ciphertext Attack against Protocols based on the RSA Encryption Standard PKCS #1. In *Crypto '98*, LNCS 1462, pages 1–12. Springer-Verlag, 1998.
5. V. Boyko, M. Peinado, and R. Venkatesan. Speedings up Discrete Log and Factoring Based Schemes via Precomputations. In *Eurocrypt '98*, LNCS 1403. Springer-Verlag, 1998.
6. R. Canetti, O. Goldreich, and S. Halevi. The Random Oracles Methodology, Revisited. In *Proc. of the 30th STOC*. ACM Press, 1998.
7. D. Coppersmith, M. Franklin, J. Patarin, and M. Reiter. Low-Exponent RSA with Related Messages. In *Eurocrypt '96*, LNCS 1070, pages 1–9. Springer-Verlag, 1996.
8. R. Cramer and V. Shoup. A Practical Public Key Cryptosystem Provably Secure against Adaptive Chosen Ciphertext Attack. In *Crypto '98*, LNCS 1462, pages 13–25. Springer-Verlag, 1998.

9. W. Diffie and M. E. Hellman. New Directions in Cryptography. In *IEEE Transactions on Information Theory*, volume IT–22, no. 6, pages 644–654, November 1976.

10. D. Dolev, C. Dwork, and M. Naor. Non-Malleable Cryptography. In *Proc. of the 23rd STOC*. ACM Press, 1991.

11. T. El Gamal. A Public Key Cryptosystem and a Signature Scheme Based on Discrete Logarithms. In *IEEE Transactions on Information Theory*, volume IT–31, no. 4, pages 469–472, July 1985.

12. S. Goldwasser and S. Micali. Probabilistic Encryption. *Journal of Computer and System Sciences*, 28:270–299, 1984.

13. J. Håstad. Solving Simultaneous Modular Equations of Low Degree. *SIAM Journal of Computing*, 17:336–341, 1988.

14. SET Secure Electronic Transaction LLC. SET Secure Electronic Transaction Specification – Book 3: Formal Protocol Definition, may 1997. Available from http://www.setco.org/.

15. D. Naccache and J. Stern. A New Cryptosystem based on Higher Residues. In *Proc. of the 5th CCCS*, pages 59–66. ACM press, 1998.

16. M. Naor and M. Yung. Public-Key Cryptosystems Provably Secure against Chosen Ciphertext Attacks. In *Proc. of the 22nd STOC*, pages 427–437. ACM Press, 1990.

17. T. Okamoto and S. Uchiyama. A New Public Key Cryptosystem as Secure as Factoring. In *Eurocrypt '98*, LNCS 1403, pages 308–318. Springer-Verlag, 1998.

18. P. Paillier. Public-Key Cryptosystems Based on Discrete Logarithms Residues. In *Eurocrypt '99*, LNCS 1592, pages 221–236. Springer-Verlag, 1999.

19. C. Rackoff and D. R. Simon. Non-Interactive Zero-Knowledge Proof of Knowledge and Chosen Ciphertext Attack. In *Crypto '91*, LNCS 576, pages 433–444. Springer-Verlag, 1992.

20. R. Rivest, A. Shamir, and L. Adleman. A Method for Obtaining Digital Signatures and Public Key Cryptosystems. *Communications of the ACM*, 21(2):120–126, February 1978.

21. RSA Data Security, Inc. Public Key Cryptography Standards – PKCS. Available from http://www.rsa.com/rsalabs/pubs/PKCS/.

22. V. Strassen. The Computational Complexity of Continued Fractions. *SIAM Journal of Computing*, 12(1):1–27, 1983.

23. Y. Tsiounis and M. Yung. On the Security of El Gamal based Encryption. In *PKC '98*, LNCS. Springer-Verlag, 1998.

Resistance Against General Iterated Attacks

Serge Vaudenay

École Normale Supérieure — CNRS
Serge.Vaudenay@ens.fr

Abstract. In this paper we study the resistance of a block cipher against a class of general attacks which we call "iterated attacks". This class includes some elementary versions of differential and linear cryptanalysis. We prove that we can upper bound the complexity of the attack by using decorrelation techniques. Our main theorem enables to prove the security against these attacks (in our model) of some recently proposed block ciphers COCONUT98 and PEANUT98, as well as the AES candidate DFC. We outline that decorrelation to the order $2d$ is required for proving security against iterated attacks of order d.

1 Introduction

Since public-key cryptography has been discovered in the late 70s, *proving* the security of cryptographic protocols has been a challenging problem. Recently, the random oracle model [2] and the generic algorithm techniques [34] have introduced new tools for validating cryptographic algorithms. Although much older, the area of symmetric cryptography did not get so many tools.

In the early 90s, Biham and Shamir [3] introduced the notion of differential cryptanalysis and Matsui [18,19] introduced the notion of linear cryptanalysis, which was a quite general model of attacks. Since then many authors tried to formalize these attacks and study their complexity in order to prove the security of block ciphers against it. Earlier work, initiated by Nyberg [23] was based on algebraic techniques.

Recently, Carter-Wegman's combinatoric notion of "universal functions" [5,42] has been adapted in context with encryption and the notion of "decorrelation bias" has been formalized [36,37]. Measurement of the decorrelation (*e.g.* by the decorrelation bias) enables to quantify the security of block ciphers against several classes of attacks. In [36,37], several real-life block cipher prototypes have been proposed, namely COCONUT98 and PEANUT98. Their decorrelation bias have been measured, and the security against basic versions of differential and linear cryptanalysis (as formalized in the present paper) has been formally proved. Similarly, [7] submitted the DFC candidate to the AES process.

In this paper, we generalize these results in a uniform approach. We introduce the notion of "iterated attack of order d" and we prove how the decorrelation bias can measure the security against any of it. Differential and linear cryptanalysis happen to be included in this class of attacks (differential attacks have an order of 2, and linear attacks have an order of 1). In particular we *prove* the security of

J. Stern (Ed.): EUROCRYPT'99, LNCS 1592, pp. 255–271, 1999.
© Springer-Verlag Berlin Heidelberg 1999

the above mentioned block ciphers against any iterated known plaintext attack of order 1.[1]

This paper is organized as follows. First we recall the previous results in decorrelation theory which are interesting for our purpose in Section 2. Our contribution starts in Section 3. We define the class of iterated attack of given order. We prove by a counterexample that decorrelation of order d is not sufficient to thwart all iterated attacks of order d. We then show how decorrelation of order $2d$ gives an upper bound on the efficiency of any iterated attacks of order d. We show how to use this result for a practical block cipher (namely, PEANUT98 or DFC). Finally, in Section 4 we investigate how to use the same techniques for combining several cryptanalysis all together and Section 5 investigates extensions of iterated attacks.

2 Previous Work

2.1 Provable Security for Block Ciphers

The notion of "provable security" is often used in public key cryptography. The area of symmetric encryption has seldom results on provable security, and with rare link with each other.

First of all, Shannon's approach [33] (1949) formalizes the notion of "perfect secrecy". It proves the security of Vernam's cipher [40] (also known as the "one-time-pad"). The drawback is that the key must be at least as long as the plaintext, used only once, and perfectly random (*i.e.* chosen with an unbiased uniform distribution).

The Wegman-Carter [42] (1981) approach enables to construct "provably secure" Message Authentication Code (MAC) algorithms by combining the notion of universal function [5] and Vernam's cipher. It has several refinements (see for instance [11,9]).

The Luby-Rackoff approach [16] (1988) uses the model of distinguishability (which was well known in the area of pseudorandomness, see [8]), also known as Turing's test, for proving that a random Feistel cipher [6] over messages of m bits is provably secure if we use it less than $2^{\frac{m}{4}}$ times. This has many refinements (*e.g.* see [28,29,30,17,22,31]). It relies the security of the cipher on the pseudorandomness of the round function, which is indeed hard to achieve (because of the key length) for real-life ciphers. We can for instance mention Knudsen's recent DEAL AES candidate [12] which is based on this construction. Here the "provable" security of DEAL relies on the assumption that DES [1] defines a family of random functions. Although this assumption does not make much sense, this provides a piece of security proof.[2]

---

[1] Iterated attacks of order 1 do not include differential attacks, but the security against differential attacks is proven by other approaches as detailed below.

[2] So far, we are not aware about any result which would formally prove that DEAL is significantly more secure than DES.

Biham and Shamir's attacks [3] gave a new breath to the area of symmetric encryption.

First of all, Lai-Massey's notion of Markov cipher [14,15,13] (1990) enables to formalize the complexity of differential cryptanalysis under the hypothesis of stochastic equivalence which assumes that all keys behave as for the average. An alternate approach due to Nyberg [23,24,25] makes links with some non-linear properties of the internal substitution boxes of the ciphers.

Finally, the Nyberg-Knudsen construction [26,27] (1992) enables to construct block ciphers which are "provably secure" against differential and linear cryptanalysis. They also gave some prototype examples of real-life ciphers which happened to be weak against more general attacks (see [10]). This construction has been successfully used by Matsui in the MISTY construction [20,21] (1996) which has no known attacks so far.

These independent results have been linked with each other through the decorrelation theory [36,37] (1998).

These notions of provable security must however be interpreted with great care, mostly because it refers to some security results against some kinds of attacks and in some sharply formalized model. It does not refer to the intuitive notion of "unbreakability" and must not be blindly trusted. The Jakobsen-Knudsen's attack [10] against the Nyberg-Knudsen ciphers [27] illustrates that security against some attacks does not provide security against other ones. It may also be possible to attack some trusted algorithms (like RSA [32]) in some real-life model (the RSA PKCS#1 standard) without mathematically breaking the algorithm, as was shown by Bleichenbacher's attack [4]. Some constructions which are proposed by the decorrelation theory happen to be vulnerable against some more general attacks as well.[3] We thus need to keep this warning in mind when dealing with "provable security".

2.2   Decorrelation Theory

In our setup, a block cipher is considered as a random permutation C over a message-block space $\mathcal{M}$. (Here the randomness comes from the random choice of the secret key.) The efficiency of a cryptanalysis can be measured by the average complexity of the algorithm over the distribution of the permutation (*i.e.* of the secret key).

Definition 1. *Given a random function F from a given set $\mathcal{M}_1$ to a given set $\mathcal{M}_2$ and an integer d, we define the "d-wise distribution matrix" $[F]^d$ of F as a $\mathcal{M}_1^d \times \mathcal{M}_2^d$-matrix where the (x, y)-entry of $[F]^d$ corresponding to the multi-points $x = (x_1, \ldots, x_d) \in \mathcal{M}_1^d$ and $y = (y_1, \ldots, y_d) \in \mathcal{M}_2^d$ is defined as the probability that we have $F(x_i) = y_i$ for $i = 1, \ldots, d$.*

---

[3] Wagner [41] recently broke the COCONUT98 cipher by a "boomerang attack" which is a kind of intermediate attack approach between differential and higher differential attacks.

Basically, each row of the d-wise distribution matrix corresponds to the distribution of the d-tuple $(F(x_1), \ldots, F(x_d))$ where $(x_1, \ldots, x_d)$ corresponds to the index of the row.

In this paper, we consider the following matrix norm over $\mathbf{R}^{\mathcal{M}^d \times \mathcal{M}^d}$ defined by

$$||A|| = \max_x \sum_y |A_{x,y}|$$

for any matrix A.[4]

Definition 2. *Let C be a random permutation over $\mathcal{M}$. We call the quantity $||[C]^d - [C^*]^d||$ the "d-wise decorrelation bias of permutation C" and we denote it $\mathrm{DecP}^d(C)$, where C^* is a uniformly distributed random permutation.*

A decorrelation bias of zero means that for any multi-point $x = (x_1, \ldots, x_d)$ the multi-point $(C(x_1), \ldots, C(x_d))$ has the same distribution of the multi-point $(C^*(x_1), \ldots, C^*(x_d))$, so that C and C^* have the same "decorrelation". Throughout the paper, C^* denotes a uniformly distributed permutation which serves as a reference (which will be called "perfect cipher"). We say that its decorrelation is "perfect". For instance, saying that a cipher C on $\mathcal{M}$ has a perfect pairwise decorrelation means that for any $x_1 \neq x_2$, the random variable $(C(x_1), C(x_2))$ is uniformly distributed among all the (y_1, y_2) pairs such that $y_1 \neq y_2$. This notion is fairly similar to the notion of universal functions which was been introduced by Carter and Wegman [5,42].

The matrix norm property (*i.e.* $||A \times B|| \leq ||A||.||B||$) implies

$$\mathrm{DecP}^d(C_1 \circ C_2) \leq \mathrm{DecP}^d(C_1).\mathrm{DecP}^d(C_2).$$

Thus we can built ciphers with arbitrarily small decorrelation bias by iterating a simple cipher as long as its own decorrelation bias is smaller than 1. The security results show that when the decorrelation bias is small, then the complexity of the attack is high.

As an example we mention the simple affine cipher defined by $C(x) = Ax + B$ where $(A, B) \in_U \mathrm{GF}(2^m)^* \times \mathrm{GF}(2^m)$ is a random key. This cipher is perfectly decorrelated to the order 2. It is the basic COCONUT cipher [36].

2.3   Security Model

In the Luby-Rackoff model [16], an attacker is an infinitely powerful Turing machine $\mathcal{A}^{\mathcal{O}}$ which has access to an oracle $\mathcal{O}$. Its aim is to distinguish if the oracle implements a cipher C or the Perfect Cipher C^* by querying it and with a limited number d of inputs. The attacker must finally answer 0 ("reject") or 1 ("accept"). We measure the ability to distinguish C from C^* by the advantage $\mathrm{Adv}_{\mathcal{A}}(C, C^*) = |p - p^*|$ where p (resp. p^*) is the probability of answering 1 if $\mathcal{O}$

---

[4] This norm is the infinity-associated matrix norm and is usually denoted $|||.|||_\infty$. Other norms have been considered, *e.g.* in [38].

implements C (resp. C^*). In this paper we focus on non-adaptive attacks *i.e.* on distinguishers illustrated on Fig. 1: here no X_i queried to the oracle depends on some previous answers $C(X_j)$. The chosen norm is well suited to this notion of

---

Parameter: a complexity n
Input: an oracle which implements a function c
    1. compute some messages $X = (X_1, \ldots, X_d)$
    2. get $Y = (c(X_1), \ldots, c(X_d))$ from the oracle
    3. depending on X and Y, output 0 or 1

Fig. 1. A Generic d-Limited Non-Adaptive Distinguisher.

non-adaptive attack as shown by the following result (taken from [36,37]).

Theorem 3. *Let d be an integer. Let C be a cipher. The best d-limited non-adaptive distinguisher $\mathcal{A}$ for C is such that*

$$\mathrm{Adv}_{\mathcal{A}}(C, C^*) = \frac{1}{2}\mathrm{DecP}^d(C).$$

Thus the decorrelation bias for the $\|.\|$ norm expresses the best possible advantage for a non-adaptive attack.

For instance, if C is the basic COCONUT cipher and $d = 2$, then the advantage of *any* non-adaptive attack which is limited to 2 queries is zero: this cipher is perfectly secure when used only twice (as one-time pad [40] is perfectly secure when used only once).

2.4   Differential and Linear Cryptanalysis

In this section we assume that $\mathcal{M} = \mathrm{GF}(2^m)$. The inner dot product $a \cdot b$ in $\mathrm{GF}(2^m)$ is the parity of the bitwise AND of a and b.

We formalize the basic notion of differential (resp. linear) cryptanalysis by the distinguisher which is characterized by a pair $(a, b) \in \mathcal{M}^2$ (and which is called a "characteristic") and which is depicted on Fig. 2 (resp. Fig. 3). Linear cryptanalysis also needs an "acceptance set" B.

These formalizations are somewhat different from the original ones. We claim that they are straightforward adaptations of the original attacks in the Luby-Rackoff model. Actually, the Biham-Shamir's original 3R, 2R and 1R attacks [3] can be considered as implicitly starting with the attack which is depicted on Fig. 2 against the same cipher with 3, 2 or 1 less round. One of the technical problems of differential cryptanalysis is that we do not have access to the explicit output of the oracle so we have to filter the outputs and isolate "good pairs" from "wrong pairs". The (theoretical) differential distinguisher against a cipher diminished by i rounds is thus more efficient than Biham-Shamir's iR basic

---

Parameters: a complexity n, a characteristic (a, b)
Input: an oracle which implements a function c
   1. for i from 1 to n do
      (a) pick uniformly a random X and query for $c(X)$ and $c(X + a)$
      (b) if $c(X + a) = c(X) + b$, stop and output 1
   2. output 0

Fig. 2. Differential Distinguisher.

Parameters: a complexity n, a characteristic (a, b), an acceptance set B
Input: an oracle which implements a function c
   1. initialize the counter value u to zero
   2. for i from 1 to n do
      (a) pick a random X with a uniform distribution and query for $c(X)$
      (b) if $X \cdot a = c(X) \cdot b$, increment the counter u
   3. if $u \in B$, output 1, otherwise output 0

Fig. 3. Linear Distinguisher.

attack, therefore a lower bound on the complexity of differential distinguishers leads to a lower bound on the complexity on these original attacks.[5] Similarly, Fig. 3 is the heart of Matsui's original attack against DES [19] when c is DES reduced to 14 rounds.

It has been shown (see [36,37]) that for any differential distinguisher we have

$$\text{Adv}_{\text{Fig.}2}(C, C^*) \leq \frac{n}{2^m - 1} + \frac{n}{2}\text{DecP}^2(C). \tag{1}$$

(In particular, the probability of the differential characteristic which usually introduces a dependency on the key in formal expressions is completely replaced by $\text{DecP}^2(C)$: the complexity analysis of the attack on average on the key uses only the decorrelation bias and does not rely on any unproven assumption such as the hypothesis of stochastic equivalence.[6]) Similarly for any linear distinguisher we have

$$\lim_{n \to +\infty} \frac{\text{Adv}_{\text{Fig.}3}(C, C^*)}{n^{\frac{1}{3}}} \leq 9.3 \left(\frac{1}{2^m - 1} + 2\text{DecP}^2(C) \right)^{\frac{1}{3}}. \tag{2}$$

---

[5] We outline that further versions and extensions of differential cryptanalysis use more tricks and escape from this model. This is why we refer to the "original" differential cryptanalysis.

[6] This does not mean that no "weak keys" exist, which is wrong in general (DFC happens to have weak keys as shown by Coppersmith). This shows that the attack does not work on average, which implies that the fraction of weak keys is negligible against the average case (indeed, weak keys of DFC consist in a fraction of 2^{-128}).

Therefore the decorrelation bias to the order 2 leads to upper bounds on the best advantages of both differential and linear attacks.

2.5   Some Constructions

In [36], two real-life block ciphers (called COCONUT98 and PEANUT98) have been proposed. They come from the general family constructions COCONUT and PEANUT.

A cipher in the COCONUT family is characterized by some parameters (m, p) where m is the message-block length and p is an irreducible polynomial of degree m in GF(2). The COCONUT98 Cipher corresponds to the parameters $m = 64$ and $p = x^{64} + x^{11} + x^2 + x + 1$. From the construction, any of COCONUT ciphers has a perfect pairwise decorrelation. Therefore from Equations (1) and (2) no differential or linear distinguisher (as formalized on Fig. 2 and 3) can be efficient.

A cipher in the PEANUT family has some parameters (m, r, d, p). Here m is the message-block length, r is the number of rounds (actually, a PEANUT cipher is an r-round Feistel cipher [6]), d is the order of constructed decorrelation, and p is a prime number greater than $2^{\frac{m}{2}}$. The PEANUT98 Cipher corresponds to $m = 64$, $r = 9$, $d = 2$ and $p = 2^{32} + 15$. It has been shown that the d-wise decorrelation bias of this function has an upper bound which is equal to

$$\left(\left(1 + 2 \left(p^d 2^{-\frac{md}{2}} - 1 \right) \right)^3 - 1 + \frac{2d^2}{2^{\frac{m}{2}}} \right)^{\lfloor \frac{r}{3} \rfloor} \tag{3}$$

This bound is well approximated by

$$\left(6d\delta + d^2 2^{1 - \frac{m}{2}} \right)^{\lfloor \frac{r}{3} \rfloor}$$

where $p = 2^{\frac{m}{2}}(1 + \delta)$. Hence for the PEANUT98 Cipher we have $\mathrm{DecP}^2(C) \leq 2^{-76}$. The AES DFC candidate is also in the PEANUT family with parameters $m = 128$, $r = 8$, $d = 2$ and $p = 2^{64} + 13$. Therefore $\mathrm{DecP}^2(C) \leq 2^{-113}$ for it (even if we remove two rounds). Equations (1) and (2) show that differential and linear distinguishers must have a high complexity against both ciphers.

2.6   Several Aspect of the Decorrelation Theory

The approach of the decorrelation theory consists of four important steps.

1. Defining the distance between $[C]^d$ and $[C^*]^d$. We have seen that we can use matrix norms. This paper uses the $|||.|||_\infty$ norm. Some other norms can be considered such as the Euclidean L_2 norm as detailed in [38]. The original concept of universal functions deals with the infinity norm (defined as the maximum of all entries). The choice of the distance is very important, because some norms seem to provide better complexity lower bounds than others.

2. Constructing simple toy random function (which we call "decorrelation modules") with low decorrelation bias. For instance, the PEANUT construction of [36,37] shows how the decorrelation of the $Ax + B \bmod p \bmod 2^{\frac{m}{2}}$ random function (when $(A, B) \in_U \{0, 1\}^m$) for a prime p greater than $2^{\frac{m}{2}}$ has a decorrelation bias which is less than $2(p^d 2^{-\frac{md}{2}} - 1)$ for $d = 2$ which is approximately 4δ for $p = 2^{\frac{m}{2}}(1 + \delta)$.

3. Constructing decorrelated ciphers: proving how the decorrelation bias of the decorrelation modules can be inherited by a larger structure. For instance, the PEANUT construction shows how the decorrelation of the previous primitive is inherited by a Feistel network [6] which uses it as a round function. (Which leads to the bound of Equation (3).)

4. Considering classes of attacks and proving how the decorrelation bias of the cipher makes a lower bound for the complexity of the attack. For instance, proving how the decorrelation to the order 2 provides security against the class of differential or linear attacks.

The present paper deals with the fourth step only.

3   Iterated Attacks of Order d

In this section we introduce the notion of "iterated attack".

3.1   Definition

Equations (1) and (2) suggest that we try to generalize them to a model of iterated attacks. Intuitively, this is an attack in which we iterate (independently) n times an elementary distinguisher which is limited to d queries. After performing one elementary distinguisher we get only one bit of information (we will extend this model for more bits in Section 5, but the results of Section 3 and 4 are only applicable with this limitation of one bit). We focus here on non-adaptive attacks.

Definition 4. *Let n and d be some integers and $\mathcal{M}$ be a set. A non-adaptive "iterated distinguisher of order d and complexity n" for a permutation on $\mathcal{M}$ is defined by*

- *a distribution $\mathcal{D}$ on $\mathcal{M}^d$ (a "plaintext distribution"),*
- *a function T from $\mathcal{M}^{2d}$ to $[0, 1]$ (a "test function"),*
- *a function A from $\{0, 1\}^n$ to $[0, 1]$ (an "acceptance function").*

The distinguisher runs as illustrated on Fig. 4.

Obviously differential and linear distinguishers as formalized on Fig. 2 and 3 are particular cases of iterated attacks (of order 2 and 1 respectively). Namely, if $d = 2$, if the distribution $\mathcal{D}$ is the distribution of $(X, X + a)$ where X has a uniform distribution, if $T((x_1, x_2), (y_1, y_2))$ is defined to be 1 if $y_2 = y_1 + b$ and 0 otherwise, and finally if $A(t_1, \ldots, t_n)$ is defined to be the product of all t_is, then we get a differential distinguisher with characteristic (a, b). Similarly, if $d = 1$, if $\mathcal{D}$ is uniform, if $T(x, y)$ is defined to be 1 if $a \cdot x = b \cdot y$ and 0 otherwise and finally if $A(t_1, \dots, t_n)$ is defined to be 1 if the sum of all t_is is in B and 0 otherwise, then we get a linear distinguisher with characteristic (a, b) and acceptance set B. Iterated attacks of order at most 2 are therefore more general than differential and linear attacks.

Parameters: a complexity n, a plaintext distribution $\mathcal{D}$, a test function T, an acceptance function A

Input: an oracle which implements a function c
1. for i from 1 to n do
   (a) pick a random $X = (X_1, \dots, X_d)$ with distribution $\mathcal{D}$
   (b) get $Y = (c(X_1), \dots, c(X_d))$ from the oracle c
   (c) pick a random $T_i \in \{0, 1\}$ with an expected value of $T(X, Y)$
2. randomly output 0 or 1 with an expected value of $A(T_1, \dots, T_n)$

Fig. 4. Non-Adaptive Iterated Attack of Order d.

When $\mathcal{D}$ is the uniform distribution, we will refer to "known plaintext iterated attacks".

3.2   A Counterexample

It is tempting to believe that a cipher resists to this model of attacks once it has a small d-wise decorrelation bias. This is wrong as the following example shows with $d = 2$. Let C be the simple $Ax + B$ cipher over $\mathrm{GF}(q)$ where $(A, B) \in_U \mathrm{GF}(q)^* \times \mathrm{GF}(q)$. It has a perfect pairwise decorrelation. Obviously, any $((x_1, x_2), (y_1, y_2))$ sample with $x_1 \neq x_2$ and such that $y_1 = C(x_1)$ and $y_2 = C(x_2)$ enables to get (A, B) as a function $f(x_1, x_2, y_1, y_2)$. Let D be a subset of distinguished values of $\mathrm{GF}(q)^* \times \mathrm{GF}(q)$ with a given cardinality denoted $q(q-1)/\mu$. We use the uniform distribution of all (X_1, X_2) pairs such that $X_1 \neq X_2$ as the plaintext distribution. We define

$$T((x_1, x_2), (y_1, y_2)) = \begin{cases} 1 \text{ if } f(x_1, x_2, y_1, y_2) \in D \\ 0 \text{ otherwise} \end{cases}$$

and

$$A(t_1, \dots, t_n) = \begin{cases} 1 \text{ if } (t_1, \dots, t_n) \neq (0, \dots, 0) \\ 0 \text{ otherwise} \end{cases}$$

The trick is that all iterations will provide the same answer for C but a random one for C^*. For the corresponding iterated attack we thus have $p = 1/\mu$ and

$$p^* = 1 - \left(1 - \frac{1}{\mu}\right)^n.$$

For $n = 2$ (two iterations only) we have an advantage of $\frac{1}{\mu}\left(1 - \frac{1}{\mu}\right)$ thus we can have a quite large $|p - p^*|$ although C is perfectly pairwise decorrelated, and that we have an iterated attack of order 2. The trick comes from the fact that the test $\mathcal{T}$ provides a same expected result for C and C^* but a totally different standard deviation, which is avoided by decorrelation to the order $2d = 4$ as shown in the next section.

This counterexample shows that decorrelation of order d is not sufficient in general to prove the security against iterated attacks of order d. In some special cases (as for differential attacks) it may however be sufficient. In the next section we show that decorrelation of order $2d$ is sufficient.

3.3   Security Result

We can however prove the security when the cipher has a good decorrelation to the order $2d$.

Theorem 5. *Let C be a cipher on a message space $\mathcal{M}$ of size M such that* $\mathrm{DecP}^{2d}(C) \leq \epsilon$ *for some given d. For any non-adaptive iterated attack of order d and complexity n which uses a distribution $\mathcal{D}$ (see Fig. 4), we have*

$$\mathrm{Adv}_{\mathrm{Fig}.4}(C, C^*) \leq 3\left(\left(2\delta + \frac{2d^2}{M} + \frac{d^3}{M(M-d)} + \frac{3\epsilon}{2}\right)n^2\right)^{\frac{1}{3}} + \frac{n\epsilon}{2}$$

where δ is the probability that for two independent random X and X' with distribution $\mathcal{D}$ there exists i and j such that $X_i = X'_j$.

In the particular case where $\mathcal{D}$ is the uniform distribution (*i.e.* if we have a known plaintext iterated attack), we have $\delta \leq \frac{d^2}{M}$ so

$$\mathrm{Adv}_{\mathrm{Fig}.4}(C, C^*) \leq 3\left(\left(\frac{4d^2}{M} + \frac{d^3}{M(M-d)} + \frac{3\epsilon}{2}\right)n^2\right)^{\frac{1}{3}} + \frac{n\epsilon}{2}.$$

This result shows that with a low decorrelation bias ϵ we need

$$n = \Omega(\min(\epsilon^{-\frac{1}{2}}, \sqrt{M}))$$

in order to get a significant advantage unless the distribution $\mathcal{D}$ has some special property. For known plaintext attacks, the attacker cannot choose this distribution so this results is meaningful. For other attacks we can wonder what happens if the attacker choose a clever distribution. We believe that the present result can be improved in further work. Actually, if the distribution is such that X_1 is always the same query we get the worse case because $\delta = 1$. Having the same query to the oracle is however a strange way for attacking it and we believe that this strategy does not provide any advantage.[7]

---

[7] We did not state a theorem in term of known plaintext attack only in order to stimulate further research in this way.

If we apply this Theorem to linear cryptanalysis ($d = 1$ and $\delta = \frac{1}{M}$) we obtain

$$\text{Adv}_{\text{Fig.}2}(C, C^*) \leq 3\left(\left(\frac{4}{M} + \frac{1}{M(M-d)} + \frac{3\epsilon}{2}\right)n^2\right)^{\frac{1}{3}} + \frac{n\epsilon}{2}.$$

This result is weaker than Equation (2). Similarly, in order to apply it to differential distinguisher ($d = 2$ and $\delta \leq \frac{4}{M}$), we need decorrelation to the order 4 although Equation (1) needs decorrelation to the order 2 only. This is the cost of more general results!

Proof. Let Z (resp. Z^*) be the probability over the distribution of X that the test accepts $(X, C(X))$ (resp. $(X, C^*(X))$), *i.e.*

$$Z = E_X(T(X, C(X))).$$

(Z depends on C.) Let p (resp. p^*) be the probability that the attack accepts, *i.e.*

$$p = E_C(A(T_1, \ldots, T_n)).$$

Since the T_is are independent and with the same expected value Z which only depends on C, we have

$$p = E_C\left(\sum_{t_1, \ldots, t_n \in \{0,1\}} A(t_1, \ldots, t_n) Z^{t_1 + \cdots + t_n}(1 - Z)^{n - (t_1 + \cdots + t_n)}\right).$$

We thus have $p = E(f(Z))$ where $f(z)$ is a polynomial of degree at most n with values in $[0, 1]$ for any $z \in [0, 1]$ entries and with the form $f(z) = \sum a_i z^i (1 - z)^{n-i}$. It is straightforward that $|f'(z)| \leq n$ for any $z \in [0, 1]$. Thus we have $|f(z) - f(z^*)| \leq n|z - z^*|$.

The crucial point in the proof is in proving that $|Z - Z^*|$ is small within a high probability. For this, we need $|E(Z) - E(Z^*)|$ and $|V(Z) - V(Z^*)|$ to be both small.

From Theorem 3 we know that $|E(Z) - E(Z^*)| \leq \frac{\epsilon}{2}$. We note that Z^2 corresponds to a another test but with $2d$ entries, hence we have $|E(Z^2) - E((Z^*)^2)| \leq \frac{\epsilon}{2}$. Hence $|V(Z) - V(Z^*)| \leq \frac{3}{2}\epsilon$. Now from Tchebichev's Inequality we have

$$\Pr[|Z - E(Z)| > \lambda] \leq \frac{V(Z)}{\lambda^2}.$$

Hence we have

$$\Pr\left[|Z - Z^*| > \frac{\epsilon}{2} + 2\lambda\right] \leq \frac{2V(Z^*) + \frac{3}{2}\epsilon}{\lambda^2}$$

thus

$$|p - p^*| \leq \frac{2V(Z^*) + \frac{3}{2}\epsilon}{\lambda^2} + n\left(\frac{\epsilon}{2} + 2\lambda\right)$$

so, with $\lambda = \left(\frac{2V(Z^*) + \frac{3}{2}\epsilon}{n} \right)^{\frac{1}{3}}$ we have

$$|p - p^*| \leq 3 \left(\left(2V(Z^*) + \frac{3\epsilon}{2} \right) n^2 \right)^{\frac{1}{3}} + \frac{n\epsilon}{2}.$$

The variance $V(Z^*)$ is expressed by

$$\sum_{\substack{x,y \\ x',y'}} \Pr_{\mathcal{D}^2}[x, x'] T(x, y) T(x', y') \left(\Pr_{C^*} \left[\begin{array}{c} x \to y \\ x' \to y' \end{array} \right] - \Pr_{C^*}[x \to y] \Pr_{C^*}[x' \to y'] \right)$$

which is maximal when $T(x, y)$ is 0 or 1 by linear programming results. Thus

$$V(Z^*) \leq \frac{1}{2} \sum_{\substack{x,y \\ x',y'}} \Pr_X[x] \Pr_X[x'] \left| \Pr_{C^*} \left[\begin{array}{c} x \to y \\ x' \to y' \end{array} \right] - \Pr_{C^*}[x \to y] \Pr_{C^*}[x' \to y'] \right|.$$

The sum over all x and x' entries with colliding entries (*i.e.* with some $x_i = x'_j$) is less than δ. The sum over all y and y' entries with colliding entries and no colliding x and x' is less than $d^2/2M$. The sum over all no colliding x and x' and no colliding y and y' is less than

$$\frac{1 - \delta}{2} \left(1 - \frac{M(M-1)\ldots(M-2d+1)}{M^2(M-1)^2\ldots(M-d+1)^2} \right)$$

which is less than $\frac{d^2}{2(M-d)}$. Thus we have $V(Z^*) \leq \delta + \frac{d^2}{2M} + \frac{d^2}{2(M-d)}$ which is equal to $\delta + \frac{d^2}{M} + \frac{d^3}{2M(M-d)}$. $\qquad\square$

3.4   Applications

PEANUT98 is a 9-round Feistel Cipher for message-blocks of size 64 which has been proposed in [36] with a constructed pairwise decorrelation such that $\mathrm{DecP}^2(\mathrm{PEANUT98}) \leq 2^{-76}$ as shown in Section 2.5. From Equation (1) we know that no differential distinguisher with a number of chosen plaintext pairs less than 2^{76} will have an advantage greater than 50%. From Equation (2) we know that no linear distinguisher with a number of known plaintext less than 2^{62} will have an advantage greater than 50%. Now from Theorem 5 we know that no known plaintext iterated attack of order 1 (*e.g.* linear attacks) with a number of known plaintext less than 2^{33} will have an advantage greater than 50%. For linear cryptanalysis, this result is weaker than Equation (2), but more general.

Similarly, DFC is immune against any known plaintext iterated attack of order 1 with a number of known plaintext less than 2^{52} in the sense that the advantage of these attacks will always be less than 50%.

All these results are applicable to the COCONUT98 Cipher as well since its pairwise decorrelation bias is even smaller (it is actually zero).

The threshold of 50% is arbitrary here. If we have an attack with low advantage α, we intuitively want to iterate it at least $1/\alpha$ times in order to get a significant success rate. The complexity is therefore increased accordingly. We thus adopted this symbolic threshold of 50%.

4 On Combining Several Attacks

When several (inefficient) attacks hold against a cipher C, it is natural to wonder whether or not we can combine their effort in order to get an efficient attack. This situation is formalized by changing a few things on Fig. 4 and we can rewrite Theorem 5 in this setting. Firstly, the test in each iteration can be changed. Secondly, n must be considered as relatively small, and d as relatively large: we use a few attacks (n) which have no real limitations (d) on the number of queries. This situation is different from the previous one where we used many attacks (many times the same one actually) of limited order d. For this reason and since we want n to express the complexity we rewrite d into n_i for the ith attack and n into r. The resulting model is illustrated on Fig. 5.

Parameters: several attacks $\mathcal{A}_1, \dots, \mathcal{A}_r$, an acceptance function A
Input: an oracle which implements a function c
    1. for i from 1 to r do in parallel
        (a) perform the attack $\mathcal{A}_i$ against c
        (b) set T_i to the result of the attack
    2. randomly output 0 or 1 with an expected value of $A(T_1, \dots, T_n)$

Fig. 5. Combined Attack.

Theorem 6. *Let C be a cipher on a message space $\mathcal{M}$ of size M. Let $\mathcal{A}_1, \dots, \mathcal{A}_r$ be r attacks on C with advantages $\mathrm{Adv}_{\mathcal{A}_1}, \dots, \mathrm{Adv}_{\mathcal{A}_r}$ respectively. For each i, we let n_i denote the number of queries from $\mathcal{A}_i$ and we let $\mathcal{A}_i^2$ denotes the following attack.*

Input: *an oracle which implements a cipher c*
    1. perform the attack $\mathcal{A}_i$ and set a to the result
    2. perform the attack $\mathcal{A}_i$ and set b to the result
    3. if $a = b = 1$ output 1 otherwise output 0

We let $\mathrm{Adv}_{\mathcal{A}_i^2}$ denote its advantage, and δ_i denote the probability that the two $\mathcal{A}_i$ attack executions query c with one input in common. For any combined attack (depicted on Fig. 5) with independent attacks, $\mathrm{Adv}_{\mathrm{Fig.5}}(C, C^*)$ is less than

$$\sum_{i=1}^{r} \left(\mathrm{Adv}_{\mathcal{A}_i} + 3 \left(2\delta_i + \frac{2n_i^2}{M} + \frac{n_i^3}{M(M-d)} + 2\mathrm{Adv}_{\mathcal{A}_i} + \mathrm{Adv}_{\mathcal{A}_i^2} \right)^{\frac{1}{3}} \right).$$

For instance, when the attacks are known plaintext attacks with a plaintext source with uniform distribution, we have $\delta_i \leq \frac{n_i^2}{M}$.

This result does not depend on the decorrelation of the cipher but only upper bound what we can best achieve when combining several attacks. The occurrence of $\mathcal{A}_i^2$ is a little frustrating but is necessary. Section 3.2 is actually a counterexample in which some attack $\mathcal{A}$ is totally inefficient (with an advantage of 0) but with a quite high $\mathrm{Adv}_{\mathcal{A}^2}$.

Proof. As for the proof of Theorem 5, the advantage can be written

$$\mathrm{Adv}_{\mathrm{Fig.5}}(C, C^*) = |E(f(Z_1, \ldots, Z_r) - f(Z_1^*, \ldots, Z_r^*))|$$

for a polynomial $f(x_1, \ldots, x_r)$ of partial degrees at most 1 and with values in $[0,1]$ whenever all entries are in $[0,1]$. All partial derivatives $f_i'(x_1, \ldots, x_r)$ are in $[-1,1]$, so we have

$$\mathrm{Adv}_{\mathrm{Fig.5}}(C, C^*) \le \sum_{i=1}^{r} E(|Z_i - Z_i^*|).$$

We have $|E(Z_i - Z_i^*)| = \mathrm{Adv}_{\mathcal{A}_i}$ and $|E(Z_i^2 - (Z_i^*)^2)| = \mathrm{Adv}_{\mathcal{A}_i^2}$. So, as in the proof of Theorem 5, we obtain

$$\mathrm{Adv}_{\mathrm{Fig.5}}(C, C^*) \le \sum_{i=1}^{r} \left(\mathrm{Adv}_{\mathcal{A}_i} + 3 \left(2V(Z_i^*) + 2\mathrm{Adv}_{\mathcal{A}_i} + \mathrm{Adv}_{\mathcal{A}^2} \right)^{\frac{1}{3}} \right).$$

and finally $V(Z_i^*) \le \delta_i + \frac{n_i^2}{M} + \frac{n_i^3}{2M(M-d)}$. □

5   Generalization

We can even generalize Theorem 5 in the case where the iterations of the attack produce an information T_i which is not necessarily binary. We outline that if the size of T_i is unlimited, then there is no possible result because the attack has unlimited computation power and it would be able to perform exhaustive search with all information from the queries.

Theorem 7. *Let C be a cipher on a message space of size M such that we have $\mathrm{DecP}^{2d}(C) \le \epsilon$ for some given d. For any non-adaptive iterated attack of order d and complexity n which uses a distribution $\mathcal{D}$ (see Fig. 4) and where we allow the T_i to be in the set $\{1, \ldots, s\}$, we have*

$$\mathrm{Adv}_{\mathrm{Fig.4}}(C, C^*) \le 3s \left(\left(2\delta + \frac{2d^2}{M} + \frac{d^3}{M(M-d)} + \frac{3\epsilon}{2} \right) n^2 \right)^{\frac{1}{3}} + \frac{ns\epsilon}{2}$$

where δ is the probability that for two independent random X and X' with distribution $\mathcal{D}$ there exists i and j such that $X_i = X_j'$.

Proof. In the proof of Theorem 5, $f(Z)$ is replaced by a polynomial $f(Z_1, \ldots, Z_s)$ in term of $Z_j = \Pr[T_i = j]$ for $j = 1, \ldots, s$. For two distributions $(z_1, \ldots, z_s)$ and $(z_1^*, \ldots, z_s^*)$, we have

$$|f(z_1, \ldots, z_s) - f(z_1^*, \ldots, z_s^*)| \le n \sum_{i=1}^{s} |z_i - z_i^*|.$$

As in the previous proof we have

$$\Pr\left[|Z_i - Z_i^*| > \frac{\epsilon}{2} + 2\lambda\right] \leq \frac{2V(Z_i^*) + \frac{3}{2}\epsilon}{\lambda^2}$$

for any λ and $V(Z_i^*) \leq \delta + \frac{d^2}{M} + \frac{d^3}{2M(M-d)}$. Hence the situation simply consists in multiplying the lower bound by s. $\qquad\square$

6   Conclusion

We showed how to unify differential and linear distinguishers in a general notion of iterated attack. We then proved that decorrelation enables to quantify the security against any iterated attack. This result happened to be applicable to a real life block cipher. Our result are however not so tight because of the use of Tchebichev's Inequality, and it is still an open problem to improve the complexity upper bounds (with Chernov's bounds?). We encourage researches in this direction.

References

1. Data Encryption Standard. *Federal Information Processing Standard Publication 46*, U. S. National Bureau of Standards, 1977.
2. M. Bellare, P. Rogaway. Random Oracles are Practical: a Paradigm for Designing Efficient Protocols. In *1st ACM Conference on Computer and Communications Security*, Fairfax, Virginia, U.S.A., pp. 62–73, ACM Press, 1993.
3. E. Biham, A. Shamir. *Differential Cryptanalysis of the Data Encryption Standard*, Springer-Verlag, 1993.
4. D. Bleichenbacher. Chosen Ciphertext Attacks Against Protocols Based on the RSA Encryption Standard PKCS#1. In *Advances in Cryptology CRYPTO'98*, Santa Barbara, California, U.S.A., Lectures Notes in Computer Science 1462, pp. 1–12, Springer-Verlag, 1998.
5. L. Carter, M. Wegman. Universal Classes of Hash Functions. *Journal of Computer and System Sciences*, vol. 18, pp. 143–154, 1979.
6. H. Feistel. Cryptography and Computer Privacy. *Scientific American*, vol. 228, pp. 15–23, 1973.
7. H. Gilbert, M. Girault, P. Hoogvorst, F. Noilhan, T. Pornin, G. Poupard, J. Stern, S. Vaudenay. Decorrelated Fast Cipher: an AES Candidate. Submitted to the Advanced Encryption Standard process. In *CD-ROM "AES CD-1: Documentation"*, National Institute of Standards and Technology (NIST), August 1998.
8. O. Goldreich, S. Goldwasser, S. Micali. How to Construct Random Functions. In *Proceedings of the 25th IEEE Symposium on Foundations of Computer Science*, Singer Island, U.S.A., pp. 464–479, IEEE, 1984.
9. S. Halevi, H. Krawczyk. MMH: Software Message Authentication in the Gbit/Second Rates. In *Fast Software Encryption*, Haifa, Israel, Lectures Notes in Computer Science 1267, pp. 172–189, Springer-Verlag, 1997.
10. T. Jakobsen, L. R. Knudsen. The Interpolation Attack on Block Ciphers. In *Fast Software Encryption*, Haifa, Israel, Lectures Notes in Computer Science 1267, pp. 28–40, Springer-Verlag, 1997.

11. H. Krawczyk. LFSR-based Hashing and Authentication. In *Advances in Cryptology CRYPTO'94*, Santa Barbara, California, U.S.A., Lectures Notes in Computer Science 839, pp. 129–139, Springer-Verlag, 1994.

12. L. R. Knudsen. DEAL - A 128-Bit Block Cipher. Presented at the SAC'97 Workshop (Invited Lecture). Submitted to the Advanced Encryption Standard process. In *CD-ROM "AES CD-1: Documentation"*, National Institute of Standards and Technology (NIST), August 1998.

13. X. Lai. *On the Design and Security of Block Ciphers*, ETH Series in Information Processing, vol. 1, Hartung-Gorre Verlag Konstanz, 1992.

14. X. Lai, J. L. Massey. A Proposal for a New Block Encryption Standard. In *Advances in Cryptology EUROCRYPT'90*, Aarhus, Denemark, Lectures Notes in Computer Science 473, pp. 389–404, Springer-Verlag, 1991.

15. X. Lai, J. L. Massey, S. Murphy. Markov Ciphers and Differential Cryptanalysis. In *Advances in Cryptology EUROCRYPT'91*, Brighton, United Kingdom, Lectures Notes in Computer Science 547, pp. 17–38, Springer-Verlag, 1991.

16. M. Luby, C. Rackoff. How to Construct Pseudorandom Permutations from Pseudorandom Functions. *SIAM Journal on Computing*, vol. 17, pp. 373–386, 1988.

17. S. Lucks. Faster Luby-Rackoff Ciphers. In *Fast Software Encryption*, Cambridge, United Kingdom, Lectures Notes in Computer Science 1039, pp. 189–203, Springer-Verlag, 1996.

18. M. Matsui. Linear Cryptanalysis Methods for DES Cipher. In *Advances in Cryptology EUROCRYPT'93*, Lofthus, Norway, Lectures Notes in Computer Science 765, pp. 386–397, Springer-Verlag, 1994.

19. M. Matsui. The First Experimental Cryptanalysis of the Data Encryption Standard. In *Advances in Cryptology CRYPTO'94*, Santa Barbara, California, U.S.A., Lectures Notes in Computer Science 839, pp. 1–11, Springer-Verlag, 1994.

20. M. Matsui. New Structure of Block Ciphers with Provable Security against Differential and Linear Cryptanalysis. In *Fast Software Encryption*, Cambridge, United Kingdom, Lectures Notes in Computer Science 1039, pp. 205–218, Springer-Verlag, 1996.

21. M. Matsui. New Block Encryption Algorithm MISTY. In *Fast Software Encryption*, Haifa, Israel, Lectures Notes in Computer Science 1267, pp. 54–68, Springer-Verlag, 1997.

22. M. Naor, O. Reingold. On the construction of pseudo-random permutations: Luby-Rackoff revisited. Presented at the Security in Communication Networks Workshop, Amalfi, Italy, 1996. Submitted for publication.
    http://www.unisa.it/SCN96/papers/Reingold.ps

23. K. Nyberg. Perfect Nonlinear *S*-Boxes. In *Advances in Cryptology EUROCRYPT'91*, Brighton, United Kingdom, Lectures Notes in Computer Science 547, pp. 378–385, Springer-Verlag, 1991.

24. K. Nyberg. Differentially Uniform Mapping for Cryptography. In *Advances in Cryptology EUROCRYPT'93*, Lofthus, Norway, Lectures Notes in Computer Science 765, pp. 55–64, Springer-Verlag, 1994.

25. K. Nyberg. Linear Approximation of Block Ciphers. In *Advances in Cryptology EUROCRYPT'94*, Perugia, Italy, Lectures Notes in Computer Science 950, pp. 439–444, Springer-Verlag, 1995.

26. K. Nyberg, L. R. Knudsen. Provable Security against a Differential Cryptanalysis. In *Advances in Cryptology CRYPTO'92*, Santa Barbara, California, U.S.A., Lectures Notes in Computer Science 740, pp. 566–574, Springer-Verlag, 1993.

27. K. Nyberg, L. R. Knudsen. Provable Security against a Differential Attack. *Journal of Cryptology*, vol. 8, pp. 27–37, 1995.

28. J. Pieprzyk. How to Construct Pseudorandom Permutations from a Single Pseudorandom Functions. In *Advances in Cryptology EUROCRYPT'90*, Aarhus, Denemark, Lectures Notes in Computer Science 473, pp. 140–150, Springer-Verlag, 1991.

29. J. Patarin. *Etude des Générateurs de Permutations Basés sur le Schéma du D.E.S.*, Thèse de Doctorat de l'Université de Paris 6, 1991.

30. J. Patarin. How to Construct Pseudorandom and Super Pseudorandom Permutations from One Single Pseudorandom Function. In *Advances in Cryptology EUROCRYPT'92*, Balatonfüred, Hungary, Lectures Notes in Computer Science 658, pp. 256–266, Springer-Verlag, 1993.

31. J. Patarin. About Feistel Schemes with Six (or More) Rounds. In *Fast Software Encryption*, Paris, France, Lectures Notes in Computer Science 1372, pp. 103–121, Springer-Verlag, 1998.

32. R. L. Rivest, A. Shamir and L. M. Adleman. A Method for Obtaining Digital Signatures and Public-key Cryptosystem. In *Communications of the ACM*, vol. 21, pp. 120–126, 1978.

33. C. E. Shannon. Communication Theory of Secrecy Systems. *Bell system technical journal*, vol. 28, pp. 656–715, 1949.

34. V. Shoup. Lower Bounds for Discrete Logarithms and Related Problems. In *Advances in Cryptology EUROCRYPT'97*, Konstanz, Germany, Lectures Notes in Computer Science 1233, pp. 256–266, Springer-Verlag, 1997.

35. S. Vaudenay. An Experiment on DES — Statistical Cryptanalysis. In *3rd ACM Conference on Computer and Communications Security*, New Delhi, India, pp. 139–147, ACM Press, 1996.

36. S. Vaudenay. Provable Security for Block Ciphers by Decorrelation. In *STACS 98*, Paris, France, Lectures Notes in Computer Science 1373, pp. 249–275, Springer-Verlag, 1998.

37. S. Vaudenay. Provable Security for Block Ciphers by Decorrelation. (Full Paper.) Submitted. Preliminary version available on
URL:ftp://ftp.ens.fr/pub/reports/liens/liens-98-8.A4.ps.Z

38. S. Vaudenay. Feistel Ciphers with L_2-Decorrelation. To appear in SAC'98, LNCS.

39. S. Vaudenay. The Decorrelation Technique Home-Page.
URL:http://www.dmi.ens.fr/~vaudenay/decorrelation.html

40. G. S. Vernam. Cipher Printing Telegraph Systems for Secret Wire and Radio Telegraphic Communications. *Journal of the American Institute of Electrical Engineers*, vol. 45, pp. 109–115, 1926.

41. D. Wagner. The Boomerang Attack. Personal communication.

42. M. N. Wegman, J. L. Carter. New Hash Functions and their Use in Authentication and Set Equality. *Journal of Computer and System Sciences*, vol. 22, pp. 265–279, 1981.

XOR and Non-XOR Differential Probabilities

Philip Hawkes[1] and Luke O'Connor[2]

[1] Qualcomm International,
Suite 410, Birkenhead Point,
Drummoyne, NSW, 2047, Australia
phawkes@qualcomm.com
[2] IBM Research Division
Zurich Research Laboratory
Säumerstrasse 4, Rüschlikon
CH-8803, Switzerland
oco@zurich.ibm.com

Abstract. Differential cryptanalysis is a well-known attack on iterated ciphers whose success is determined by the probability of predicting sequences of differences from one round of the cipher to the next. The notion of difference is typically defined with respect to the group operation(s) used to combine the subkey in the round function F. For a given round operation π of F, such as an S-box, let $DP_\otimes(\pi)$ denote the probability of the most likely non-trivial difference for π when differences are defined with respect to $\otimes$. In this paper we investigate how the distribution of $DP_\otimes(\pi)$ varies as the group operation $\otimes$ is varied when π is a uniformly selected permutation. We prove that $DP_\otimes(\pi)$ is maximised with high probability when differences are defined with respect to XOR.

1 Introduction

Differential cryptanalysis (DC) is a well-known chosen-plaintext attack based on predicting how certain changes or differences in the plaintext propagate through a cipher. DC was well publicized by Biham and Shamir [3] as a tool for the cryptanalysis of DES-like ciphers. Biham and Shamir defined the *difference* $\Delta(X, X^*)$ between two n-bit blocks X, X^* by $\Delta(X, X^*) = X \oplus X^*$ where $\oplus$ denotes the bit-wise exclusive-OR (XOR) operation. To extend the application of DC to other ciphers Lai, Massey and Murphy [14] adapted the definition of differences to $\Delta(X, X^*) = X \otimes (X^*)^{-1}$, where $\otimes$ is an Abelian group operation and $(X^*)^{-1}$ is the group inverse of X^*. The choice of difference used to analyse a cipher is usually selected so that the subkey Z is cancelled by the difference operator:

$$\Delta(X, X^*) = \Delta((X \otimes Z), (X^* \otimes Z)) = X \otimes (X^*)^{-1}. \qquad (1)$$

Consequently, the choice of operation used to define differences is typically defined by the group operations(s) used to combine the key into the cipher. Commonly used group operations include XOR $(\mathbb{Z}_2^n, \oplus)$, modular addition $(\mathbb{Z}_{2^n}, \boxplus)$, and modular multiplication $(\mathbb{Z}_{2^n+1}^*, \odot)$, where $(2^n + 1)$ is prime. In general the

J. Stern (Ed.): EUROCRYPT'99, LNCS 1592, pp. 272–285, 1999.
© Springer-Verlag Berlin Heidelberg 1999

n	4	5	6	7	8
av. max. $\boxplus$	0.2771	0.1617	0.0919	0.0515	0.0284
av. max. $\odot$	0.2764	-	-	-	0.0283
av. max. $\oplus$	0.4186	0.2487	0.1426	0.0806	0.0443

Table 1. The average maximum probability for differential approximations to randomly selected bijections $\pi : \mathbb{Z}_2^n \rightarrow \mathbb{Z}_2^n$ defined for the operations $\otimes \in \{ \boxplus, \oplus, \odot \}$, where $4 \leq n \leq 8$. Note that differences for $\odot$ are only defined for n when $2^n + 1$ is prime.

inputs x_1, x_2 to $\odot$ in a cipher will be elements of $\mathbb{Z}_2^n$ rather than $\mathbb{Z}_{2^n+1}^*$, and when evaluating $x_1 \odot x_2$ we first map x_i to 2^n if x_i is zero; also $x_1 \odot x_2$ is mapped to zero if it is equal to 2^n.

The purpose of this paper is to examine how the probability of differential approximations for permutations π vary as the group operation $\otimes$ used to define differences is varied. The study of permutations can be justified on two grounds. First, many blocks ciphers make use of permutations: in some cases these permutations are 'small', often referred to as S-boxes if implemented as tables, such as in SAFER K-64 [17], TWOFISH [21], CRYPTON [16], E2 [12] and Rijndael [5] (all use 8-bit permutations), while other ciphers use larger permutations such as IDEA [13] (subkey multiplication is equivalent to a 16-bit permutation look-up) and DFC [7] (64-bit permutation). The second reason to study permutations is that a block cipher implements a permutation π for any fixed key, and the cipher itself then represents a family of permutations. By studying the properties of permutations we can examine how, for example, permutations generated by an iterative block structure differ from truly random permutations.

Our research was initially motivated by the results presented in Table 1, which shows the average maximum differential approximation to several thousand n-bit permutations, $4 \leq n \leq 8$, with respect to the group operations $\oplus$, $\boxplus$ and $\odot$. In all cases $DP_{\otimes}(\pi)$ was maximised for XOR differences. For example, the column for $n = 8$ indicates that the best approximation for 8-bit mappings with respect to $\otimes \in \{\odot, \boxplus\}$ will have a probability between $7/256$ and $8/256$, while the corresponding probability for XOR differences was between $10/256$ and $12/256$. Experiments also showed that XOR differences yielded higher probability differentials for the S-boxes of DES than differences with respect to $\boxplus$. While this phenomenon is quite likely to be known by some researchers[1], this is the first paper which analyses it mathematically.

We first present our main results and then discuss their implications. We will consider all abelian groups of order 2^n, and to this end, let $(\mathbb{Z}_2^n, \otimes)$ be an abelian group of order 2^n with identity element I. For $\alpha, \beta \in \mathbb{Z}_2^n \setminus \{I\}$ and an n-bit permutation π we define

---

[1] For example, this observation was stated by M. Dichtl during a seminar presented at Isaac Newton Institute, 1996.

n	8	16	32	64	128	256	512	1024
B_n	4.6	7.2	11.7	20.8	34.3	60.4	108.1	195.6
$2 \cdot \lceil B_n \rceil$	10	16	24	42	70	122	218	392

Table 2. The values of $B_n = \ln N^2 / \ln\ln N^2$, $N = (2^n - 1)$ for several n.

$$DP_\otimes(\pi, \alpha, \beta) = \frac{1}{2^n} \cdot \sum_{\substack{X, X^* \\ \Delta(X, X^*)=\alpha}} [\Delta(\pi(X), \pi(X^*)) = \beta\}]$$

where $[\cdot]$ is a predicate that evaluates to 0 or 1. Thus $DP_\otimes(\pi, \alpha, \beta)$ is the probability that an input difference of α leads to an output difference of β in π when differences defined with respect to $\otimes$. Further, we define $DP_\otimes(\pi) = \max_{\alpha, \beta \neq I} DP_\otimes(\pi, \alpha, \beta)$ to be the highest probability difference in π with respect to $\otimes$. One of the main results of this paper is to prove that asymptotically, for uniformly selected π,

$$\Pr\left(\frac{n \ln 2}{2^{n-1} \ln n} \leq DP_\otimes(\pi) < \frac{n}{2^{n-1}}\right) \sim 1, \qquad \otimes = \oplus, \tag{2}$$

$$\Pr\left(DP_\otimes(\pi) < \frac{n \ln 2}{2^{n-1} \ln n}\right) \sim 1, \qquad \otimes \in \{\boxplus, \odot\}. \tag{3}$$

Equivalently, the fraction of n-bit permutations that do not satisfy the bounds of (2) and (3) tends to 0 and n increases. Our results concentrate on a comparison between $\otimes = \oplus$ and $\otimes \in \{\boxplus, \odot\}$, since the latter two group operations are the most pertinent to cryptography. The $(n \ln 2)/(2^{n-1} \ln n)$ term in (2) and (3) is derived as an asymptotic estimate of $2B_n/2^n$ where $B_n = \ln N^2 / \ln(\ln N^2))$ and $N = 2^n - 1$ is the number of non-trivial differences. For smaller n, $2B_n$ can be used in (2) and (3), and some relevant values of B_n are given in Table 2. For 8-bit permutations the critical value is $B_8 = 10$, meaning that XOR approximations are likely[2] to occur with probability at least $10/256$ while approximations based on $\otimes \in \{\boxplus, \odot\}$ with probability less than $10/256$. The general conclusion is that it is very likely that selecting a permutation π at random will yield higher probability XOR difference approximations than differences defined with respect to the groups $\boxplus$ and $\odot$. [3]

The bounds of (2) and (3) indicate that with high probability the best DC XOR approximation to a 64-bit permutation lies in the interval $[2^{-58.6}, 2^{-57}]$,

---

[2] We note that the authors of TWOFISH were able to find 8-bit permutations with best XOR difference approximation of at most $10/256$ in 'a few tens of hours' [21, p.24]. These permutations were composed to form the basis of the S-boxes for TWOFISH, where for a majority of the keys the best XOR approximation has probability $12/256$.

[3] We note that it is always possible to pick a 'cooked' permutation π for which XOR differences have lower probability than $\otimes \in \{\boxplus, \odot\}$ differences, such as $\pi(x) = x \oplus c$ for some group element c. We simply assert that this event is unlikely to happen if π is selected randomly or in some unbiased manner.

while for a 128-bit permutation the interval is $[2^{-121.9}, 2^{-120}]$. Thus if we assume that 48-round DES acts as random 64-bit permutation, the best XOR approximation will occur with much higher probability than suggested by extending the 2-round iterative characteristic used for the DC of 16-round DES. While we acknowledge that it may be computationally infeasible to find such a high probability characteristic, the bounds of (2) and (3) strongly suggest that far more probable DC approximations are available than indicated by the round-by-round approximation approach based on characteristics.

We are hesitant to apply our results in general to existing ciphers, say by changing $\oplus$ operations to $\boxplus$ operations and claiming improved security against DC. This is certainly not the case for DES [4]. We believe that to fully take advantage of our results would require the design of a new cipher, and this is not the subject of this paper. We hope that our present results will form the basis for further research into the most appropriate group operation(s) to be used in the design and analysis of block ciphers against DC. We note however that Adams [1] has already used our results to suggest the security of the CAST-256 algorithm. We also note that in general the designer cannot force the cryptanalyst to use differences defined with respect to a given group operation $\otimes$. A case in point is a DC of RC5 [11] where the natural choice of difference was based on $\boxplus$, but differences with respect to $\oplus$ were used regardless. On the other hand, XOR differences give high probability approximations to the two S-boxes used in SAFER K-64, but the use of other non-XOR operations such as the Pseudo Hadamard Transform appears to have successfully thwarted on DC based on XOR differences alone.

It remains now to prove (2) and (3). As a first step we determine that the distribution of $DP_{\otimes}(\pi, \alpha, \beta)$, asymptotically follows the Poisson distribution $\Pr(X = t) = e^{-\mu} \cdot \mu^t / t!$, for $t \geq 0$. When both group elements α, β have order 2, the Poisson parameter is $\mu = 1/2$, while it is $\mu = 1$ for any other pair of elements with orders both distinct from 2. Note that all elements of $(\mathbb{Z}_2^n - \{0\}, \oplus)$ have order 2, while almost all elements of $(\mathbb{Z}_2^n - \{I\}, \otimes)$, $\otimes \neq \oplus$, have order greater than 2, which will be shown to cause the higher XOR approximations. Also similar comments apply if α is a difference with respect to $\otimes_1$, and β is a difference with respect to $\otimes_2$, Such differences have been called a quasi-differentials [18], and naturally arise in the DC of SAFER [17] which uses both $\oplus$ and $\boxplus$ to mask the inputs and outputs to its S-boxes.

The upper bound in (2) is from [19], while given $Y_k = \sum_{\alpha, \beta} \Pr(DP_{\otimes}(\pi, \alpha, \beta) = k)$, the upper bound in (3) can be proven directly from $\Pr(DP_{\otimes}(\pi) \leq t) \leq (1 - \sum_{k \geq t} \mathbf{E}[Y_k])$ when $Y_k = o(2^{2n})$. The lower bound in (2) is harder to prove. Note that Y_k defined above is the expected number of entries in the difference table of size k. Our approach is to find a value of k for which $\mathbf{E}[Y_k] \geq n$ and $\mathbf{Var}[Y_k] \sim n$, from which it follows via Chebychev's inequality that an entry of size k exists with probability tending to 1 with n. As it turns out, $k = B_n$ satisfies these conditions. Even though we work with expectations we note that that bounds in (2) and (3) are not expectations or for the average case.

The paper is set out as follows. In §2 we introduce notation and reduce the problem of enumerating $2^n \cdot DP_\otimes(\pi, \alpha, \beta)$ to a counting problem on graphs. This counting problem is combined with the inclusion-exclusion principle to obtain the distribution of probabilities for a differential approximation. The distribution of values for individual entries are shown to be asymptotically Poisson in §2. In §3 the bound given in (2) and (3) are proven. Our conclusions are presented in §4 and several proofs are delegated to the appendix in §5.

2   An Equivalent Graph Theory Problem

We let $\Pi^{(n)}$ denote the set of n-bit permutations, and write $\pi \in_R \Pi^{(n)}$ to denote a uniformly selected n-bit permutation. The problem of determining the distribution of $DP_\otimes(\pi)$ can be considered as an enumeration problem: count the number of edge-preserving mappings between two appropriately defined directed graphs, given below. Recall that the set of n-bit blocks is denoted $\mathbb{Z}_2^n$ and can represented by the set $\{0, 1, \ldots, 2^n - 1\}$.

Definition 1. For a group $(\mathbb{Z}_2^n, \otimes)$ of order 2^n and a non-trivial (non-identity) difference $\alpha \in \mathbb{Z}_2^n$ there is an associated directed graph $D_\alpha = (V, E_\alpha)$, $|V| = 2^n$, where each vertex $v \in V$ has a unique label $l(v) \in \mathbb{Z}_2^n$. Then any group element X is uniquely associated with the vertex $u \in V$ such that $l(u) = X$. The directed edge set of D_α is defined as $E_\alpha = \{(u, v) \mid l(u) \otimes (l(v))^{-1} = \alpha\}$, meaning that (u, v) is an edge when $X = l(u)$ and $v = l(X^*)$ and $\Delta(X, X^*) = \alpha$. We call D_α the *difference graph* of α with respect to $\otimes$.   □

As a result of the group property, every vertex of D_α and D_β has indegree and outdegree one. Consequently, the arcs of D_α and D_β form cycles. Further, D_α consists of $\frac{2^n}{\text{ord } \alpha}$ labeled disjoint cycles of length ord α, which follows from Lagrange's Theorem since the cycles correspond to cosets. Let $D_\alpha = (V, E_\alpha)$ and $D_\beta = (V, E_\beta)$ be the difference directed graphs representing any two differences $\alpha, \beta \in \mathbb{Z}_2^n$. For a permutation $\pi \in \Pi^{(n)}$ we define

$$d_{\otimes, \pi}(D_\alpha, D_\beta) = \# \{(u, v) \in E_\alpha \mid (u^*, v^*) \in E_\beta, \ l(u^*) = \pi(l(u)), l(v^*) = \pi(l(v))\}.$$

Thus $2^n \cdot DP_\otimes(\pi, \alpha, \beta) = d_{\otimes, \pi}(D_\alpha, D_\beta)$, and this value depends on the number of edges mapped between the two distance graphs.

Example 2. Consider $(\mathbb{Z}_2^3, \boxplus)$, the group of addition modulo 8. The directed graphs D_1, D_2 representing the differences $\Delta(X, X^*) = 1$ and $\Delta(X, X^*) = 2$ are shown in Figure 1. Notice that the arcs of D_1 and D_2 form cycles of length 8 and 4 respectively, as ord $1 = 8$ and ord $2 = 4$ with respect to $\boxplus$. Let $\pi \in \Pi^{(3)}$ be the permutation $(3, 0, 7, 1, 2, 5, 4, 6)$ where $\pi(0) = 3, \pi(1) = 0, \pi(2) = 7$ and so on. Then the only arcs of D_1 mapped by π to arcs of D_2 are the arcs labeled by $(3, 2)$ and $(7, 6)$ of D_1 which are mapped to the arcs labeled by $(1, 7)$ and $(6, 4)$ respectively of D_2. Consequently $DP_\boxplus(\pi, 1, 2) = d_{\boxplus, \pi}(D_1, D_2) = 2$.   □

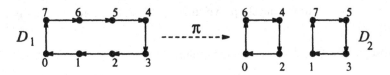

Fig. 1. The directed graphs D_1 and D_2 representing the two differences $\Delta(X, X^*) = 1$ and $\Delta(X, X^*) = 2$ using the 3-bit $\boxplus$ operation to define the differences.

Theorem 3. For any Abelian group G and elements $\alpha, \beta, \in G$, the probability $\Pr(2^n \cdot DP_\otimes(\pi, \alpha, \beta) = t)$ only depends on t, ord $G = \#G$, $a =$ ord α and $b =$ ord β. For $a = 2^r$, $b = 2^s$, $1 \leq r \leq n$, $1 \leq s \leq n$, and $0 \leq t \leq 2^n$, define

$$p_t(\#G, a, b) \stackrel{\text{def}}{=} \Pr\left(2^n \cdot DP_\otimes(\pi, \alpha, \beta) = t \mid \pi \in_R \Pi^{(n)}\right). \tag{4}$$

Our main goal is to show that $p_t(\#G, a, b)$ asymptotically follows the Poisson distribution. To show this we need to consider the the distribution of (element) orders in $(\mathbb{Z}_2^n, \otimes)$. In the group $(\mathbb{Z}_2^n, \oplus)$, all the nonzero elements have order 2, and the resulting directed graphs D_α consist of 2^{n-1} cycles of length 2. However, in the group $(\mathbb{Z}_2^n, \boxplus)$ there are 2^{a-1} elements of order 2^a, $1 \leq a \leq n$, and the identity (0) has order one. For $2^n + 1$ prime, the groups $(\mathbb{Z}_2^n, \odot)$ and $(\mathbb{Z}_2^n, \boxplus)$ are isomorphic, and thus have the same distribution of orders.

Corollary 4. Let $\otimes \in \{\boxplus, \odot\}$. Then there are 2^{2a-2} pairs of group elements α, β for which ord $\alpha =$ ord $\beta = 2^a$, $1 \leq a \leq n$, and 2^{a+b-2} pairs for which $\{$ord $\alpha,$ ord $\beta\} = \{2^a, 2^b\}$, $1 \leq a < b \leq n$.

To bound the value of $DP_\otimes(\pi)$, we need only determine $p_t(2^n, a, b)$ for $a = 2^r$, $b = 2^s$, $1 \leq r \leq n$, $1 \leq s \leq n$, and $0 \leq t \leq 2^n$, and apply Corollary 4. We now cast determining $p_t(2^n, a, b)$ to an enumeration problem in terms of the *inclusion-exclusion principle (IEP)* (see for example Hall [9]).

Let α and β be elements of $(\mathbb{Z}_2^n, \otimes)$, and let $D_\alpha = (V, E_\alpha)$ and $D_\beta = (V, E_\beta)$ be their respective (difference) graphs. For each edge $uv \in E_\alpha$ define A_{uv} as $A_{uv} = \{\pi \in \Pi^{(n)} \mid (\pi(u), \pi(v)) \in E_\beta\}$, which is the set of permutations π that preserve the edge uv of D_α in D_β. Then, by the inclusion-exclusion principle, the number of permutations π that preserve exactly t edges from D_α in D_β is

$$P_t = \sum_{i=0}^{j-t}(-1)^i\binom{t+i}{i}S_{t+i}, \qquad S_k = \sum_{\substack{\mathcal{Y} \subseteq E_\alpha \\ |\mathcal{Y}|=k}}\left|\bigcap_{uv \in \mathcal{Y}} A_{uv}\right|, \tag{5}$$

and it follows that $p_t(2^n, a, b) = P_t/(2^n!)$. In the case of XOR differences ($\otimes = \oplus$) it is known [19] that

$$P_{2t} \sim \binom{2^{n-1}}{t}^2 \cdot 2^t \cdot t! \cdot \frac{(2^{n-1} - t)!}{e^{1/2}}. \tag{6}$$

In this case we can immediately prove that $p_{2t}(2^n, 2, 2)$ is asymptotically Poisson distributed.

Lemma 5. If $t \in o(2^{n/2})$ as $n \to \infty$, then

$$p_{2t}(2^n, 2, 2) = \frac{e^{-\frac{1}{2}}}{2^t \cdot t!} \cdot (1 + O((t+1)^2/2^n)).$$

Proof. From [20] we have that

$$p_{2t}(2^n, 2, 2) = \frac{1}{2^{n!}} \cdot \binom{2^{n-1}}{t}^2 \cdot 2^t \cdot t! \cdot \Phi(2^{n-1} - t) \tag{7}$$

where $\Phi(2^{n-1} - t) = (2^n - 2t)! \cdot e^{-1/2} \cdot (1 + O(1/(2^n - 2t)))$. If $t \in o(\sqrt{2^n})$ then it can be shown that

$$\binom{2^{n-1}}{t} = \frac{(2^{n-1})^t}{t!} \cdot (1 + O(t^2/2^{n-1})) = \frac{(2^n)^t}{2^t \cdot t!} \cdot (1 + O(t^2/2^n)),$$

$$\Rightarrow \binom{2^{n-1}}{t}^2 = \left(\frac{(2^n)^t}{2^t \cdot t!}\right)^2 \cdot (1 + O(t^2/2^{n-1})),$$

as $(1 + O(t^2/2^n))^2 = 1 + 2 \cdot O(t^2/2^n) + O(t^4/2^{2n}) = 1 + O(t^2/2^n)$. Substituting these approximations into (7) yields the theorem. □

In this case determining an exact expression for P_t is assisted by the fact that ord $\alpha = $ ord $\beta = 2$, and the sets A_{uv} are 'independent' in the sense that uv is the only edge incident on u and v. For a general group operation $\otimes \neq \oplus$, most groups elements α will have ord $\alpha > 2$, and hence induce a difference graph for which there exist sets $A_{u_1 v_1}, A_{u_2, v_2}$ and $v_1 = u_2$. Dependence between the A_{uv} sets considerably complicates the expressions for P_t. The following expression for $p_t(2^n, 2, 4)$ taken from [10], which also gives an involved formula for $p_t(2^n, 4, 4)$.

Lemma 6. For $n \geq 2$, and $0 \leq t \leq 2^{n-1}$,

$$p_t(2^n, 2, 4) = \frac{1}{2^n} \cdot \sum_{i=0}^{2^{n-1} - t} (-1)^k \binom{t+i}{i} S_{t+i},$$

where for $0 \leq k \leq 2^{n-1}$,

$$S_k = \left(\sum_{i=\lceil k/2 \rceil}^{\min(k, 2^{n-2})} \binom{2^{n-2}}{i} \cdot \binom{i}{k-i} \cdot 2^{3i}\right) \cdot \binom{2^{n-1}}{k} \cdot k! \cdot (2^n - 2k)!. \tag{8}$$

For general $a, b > 4$ the expression for $S_k = S_k^{(n)}(a, b)$ becomes increasing difficult to determine exactly, and we therefore consider an asymptotic approximation. We denote $\pi(\mathcal{Y}) = \{(u^*, v^*) \mid l(u^*) = \pi(l(u)), l(v^*) = \pi(l(v)), (u, v) \in \mathcal{Y}\}$, so that we can represent $\cap_{uv \in \mathcal{Y}} A_{uv} = \{\pi \mid \pi(\mathcal{Y}) \subseteq E_\beta\}$. Observe that S_k is defined in terms of *preserved edges* , but it may be further decomposed into terms of *preserved vertices* . Observe that a set of k edges is incident on at least k vertices (a cycle) and at most $2k$ vertices (disjoint edges). Let $p(\mathcal{Y})$ be the number of vertices which are incident to the edges of $\mathcal{Y}$, where $k \leq p(\mathcal{Y}) \leq 2k$. For $k \leq j \leq 2k$, define

$$\phi(k,j) = \sum_{\substack{\mathcal{Y} \subseteq E_\alpha, \\ |\mathcal{Y}|=k,\ p(\mathcal{Y})=j}} |\{\pi \mid \pi(\mathcal{Y}) \subseteq E_\beta\}|,$$

such that S_k can be expressed as $S_k = \sum_{j=k}^{2k} \phi(k,j)$. As it turns out, $S_k \sim \phi(k,2k)$, meaning that S_k is dominated by the term mapping disjoint edges D_α to edges to disjoint edges in D_β. In [10] it was proven that for $k = o(2^{n/2})$, $\phi(k,2k) = \frac{N!}{k!} \cdot (1+o(1))$, which leads to the next theorem.

Theorem 7. Suppose that $n \geq 0$, $a = 2^r$, $b = 2^s$, $1 \leq r \leq n$ and $2 \leq s \leq n$. Then $S_k = \frac{2^{n!}}{k!} \cdot \left(1 + O\left(k^2/2^n\right)\right)$ for $k \in o(2^{n/2})$ as $n \to \infty$.

The proof of Theorem 7 is involved and lengthy, and the reader is referred to [10] for details. It still remains to derive an expression for general $p_t(2^n, a, b)$ from P_t and S_t. Our results are based on the following adaptation of a theorem by Bender [2].

Theorem 8. Suppose there is a function $A(n)$ and a value $\lambda \geq 0$, such that

$$S_k = A(n) \cdot \frac{\lambda^k}{k!} \cdot (1 + O(f(k)/g(n))),$$

and $f(k) \in o(g(n))$ for $0 \leq k \leq l(n)$, where $l(n)$ goes to infinity with n. Let $j = l(n) - t$ and define $f^*(t) = \sum_{i=0}^{j-1} f(t+i) \cdot \lambda^i/i!$. If $m(n)$ is a function such that $l(n) - m(n)$ goes to infinity with n, then for each t, $0 \leq t \leq m(n)$,

$$P_t = A(n) \cdot e^{-\lambda} \cdot \frac{\lambda^t}{t!} \cdot (1 + O(f^*(t)/g(n))). \tag{9}$$

By applying this theorem we are able to show that $p_t(2^n, a, b)$ is asymptotically Poisson.

Corollary 9. Provided $a > 2$ or $b > 2$, and $t \in o(2^{n/2}/2)$,

$$p_t(2^n, a, b) = \frac{e^{-1}}{t!} \cdot (1 + O((t+1)^2/2n)).$$

Proof. Theorem 7 proves that $S_k = 2^{n!}/k! \cdot (1 + O(k^2/2^n))$ for $k = o(2^{n/2})$. Theorem 8 can now be applied with $A(n) = 2^n!$, $\lambda = 1$, $l(n) = o(2^{n/2})$, $f(k) = k^2$, $g(n) = 2^n$, $f^*(t) = O((t+1)^2)$ and $m(n) = o(2^{n/2})$. $\square$

The main result of this section can now be stated, which we call the *asymptotic Poisson approximation (PA)* to $DP_\otimes(\pi, \alpha, \beta)$.

Theorem 10. Let $(\mathbb{Z}_2^n, \otimes)$ be an Abelian group of order 2^n and $\alpha, \beta \in \mathbb{Z}_2^n$ be non-trivial differences. If $\pi \in_R \Pi^{(n)}$ and $t = o(2^{n/2})$

$$\Pr\left(DP_\otimes(\pi, \alpha, \beta) = \tfrac{t}{2^{n-1}}\right) \sim e^{-\frac{1}{2}} \cdot \left(\tfrac{1}{2}\right)^t / t! \quad \text{if ord } \alpha = \text{ord } \beta = 2,$$
$$\Pr\left(DP_\otimes(\pi, \alpha, \beta) = \tfrac{t}{2^n}\right) \sim e^{-1}/t! \qquad \text{otherwise.}$$

Let $\mathbf{E}[X]$, $\mathbf{Var}[X] = \mathbf{E}[X^2] - (\mathbf{E}[X])^2$ and $\sigma[X] = \sqrt{\mathbf{Var}[X]}$ denote the expectation, variance and standard deviation of the random variable X. It is known that if the distribution of values for X is Poisson, then $\mathbf{Var}[X] = \mathbf{E}[X] = \mu$. Then, for example, a little algebraic manipulation reveals that the distribution of values for $DP_\otimes(\pi, \alpha, \beta)$ has $\mathbf{E}[DP_\otimes(\pi, \alpha, \beta)] \sim 1/2^n$ and $\sigma[DP_\otimes(\pi, \alpha, \beta)] \sim \eta/2^n$, where $\eta = \sqrt{2}$ if ord $\alpha = $ ord $\beta = 2$ and $\eta = 1$ otherwise. This indicates that the probabilities for a differential approximation $\Delta X = \alpha \rightarrow \Delta\pi(X) = \beta$ where ord $\alpha = $ ord $\beta = 2$ are distributed $\sqrt{2}$ times as far from $1/2^n$ as the probabilities for other differential approximations. Consequently, differential approximations for which ord $\alpha = $ ord $\beta = 2$ are more likely to have higher probabilities.

3   Bounding the Maximum Difference Table Entry

In this section we use the PA to obtain bounds on $DP_\otimes(\pi)$ that hold asymptotically with probability one. The distribution of differences with respect to $\oplus$ is approximated using a Poisson distribution with $\mu = \frac{1}{2}$, as all non-trivial elements have order two. The distribution of differences for $\otimes \in \{\boxplus, \odot\}$ is approximated using a Poisson distribution with $\mu = 1$, as there is only one pair (α, β) with ord $\alpha = $ ord $\beta = 2$. We determine the expectation and variance of $\theta_t(\otimes, \pi)$, defined to be

$$\theta_t(\otimes, \pi) = \frac{1}{(2^n - 1)^2} \sum_{\alpha \neq I} \sum_{\beta \neq I} [2^n \cdot DP_\otimes(\alpha, \beta) = t] \tag{10}$$

which is the fraction input/output differences that map exactly t pairs, $0 \le t \le 2^n$.

Corollary 11. For $\pi \in_R \Pi_{2^n}$, $\mathbf{E}[\theta_{2t}(\oplus, \pi)] \sim e^{-\frac{1}{2}} \cdot \left(\frac{1}{2}\right)^t / t!$ and $\mathbf{E}[\theta_t(\otimes, \pi)] \sim e^{-1}/t!$ uniformly for $t = o(2^{n/2})$ where $\otimes \in \{\boxplus, \odot\}$.

This information is sufficient for obtaining upper bounds on $DP_\otimes(\pi)$ for $\otimes \in \{\oplus, \boxplus, \odot\}$. However, to obtain our lower bound on the maximum entry in differences tables with respect to $\oplus$, the variance of $\theta_{2t}(\oplus, \pi)$ is required. We have not attempted to determine the variance in $\theta_t(\otimes, \pi)$ for $\otimes \in \{\boxplus, \odot\}$ as the counting problem is very complex, and this variance is not required for the results of this paper. See [10] for a proof of the next lemma.

Lemma 12. For $\pi \in_R \Pi_{2^n}$ and $t = o(2^{n/2})$

$$\mathbf{Var}[\theta_{2t}(\oplus, \pi)] \sim \frac{1}{(2^n - 1)^2} \cdot e^{-\frac{1}{2}} \cdot \left(\frac{1}{2}\right)^t / t! \cdot \left(1 - e^{-\frac{1}{2}} \cdot \left(\frac{1}{2}\right)^t / t!\right).$$

For nontrivial α, β define $\Psi_{\alpha,\beta}^{(t)}$, $0 \leq t \leq 2^{n-1}$, where $\Psi_{\alpha,\beta}^{(t)} = 1$ if $2^n \cdot DP_\oplus(\pi, \alpha, \beta) = 2t$ and $\Psi_{\alpha,\beta}^{(t)} = 0$ otherwise. It follows that $\Psi^{(t)} = \sum_{\alpha,\beta \neq I} \Psi_{\alpha,\beta}^{(t)} = (2^n - 1)^2 \cdot \theta_{2t}(\oplus, \pi)$. Note that $\mathbf{E}[\Psi^{(t)}] = (2^n - 1)^2 \cdot \mathbf{E}[\theta_{2t}(\oplus, \pi)] \sim (2^n - 1)^2 \cdot e^{-\frac{1}{2}} \cdot \frac{1}{2}^t / t!$ for $t = o(2^{n/2})$. Similarly,

$$\mathbf{Var}\left[\Psi^{(t)}\right] = (2^n - 1)^4 \cdot \mathbf{Var}[\theta_{2t}(T_{\oplus,\pi})] \tag{11}$$

$$\sim (2^n - 1)^2 \cdot \frac{e^{-\frac{1}{2}}}{2^t \cdot t!} \cdot \left(1 - \frac{e^{-\frac{1}{2}}}{2^t \cdot t!}\right) \tag{12}$$

$$\sim \mathbf{E}[\Psi^{(t)}] \cdot \left(1 - \frac{e^{-\frac{1}{2}}}{2^t \cdot t!}\right) \tag{13}$$

drawing on the result of Lemma 12. Define B_n as $B_n = \ln N^2 / \ln \ln N^2$, where $N = (2^n - 1)$, and observe that the Poisson approximation (Corollary 10) holds for $0 \leq t \leq 2B_n$ since $2B_n = o(2^{n/2})$. The next two lemmas are proved using the previous variance calculations in the Appendix.

Lemma 13. If $\pi \in_R \Pi^{(n)}$, then $\Pr\left(B_n/2^{n-1} \leq DP_\oplus(\pi) < n/2^{n-1}\right) \sim 1$.

Lemma 14. If $\pi \in_R \Pi^{(n)}$, then $\Pr\left(DP_\otimes(\pi) < B_n/2^{n-1}\right) \sim 1$, where $\otimes \in \{\boxplus, \odot\}$.

Asymptotically B_n tends to $(2n \ln 2)/\ln n$, which when applied to the previous two lemmas, determines the bounds given in (2) and (3). Statements concerning the best differential approximation of a randomly selected permutation can now be made. For example, the probability of the best approximation with respect to XOR differences is in the range $[2^{-58.6}, 2^{-57}]$ for a random 64-bit permutation and in the range $[2^{-121.9}, 2^{-120}]$ for a random 128-bit approximation. The values $2^{-58.6}$ and $2^{-121.9}$ are also upper bounds on the probability of approximations with respect to $\otimes \neq \oplus$ for random 64-bit and 128-bit permutations respectively. Further bounds on the maximum entry can be obtained for difference tables with respect to other group operations, and these bounds will rely primarily on the fraction of entries in the difference table for which both elements have order 2.

Finally, Lemma 13 and Lemma 14 combine to confirm our initial observation that in general XOR differences yield higher probability approximations than differences with respect to modular addition and modular multiplication.

Corollary 15. If $\pi \in_R \Pi^{(n)}$, then $\Pr\left(DP_\oplus(\pi) > DP_\otimes(\pi)\right) \sim 1$, for $\otimes \in \{\boxplus, \odot\}$.

4 Conclusion

We have shown that with high probability, XOR differences yield better differential approximations than differences with respect $\otimes \in \{\boxplus, \odot\}$. Furthermore, we determined asymptotic approximations to the difference distribution of three group operations $\otimes \in \{\oplus, \boxplus, \odot\}$, and bound the probability of the most likely difference. Further bounds on the maximum entry can be obtained for difference tables with respect to other group operations, and these bounds will rely primarily on the fraction of entries in the difference table for which both elements have order 2. The Poisson approximation (Corollary 10) can also be applied to quasi-differentials and the maximum probability can be similarly bounded.

We have concentrated on the three groups defined by $\oplus$, $\boxplus$ and $\odot$, but the other groups can be considered using the same analysis. The Poisson approximation of $DP_{\otimes}(\alpha, \beta)$ holds for all group elements with order at least 2. On the other hand, the bounds on $DP_{\otimes}(\pi)$ depend on the distribution of group elements. Bounding $DP_{\otimes}(\pi)$ for a given group $(G, \otimes)$ requires knowledge of how element orders are distributed within G. Our results in this paper are based on all non-identity elements of $(\mathbb{Z}_2^n, \oplus)$ having order 2, and the element orders of $(\mathbb{Z}_{2^n}, \boxplus)$ and $(\mathbb{Z}_{2^n+1}^*, \odot)$ being determined as in Corollary 4.

The distribution of entries in difference tables has previously been predicted using a "balls-in-bins" model [15], summarized as follows. In modeling differences tables with respect to XOR, we let the "balls" represent the *unordered* pairs of difference α and let the "bins" represent the possible non-trivial output differences. If the 2^{n-1} input pairs of input difference α (the "balls") can be allocated randomly and independently to any of the $(2^n - 1)$ "bins", then the resulting distribution approaches a Poisson distribution with parameter $\mu = \frac{2^{n-1}}{2^n-1} \sim \frac{1}{2}$. In modeling differences tables with respect to $\otimes \neq \oplus$, we let the "balls" represent the *ordered* pairs of difference α and let the "bins" represent the possible non-trivial output differences. If the input pairs of input difference α (the "balls") can be allocated randomly and independently to any of the $(2^n - 1)$ "bins", then the resulting distribution approaches a Poisson distribution with parameter $\mu = \frac{2^n}{2^n-1} \sim 1$. Our results add validity to the "balls-in-bins" approach for predicting $DP_{\otimes}(\pi)$.

Acknowledgments

We would like to thank the diligent referees for their comments on this work, especially Eli Biham.

5   Appendix

Lemma 13 If $\pi \in_R \Pi^{(n)}$, then $\Pr\left(B_n/2^{n-1} DP_{\oplus}(\pi) < n/2^{n-1}\right) \sim 1$, where $\otimes \in \{\boxplus, \odot\}$.

Proof. O'Connor [19] proved that $\Pr\left(DP_{\oplus}(\pi) \geq n/2^{n-1}\right) = o(1)$. Denote $\Psi = \Psi^{(B_n)}$, and observe that $\mathbf{Var}[\Psi] \sim \mathbf{E}[\Psi]$ as B_n increases with n. Chebychev's inequality (see for example [6]) is applied to show that

$$\Pr(DP_{\oplus}(\pi) < 2B_n) \leq \Pr(\Psi = 0) \leq \Pr\left(|\Psi - \mathbf{E}[\Psi]| \geq \mathbf{E}[\Psi]\right) \leq \frac{\mathbf{Var}[\Psi]}{(\mathbf{E}[\Psi])^2}$$

which is asymptotic to $\frac{1}{\mathbf{E}[\Psi]}$. The expected number of entries $2B_n$ in the differences tables with respect to $\oplus$ is equal to

$$\mathbf{E}[\Psi] \;=\; (2^n - 1)^2 \cdot \mathbf{E}[\theta_{2B_n}(\oplus, \pi)] \;\sim\; N^2 \cdot \frac{e^{-\frac{1}{2}}}{2^{B_n} \cdot B_n!}$$

By applying Stirling's formula for $n!$ (see, for example [8, page 213]),

$$B_n! \sim \left(\frac{B_n}{e}\right)^{B_n} \cdot \sqrt{2\pi B_n} \;=\; \frac{(\ln N^2)^{\ln N^2 / \ln \ln N^2}}{(e \cdot \ln \ln N^2)^{B_n}} \cdot \sqrt{2\pi B_n},$$

where $(\ln N^2)^{\ln N^2 / \ln \ln N^2} = (e^{\ln \ln N^2})^{\ln N^2 / \ln \ln N^2} = e^{\ln N^2} = N^2$. Consequently,

$$\Pr(DP_\oplus(\pi) < 2B_n) \leq \frac{1}{\mathbf{E}[\Psi]} \tag{14}$$

$$\sim \frac{1}{N^2} \cdot \frac{2^{B_n}}{e^{-\frac{1}{2}}} \cdot \frac{N^2 \cdot \sqrt{2\pi B_n}}{(e \cdot \ln \ln N^2)^{B_n}} \tag{15}$$

$$= e^{\frac{1}{2}} \cdot \frac{\sqrt{2\pi B_n}}{((e/2) \cdot \ln \ln N^2)^{B_n}} = o(1), \tag{16}$$

as $\sqrt{2\pi B_n} = o\left(((e/2) \cdot \ln \ln N^2)^{B_n}\right)$. Therefore, the probability that the maximum entry is either less than $2B_n$ or greater than or equal to $2n$ is $o(1)$, and the lemma is proved. $\square$

Lemma 14 If $\pi \in_R \Pi^{(n)}$, then $\Pr\left(DP_\otimes(\pi) < B_n / 2^{n-1}\right) \sim 1$, where $\otimes \in \{\boxplus, \odot\}$.

Proof. Assume $\otimes \neq \oplus$. Let $\Omega^{(t)} = (2^n - 1)^2 \cdot \theta_t(\otimes\pi)$ denote the number of entries t in the differences table with respect to $\otimes$, and in particular denote $\Omega = \Omega^{(2B_n)}$. Recall that $\mathbf{E}[\Omega] = (2^n - 1)^2 \cdot \mathbf{E}[\theta_{2B_n}(\otimes, \pi)] \sim N^2 \cdot e^{-1}/(2B_n)!$. By applying Stirling's formula for $n!$,

$$(2B_n)! \sim \left(\frac{2B_n}{e}\right)^{2B_n} \cdot \sqrt{2\pi \cdot (2B_n)} \;=\; \left(\frac{2\ln N^2}{e \cdot \ln \ln N^2}\right)^{2B_n} \cdot 2\sqrt{\pi B_n} \tag{17}$$

$$= \frac{(\ln N^2)^{2\ln N^2 / \ln \ln N^2}}{((e/2) \cdot \ln \ln N^2)^{2B_n}} \cdot 2\sqrt{\pi B_n}, \tag{18}$$

where $(\ln N^2)^{2\ln N^2/\ln\ln N^2} = (e^{\ln\ln N^2})^{2\ln N^2/\ln\ln N^2} = e^{2\ln N^2} = N^2$. Thus

$$\mathbf{E}[\Omega] \sim N^2 \cdot \frac{e^{-1}}{N^4} \cdot \left((e/2)\cdot\ln\ln N^2\right)^{2\ln N^2/\ln\ln N^2} \cdot \frac{1}{2\sqrt{\pi B_n}}$$

$$= \frac{e^{-1}}{2\sqrt{\pi B_n}} \cdot \left(\frac{\left((e/2)\cdot\ln\ln N^2\right)^{2/\ln\ln N^2}}{(N^2)^{1/\ln N^2}}\right)^{\ln N^2}$$

$$= \frac{e^{-1}}{2\sqrt{\pi B_n}} \cdot \left(\underbrace{\frac{\left((e/2)\cdot\ln\ln N^2\right)^{2/\ln\ln N^2}}{e}}_{y(N)}\right)^{\ln N^2},$$

and we can show that $y(N) \leq 1$. Therefore,

$$\mathbf{E}[\Omega] \sim \frac{e^{-1}}{2\sqrt{\pi B_n}} \cdot y(N)^{\ln N^2} = o(1)$$

as B_n increases with n. Now, for $2B_n = o(2^{n/2})$, $\mathbf{E}[\Omega^{(t)}] \leq \mathbf{E}[\Omega]/(2B_n)^{t-2B_n}$. (The value of $\mathbf{E}[\Omega^{(t)}]$ is insignificant for $t \neq o(2^{n/2})$). Therefore, the expected number of entries greater than or equal to $2B_n$ in a difference table with respect to $\otimes$ is

$$\sum_{t\geq 2B_n}\mathbf{E}[\Omega^{(t)}] \leq \sum_{t\geq 2B_n}\frac{1}{(2B_n)^{t-2B_n}}\cdot\mathbf{E}[\Omega] \tag{19}$$

$$= \mathbf{E}[\Omega]\cdot\sum_{i\geq 0}\frac{1}{(2B_n)^i} \tag{20}$$

$$= \frac{\mathbf{E}[\Omega]}{1-1/(2B_n)} \tag{21}$$

$$\sim \mathbf{E}[\Omega] = o(1). \tag{22}$$

Note that the probability that $DP_\otimes(\pi) \geq B_n/2^{n-1}$ is less than the expected number of entries of size $t \geq 2B_n$. Therefore, $\Pr(DP_\otimes(\pi) \geq B_n/2^{n-1}) = o(1)$ as $n \to \infty$. □

References

1. C. M. Adams. The CAST-256 Encryption Algorithm. NIST Advanced Encryption Standard (AES) submission, description available at http://www.entrust.com/resources/pdf/cast.pdf.
2. E. A. Bender. Asymptotic methods in enumeration. *SIAM Review*, 16(4):485–515, 1974.
3. E. Biham and A. Shamir. Differential cryptanalysis of DES-like cryptosystems. *Journal of Cryptology*, 4(1):3–72, 1991.
4. E. Biham and A. Shamir. *Differential cryptanalysis of Data Encryption Standard*. Springer–Verlag, 1993.

5. J. Daemen and V. Rijmen. AES proposal: Rijndael. NIST Advanced Encryption Standard submission, description available at http://www.esat.kuleuven.ac.be/~rijmen/rijndael.
6. W. Feller. *An Introduction to Probability Theory and its Applications*. New York: Wiley, 3rd edition, Volume 1, 1968.
7. H. Gilbert, M. Girault, P. Hoogvorst, F. Noilhan, T. Pornin, G. Poupard, J. Stern, and S. Vaudenay. Decorrelated Fast Cipher. NIST Advanced Encryption Standard (AES) submission, description available http://www.dmi.ens.fr/~vaudenay/dfc.html.
8. R. P. Grimaldi. *Discrete and Combinatorial Mathematcis: An Applied Introduction*. Addison Wesley Publishing Company, 1989.
9. M. Hall. *Combinatorial Theory*. Blaisdell Publishing Company, 1967.
10. P. Hawkes and L. J. O'Connor. Aymptotic bounds on differential probabilities. Technical Report RZ 3018, IBM Research Report, May, 1998. Available from http://www.research.ibm.com.
11. B. S Kaliski and L. Y. Yiqun. On differential and linear cryptanalysis of the RC5 algorithm. *Advances in Cryptology, CRYPTO 95, Lecture Notes in Computer Science, vol. 963, D. Coppersmith eds., Springer-Verlag*, pages 171–184, 1995.
12. M. Kanda, S. Moriai, A. Kazumaro, H. Ueda, M. Ohkubo, Y. Takashima, K. Ohta, and T. Matsumoto. Specification of E2 - a 128-bit block cipher. NIST Advanced Encryption Standard submission, description available at http://titan.isl.ntt.co.jp/e2.
13. X. Lai. *On the design and security of block ciphers*. ETH Series in Information Processing, editor J. Massey, Hartung-Gorre Verlag Konstanz, 1992.
14. X. Lai and J. L. Massey. A proposal for a new block encryption standard. In *Advances in Cryptology, EUROCRYPT 90, Lecture Notes in Computer Science, vol. 473, I. B. Damgård ed., Springer-Verlag*, pages 389–404, 1991.
15. J. Lee, H. M. Heys, and S. E. Tavares. Resistance of a CAST-like encryption algorithm to linear and differential cryptanalysis. *Designs, Codes and Cryptography*, 12(3):267–282, 1997.
16. C. H. Lim. Specification and analysis of CRYPTON version 1.0. NIST Adavanced Encryption Standard (AES) submission, description available at http://crypt.future.co.kr/~chlim/crypton.html.
17. J. L. Massey. SAFER: a byte-oriented ciphering algorithm. *Fast Software Encryption, Lecture Notes in Computer Science, vol. 809, R. Anderson ed., Springer-Verlag*, pages 1–17, 1993.
18. J. L. Massey. SAFER K-64: one year later. *Fast Software Encryption, Lecture Notes in Computer Science, vol. 1008, B. Preneel ed., Springer-Verlag*, pages 212–241, 1994.
19. L. J. O'Connor. On the distribution of characteristics in bijective mappings. *Advances in Cryptology, EUROCRYPT 93, Lecture Notes in Computer Science, vol. 765, T. Helleseth ed., Springer-Verlag*, pages 360–370, 1994.
20. L. J. O'Connor. On the distribution of characteristics in bijective mappings. *Journal of Cryptology*, 8(2):67–86, 1995.
21. B. Schneier, J. Kelsey, D. Whiting, D. Wagner, C. Hall, and N. Ferguson. Twofish: a 128-bit block cipher. NIST Advanced Encryption Standard (AES) submission, description available http://www.counterpane.com/twofish.html.

S-boxes with Controllable Nonlinearity

Jung Hee Cheon, Seongtaek Chee, and Choonsik Park

Electronics and Telecommunications Research Institute,
161 Kajong-Dong,Yusong-Gu, Taejon, 305-350, ROK
{jhcheon, chee, csp}@etri.re.kr

Abstract. In this paper, we give some relationship between the nonlinearity of rational functions over $\mathbb{F}_{2^n}$ and the number of points of associated hyperelliptic curve. Using this, we get a lower bound on nonlinearity of rational-typed vector Boolean functions over $\mathbb{F}_{2^n}$. While the previous works give us a lower bound on nonlinearity only for special-typed monomials, our result gives us general bound applicable for all rational fuctions defined over $\mathbb{F}_{2^n}$. As an application of our results, we get a lower bound on nonlinearity of $n \times kn$ S-boxes.

1 Introduction

One of the powerful attack for block ciphers is linear cryptanalysis which was developed by Matsui[5] in 1993. The basic idea of linear cryptanalysis is to find a linear relation among the plain text, cipher text and key bits. Such a relation usually occurs by a low nonlinearity of substitutions in block ciphers.

Nonlinearity for Boolean functions was well-established [9]. However, it is very difficult to analyze nonlinearity for vector Boolean functions, in general. Some results on nonlinearity of vector Boolean functions were found in [2,6,7]. But the results are only concerned with the special types of monomials over $\mathbb{F}_{2^n}$.

In this paper, we derive a novel relationship between the nonlinearity of a rational function over $\mathbb{F}_{2^n}$ and the number of points of hyperelliptic curve over that field. And, using such a relationship we obtain a lower bound on nonlinearity of rational-typed vector Boolean functions over $\mathbb{F}_{2^n}$. Furthermore, we give a lower bound on nonlinearity of S-box constructed by concatenating two or more S-boxes over $\mathbb{F}_{2^n}$. Similar method has been used in the CAST algorithm [1], in which 8×32 S-boxes were constructed by selecting 32 bent Boolean functions over $\mathbb{F}_{2^8}$. In that case, their S-boxes has been believed to be highly nonlinear, but nobody gave lower bound on the nonlinearity. It has been known that it might be very difficult to prove the lower bound on the nonlinearity of such S-boxes [8].

2 Preliminaries

2.1 Nonlinearity

We consider a vector Boolean function $F : \mathbb{F}_{2^n} \to \mathbb{F}_{2^n}$. Let $b = (b_1, b_2, \cdots, b_n)$ be a nonzero element of $\mathbb{F}_{2^n}$. We denote by $b \cdot F$ the Boolean function which is the

J. Stern (Ed.): EUROCRYPT'99, LNCS 1592, pp. 286–294, 1999.

linear combination $b_1 f_1 + b_2 f_2 + \cdots + b_n f_n$ of the coordinate Boolean functions $f_1, f_2, \cdots, f_n$ of F.

Definition 1. *The nonlinearity of F, $\mathcal{N}(F)$, is defined as*

$$\mathcal{N}(F) = \min_{b \neq 0} \min_{A \in \Gamma} \#\{x : A(x) \neq b \cdot F(x)\},$$

where Γ is the set of all affine functions over $\mathbb{F}_{2^n}$.

If we define $\mathcal{L}(F, a, b) = \#\{x : a \cdot x = b \cdot F(x)\}$, then we have

$$\mathcal{N}(F) = 2^{n-1} - \max_{b \neq 0} \max_{a} |2^{n-1} - \mathcal{L}(F, a, b)|. \tag{1}$$

Observe that nonlinearity of arbitrary vector Boolean functions is upper-bounded as

$$\mathcal{N}(F) \leq 2^{n-1} - 2^{\frac{n}{2}-1}.$$

and the equality holds for only bent functions.

The nonlinearity for special types of F, usually monomials, are investigated by Nyberg [7].

Theorem 2.

1. *Let $F(x) = x^{2^k+1}$.*

   (a) If n/s is odd for $s = \gcd(n, k)$, then

   $$\mathcal{N}(F) = 2^{n-1} - 2^{(n+s)/2-1}. \tag{2}$$

   (b) If n is odd and $\gcd(n, k) = 1$, then

   $$\mathcal{N}(F^{-1}) = 2^{n-1} - 2^{(n-1)/2}. \tag{3}$$

2. *For $F(x) = x^{-1}$,*

   $$\mathcal{N}(F) \geq 2^{n-1} - 2^{n/2}. \tag{4}$$

2.2   Hyperelliptic Curves

In this section, we introduce a hyperelliptic curve and the Weil theorem which have important roles in proving our main theorem. A hyperelliptic curve C over $\mathbb{F}_{2^n}$ is an equation of the form

$$C : y^2 + h(x)y = f(x), \tag{5}$$

where $f(x), h(x) \in \mathbb{F}_{2^n}[x]$ with $2 \deg h(x) + 1 \leq \deg f(x)$, and there are no solutions x, y in the algebraic closure of $\mathbb{F}_{2^n}$, which simultaneously satisfy the equation $y^2 + h(x)y = f(x)$ and the partial derivative equations $2y + h(x) = 0$ and $h'(x)y - f'(x) = 0$. When a curve C has no solutions which satisfies the three equations, we say that C is nonsingular. Otherwise, we say that C is sigular.

We define the set of $\mathbb{F}_{2^n}$-rational points on C, denoted $C(\mathbb{F}_{2^n})$, the set of all points $(x, y) \in \mathbb{F}_{2^n} \times \mathbb{F}_{2^n}$ that satisfies the equation (5) of the curve C, together with a special point at infinity, denoted O.

For the number $\#C(\mathbb{F}_{2^n})$ of the $\mathbb{F}_{2^n}$-rational points on C, we have the following nontrivial bound [4].

Theorem 3 (Weil). *For any hyperelliptic curve C over $\mathbb{F}_{2^n}$, we have*

$$|\#C(\mathbb{F}_{2^n}) - 2^n - 1| \leq 2g\sqrt{2^n}, \tag{6}$$

where g is the genus of the hyperelliptic curve C.

By the Riemann-Hurwitz formula, we have $g = \lfloor \frac{d-1}{2} \rfloor$ for the degree d of f(See [4, p332]). When a curve given by the equation (5) is singular, the theorem does not hold. In this case, we have the following, using the theory of desingularization of algebraic curves(See [4, p.358]).

$$|\#C(\mathbb{F}_{2^n}) - 2^n - 1| \leq 2g\sqrt{2^n} - g + \frac{(d-1)(d-3)}{2}, \tag{7}$$

where g is the genus of the singular curve C and d is the degree of f. Since the genus g is less than $\lfloor \frac{d-1}{2} \rfloor$, we can get the same inequality for a singular curve under some condition.

Corollary 4. *Let C be a curve given by the irrduducible equation $y^2 + h(x)y = f(x)$, which satisfies $\deg f \geq \max\{2\deg h + 1, 3\}$. Assume that C is nonsingular or $d = \deg f \leq 2^{n/4+1} + 2$. Then we have*

$$|\#C(\mathbb{F}_{2^n}) - 2^n - 1| \leq 2\lfloor \frac{d-1}{2} \rfloor \sqrt{2^n}. \tag{8}$$

Proof. If C is nonsingular, we have $g = \lfloor \frac{d-1}{2} \rfloor$. Hence the corollary is proved. If C is singular, we have $g \leq \lfloor \frac{d-1}{2} \rfloor - 1$ so that

$$|\#C(\mathbb{F}_{2^n}) - 2^n - 1| \leq (2\sqrt{2^n} - 1)(\lfloor \frac{d-1}{2} \rfloor - 1) + \frac{(d-1)(d-3)}{2}.$$

The right-hand side is less than or equal to $2\lfloor \frac{d-1}{2} \rfloor \sqrt{2^n}$ if $d^2 - 5d + 7 \leq 4\sqrt{2^n}$. Hence the corollary follows for $3 \leq d \leq 2^{n/4+1} + 2$.

3   Nonlinearity of Rational Functions over $\mathbb{F}_{2^n}$

In this section, we get a lower bound on nonlinearity of rational functions over a finite field, using the bound on the numbers of points of hyperelliptic curves over that field. We consider a rational function of the form $F(x) = P(x)/Q^2(x)$ for $P(x), Q(x) \in \mathbb{F}_{2^n}[x]$ where we may define $F(\alpha)$ to be any elements of $\mathbb{F}_{2^n}$ for a root α of $Q(x)$.

First, we introduce a lemma. We donote by $Tr(\cdot)$, an absolute trace mapping [3].

Lemma 5. *The following polynomial equation of one variable x*

$$x^2 + ax + b = 0, \quad a \neq 0, \quad b \in \mathbb{F}_{2^n} \tag{9}$$

is reducible over $\mathbb{F}_{2^n}$ if and only if $Tr(\frac{b}{a^2}) = 0$.

Proof. If we replace by ax, x of the equation (5) and divide the equation by a^2, we get $x^2 + x + b/a^2 = 0$. Hence $x^2 + ax + b = 0$ is reducible over $\mathbb{F}_{2^n}$ if and only if $x^2 - x = b/a^2$ has a root in $\mathbb{F}_{2^n}$. By Hilbert theorem 90 [3], it is equivalent to $Tr(b/a^2) = 0$.

By using the above lemma, we can derive the following theorem.

Theorem 6. *Let $P(x), Q(x), G(x) \in \mathbb{F}_{2^n}[x]$, $F(x) = P(x)/Q^2(x)$ where $G(x)$ is a permutation. Suppose that $C_{a,b} : y^2 + Q(x)y = aQ^2(x)G(x) + bP(x)$ is nonsingular for each $a, b \neq 0$ in $\mathbb{F}_{2^n}$, or $d = \max\{2\deg Q + \deg G, \deg P\} \leq (2^{n/2+2} - 2)^{1/2} + 2$. If $Q(x)$ has r distinct roots in $\mathbb{F}_{2^n}$ and $\gcd(P(x), Q(x))$ has s distinct roots in $\mathbb{F}_{2^n}$, then the nonlinearity of $F \circ G^{-1}$ is lower-bounded as follows :*

$$\mathcal{N}(F \circ G^{-1}) \geq 2^{n-1} - \lfloor \frac{d-1}{2} \rfloor 2^{n/2} - r + \frac{s}{2}.$$

Proof. Choose a basis B of $\mathbb{F}_{2^n}$ over $\mathbb{F}_2$ and take its dual basis $\hat{B}$. Represent binary vectors in $\mathbb{F}_{2^n}$, a and b by the basis B, and $G(x)$ and $F(x)$ by its dual basis $\hat{B}$. Then we have

$$a \cdot G(x) = Tr(aG(x)), \quad b \cdot F(x) = Tr(bF(x)).$$

Hence

$$\begin{aligned}
\mathcal{L}(F \circ G^{-1}, a, b) &= \#\{x | a \cdot x = b \cdot F(G^{-1}(x))\} \\
&= \#\{x | Tr(aG(x)) = Tr(bF(x))\} \\
&= \#\{x | Tr(aG(x) + bF(x)) = 0\}
\end{aligned}$$

Let $\alpha_1, \alpha_2, \cdots, \alpha_r$ be r distinct roots of $Q(x)$. If $\alpha \neq \alpha_i$ for all i, $C_{a,b}$ has two distinct points whose x-coordinate is α, whenever the equation of y, $y^2 + Q(\alpha)y - (aQ^2(\alpha)G(\alpha) + bP(\alpha))$, is reducible. Also, $C_{a,b}$ has one point whose x-coordinate is α_i, whenever the equation of y, $y^2 - bP(\alpha_i)$, is reducible. Considering the infinity point O, we have

$$\begin{aligned}
&\#C_{a,b}(\mathbb{F}_{2^n}) - 1 \\
&= 2 \cdot \#\{x | Tr(\frac{aQ^2(x)G(x) + bP(x)}{Q(x)^2}) = 0, Q(x) \neq 0\} + \sum_i \#\{y | y^2 = bP(\alpha_i)\} \\
&= 2 \cdot \#\{x | Tr(aG(x)) = Tr(bF(x)), Q(x) \neq 0\} + \sum_i \#\{y | y^2 = bP(\alpha_i)\} \qquad (10) \\
&= 2\mathcal{L}(F \circ G^{-1}, a, b) - 2\#\{i | Tr(aG(\alpha_i)) = Tr(bF(\alpha_i))\} + \sum_i \#\{y | y^2 = bP(\alpha_i)\}.
\end{aligned}$$

The first equality follows from lemma 5. Observe that all curves $C_{a,b}$ for $a, b \neq 0$ satisfy the assumption of Corollary 4 and the degree of the equation $C_{a,b}$ at x is less than or equal to d. Hence we have

$$|\#C_{a,b}(\mathbb{F}_{2^n}) - 2^n - 1| \leq 2\lfloor \frac{d-1}{2} \rfloor \sqrt{2^n}. \qquad (11)$$

Combining it with the identity (10), we have

$$|2^{n-1} - \mathcal{L}(F \circ G^{-1}, a, b)|$$

$$\leq \lfloor \frac{d-1}{2} \rfloor \sqrt{2^n} + |\#\{i|Tr(aG(\alpha_i)) = Tr(bF(\alpha_i))\} - \frac{1}{2} \sum_i \#\{y|y^2 = bP(\alpha_i)\}|.$$

If we take the maximum through all $a, b \neq 0 \in \mathbb{F}_{2^n}$, we have

$$\max_{a,b \neq 0} |2^{n-1} - \mathcal{L}(F \circ G^{-1}, a, b)| \leq \lfloor \frac{d-1}{2} \rfloor \sqrt{2^n} + r - \frac{s}{2}.$$

Hence we have

$$\mathcal{N}(F \circ G^{-1}) \geq 2^{n-1} - \lfloor \frac{d-1}{2} \rfloor 2^{n/2} - r + \frac{s}{2}.$$

Observe that $C(F, a, b)$ is singular if and only if $Q(x) = 0$, $Q'(x)y = bP'(x)$ and $y^2 = bP(x)$ has a common solution. Hence $C(F, a, b)$ is non-singluar for any nonzero $b \in \mathbb{F}_{2^n}$ if $F(x)$ satisfies the following condition:

For any root of $Q(x) = 0$ in the algebraic closure of $\mathbb{F}_{2^n}$,

$$(\frac{Q'(\alpha)}{P'(\alpha)})^2 P(\alpha) \notin \mathbb{F}_{2^n}^*. \tag{12}$$

If we use Theorem 6, we can obtain the following useful results.

Corollary 7.

1. *For any polynomial $F(x) \in \mathbb{F}_{2^n}[x]$ of degree $d \geq 3$,*

$$\mathcal{N}(F) \geq 2^{n-1} - \lfloor \frac{d-1}{2} \rfloor 2^{n/2}.$$

2. *For any polynomial $H(x) \in \mathbb{F}_{2^n}[x]$ of degree m and a positive integer k, $F(x) = \frac{H(x)}{x^{2k}-1}$ has a lower bound on its nonlinearity as follows:*

$$\mathcal{N}(F) \geq 2^{n-1} - \lfloor \frac{d-1}{2} \rfloor 2^{n/2} - \frac{1}{2},$$

where $d = \max\{2k + 1, m + 1\}$.

Proof. 1. We take $G(x) = x$, $Q(x) = 1$ and $P(x) = F(x)$ in Theorem 6. Then a curve $C_{a,b} : y^2 + y = ax + bF(x)$ is irreducible and nonsingular for each $a, b \neq 0$. Since the degree of each curve $C_{a,b}$ at x is d, we have the above assertion.

2. We take $G(x) = x$, $Q(x) = x^k$ and $P(x) = xH(x)$ in Theorem 6. Then a curve $C_{a,b} : y^2 + x^k y = ax^{2k+1} + bxH(x)$ is irreducible and nonsingular for each $a, b \neq 0$. If we take $d = \max\{2k + 1, m + 1\}$, the degree of each curve $C_{a,b}$ is less than or equal to d, which completes the proof.

We can extend the above corollary to the composite function cases.

Corollary 8. *Assume that e, f be integers satisfying $ef \equiv 1 \bmod (2^n - 1)$.*

1. *For any polynomial $F(x) \in \mathbb{F}_{2^n}[x]$ of degree $m \geq 3$,*

$$\mathcal{N}(F(x^f)) \geq 2^{n-1} - \lfloor \frac{d-1}{2} \rfloor 2^{n/2},$$

where $d = \max\{e, m\}$.

2. *For any polynomial $H(x) \in \mathbb{F}_{2^n}[x]$ of degree m and a positive integer k, let $F(x) = \frac{H(x)}{x^{2k}-1}$.*

$$\mathcal{N}(F(x^f)) \geq 2^{n-1} - \lfloor \frac{d-1}{2} \rfloor 2^{n/2} - \frac{1}{2},$$

where $d = \max\{2k + e, m + 1\}$.

Proof. 1. Take $G(x) = x^e$, $Q(x) = 1$ and $P(x) = F(x)$ in Theorem 6. Then for a curve $C_{a,b} : y^2 + y = ax^e + bF(x)$ the similar assertions as the proof of Corollary 7 hold.

2. Take $G(x) = x^e$, $Q(x) = x^k$ and $P(x) = xH(x)$ in Theorem 6. Then for a curve $C_{a,b} : y^2 + x^k y = ax^{2k+e} + bxH(x)$ the similar assertions as the proof of Corollary 7 hold.

By applying Corollary 7, we get some useful results. See the example.

Example 9. 1. For $F(x) = x^3 + x^5 + x^6 \in \mathbb{F}_{2^n}[x]$,

$$\mathcal{N}(F) \geq 2^{n-1} - 2^{n/2+1}.$$

2. For $F(x) = x^{-1} + x^3 \in \mathbb{F}_{2^n}[x]$,

$$\mathcal{N}(F) \geq 2^{n-1} - 2^{n/2+1} - \frac{1}{2}.$$

Furthermore we can get rid of the last term '1/2' if n is odd.

3. For $F(x) = x^{-3} + x^{-1} \in \mathbb{F}_{2^n}[x]$,

$$\mathcal{N}(F) \geq 2^{n-1} - 2^{n/2+1} - \frac{1}{2}.$$

Furthermore we can get rid of the last term '1/2' if n is even.

If we apply Corollary 8, we can obtain lower bounds on nonlinearity of some monomials whose nonlinearity has not analyzed theoretically yet.

Example 10. Consider $\mathbb{F}_{2^7}$. Let $F(x) = x^3$ and $G(x) = (x^5)^{-1} = x^{51}$. Then we have $F \circ G^{-1}(x) = x^{28}$. By the above statement, we have

$$\mathcal{N}(x^{28}) \geq 2^{n-1} - 2^{n/2+1}.$$

Since nonlinearity preserves under composition with linear functions like x^2, x^7 has the same nonlinearity with $(x^7)^4$. Hence we have

$$\mathcal{N}(x^7) \geq 2^{n-1} - 2^{n/2+1}.$$

We can apply Theorem 6 directly to get a lower bound on nonlinearity of some rational functions,

Example 11. 1. For any irreducible polynomial $H(x)$ of degree d, we have

$$\mathcal{N}(1/H) \geq 2^{n-1} - d \cdot 2^{n/2}.$$

2. For $F(x) = x^{-3}(x-1)^{-1}($ we assume $F(0) = F(1) = 0)$,

$$\mathcal{N}(F) \geq 2^{n-1} - 3 \cdot 2^{n/2} - 1.$$

Table 1 shows the tightness of our lower bound on nonlinearity. The third columnn shows the lower bound obtained by Theorem 6 and the fourth column shows the exact value of nonlinearity calculated by computational experiment. Note that S-boxes in Table 1 may not be a permutation. In order to apply them for block cipher, the other properties such as differential probability should be investigated.

Table 1. Lower bound on Nonlinearity and its Exact Value

Function	n	Our Lower Bound	Exact Value
$x^3 + x^5 + x^6$	7	48	48
	8	96	96
$x^{-1} + x^3$	7	41	46
	8	96	100
$x^{-3} + x^{-1}$	7	41	46
	8	96	97

4   Nonlinearity of $n \times kn$ S-boxes

In this section. we derive nonlinearity of $n \times kn$ S-box constructed by concatenating k $n \times n$ S-boxes over $\mathbb{F}_{2^n}$. At first, we present a proposition to relate nonlinearity of $n \times kn$ S-box to that of $n \times n$ S-box.

Proposition 12. *Let* $F : \mathbb{F}_{2^n} \to \mathbb{F}_{2^{kn}}$ *be a vector Boolean fuctions with* $F = (F_1, F_2, \cdots, F_k)$ *for* $F_i : \mathbb{F}_{2^n} \to \mathbb{F}_{2^n}$. *Then we have*

$$\mathcal{N}(F) = \min_{(c_1, c_2, \cdots, c_k) \in \mathbb{F}_{2^{kn}}^*} \mathcal{N}(c_1 F_1 + c_2 F_2 + \cdots + c_k F_k),$$

where the sum and product are the field operations in $\mathbb{F}_{2^{kn}}$.

Proof. Choose a basis B of $\mathbb{F}_{2^n}$ over $\mathbb{F}_2$ and take its dual basis $\hat{B}$. Let us represent by the basis B the left sides of all inner products and by its dual basis $\hat{B}$ their right sides. For any nonzero $b = (c_1, c_2, \cdots, c_k)$ with $c_i \in \mathbb{F}_{2^n}$, we have

$$\begin{aligned}
\mathcal{L}(F, a, b) &= \#\{x \mid a \cdot x = b \cdot F(x)\} \\
&= \#\{x \mid Tr(ax + bF(x)) = 0\} \\
&= \#\{x \mid Tr(ax + c_1 F_1(x) + \cdots + c_k F_k(x)) = 0\} \\
&= \mathcal{L}(c_1 F_1 + \cdots + c_k F_k, a, 1).
\end{aligned}$$

where 1 is a binary vector representing an identity element by the basis B.

Conversely, for any nonzero $(c_1, c_2, \cdots, c_k) \in \mathbb{F}_{2^{kn}}, c_i \in \mathbb{F}_{2^n}$ and a nonzero $b_0 \in \mathbb{F}_{2^n}$, there exists a nonzero $b \in \mathbb{F}_{2^{kn}}$ such that $\mathcal{L}(c_1 F_1 + \cdots + c_k F_k, a, b_0) = \mathcal{L}(F, a, b)$, which completes the proof.

By the above proposition, we can apply Theorem 6 to get a lower bound on nonlinearity of $n \times kn$ S-box. For example, consider an $n \times 2n$ S-box $F = (F_1, F_2)$ where $F_1(x) = x^{-1}$ and $F_2(x) = x^3$ are S-boxes over $\mathbb{F}_{2^n}$. Then

$$\begin{aligned}
\mathcal{N}(F) &= \min_{(c_1, c_2) \neq 0} \mathcal{N}(c_1 x^{-1} + c_2 x^3) \\
&= \min\{\min_{c_i \neq 0} \mathcal{N}(c_1 x^{-1} + c_2 x^3), \mathcal{N}(x^{-1}), \mathcal{N}(x^3)\} \\
&\geq 2^{n-1} - 2^{n/2+1} + \frac{1}{2}.
\end{aligned}$$

The first equality follows from Proposition 12 and the last inequality follows from Corollary 7.

Similarly, we can get lower bounds on nonlinearity of various n-by-kn boxes. We present some of them in Table 2. The second column shows a lower bound of nonlinearity of the S-boxes in the first column for even or odd n. The third and fourth column shows the exact value of nonlinearity calculated by computational experiment.

In Table 2, every rational function such as x^{-1} and x^3 is a vector Boolean function from $\mathbb{F}_{2^n}$ to $\mathbb{F}_{2^n}$. Note that all functions are permutations for odd n, but only x^{-1} and x^7 are permutations for $n = 8$. If we combine our result with Theorem 17 in [8], we can also construct highly nonlinear $kn \times kn$ S-boxes.

Acknowledgement
We would like to thank Sang Geun Hahn for his helpful comments. We are especially grateful to Sangjoon Park who pointed us to the problem of relating hyperelliptic curves to nonlinearity of Boolean functions.

References

1. C. Adams and S. E. Tavares, "Designing S-boxes for Ciphers Resistant to Differential Cryptanalysis," Proc. of SPRC'93, 1993.

Table 2. Lower Bounds of Nonlinearity of n-by-kn S-boxes

S-box	Lower Bound of Nonlinearity	$n = 7$	$n = 8$
(x^{-1}, x^3)	$2^{n-1} - 2^{n/2+1} - \frac{1}{2}$	41	96
(x^{-1}, x^{-3})	$2^{n-1} - 2^{n/2+1} - \frac{1}{2}$	41	96
(x^3, x^5)	$2^{n-1} - 2^{n/2+1}$	42	96
(x^{-3}, x^{-5})	$2^{n-1} - 3 \cdot 2^{n/2} - \frac{1}{2}$	30	80
(x^{-3}, x^{-1}, x^3)	$2^{n-1} - 3 \cdot 2^{n/2} - \frac{1}{2}$	30	80
(x^{-1}, x^3, x^5)	$2^{n-1} - 3 \cdot 2^{n/2} - \frac{1}{2}$	30	80
(x^3, x^5, x^7)	$2^{n-1} - 3 \cdot 2^{n/2}$	31	80
$(x^{-3}, x^{-1}, x^3, x^5)$	$2^{n-1} - 4 \cdot 2^{n/2} - \frac{1}{2}$	19	64
(x^{-1}, x^3, x^5, x^7)	$2^{n-1} - 4 \cdot 2^{n/2} - \frac{1}{2}$	19	64
(x^3, x^5, x^7, x^9)	$2^{n-1} - 4 \cdot 2^{n/2}$	19	64

2. T. Beth and D. Ding, "On Almost Perfect Nonlinear Permutations," Proc. of Eurocrypt'93, pp. 65 – 76, Springer-Verlag, 1994.
3. R. Lidl and H. Niederreiter, *Introduction to Finite Fields and their Applications*, Cambridge University Press, 1986.
4. D. Lorenzini, *An Invitation to Arithmetic Geometry*, American Mathematical Society, 1996.
5. M. Matsui, "Linear Cryptanalysis Method for DES cipher," Proc. of Eurocrypt'93, pp.386 – 397, Springer-Verlag, 1993.
6. K. Nyberg, "On the Construction of Highly Nonlinear Permutation," Proc. of Eurocrypt'92, pp. 92 – 98, Springer-Verlag, 1993.
7. K. Nyberg, "Differentially Uniform Mappings for Cryptography," Proc. of Eurocrypt'93, pp. 55 – 64, Springer-Verlag, 1994.
8. K. Nyberg, "S-Boxes and Round Functions with Controllable Linearity and Differential Uniformity," Proc. of the Second Fast Software Encryption, pp. 111 – 130, Springer-Verlag, 1994.
9. J. Seberry, X. -M. Zhang and Y. Zheng, "Nonlinearly Balanced Functions and Their Propagation Characteristics," Proc. of Crypto'93, pp. 49 – 60, Springer-Verlag, 1993.
10. J. Silverman, *The Arithmetic of Elliptic Curves*, Springer-Verlag, 1992.

Secure Distributed Key Generation
for Discrete-Log Based Cryptosystems

Rosario Gennaro[1], Stanisław Jarecki[2], Hugo Krawczyk[3], and Tal Rabin[1]

[1] IBM T.J.Watson Research Center, PO Box 704, Yorktown Heights, NY 10598, USA
{rosario,talr}@watson.ibm.com
[2] MIT Laboratory for Computer Science, 545 Tech Square, Cambridge, MA 02139,
USA, stasio@theory.lcs.mit.edu
[3] Department of Electrical Engineering, Technion, Haifa 32000, Israel, and
IBM T.J. Watson Research Center, New York, USA
hugo@ee.technion.ac.il

Abstract. Distributed key generation is a main component of threshold cryptosystems and distributed cryptographic computing in general. Solutions to the distributed generation of private keys for discrete-log based cryptosystems have been known for several years and used in a variety of protocols and in many research papers. However, these solutions fail to provide the full security required and claimed by these works. We show how an active attacker controlling a small number of parties can bias the values of the generated keys, thus violating basic correctness and secrecy requirements of a key generation protocol. In particular, our attacks point out to the places where the proofs of security fail.

Based on these findings we designed a distributed key generation protocol which we present here together with a rigorous proof of security. Our solution, that achieves optimal resiliency, can be used as a drop-in replacement for key generation modules as well as other components of threshold or proactive discrete-log based cryptosystems.

Keywords: Threshold Cryptography. Distributed Key Generation. VSS. Discrete Logarithm.

1 Introduction

Distributed key generation is a main component of threshold cryptosystems. It allows a set of n servers to jointly generate a pair of public and private keys according to the distribution defined by the underlying cryptosystem without having to ever compute, reconstruct, or store the secret key in any single location and without assuming any trusted party (dealer). While the public key is output in the clear, the private key is maintained as a (virtual) secret shared via a threshold scheme. In particular, no attacker can learn anything about the key as long as it does not break into a specified number, $t+1$, of servers. This shared private key can be later used by a threshold cryptosystem, e.g., to compute signatures or decryptions, without ever being reconstructed in a single location. For discrete–log based schemes, distributed key generation amounts to generating a

J. Stern (Ed.): EUROCRYPT'99, LNCS 1592, pp. 295–310, 1999.

secret sharing of a random, uniformly distributed value x and making public the value $y = g^x$. We refer to such a protocol as DKG.

A DKG protocol may be run in the presence of a malicious adversary who corrupts a fraction (or threshold) of the players and forces them to follow an arbitrary protocol of his choice. Informally, we say that a DKG protocol is secure if the output of the non-corrupted parties is correct (i.e. the shares held by the good players define a unique uniformly distributed value x and the public value y satisfies $y = g^x$), and the adversary learns no information about the chosen secret x beyond, of course, what is learned from the public value y.

Solutions to the shared generation of private keys for discrete-log based threshold cryptosystems [DF89] have been known and used for a long time. Indeed, the first DKG scheme was proposed by Pedersen in [Ped91a]. It then appeared, with various modifications, in several papers on threshold cryptography, e.g., [CMI93, Har94, LHL94, GJKR96, HJJ+97, PK96, SG98], and distributed cryptographic applications that rely on it, e.g., [CGS97]. Moreover, a secure DKG protocol is an important building block in other distributed protocols for tasks different than the generation of keys. One example is the generation of the randomizers in discrete-log based signature schemes (for example the r value in a (r, s) DSS signature as in [GJKR96]). Another example is the generation of the refreshing polynomial in proactive secret sharing and signature schemes [HJKY95, HJJ+97, FGMY97].

The basic idea in Pedersen's DKG protocol [Ped91a] (as well as in the subsequent variants) is to have n parallel executions of Feldman's verifiable secret sharing (VSS) protocol [Fel87] in which each player P_i acts as a dealer of a random secret z_i that he picks. The secret value x is taken to be the sum of the properly shared z_i's. Since Feldman's VSS has the additional property of revealing $y_i = g^{z_i}$, the public value y is the product of the y_i's that correspond to those properly shared z_i's.

In this paper we show that, in spite of its use in many protocols, Pedersen's DKG cannot guarantee the correctness of the output distribution in the presence of an adversary. Specifically, we show a strategy for an adversary to manipulate the distribution of the resulting secret x to something quite different from the uniform distribution. This flaw stresses a well-known basic principle for the design of cryptographic protocols, namely, that secure components can turn insecure when composed to generate new protocols. We note that this ability of the attacker to bias the output distribution represents a flaw in several aspects of the protocol's security. It clearly violates the basic correctness requirement about the output distribution of the protocol; but it also weakens the secrecy property of the solution. Indeed, the attacker acquires in this way some a-priori knowledge on the secret which does not exist when the secret is chosen truly at random. Moreover, these attacks translate into flaws in the attempted proofs of these protocols; specifically, they show that simulation arguments (à la zero-knowledge) as used to prove the secrecy of these protocols must fail.

In contrast to the above, we present a protocol that enjoys a full proof of security. We first present the formal requirements for a secure solution of the

DKG problem, then present a particular DKG protocol and rigorously prove that it satisfies the security requirements. In particular, we show that the output distribution of private and public keys is as required, and prove the secrecy requirement from the protocol via a full simulation argument. Our solution is based on ideas similar to Pedersen's DKG (in particular, it also uses Feldman's VSS as a main component), but we are careful about designing an initial *commitment phase* where each player commits to its initial choice z_i in a way that prevents the attacker from later biasing the output distribution of the protocol. For this commitment phase we use another protocol of Pedersen, i.e., Pedersen's VSS (verifiable secret sharing) protocol as presented in [Ped91b]. Very importantly, our solution preserves most of the efficiency and simplicity of the original DKG solution of [Ped91a], in particular it has comparable computational complexity and the same optimal threshold of $t < n/2$.

Organization: In Section 2 we present the basic communication and adversarial models for our protocols. In Section 3 we describe previously proposed solutions to the DKG problem and show where they fail. In Section 4 we present our solution and its full analysis; we also discuss some other applications of our protocol. Finally, in Section 5 we discuss an enhanced (and more realistic) security model under which our solution works as well.

2 Preliminaries

Communication Model. We assume that our computation model is composed of a set of n *players* $P_1, \ldots, P_n$ that can be modeled by polynomial-time randomized Turing machines. They are connected by a complete network of private (i.e. untappable) point-to-point channels. In addition, the players have access to a dedicated broadcast channel.

For simplicity of the discussion that follows, we assume a *fully synchronous* communication model, i.e. that messages of a given round in the protocol are sent by all players simultaneously, and that they are simultaneously delivered to their recipients. This model is not realistic enough for many applications, but it is often assumed in the literature; moreover, our attacks against known DKG protocols (Section 3) work even in this simplified setting.

In Section 5 we introduce a more realistic, *partially synchronous* communication model. Our solution to the DKG problem (Section 4) and its security proof work in this strictly stronger adversarial model.

The Adversary. We assume that an adversary, $\mathcal{A}$, can corrupt up to t of the n players in the network, for any value of $t < n/2$ (this is the best achievable threshold – or resilience – for solutions that provide both secrecy and robustness). We consider a malicious adversary that may cause corrupted players to divert from the specified protocol in *any* way. We assume that the computational power of the adversary is adequately modeled by a probabilistic polynomial time Turing machine. Our adversary is *static*, i.e. chooses the corrupted players at the beginning of the protocol (see section 4.2 for a reference to a recent extension of our results to the non-static – or *adaptive* – adversary setting).

3   Distributed Key Generation in DLog-Based Schemes

In this section we define the minimal requirements for a secure distributed key generation protocol. We show how previous solutions fail to satisfy these requirements. We also discuss the applicability of our attacks to other existing distributed protocols.

3.1   Requirements of a Secure DKG Protocol

As we mentioned in the introduction, distributed generation of keys in a discrete–log based scheme amounts to generating a secret sharing of a random, uniformly distributed value x and making public the value $y = g^x$. Specifically, in a discrete–log based scheme with a large prime p and an element g of order q in $\mathbb{Z}_p^*$ where q is a large prime dividing $p - 1$, the distributed protocol DKG performed by n players $P_1, \ldots, P_n$ generates private outputs $x_1, \ldots, x_n$, called *the shares*, and a public output y. The protocol is called *t-secure* (or secure with threshold t) if in the presence of an attacker that corrupts at most t parties the following requirements for correctness and secrecy are satisfied:

Correctness:

(C1) All subsets of $t + 1$ shares provided by honest players define the same unique secret key x.

(C2) All honest parties have the same value of public key $y = g^x \bmod p$, where x is the unique secret guaranteed by (C1).

(C3) x is uniformly distributed in $\mathbb{Z}_q$ (and hence y is uniformly distributed in the subgroup generated by g).

Secrecy: No information on x can be learned by the adversary except for what is implied by the value $y = g^x \bmod p$.

More formally, we state this condition in terms of simulatability: for every (probabilistic polynomial-time) adversary $\mathcal{A}$, there exists a (probabilistic polynomial-time) simulator SIM, such that on input an element y in the subgroup of $\mathbb{Z}_p^*$ generated by g, produces an output distribution which is polynomially indistinguishable from $\mathcal{A}$'s view of a run of the DKG protocol that ends with y as its public key output, and where $\mathcal{A}$ corrupts up to t parties.

The above is a minimal set of requirements needed in all known applications of such a protocol. In many applications a stronger version of (C1) is desirable, which reflects two additional aspects: (1) It requires the existence of an *efficient procedure* to build the secret x out of $t+1$ shares; and (2) it requires this procedure to be *robust*, i.e. the reconstruction of x should be possible also in the presence of malicious parties that try to foil the computation. We note that these added properties are useful not only in applications that require explicit reconstruction of the secret, but also in applications (such as threshold cryptosystems) that use the secret x in a distributed manner (without ever reconstructing it) to compute some cryptographic function, e.g. a signature. Thus, we formulate (C1')
as follows:

(C1') There is an efficient procedure that on input the n shares submitted by the players and the public information produced by the DKG protocol, outputs the unique value x, even if up to t shares are submitted by faulty players.

3.2 The Insecurity of a Common DKG Protocol

The Joint-Feldman Protocol. Feldman [Fel87] presents a *verifiable secret sharing* (VSS) protocol, denoted by Feldman-VSS, that allows a *trusted* dealer to share a key x among n parties in a way that the above security properties are achieved (with the exception that the protocol assumes the dealer never to be corrupted by the attacker). Based on this protocol, Pedersen [Ped91a] proposes the first distributed solution to this problem, i.e. the first DKG protocol. It specifies the run of n parallel executions of Feldman-VSS as follows. Each player P_i selects a random secret $z_i \in \mathbb{Z}_q$ and shares it among the n players using Feldman-VSS. This defines the set *QUAL* of players that shared their secrets properly. The random secret x is set to be the sum of the properly shared secrets and each player can compute his share of x by locally summing up the shares he received. The value y can be computed as the product of the public values $y_i = g^{z_i} \bmod p$ generated by the proper executions of the Feldman-VSS protocols. Similarly, the verification values $A_1, \ldots, A_t$ necessary for robust reconstruction of x in Feldman-VSS, can be computed as products of the corresponding verification values generated by each properly executed VSS protocol.

In Figure 1 we present a simplified version of the protocol proposed in [Ped91a], which we call Joint-Feldman. By concentrating on the core of the protocol we are able to emphasize the central weakness in its design. We also show that several variants of this core protocol (including the full protocol from [Ped91a] and other modifications [HJKY95, HJJ+97]) are also insecure.

An Attack Against Joint-Feldman. We show how an adversary can influence the distribution of the result of Joint-Feldman to a non-uniform distribution.

It can be seen, from the above description of the protocol that the determining factor for what the value x will be, is the definition of the set *QUAL*. The attack utilizes the fact that the decision whether a player is in *QUAL* or not, even given the fully synchronous communication model, occurs after the adversary has seen the values y_i of all players. The values y_i are made public in Step 1 and the disqualification of players occurs in Steps 2-3. Using this timing discrepancy, the attacker can affect the distribution of the pair (x, y).

More specifically the attack works as follows. Assume the adversary wants to bias the distribution towards keys y whose last bit is 0. It assumes two faulty players, say P_1 and P_2. In Step 1, P_1 gives players $P_3, \ldots, P_{t+2}$ shares which are inconsistent with his broadcast values, i.e. they do not pass the test of Step 2. The rest of the players receive consistent shares. Thus, in Step 2 there will be t complaints against P_1, yet t complaints are not sufficient for disqualification. Now, at the end of Step 1 the adversary computes $\alpha = \prod_{i=1}^{n} y_i$ and $\beta = \prod_{i=2}^{n} y_i$. If α ends with 0 then P_1 will do nothing and continue the protocol as written. If α ends with 1 then the adversary forces the disqualification of P_1 in Step 3.

Protocol Joint-Feldman

1. Each player P_i chooses a random polynomial $f_i(z)$ over $\mathbf{Z}_q$ of degree t:

$$f_i(z) = a_{i0} + a_{i1}z + \ldots + a_{it}z^t$$

   P_i broadcasts $A_{ik} = g^{a_{ik}} \bmod p$ for $k = 0, \ldots, t$. Denote a_{i0} by z_i and A_{i0} by y_i. Each P_i computes the shares $s_{ij} = f_i(j) \bmod q$ for $j = 1, \ldots, n$ and sends s_{ij} secretly to player P_j.

2. Each P_j verifies the shares he received from the other players by checking for $i = 1, \ldots, n$:

$$g^{s_{ij}} = \prod_{k=0}^{t} (A_{ik})^{j^k} \bmod p \tag{1}$$

   If the check fails for an index i, P_j broadcasts a *complaint* against P_i.

3. If more than t players complain against a player P_i, that player is clearly faulty and he is disqualified. Otherwise P_i reveals the share s_{ij} matching Eq. 1 for each complaining player P_j. If any of the revealed shares fails this equation, P_i is disqualified. We define the set $QUAL$ to be the set of non-disqualified players.

4. The public value y is computed as $y = \prod_{i \in QUAL} y_i \bmod p$. The public verification values are computed as $A_k = \prod_{i \in QUAL} A_{ik} \bmod p$ for $k = 1, \ldots, t$. Each player P_j sets his share of the secret as $x_j = \sum_{i \in QUAL} s_{ij} \bmod q$. The secret shared value x itself is not computed by any party, but it is equal to $x = \sum_{i \in QUAL} z_i \bmod q$.

Fig. 1. An insecure solution for distributed generation of secret keys

---

This is achieved by asking P_2 to also broadcast a complaint against P_1, which brings the number of complaints to $t+1$. This action sets the public value y to β which ends with 0 with probability $1/2$. Thus effectively the attacker has forced strings ending in 0 to appear with probability $3/4$ rather than $1/2$.

Why the Simulation Fails. An attempt to prove this protocol secure would use a simulation argument. Following is an explanation of why such a simulator would fail. Consider a simulator S which receives the value y and needs to "hit" this value. That is, S needs to generate a transcript which is indistinguishable from an actual run of the protocol that outputs y as the public key, and where the adversary controls up to t players, say $P_1, \ldots, P_t$. The simulator has enough information to compute the values $z_1, \ldots, z_t$ that the adversary has shared in Step 1. Now S needs to commit itself to the values shared by the good players. However, the attack described in the paragraph above can be easily extended to a strategy that allows the adversary to decide in Steps 2-3 on the set Q of faulty players whose values will be considered in the final computation (i.e. $QUAL = Q \cup \{t+1, \ldots, n\}$). Consequently, in Step 1, the simulator S does not know how to pick the good players' values $y_{t+1}, ..., y_n$ so that $(\prod_{i \in Q} y_i) \cdot (y_{t+1} \cdot ... \cdot y_n) = y \bmod p$, as S still does not know the set Q. Since the number of possible sets Q that the adversary can choose is exponential in t, then S has no effective strategy to simulate this computation in polynomial time.

Other Insecure Variants of the Joint-Feldman Protocol. The many variants and extensions of the Joint-Feldman protocol which have appeared in the literature are also insecure. They all fail to achieve the correctness property (C3) and the secrecy requirement as presented in Section 3.1. The variants include: signatures on shares, commitments to y_i, committing encryption on broadcast channel, committing encryption with reconstruction, and "stop, kill and rewind". Due to space limitations, we invite the reader to the on-line appendix to this paper [GJKR99] for the description of these variants and their flaws.

4   The New Protocol

Our solution enjoys the same flavor and simplicity as the Joint-Feldman protocol presented in Figure 1, i.e. each player shares a random value and the random secret is generated by summing up these values.

But we use a different sharing and then introduce methods to extract the public key. We start by running a commitment stage where each player P_i commits to a t-degree polynomial (t is the scheme's threshold) $f_i(z)$ whose constant coefficient is the random value, z_i, contributed by P_i to the jointly generated secret x. We require the following properties from this commitment stage: First, the attacker cannot force a commitment by a (corrupted) player P_j to depend on the commitment(s) of any set of honest players. Second, for any player P_i that is not disqualified during this stage, there is a unique polynomial f_i committed to by P_i and this polynomial is recoverable by the honest players (this may be needed if player P_i misbehaves at a later stage of the protocol). Finally, for each honest player P_i and non-disqualified player P_j, P_i holds the value $f_i(j)$ at the end of the commitment stage.

To realize the above commitment stage we use the information-theoretic verifiable secret sharing (VSS) protocol due to Pedersen [Ped91b], and which we denote by Pedersen-VSS. We show that at the end of the commitment stage the value of the secret x is determined and no later misbehavior by any party can change it (indeed, if a non-disqualified player misbehaves later in the protocol his value z_i is publicly reconstructed by the honest players). Most importantly, this guarantees that no bias in the output x or y of the protocol is possible, and it allows us to present a full proof of security based on a careful simulation argument. After the value x is fixed we enable the parties to efficiently and securely compute $g^x \bmod p$.

In the next subsection we present the detailed solution and its analysis. But first we expand on Pedersen's VSS protocol.

Pedersen's VSS. As said, we use the protocol Pedersen-VSS introduced in [Ped91b] as a central tool in our solution. For lack of space we do not explicitly describe Pedersen-VSS here, however its description is implicit in step 1 of

Figure 2. We note that this protocol uses, in addition to the parameters p, q, g which are inherent to the DKG problem, an element h in the subgroup of $\mathbb{Z}_p^*$ generated by g. It is assumed that the adversary cannot find the discrete logarithm of h relative to the base g. In section 4.2 we discuss how this value of h can be generated in the context of our DKG solution. Some of the main properties of Pedersen-VSS are summarized in the next Lemma and used in the analysis of our DKG solution in the next subsection.

Lemma 1. [Ped91b] Pedersen-VSS *satisfies the following properties in the presence of an adversary that corrupts at most t parties and which cannot compute* $dlog_g h$:

1. *If the dealer is not disqualified during the protocol then all honest players hold shares that interpolate to a unique polynomial of degree t. In particular, any $t + 1$ of these shares suffice to efficiently reconstruct (via interpolation) the secret s.*
2. *The protocol produces information (the public values C_k and private values s_i') that can be used at reconstruction time to test for the correctness of each share; thus, reconstruction is possible, even in the presence of malicious players, from any subset of shares containing at least $t + 1$ correct shares.*
3. *The view of the adversary is independent of the value of the secret s, and therefore the secrecy of s is unconditional.*

4.1 Secure DKG Protocol

Our secure solution to the distributed generation of keys follows the above ideas and is presented in detail in Figure 2. We denote this protocol as DKG.
The security properties of this solution are stated in the next Theorem.

Theorem 2. *Protocol DKG from Figure 2 is a secure protocol for distributed key generation in discrete-log based cryptosystems, namely, it satisfies the correctness and secrecy requirements of Section 3.1 with threshold t, for any $t < n/2$.*

Proof of Correctness. We first note that all honest players in the protocol compute the same set $QUAL$ since the determination of which players are to be disqualified depends on public broadcast information which is known to all (honest) players.

(C1) At the end of Step 2 of the protocol it holds that if $i \in QUAL$ then player P_i has successfully performed the dealing of z_i under Pedersen-VSS. From part 1 of Lemma 1 we know that all honest players hold shares (s_{ij}) which interpolate to a unique polynomial with constant coefficient equal to z_i. Thus, for any set $\mathcal{R}$ of $t + 1$ correct shares, $z_i = \sum_{j \in \mathcal{R}} \gamma_j \cdot s_{ij} \mod q$ where γ_j are appropriate Lagrange interpolation coefficients for the set $\mathcal{R}$. Since each honest party P_j computes its share x_j of x as $x_j = \sum_{i \in QUAL} s_{ij}$, then we have that for the set of shares $\mathcal{R}$:

$$x = \sum_{i \in QUAL} z_i = \sum_{i \in QUAL} \left(\sum_{j \in \mathcal{R}} \gamma_j \cdot s_{ij} \right) = \sum_{j \in \mathcal{R}} \gamma_j \cdot \left(\sum_{i \in QUAL} s_{ij} \right) = \sum_{j \in \mathcal{R}} \gamma_j x_j$$

Protocol DKG

Generating x:

1. Each player P_i performs a **Pedersen-VSS** of a random value z_i as a dealer:
   (a) P_i chooses two random polynomials $f_i(z), f_i'(z)$ over $\mathbb{Z}_q$ of degree t:

   $$f_i(z) = a_{i0} + a_{i1}z + \ldots + a_{it}z^t \qquad f_i'(z) = b_{i0} + b_{i1}z + \ldots + b_{it}z^t$$

   Let $z_i = a_{i0} = f_i(0)$. P_i broadcasts $C_{ik} = g^{a_{ik}}h^{b_{ik}} \bmod p$ for $k = 0, \ldots, t$. P_i computes the shares $s_{ij} = f_i(j), s_{ij}' = f_i'(j) \bmod q$ for $j = 1, \ldots, n$ and sends s_{ij}, s_{ij}' to player P_j.
   (b) Each player P_j verifies the shares he received from the other players. For each $i = 1, \ldots, n$, P_j checks if

   $$g^{s_{ij}}h^{s_{ij}'} = \prod_{k=0}^{t}(C_{ik})^{j^k} \bmod p \qquad (2)$$

   If the check fails for an index i, P_j broadcasts a *complaint* against P_i.
   (c) Each player P_i who, as a dealer, received a complaint from player P_j broadcasts the values s_{ij}, s_{ij}' that satisfy Eq. 2.
   (d) Each player marks as *disqualified* any player that either
   - received more than t complaints in Step 1b, or
   - answered to a complaint in Step 1c with values that falsify Eq. 2.

2. Each player then builds the set of non-disqualified players *QUAL*. (We show in the analysis that all honest players build the same set *QUAL* and hence, for simplicity, we denote it with a unique global name.)
3. The distributed secret value x is not explicitly computed by any party, but it equals $x = \sum_{i \in QUAL} z_i \bmod q$. Each player P_i sets his share of the secret as $x_i = \sum_{j \in QUAL} s_{ji} \bmod q$ and the value $x_i' = \sum_{j \in QUAL} s_{ji}' \bmod q$.

Extracting $y = g^x \bmod p$:

4. Each player $i \in QUAL$ exposes $y_i = g^{z_i} \bmod p$ via **Feldman VSS**:
   (a) Each player P_i, $i \in QUAL$, broadcasts $A_{ik} = g^{a_{ik}} \bmod p$ for $k = 0, \ldots, t$.
   (b) Each player P_j verifies the values broadcast by the other players in *QUAL*, namely, for each $i \in QUAL$, P_j checks if

   $$g^{s_{ij}} = \prod_{k=0}^{t}(A_{ik})^{j^k} \bmod p \qquad (3)$$

   If the check fails for an index i, P_j *complains* against P_i by broadcasting the values s_{ij}, s_{ij}' that satisfy Eq. 2 but do not satisfy Eq. 3.
   (c) For players P_i who receive at least one valid complaint, i.e. values which satisfy Eq. 2 and not Eq. 3, the other players run the reconstruction phase of **Pedersen-VSS** to compute z_i, $f_i(z)$, A_{ik} for $k = 0, \ldots, t$ in the clear. For all players in *QUAL*, set $y_i = A_{i0} = g^{z_i} \bmod p$. Compute $y = \prod_{i \in QUAL} y_i \bmod p$.

Fig. 2. Secure distributed key generation in discrete–log based systems

Since this holds for *any* set of $t+1$ correct shares then x is uniquely defined.

(C1') The above argument in (C1) shows that the secret x can be efficiently reconstructed, via interpolation, out of any $t+1$ correct shares. We need to show that we can tell apart correct shares from incorrect ones. For this we show that for each share x_j, the value g^{x_j} can be computed from publicly available information broadcast in Step 4a:

$$g^{x_j} = g^{\sum_{i \in QUAL} s_{ij}} = \prod_{i \in QUAL} g^{s_{ij}} = \prod_{i \in QUAL} \prod_{k=0}^{t} (A_{ik})^{j^k} \bmod p$$

where the last equality follows from Eq. 3. Thus the publicly available value g^{x_j} makes it possible to verify the correctness of share x_j at reconstruction time.

(C2) The value y is computed (by the honest players) as $y = \prod_{i \in QUAL} y_i \bmod p$, where the values of y_i are derived from information broadcast in the protocol and thus known to all honest players. We need to show that indeed $y = g^x$ where $x = \sum_{i \in QUAL} z_i$. We will show that for $i \in QUAL$, $y_i = g^{z_i}$, and then $y = \prod_{i \in QUAL} y_i = \prod_{i \in QUAL} g^{z_i} = g^{\sum_{i \in QUAL} z_i} = g^x$. For parties $i \in QUAL$ against whom a valid complaint has been issued in Step 4b value z_i is publicly reconstructed and y_i set to $g^{z_i} \bmod p$ (the correct reconstruction of z_i is guaranteed by Lemma 1 (part 2)). Now we need to show that for P_i, $i \in QUAL$, against whom a valid complaint has not been issued, the value y_i is set to A_{i0}. Values $A_{ik}, k = 0, \ldots, t$ broadcast by player P_i in Step 4a define a t-degree polynomial $\hat{f}_i(z)$ in $\mathbb{Z}_q$. Since we assume that no valid complaint was issued against P_i then Eq. 3 is satisfied for all honest players, and thus $\hat{f}_i(z)$ and $f_i(z)$ have at least $t+1$ points in common, given by the shares s_{ij} held by the uncorrupted players P_j. Hence they are equal, and in particular $A_{i0} = g^{f_i(0)} = g^{z_i}$.

(C3) The secret x is defined as $x = \sum_{i \in QUAL} z_i$. Note that as long as there is one value z_i in this sum that is chosen at random and independently from other values in the sum, we are guaranteed to have uniform distribution of x. Also note that the secret x and the components z_i in the sum are already determined at the end of Step 2 of DKG (since neither the values z_i nor the set $QUAL$ change later). Let P_i be a non-corrupted player; in particular, $i \in QUAL$. At the end of Step 1 of the protocol z_i exists only as a value dealt by P_i using Pedersen-VSS. By virtue of part 3 of Lemma 1 the view (and thus actions) of the adversary are independent of this value z_i and hence the secret x is uniformly distributed (as z_i is).

Proof of Secrecy. We provide a simulator SIM for the DKG protocol in Figure 3. Here we show that the view of the adversary $\mathcal{A}$ that interacts with SIM on input y is the same as the view of $\mathcal{A}$ that interacts with the honest players in a regular run of the protocol that outputs the given y as the public key.

In the description and analysis of the simulator we assume, without loss of generality, that the adversary compromises players $P_1, \ldots, P_{t'}$, where $t' \leq t$. We denote the indices of the players controlled by the adversary by $\mathcal{B} = \{1, \ldots, t'\}$, and the indices of the players controlled by the simulator by $\mathcal{G} = \{t'+1, \ldots, n\}$.

In a regular run of protocol DKG, $\mathcal{A}$ sees the following probability distribution of data produced by the uncorrupted parties:

- Values $f_i(j), f_i'(j), i \in \mathcal{G}, j \in \mathcal{B}$, uniformly chosen in $\mathbb{Z}_q$ (and denoted as s_{ij}, s_{ij}', resp.).
- Values $C_{ik}, A_{ik}, i \in \mathcal{G}, k = 0, \ldots, t$ that correspond to (exponents of) coefficients of randomly chosen polynomials and for which the Eqs. (2) and (3) are satisfied for all $j \in \mathcal{B}$.

Algorithm of simulator SIM

We denote by $\mathcal{B}$ the set of players controlled by the adversary, and by $\mathcal{G}$ the set of honest parties (run by the simulator). Wlog, $\mathcal{B} = \{1, \ldots, t'\}$ and $\mathcal{G} = \{t'+1, \ldots, n\}, t' \leq t$.

Input: public key y

1. Perform Steps 1a-1d,2 on behalf of the uncorrupted players $P_{t'+1}, \ldots, P_n$ exactly as in protocol DKG. This includes receiving and processing the information sent privately and publicly from corrupted players to honest ones. At the end of Step 2 the following holds:
   - The set $QUAL$ is well-defined. Note that $\mathcal{G} \subseteq QUAL$ and that polynomials $f_i(z), f_i'(z)$ for $i \in \mathcal{G}$ are chosen at random.
   - The adversary's view consists of polynomials $f_i(z), f_i'(z)$ for $i \in \mathcal{B}$, the shares $(s_{ij}, s_{ij}') = (f_i(j), f_i'(j))$ for $i \in QUAL, j \in \mathcal{B}$, and all the public values C_{ik} for $i \in QUAL, k = 0, \ldots, t$.
   - SIM knows all polynomials $f_i(z), f_i'(z)$ for $i \in QUAL$ (note that for $i \in QUAL \cap \mathcal{B}$ the honest parties, and hence SIM, receive enough consistent shares from the adversary that allow SIM to compute all these parties' polynomials). In particular, SIM knows all the shares s_{ij}, s_{ij}', the coefficients a_{ik}, b_{ik} and the public values C_{ik}.
2. Perform the following computations:
   - Compute $A_{ik} = g^{a_{ik}}$ for $i \in QUAL \setminus \{n\}, k = 0, \ldots, t$
   - Set $A_{n0}^* = y \cdot \prod_{i \in (QUAL \setminus \{n\})} (A_{i0})^{-1} \bmod p$
   - Assign $s_{nj}^* = s_{nj} = f_n(j)$ for $j = 1, \ldots, t$
   - Compute $A_{nk}^* = (A_{n0}^*)^{\lambda_{k0}} \cdot \prod_{i=1}^{t} (g^{s_{ni}^*})^{\lambda_{ki}}$ for $k = 1, \ldots, t$, where λ_{ki}'s are the Lagrange interpolation coefficients.

   (a) Broadcast A_{ik} for $i \in \mathcal{G} \setminus \{n\}$, and A_{nk}^* for $k = 0, \ldots, t$

   (b) Perform for each uncorrupted player the verifications of Eq. 3 on the values $A_{ik}, i \in \mathcal{B}$, broadcast by the players controlled by the adversary. If the verification fails for some $i \in \mathcal{B}, j \in \mathcal{G}$, broadcast a complaint (s_{ij}, s_{ij}'). (Notice that the corrupted players can publish a valid complaint only against one another.)

   (c) Perform Step 4c of the protocol on behalf of the uncorrupted parties, i.e. perform reconstruction phase of Pedersen-VSS to compute z_i and y_i in the clear for every P_i against whom a valid accusation was broadcast in the previous step.

Fig. 3. Simulator for the shared key generation protocol DKG

Since here we are interested in runs of DKG that end with the value y as the public key output of the protocol, we note that the above distribution of values is induced by the choice (of the good players) of polynomials $f_i(z), f_i'(z), i \in \mathcal{G}$, uniformly distributed in the family of t-degree polynomials over $\mathbb{Z}_q$ subject to the condition that

$$\prod_{i \in QUAL} A_{i0} = y \bmod p . \tag{4}$$

In other words, this distribution is characterized by the choice of polynomials $f_i(z), f_i'(z)$ for $i \in (\mathcal{G} \setminus \{n\})$ and $f_n'(z)$ as random independent t-degree polynomials over $\mathbb{Z}_q$, and of $f_n(z)$ as a uniformly chosen polynomial from the family of t-degree polynomials over $\mathbb{Z}_q$ that satisfy the constraint $f_n(0) = dlog_g(y) - \sum_{i \in (QUAL \setminus \{n\})} f_i(0) \bmod q$. (This last constraint is necessary and sufficient to guarantee Eq. (4).) Note that, using the notation of values computed by SIM in Step 2 of the simulation, the last constraint can be denoted as $f_n(0) = dlog_g(A_{n0}^*)$.

We show that the simulator SIM outputs a probability distribution which is *identical* to the above distribution. First note that the above distribution depends on the set $QUAL$ defined at the end of Step 2 of the protocol. Since all the simulator's actions in Step 1 of the simulator are identical to the actions of honest players interacting with $\mathcal{A}$ in a real run of the protocol, thus we are assured that the set $QUAL$ is defined at the end of this simulation step identically to its value in the real protocol. We now describe the output distribution of SIM in terms of t-degree polynomials f_i^* and $f_i'^*$ corresponding to the choices of the simulator when simulating the actions of the honest players and defined as follows: For $i \in \mathcal{G} \setminus \{n\}$, set f_i^* to f_i and $f_i'^*$ to f_i'. For $i = n$, define f_n^* via the values[1] $f_n^*(0) = dlog_g(A_{n0}^*)$ and $f_n^*(j) = s_{nj}^* = f_n(j), j = 1, \dots, t$. Finally, the polynomial $f_n'^*$ is defined via the relation: $f_n^*(z) + d \cdot f_n'^*(z) = f_n(z) + d \cdot f_n'(z) \bmod q$, where $d = dlog_g(h)$. It can be seen that by this definition that the values of these polynomials evaluated at the points $j \in \mathcal{B}$ coincide with the values $f_i(j), f_i'(j)$ which are seen by the corrupted parties in Step 1 of the protocol. Also, the coefficients of these polynomials agree with the exponentials C_{ik} published by the simulated honest parties in Step 1 of the protocol (i.e. $C_{ik} = g^{a_{ik}^*} h^{b_{ik}^*}$ where a_{ik}^* and b_{ik}^* are the coefficients of polynomials $f_i^*(z), f_i'^*(z)$, respectively, for $i \in \mathcal{G}$), as well as with the exponentials $A_{ik}, i \in \mathcal{G} \setminus \{n\}$ and A_{nk}^* published by the simulator in Step 2a on behalf of the honest parties (i.e. $A_{ik} = g^{a_{ik}^*}$, $i \in \mathcal{G} \setminus \{n\}$ and $A_{nk}^* = g^{a_{nk}^*}, k = 0, \dots, t$) corresponding to the players' values in Step 4a of the protocol. Thus, these values pass the verifications of Eq. (2) and (3) as in the real protocol.

It remains to be shown that polynomials f_i^* and $f_i'^*$ belong to the right distribution. Indeed, for $i \in \mathcal{G} \setminus \{n\}$ this is immediate since they are defined identically to f_i and f_i' which are chosen according to the uniform distribution.

---

[1] Note that in this description we use discrete log values unknown to the simulator; this provides a mathematical description of the output distribution of SIM useful for our analysis but does not require or assume that SIM can compute these values.

For f_n^* we see that this polynomial evaluates in points $j = 1, \ldots, t$ to random values (s_{nj}) while at 0 it evaluates $dlog_g(A_{n0}^*)$ as required to satisfy Eq. 4. Finally, polynomial $f_n'^*$ is defined (see above) as $f_n'^*(z) = d^{-1} \cdot (f_n(z) - f_n^*(z)) + f_n'(z)$ and since $f_n'(z)$ is chosen in Step 1 as a random and independent polynomial then so is $f_n'^*(z)$.

4.2 Remarks

Efficiency. We point out that our secure protocol does not lose much in efficiency with respect to the previously known insecure Joint-Feldman protocol. Instead of Feldman-VSS, each player performs Pedersen-VSS (Steps 1-3), which takes the same number of rounds and demands at most twice more local computation. The extraction of the public key in Step 4 adds only two rounds (one if no player is dishonest) to the whole protocol. We point out that all the long modular exponentiations needed during this extraction have already been computed during the Pedersen-VSS phase, thus Step 4 is basically "for free" from a computational point of view.

Generation of h. The public value h needed to run Pedersen's VSS can be easily generated jointly by the players. Indeed it is important that nobody knows the discrete log of h with respect to g. The procedure for generating h consists of a generic distributed coin flipping protocol which generates a random value $r \in Z_p^*$. To generate a random element in the subgroup generated by g it will be enough to set $h = r^k \bmod p$ where $k = (p-1)/q$. If q^2 does not divide $p-1$ (which is easily checkable) then h is an element in the group generated by g.

4.3 Other Applications of a DKG Protocol

DKG protocols have more applications than just key generation. We sketch here two of these applications where previous flawed DKG protocols were used and for which our solution can serve as a secure plug-in replacement.

Randomizers in ElGamal/DSS Threshold Signatures. Signature schemes based on variants of the ElGamal scheme [ElG85], such as DSS, usually consist of a pair (r, s) where $r = g^k$ for a random value $k \in Z_q$. Several robust threshold versions of such signature schemes have been proposed in the literature [CMI93, GJKR96, PK96]. In these schemes the public value r and the sharing of the secret value k is jointly generated by the players running a DKG protocol. Clearly, in order for the resulting threshold scheme to be identical to the centralized case, r must be uniformly distributed in the group generated by g. However, each of these papers uses a version of the Joint-Feldman protocol which allows an adversary to bias the distribution of r. Our DKG protocol fixes this problem.

Refresh Phase in Proactive Secret Sharing and Signature Schemes. Proactive secret sharing [HJKY95] and signature schemes [HJJ+97] were introduced to cope with mobile adversaries who may corrupt more than t servers during the lifetime of the secret. In these protocols time is divided into stages, with an assumption that the adversary may corrupt at most t servers in each stage. However in different stages the adversary can control different players. In order to cope with such adversaries the basic idea of proactive secret sharing is to "refresh" the shares at the beginning of each stage so that they will be independent from shares in previous stages, except for the fact that they interpolate to the same secret. This is achieved by the players jointly creating a random polynomial $f(z)$ of degree t with free term 0 such that each player P_i holds $f(i)$. If the share of player P_i at the previous stage was s_i, the new share will be $s_i + f(i)$. In order to generate $f(z)$ the players run a variation of Joint-Feldman where each player shares value $z_i = 0$. The polynomial $f(z)$ is the sum of the polynomials $f_i(z)$ picked by each player (see Figure 1). It should be clear that the same attack described in Section 3.2 to bias the free term of $f(z)$ can be carried out to bias its any other coefficient. The result is that the polynomial $f(z)$ generated by this refresh phase is not truly random, which implies that shares from different stages are not independent. Our DKG protocol fixes this problem as well.

5   Enhanced Security: Partially Synchronous Model

In the design of distributed cryptographic protocols it is often assumed that the message delivery is fully synchronous (see Section 2). This assumption is unrealistic in many cases where only *partially synchronous* message delivery is provided (e.g. the Internet). By *partially synchronous* communication model we mean that the messages sent on either a point-to-point or the broadcast channel are received by their recipients within some fixed time bound. A failure of a communication channel to deliver a message within this time bound can be treated as a failure of the sending player. While messages arrive in this partially synchronous manner, the protocol as a whole proceeds in synchronized rounds of communication, i.e. the honest players start a given round of a protocol at the same time. To guarantee this round synchronization, and for simplicity of discussion, we assume that the players are equipped with synchronized clocks.

Notice that in a partially synchronous communication model all messages can still be delivered relatively fast, in which case, in every round of communication, the malicious adversary can wait for the messages of the uncorrupted players to arrive, then decide on his computation and communication for that round, and still get his messages delivered to the honest parties on time. Therefore we should always assume the worst case that the adversary speaks last in every communication round. In the cryptographic protocols literature this is also known as a *rushing* adversary.

Clearly the fully synchronous communication model is strictly stronger than the partially synchronous one, thus the previously existing DKG protocols which we recalled in Section 3 remain *insecure* also in this model. In fact, the relaxation of the model allows stronger attacks against many of the Joint-Feldman variants. For example, the adversary could choose the z_i's of the dishonest players dependent on the ones chosen by the honest ones (while in the fully synchronous model he is restricted to deciding whether the previously decided z_i's of the dishonest players will be "in" or "out" of the computation).

In contrast, the DKG protocol we propose in this paper is *secure* even in this more realistic partially synchronous communication setting. Intuitively, this is because the first stage involves an information-theoretic VSS of the z_i values. Thus the adversary has *no* information about these values and he has to choose the z_i's of the dishonest players in an independent fashion even if he speaks last at each round. When the values $y_i = g^{z_i}$ are revealed, it is too late for the adversary to try to do something as at that point he is committed to the z_i's which are recoverable by the honest players. A formal proof of security of our protocol in this stronger model is identical to the proof presented in Section 4.1. Indeed, it can be easily verified that the proof of security carries over to the partially synchronous communication model basically unchanged.

Extension to Adaptive Adversary. Recently, [CGJ+99] showed a modification of our DKG protocol which is secure against an *adaptive* adversary. In this model the attacker can make its decision of what parties to corrupt at any point during the run of the protocol (while in our model the corrupted parties are fixed in advance before the protocol starts). The only modification to our protocol introduced in [CGJ+99] is in the y-extracting step (Step 4), where they replace our method of publishing $y_i = A_{i0} = g^{z_i}$ values via Feldman-VSS with the following: Each player broadcasts a pair $(A_{i0}, B_{i0}) = (g^{a_{i0}}, h^{b_{i0}})$ s.t. $A_{i0} \cdot B_{i0} = C_{i0} \bmod p$, and proves in zero-knowledge that he knows the discrete logs $DLOG_g(A_{i0})$ and $DLOG_h(B_{i0})$. Proving this ensures that $y_i = g^{z_i}$. If a player fails the proof then his shared value z_i is reconstructed via the Pedersen-VSS reconstruction, as in our DKG protocol.

This modification turns out to suffice to make the protocol secure against an adaptive adversary because it allows the construction of a simulator that, at any point in the simulation, has at most a single "inconsistent player". Namely, there is at most one player that if corrupted will make the simulation fail, while all other corruptions can be handled successfully by the simulator. The way the simulator proceeds is by choosing this "inconsistent player" at random and hoping the attacker will not corrupt him. If it does, the simulation rewinds to a previous state, a new choice of inconsistent player is made, and the simulation continues. It is shown in [CGJ+99] that this brings to the successful end of the simulation in expected polynomial-time.

Acknowledgments. We thank Don Beaver for motivational discussions on this problem.

References

[CGJ+99] R. Canetti, R. Gennaro, S. Jarecki, H. Krawczyk, and T. Rabin. Adaptive Security for Threshold Cryptosystems. Mansuscript, 1999.

[CGS97]   R. Cramer, R. Gennaro, and B. Schoenmakers. A secure and optimally efficient multi-authority election scheme. In *Advances in Cryptology — Eurocrypt '97*, pages 103–118. LNCS No. 1233.

[CMI93]   M. Cerecedo, T. Matsumoto, and H. Imai. Efficient and secure multiparty generation of digital signatures based on discrete logarithms. *IEICE Trans. Fundamentals*, E76-A(4):532–545, 1993.

[DF89]   Yvo Desmedt and Yair Frankel. Threshold cryptosystems. In *Advances in Cryptology — Crypto '89*, pages 307–315. LNCS No. 435.

[ElG85]   T. ElGamal. A Public Key Cryptosystem and a Signature Scheme Based on Discrete Logarithms. *IEEE Trans. Info. Theory*, IT 31:469–472, 1985.

[Fel87]   P. Feldman. A Practical Scheme for Non-Interactive Verifiable Secret Sharing. In *Proc. 28th FOCS*, pages 427–437.

[FGMY97]   Y. Frankel, P. Gemmell, P. Mackenzie, and M. Yung. Optimal resilience proactive public-key cryptosystems. In *Proc. 38th FOCS*, pages 384–393. IEEE, 1997.

[GJKR96]   R. Gennaro, S. Jarecki, H. Krawczyk, and T. Rabin. Robust threshold DSS signatures. In *Advances in Cryptology — Eurocrypt '96*, pages 354–371. LNCS No. 1070.

[GJKR99]   R. Gennaro, S. Jarecki, H. Krawczyk, and T. Rabin. Secure Distributed Key Generation for Discrete-Log Based Cryptosystems http://www.research.ibm.com/security/dkg.ps

[Har94]   L. Harn. Group oriented (t, n) digital signature scheme. *IEE Proc.-Comput.Digit.Tech*, 141(5):307–313, Sept 1994.

[HJJ+97]   A. Herzberg, M. Jakobsson, S. Jarecki, H. Krawczyk, and M. Yung. Proactive public key and signature systems. In *1997 ACM Conference on Computers and Communication Security*, 1997.

[HJKY95]   A. Herzberg, S. Jarecki, H. Krawczyk, and M. Yung. Proactive secret sharing, or: How to cope with perpetual leakage. In *Advances in Cryptology — Crypto '95*, pages 339–352. LNCS No. 963.

[LHL94]   C.-H. Li, T. Hwang, and N.-Y. Lee. (t, n) threshold signature schemes based on discrete logarithm. In *Advances in Cryptology — Eurocrypt '94*, pages 191–200. LNCS No. 950.

[Ped91a]   T. Pedersen. A threshold cryptosystem without a trusted party. In *Advances in Cryptology — Eurocrypt '91*, pages 522–526. LNCS No. 547.

[Ped91b]   T. Pedersen. Non-interactive and information-theoretic secure verifiable secret sharing. In *Advances in Cryptology — Crypto '91*, pages 129–140. LNCS No. 576.

[PK96]   C. Park and K. Kurosawa. New ElGamal Type Threshold Digital Signature Scheme. *IEICE Trans. Fundamentals*, E79-A(1):86–93, January 1996.

[Sch91]   C. P. Schnorr. Efficient signature generation by smart cards. *Journal of Cryptology*, 4:161–174, 1991.

[SG98]   V. Shoup and R. Gennaro. Securing threshold cryptosystems against chosen ciphertext attack. In *Advances in Cryptology — Eurocrypt '98*, pages 1–16. LNCS No. 1403.

[Sha79]   A. Shamir. How to Share a Secret. *Communications of the ACM*, 22:612–613, 1979.

Efficient Multiparty Computations
Secure Against an Adaptive Adversary

Ronald Cramer[1], Ivan Damgård[2], Stefan Dziembowski[2], Martin Hirt[1], and
Tal Rabin[3]

[1] ETH Zurich[†]
{cramer,hirt}@inf.ethz.ch
[2] Aarhus University, BRICS[‡]
{ivan,stefand}@daimi.aau.dk
[3] IBM T.J.Watson Research Center
talr@watson.ibm.com

Abstract. We consider verifiable secret sharing (VSS) and multiparty
computation (MPC) in the secure-channels model, where a broadcast
channel is given and a non-zero error probability is allowed. In this model
Rabin and Ben-Or proposed VSS and MPC protocols secure against an
adversary that can corrupt any minority of the players. In this paper, we
first observe that a subprotocol of theirs, known as weak secret sharing
(WSS), is not secure against an adaptive adversary, contrary to what was
believed earlier. We then propose new and adaptively secure protocols
for WSS, VSS and MPC that are substantially more efficient than the
original ones. Our protocols generalize easily to provide security against
general Q^2-adversaries.

1 Introduction

Since the introduction of multiparty computation [Yao82, GMW87], its design
and analysis has attracted many researchers, and has generated a large body
of results. The problem stated very roughly is the following: Consider a set of
players each holding a private input, who wish to compute some agreed upon
function of their inputs in a manner which would preserve the secrecy of their
inputs. They need to carry out the computation even if some of the players
may become corrupted and actively try to interfere with the computation. Solutions to this problem have been given in various models and under different
computational assumptions.

One of the major components of the model is the type of adversary which
is assumed. The adversary is the entity which corrupts a set (of size up to t)
of players during the execution of the protocol and takes control of their actions. Two types of adversaries have been considered in the literature (barring

[†] Supported by the Swiss National Science Foundation (SNF), SPP 5003-045293.
[‡] Basic Research in Computer Science, center of the Danish National Research Foundation.

J. Stern (Ed.): EUROCRYPT'99, LNCS 1592, pp. 311–326, 1999.

slight variations): *static adversaries* and *adaptive adversaries*. The static adversary needs to choose the set of corrupted players before the execution of the protocol. The adaptive adversary on the other hand can choose the players during the execution of the protocol. It has been stated that the protocols of [BGW88, CCD88, RB89, Bea91] are secure against an adaptive adversary under the assumption that the players communicate via secure private channels.[1] In all these results the protocols are information theoretically secure. This has led many to believe that if a protocol is designed which is information theoretically secure and is executed in a model with private channels then the resulting protocol is immediately secure against an adaptive adversary. In the attempt to further our understanding of the power of these different adversaries we present an example of a natural protocol (which appears in [RB89]) which is information theoretically secure against a static adversary but fails against an adaptive adversary.

Another important goal in the design of these protocols is to provide protocols which are simple, so that they could actually be implemented in practice. For the case where the adversary can corrupt at most a third of the players reasonable protocols have been proposed [BGW88], but for the case where the adversary can corrupt a half of the players the existing solutions were quite cumbersome [RB89, Bea91]. In this paper we present solutions for multiparty computation (and for verifiable secret sharing) which are much more efficient than any existing protocol for the case where the adversary can corrupt up to a minority of the players.

More specifically we obtain a protocol for VSS which for probability of error $2^{-k+O(\log n)}$ with n players, requires $O((k + \log n)n^3)$ bits of communication as opposed to $\Omega((k + \log n)k^2 n^4)$ bits required by existing protocols. This improvement is based in part on a more efficient implementation of *information checking protocol*, a concept introduced in [RB89] which can be described very loosely speaking as a kind of unconditionally secure signature scheme. Our implementation is linear meaning that for two values that can be verified by the scheme, any linear combination of them can also be verified with no additional information. This means that linear computations can be done non-interactively when using our VSS in MPC, contrary to the implementation of [RB89] (this property was also obtained in [Bea91], but with a less efficient information checking implementation).

An essential tool in MPC (provided in both [RB89] and [Bea91]) is a protocol that allows a player who has committed, in some manner, to values a, b, and c to show that $ab = c$ without revealing extra information. We provide a protocol for this purpose giving error probability 2^{-k} which is extremely simple. It allows a multiplication step in the MPC protocol to be carried out at cost equivalent to $O(n)$ VSS's, where all earlier protocols required $O(kn)$ VSS's.

Using methods recently developed in [CDM99], our protocols generalize easily to provide security against general Q^2-adversaries [HM97].

---

[1] The transformation of such protocols to the public channel model is outside the scope of this paper, but the interested reader can refer to [BH92, CFGN96].

Outline

We first show that the weak secret sharing (WSS) scheme of [RB89, Rab94] is not adaptively secure (Section 3). In Section 4, we propose an efficient implementation of information checking, and in Section 5, a scheme for verifiable secret sharing (VSS) is developed. Based on these protocols, in Section 6 an efficient protocol for multiparty computation (MPC) is presented. Finally, in Section 7 an efficient protocol secure against general (non-threshold) adversaries is sketched.

2 Model and Definitions

In this paper, we consider the *secure-channels model with broadcast*, i.e. there are n players $P_1, \ldots, P_n$ who are pairwise connected with perfectly private and authenticated channels, and there is a broadcast channel. There is a *central adversary* with unbounded computing power who actively corrupts up to t players where $t < n/2$. To actively corrupt a player means to take full control over that player, i.e. to make the player (mis)behave in an arbitrary manner. The adversary is assumed to be *adaptive* (or dynamic), this means that he is allowed to corrupt players during the protocol execution (and his choice may depend on data seen so far), in contrast to a static adversary who only corrupts players before the protocol starts. The security of the presented protocols is unconditional with some *negligible error probability*, which is expressed in terms of a *security parameter* k. The protocols operate in a finite field $K = GF(q)$, where $q > \max(n, 2^k)$.

2.1 Definition of Information Checking

Information checking (IC) is an information theoretically secure method for authenticating data. An IC scheme consists of three protocols:

Distr(D, INT, R, s) is initiated by the dealer D. In this phase D hands the secret s to the intermediary INT and some auxiliary data to both INT and the recipient R.

AuthVal(INT, R, s) is initiated by INT and carried out by INT and R. In this phase INT ensures that in the protocol RevealVal R (if honest) will accept s, the secret held by INT.

RevealVal(INT, R, s') is initiated by INT and carried out by INT and R. In this phase R receives a value s' from INT, along with some auxiliary data, and either accepts s' or rejects it.

The IC scheme has the following properties:

Correctness:

A. If D, INT, and R are uncorrupted, and D has a secret s then R will accept s in phase RevealVal.

B. If INT and R are honest then after the phases Distr and AuthVal INT knows a value s such that R will accept s in the phase RevealVal (except with probability 2^{-k}).

C. If D and R are uncorrupted, then in phase RevealVal with probability at least $1 - 2^{-k}$, player R will reject every value s' different from s.

Secrecy:

D. The information that D hands R in phase Distr is distributed independently of the secret s. (Consequently, if D and INT are uncorrupted, and INT has not executed the protocol RevealVal, R has no information about the secret s.)

Definition 1. *An IC scheme is a triple* (Distr, AuthVal, RevealVal) *of protocols that satisfy the above properties A. to D.*

2.2 Definition of WSS

An intuitive explanation for a *weak secret-sharing (WSS) scheme* is that it is a distributed analog of a computational commitment. A WSS scheme for sharing a secret $s \in K$ consists of the two protocols Sh and Rec. WSS exhibits the same properties, i.e. it binds the committer to a single value after the sharing phase Sh (this is equivalent to the commitment stage in the computational setting), yet the committer can choose not to expose this value in the reconstruction phase Rec (which is equivalent to the exposure of the commitments). WSS satisfies the following properties, with an allowed error probability 2^{-k}:

- *Termination*: If the dealer D is honest then all honest players will complete Sh, and if the honest players invoke Rec, then each honest player will complete Rec.
- *Secrecy*: If the dealer is honest and no honest player has yet started Rec, then the adversary has no information about the shared secret s.
- *Correctness*: Once all currently uncorrupted players complete protocol Sh, there exists a *fixed* value, $r \in K \cup \{NULL\}$, such that the following requirements hold:
  1. If the dealer is uncorrupted throughout protocols Sh and Rec then r is the shared secret, i.e. $r = s$, and each uncorrupted player will outputs r at the end of protocol Rec.
  2. If the dealer is corrupted then each uncorrupted player outputs either r or NULL upon completing protocol Rec.

Definition 2. *A t-secure WSS scheme for sharing a secret $s \in K$ is a pair* (Sh, Rec) *of two protocols that satisfy the above properties even in the presence of an active adversary who corrupts up to t players.*

2.3   Definition of VSS

An important protocol, which is widely used for multiparty computation, is verifiable secret sharing (VSS) [CGMA85]. In essence a VSS scheme allows a *dealer* to share a secret among n players in such a way that the adversary that corrupts at most t of the players, obtains no information about the secret. Furthermore, the secret can be efficiently reconstructed, even if the corrupted players try to disrupt the protocol. A more formal definition is the following:

A pair (Sh, Rec) of protocols is a *verifiable secret-sharing (VSS) scheme* if it satisfies a stronger correctness property, with an allowed error probability 2^{-k}:

- *Correctness*: Once all currently uncorrupted players complete protocol Sh, there exists a *fixed* value, $r \in K$, such that the following requirements hold:
  1. If the dealer is uncorrupted throughout protocol Sh then r is the shared secret, i.e. $r = s$, and each uncorrupted player outputs r at the end protocol Rec.
  2. If the dealer is corrupted then each uncorrupted player outputs r upon completing protocol Rec.

Definition 3. *A t-secure VSS scheme for sharing a secret $s \in K$ is a pair* (Sh, Rec) *of two protocols that satisfy the termination and the secrecy property of WSS, and the above, stronger, correctness property, even in the presence of an active adversary who corrupts up to t players.*

2.4   Definition of MPC

The goal of multiparty computation (MPC) is to evaluate an agreed function $g : K^n \rightarrow K$, where each player provides one input and receives the output. The privacy of the inputs and the correctness of the output is guaranteed even if the adversary corrupts any t players. For a formal definition for security see [GL90, MR91, Bea91, Can98, MR98].

3   Adaptive Security of WSS in [RB89]

In this section we describe a protocol which is secure against a static adversary yet fail against an adaptive one. The example captures nicely the power of the adaptive adversary to delay decisions and due to that cause different values to be computed during the protocol. The protocol which we examine is the weak secret-sharing scheme (WSS) of Rabin and Ben-Or [RB89, Rab94]. The attack will only work when $t > n/3$. It is important to note that this attack applies only to the WSS protocol of [RB89] as a stand-alone protocol, and does not apply to their VSS scheme, although it uses the WSS as a subprotocol.

In order to explain the attack we present a simplified protocol of the [RB89] protocol which assumes digital signatures. It is in essence the same protocol but with many complicating (non relevant) details omitted.

WSS Share (Sh)

The dealer chooses a random polynomial $f(x)$ of degree t, such that $f(0) = s$ the secret to be shared, and sends the share $s_i = f(i)$ with his signature for s_i to each player P_i.

WSS Reconstruct (Rec)

1. Every player reveals his share s_i and the signature on s_i.
2. If *all* properly signed shares $s_{i1}, \ldots, s_{ik}$ for $k \geq t$ interpolate a single polynomial $f'(x)$ of degree at most t, then the secret is taken to be $f'(0)$, otherwise no secret is reconstructed.

The definition of WSS requires that at the end of Sh a single value $r \in K \cup \{\text{NULL}\}$ is set so that only that value (or NULL) will be reconstructed in Rec.

Clearly, if the adversary is static then the value r is set to the value interpolated through the shares held by the uncorrupted players. This value is well defined. If there exists a polynomial $f'(x)$ of degree t then $r = f'(0)$ otherwise r is NULL. During reconstruction if r was NULL then the players will set the output to NULL as all the shares of the good players will be considered in the interpolation and possibly some additional shares from the corrupted players. If r was not NULL then either the additional shares provided by the faulty players satisfy the polynomial $f'(x)$ in which case r will be reconstructed. But the adversary can decide to foil the reconstruction by having the corrupted players supply shares which do not match $f'(x)$, but this will only cause the players to output NULL but not another value $r' \neq r$.

Yet, we will show that under an adaptive adversary this requirement does not hold in the above described protocol. The attack for $n = 2t + 1$ proceeds as follows: In the protocol Sh the adaptive adversary corrupts the dealer causing him to deviate from the protocol. The dealer chooses two polynomials $f_1(x)$ and $f_2(x)$ both of degree at most t, where $f_1(0) \neq f_2(0)$, and $f_1(i) = f_2(i)$ for $i = 1, 2, 3$. For $i = 1, \ldots, 3$, player P_i receives the value $f_1(i)$ $(=f_2(i))$ as his share, for $i = 4, \ldots, t + 2$, player P_i receives $f_1(i)$, and for $i = t + 3, \ldots, 2t + 1$, player P_i receives $f_2(i)$ as his share. All shares are given out with valid signatures.

In Rec the adversary can decide whether to corrupt $P_4, \ldots, P_{t+2}$ thus forcing the secret to be $f_2(0)$, or to corrupt $P_{t+3}, \ldots, P_{2t+1}$ and thus force the secret to be $f_1(0)$. Hence it is clear that at the end of Sh there is not a *single* value which can be reconstructed in Rec. The decision on which value to reconstruct can be deferred by the adversary until the reconstruction protocol Rec is started.

Therefore the basic problem with stand-alone WSS is that it is not ensured that all honest players are on the same polynomial immediately after distribution. But when using it inside the VSS of [RB89], this property is ensured as a side effect of the VSS distribute protocol, hence the VSS protocol works correctly.

4   The Information Checking Protocol

In this section we present protocols that satisfy Definition 1 for information checking (cf. Section 2.1). They provide the same functionality as the check vector protocol from [RB89, Rab94] and the time capsule protocol from [Bea91]. However, our implementation of information checking also possesses an additional linearity property which will be utilized later in the paper.

The basic idea for the construction will be that the secret and the verification information will all lie on a polynomial of degree 1 (a line), where the secret will be the value at the origin. The dealer D hands to the intermediary INT two points on this line, and hands to the recipient R one point at a constant, but secret evaluation point α. This α is known to both D and R, but is unknown to INT. We will say that R will accept the secret which INT gives him only if the point which R holds lies on the line defined by the two points he receives from INT.

A general remark before we begin describing our protocols: In the following we adopt (for ease of exposition) the convention that whenever a player expects to receive a message from another player in the next step, and no message arrives, he assumes that some fixed default value was received. Thus we do not have to treat separately the case where no message arrives.

Definition 4. *A vector $(x, y, z) \in K^3$ is 1_α-consistent if there exists a degree 1 polynomial w over K such that $w(0) = x$, $w(1) = y$, $w(\alpha) = z$.*

Protocol Distr(D, INT, R, s):

The dealer D chooses a random value $\alpha \in K \setminus \{0, 1\}$ and additional random values $y, z \in K$ such that (s, y, z) is 1_α-consistent, and in addition he chooses a random 1_α-consistent vector (s', y', z'). D sends s, s', y, y' to the intermediary INT and z, z' to the recipient R.

Protocol Distr (together with RevealVal below) ensures ensures all properties except Property B. Adding the next protocol ensures this as well, without affecting A, C and D.

Protocol AuthVal(INT, R, s):

1. INT chooses a random element $d \in K$ and broadcasts $d, s' + ds, y' + dy$. If D observes that these values are incorrect, he broadcasts s, y. This counts as claiming that INT is corrupt. In this case the protocol ends here, and the broadcasted values will be used in the following. R will adjust his value for z, such that (s, y, z) is 1_α-consistent.
2. R checks if $(s' + ds, y' + dy, z' + dz)$ is 1_α-consistent. He broadcasts accept or reject accordingly. If D observes that R has acted incorrectly, he broadcasts z, α. This counts as claiming that R is corrupt. In this case the protocol ends here, and the broadcasted values will be used in the following. INT will adjust his value for y, such that (s, y, z) is 1_α-consistent.

3. If R rejected (and D did not claim him faulty) in the previous step, D must broadcast s, y, and the broadcasted values will be used in the following. R will adjust his value for z, such that (s, y, z) is 1_α-consistent.

Protocol RevealVal(INT, R, s):

1. INT broadcasts (s, y).
2. R verifies that (s, y, z) is 1_α-consistent and broadcasts accept or reject accordingly.

Lemma 1. *The protocols* (Distr, AuthVal, RevealVal) *described above satisfy Definition 1 for information checking (Section 2.1).*

Proof. We show that each property is satisfied:

A. It is clear that if all parties are honest, R will accept, and D will never broadcast any values.
B. The property is trivial in the cases where D broadcasts s, y or z, α. So it is enough to show that if D sends an inconsistent (s, y, z) initially, then R rejects with high probability. However, if for $e \neq d$, both $(s'+ds, y'+dy, z'+dz)$ and $(s'+es, y'+ey, z'+ez)$ are 1_α-consistent, then their difference and hence also (s, y, z) is 1_α-consistent. By the random choice of d it follows that R will accept with probability at most $1/|K|$ whenever (s, y, z) is inconsistent.
C. This property will follow from the fact that INT does not know α. Actually, we will show it holds, even if D uses the same α in all invocations of the protocol. We will exploit this property later. First note that INT learns no information on α from the Distr, AuthVal protocols: what he gets in Distr has distribution independent of α. In AuthVal, if he sends correct values, he knows in advance they will be accepted; if he doesn't, he knows that D will complain. Note also that this holds even if we consider many invocations of the authentication protocol together. Thus, all INT knows about α a priori is that it can be any value different from $0, 1$, and all candidates are equally likely.

Consider now the position of INT just before the opening of the first s-value. If he sends the correct s, y, or changes one of the values, he knows in advance R's reaction and so learns nothing new. If he sends s', y' where $s' \neq s, y' \neq y$, then R will accept if (s', y', z) is 1_α-consistent. We know that (s, y, z) is 1_α-consistent by its definition, thus so is $(s - s', y - y', 0)$. This gives a non-trivial degree 1 equation from which α can be computed. In other words, INT must guess α to have R accept a false value. He can do this with probability at most $1/(|K| - 2)$. On the other hand, if R rejects, all INT knows is that the solution to the equation is not the right value, so it can be excluded.

It follows by induction that if at most ℓ values are opened, at least $|K| - \ell - 2$ candidates for α remain from the point of view of INT, and no false values have been accepted, except with probability at most $\ell/(|K| - \ell - 2)$. In the application to VSS, ℓ will be linear in n, so the error probability is at most $2^{-k+O(\log n)}$.

D. If D and INT remain honest and R is corrupt, we must show that R does not learn s ahead of time. Observe that in the authentication protocol, R learns $z, z', d, s' + ds, y' + dy$. Note that since D and INT are honest, R knows in advance that $(s' + ds, y' + dy, z' + dz)$ will be 1_α-consistent. He can therefore compute $y' + dy$ from $z, z', d, s' + ds$, and this value can be deleted from his view without loss of generality. However, it is clear that $z, z', d, s' + ds$ has distribution independent of s.

Linearity of the IC Protocol

In our multiparty computation protocol we would like to be able to authenticate a linear combination of two values. The setting is as follows: D, R and INT have executed both protocols Distr and AuthVal for two different values s_1 and s_2. Now they wish to reveal a linear combination of these two secrets without exposing s_1 and s_2 and without carrying out any additional verification. This can be achieved if for both invocations of the IC protocol the dealer chooses *the same value* α as the random evaluation point which he gives to R. Then all the properties of the protocol still hold with the addition that the appropriate linear combination of the verification data yields a verification for the linear combination of s_1 and s_2.

IC-Signatures

In the sequel we will want to use the information checking protocol as semi "digital signatures". When a person receives a digital signature from a signer, he can later show it to anyone and have that person verify that it is in fact a valid signature. This property can be easily achieved with information checking, by carrying out the protocol with all players as explained bellow. We do not achieve all properties of digital signatures, but enough in order to achieve our goals.

The IC-Signatures will be given in the following way. Protocol Distr will be carried out by the dealer D with intermediary INT and the receiver being each player $P_1, \ldots, P_n$, each with respect to the same value s. Next, the AuthVal protocol will be performed by INT and each player P_i. Then, in protocol RevealVal, INT will broadcast s and the authentication information, and if $t + 1$ players accept the value s then we shall say that the "signature" has been confirmed. We shall call these signatures IC-signatures. These signature enable D to give INT a "signature" which only INT can use to convince the other players about the authenticity of a value received from the dealer. Thus, we use these IC-signatures as signatures given specifically from D to INT, and we denote such a signature as $\sigma_s(D, INT)$.

5   Verifiable Secret Sharing

We now present our simplified VSS protocol. The protocol is based on the bivariate solution of Feldman [FM88, BGW88] (omitting the need for error correcting codes). The protocol will use our new variant of information checking which will provide us with high efficiency.

Definition 5. *A vector* $(e_0, \ldots, e_{n-1}) \in K^n$ *is* t-*consistent if there exists a polynomial* $w(x)$ *of degree at most* t *such that* $w(i) = e_i$ *for* $0 \le i < n$.

The intuition behind the construction is that the secret will be shared using an $n \times n$ matrix of values, where each row and column is t-consistent, and where row and column i is given to player P_i. Thus, for $i \ne j$, P_i and P_j share two values in the matrix. The dealer will commit himself to all the values by signing each entry in the matrix. The row determines by simple interpolation a share of a single variate polynomial. Thus, de facto the dealer has given player P_i a signed share, s_i. The players can now check consistency of the matrix by comparing values between them and expose inconsistent behavior by the dealer using the signatures. Hence we are guaranteed that all the values held by (yet) uncorrupted players are consistent and define a single secret.[2] In order to also have the share of player P_i signed (implicitly) by the other players, player P_i gets the share b_{ij} in his row signed by player P_j. Now this in return will prevent the adversary from corrupting the secret at reconstruction time.

VSS Share (Sh)

1. The dealer D chooses a random bivariate polynomial $f(x, y)$ of degree at most t in each variable, such that $f(0, 0) = s$. Let $s_{ij} = f(i, j)$. The dealer sends to player P_i the values $a_{1i} = s_{1i}, \ldots, a_{ni} = s_{ni}$ and $b_{i1} = s_{i1}, \ldots, b_{in} = s_{in}$. For each value a_{ji}, b_{ij} D attaches a digital signature $\sigma_{a_{ji}}(D, P_i), \sigma_{b_{ij}}(D, P_i)$.
2. Player P_i checks that the two sets $a_{1i}, \ldots, a_{ni}$ and $b_{i1}, \ldots, b_{in}$ are t-consistent. If the values are not t-consistent, P_i broadcasts these values with D's signature on them. If a player hears a broadcast of inconsistent values with the dealer's signature then D is disqualified and execution is halted.
3. P_i sends a_{ji} and a signature which he generates on a_{ij}, $\sigma_{a_{ji}}(P_i, P_j)$ privately to P_j.
4. Player P_i compares the value a_{ij} which he received from P_j in the previous step to the values b_{ij} received from D. If there is an inconsistency, P_i broadcasts $b_{ij}, \sigma_{b_{ij}}(D, P_i)$.
5. Player P_i checks if P_j broadcasted a value $b_{ji}, \sigma_{b_{ji}}(D, P_j)$ which is different than the value a_{ji} which he holds. If such a broadcast exists then P_i broadcasts $a_{ji}, \sigma_{a_{ji}}(D, P_i)$.
6. If for an index pair (i, j) a player hears two broadcasts with signatures from the dealer on different values, then D is disqualified and execution is halted.

VSS Reconstruct (Rec)

1. Player P_i broadcasts the values $b_{i1}, \ldots, b_{in}$ with the signature for value b_{ij} which he received from player P_j. (If he did not receive a signature from P_j in the protocol Sh then he had already broadcasted that value with a signature from D.)

---

[2] So far, this results in a WSS which is secure against an adaptive adversary.

2. Player P_i checks whether player P_j's shares broadcasted in the previous step are t-consistent and all the signatures are valid. If not then P_j is disqualified.
3. The values of all non-disqualified player are taken and interpolated to compute the secret.

Theorem 1. *The above protocols* (Sh, Rec) *satisfy Definition 3 for VSS protocols.*

Proof. We prove that each required property is satisfied:

Secrecy. Observe that in Steps 2–6, the adversary learns nothing that he was not already told in Step 1. Thus the claim follows immediately from the properties of a bi-variate polynomial of degree t and the properties of the information checking.

Termination. From examining the protocol it is clear that the dealer D can be disqualified only if the data which he shared is inconsistent, assuming that the players cannot forge any of the dealers signatures, of which there are $O(n)$. Thus, an honest dealer will be disqualified at most with probability $O(2^{-k+\log n})$.

Correctness. First we will show that a fixed value r is defined by the distribution. Define r to be the secret which interpolates through the shares held by the set of the first $t+1$ players who have not been corrupted during Sh. Their shares are trivially t-consistent, and with probability at least $1-O(2^{-k+\log n})$, there are correct signatures for these shares, and thus they define uniquely an underlying polynomial $f'(x, y)$ as well as a secret $r = f'(0, 0)$. Let us look at another uncorrupted player outside this set. He has corroborated his shares with all these $t + 1$ players and has not found an inconsistency with them. Moreover, this player has also verified that his row and column are t-consistent. Hence, when this player's shares are added to the initial set of players' shares the set remains t-consistent, thus defining the same polynomial f' and secret r. Now we examine the two correctness conditions:
1. It is easy to see that if D is uncorrupted then this value $r = s$.
2. A value different than r will be interpolated (or the reconstruction will fail) only if a corrupted player would be able to introduce values which are inconsistent with the values held by the honest players. A corrupted player succeeded doing it only when he was not disqualified in Step 2. of the reconstruction procedure. This means that he was able to produce a set of n values which are t-consistent, and for each value to have a signature from the appropriate player to which it relates. Clearly, $t+1$ of these signatures must be from still uncorrupted players. We have already shown that these players' shares lie on $f'(x, y)$, thus if the corrupted player's shares are t-consistent they must lie on $f'(x, y)$ as well. Therefore the adversary cannot influence the value of the revealed secret.

$\square$

Efficiency. By inspection of the VSS distribution protocol Sh, one finds that n^2 field elements are distributed from D, and each of these are authenticated using Distr and AuthVal a constant number of times. Executing Distr and AuthVal requires communicating a constant number of field elements for each player, and so we find that the total communication is $O((k + \log n)n^3)$ bits, for an error probability of $2^{-k+O(\log n)}$.

6   Multiparty Computation

Based on the VSS scheme of the previous section, we now build a multiparty computation protocol. Based on the [BGW88] paradigm it is known that it is sufficient to devise methods for adding and multiplying two shared numbers.

Note that in our case (contrary to e.g. [BGW88]) a VSS of a value a consists not only of the shares $a_1, \ldots, a_n$ where a_i is held (in fact implicitly) by P_i, it is explicitly held by P_i via the subshares $a_{i1}, \ldots, a_{in}$ where a_{ij} is held also by player P_j, and P_i has a IC-signature from P_j on that value. This structure and the IC-signatures are required for the reconstruction. Thus, if we wish to compute the sum/multiplication of two secrets we need to have the resultant in this same form.

We will prove the following theorem in the next two subsections.

Theorem 2. *Assume the model with a complete network of private channels between n players and a broadcast channel. Let C be any arithmetic circuit over the field K, where $|K| > max(n, \log k)$ and k is a security parameter. Then there is a multiparty computation protocol for computing C, secure against any adaptive adversary corrupting less than $n/2$ of the players. The complexity of this protocol is $O(n^2|C|)$ VSS protocols with error probability $2^{-k+O(\log n)}$, where $|C|$ is the number of gates in C. This amounts to $O(|C|kn^5)$ bits of communication.*

6.1   Addition

Addition is straightforward: For two secrets a and b shared with (implicit) shares $a_1, \ldots, a_n$ and $b_1, \ldots, b_n$, all the subshares, and their appropriate IC-signatures, each player P_i needs to add his two (implicit) shares a_i and b_i which means that he needs to hold a IC-signature from P_j for $a_{ij} + b_{ij}$. But this is immediately achieved as the sum of two IC-signatures results in an IC-signature for the sum of the values signed. Thus, we have computed the addition of two shared secrets.

6.2   Multiplication

Multiplication is slightly more involved. Assume that we have two secrets a and b with (implicit) shares $a_1, \ldots, a_n$ and $b_1, \ldots, b_n$ and all the subshares and their appropriate IC-signatures. We apply the method from [GRR98]. This method calls for every player to multiply his shares of a, resp. b and to share the result of this using VSS. This results in n VSS's and a proper sharing of the result c can be computed as a fixed linear combination of these (i.e. each player computes a linear combination of his shares from the n VSS's). Since our VSS is linear, like the one used in [GRR98], the same method will work for us, provided we can show that player P_i can share a secret c_i using VSS, such that it will hold that $c_i = a_i b_i$ and to prove that he has done so properly. If P_i fails to complete this process the simplest solution for recovery is to go back to the start of the computation, reconstruct the inputs of P_i, and redo the computation, this time simulating P_i openly. This will allow the adversary to slow down the computation by at most a factor linear in n.

In order to eliminate subindices let us recap our goal stated from the point of view of a player D. He needs to share a secret c using VSS which satisfies that $c = ab$. The value a is shared via subshares $a_1, \ldots, a_n$ (lying on a polynomial f_a, say) where a_i is held by player P_i and D holds an IC-signature of this value from P_i. The same holds for the value b (with a polynomial f_b).

1. D shares the value $c = ab$ using the VSS Share protocol. Let f_c be the polynomial defined by this sharing.[3]
2. D chooses a random $\beta \in K$ and he shares β and βb. The sharing of these values is very primitive. D chooses a polynomial $f_\beta(x) = \beta_t x^t + \ldots + \beta_1 x + \beta$ and gives player P_i the value $f_\beta(i)$ and an IC-signature on this value. A player complains if he did not receive a share and a signature, and the dealer exposes these values. The same is done for βb (with a polynomial $f_{\beta b}$).
3. The players jointly generate, using standard techniques, a random value r, and expose it.
4. D broadcast the polynomial $f_1(x) = r f_a(x) + f_\beta(x)$.
5. Player P_i checks that the appropriate linear combination of his shares lies on this polynomial, if it does not he exposes his signed share $f_\beta(i)$ and requires the dealer to expose the IC-signature which the dealer holds generated by P_i for the value a_i. If the dealer fails then D is disqualified.
6. If the dealer has not been disqualified each player locally computes $r_1 = f_1(0)$.
7. D broadcasts the polynomial $f_2(x) = r_1 f_b(x) - f_{\beta b}(x) - r f_c(x)$.
8. Each player checks that the appropriate linear combination of his shares lies on this polynomial, if it does not he exposes his signed share $f_{\beta b}(i)$ and $f_c(i)$ and requires the dealer to expose the IC-signature which the dealer holds generated by P_i for the value b_i. If the dealer fails then D is disqualified.
9. If D has not been disqualified P_i verifies that $f_2(0) = 0$, and accepts the sharing of c, otherwise D is disqualified.

The security of the protocol is guaranteed by the following lemma.

Lemma 2. *Executing the above protocol for sharing $c = ab$ does not give the adversary any information that he did not know before.*

Proof. Wlog we can assume that the dealer is honest. Thus all the values revealed during the protocol look random to the adversary (except for the polynomial f_2 which is a random polynomial such that $f_2(0) = 0$). Therefore the adversary learns nothing. □

---

[3] Note that f_c is not the bivariate polynomial directly constructed by D rather it is the univariate polynomial defined by the implicit shares of c.

Lemma 3. *If $c \neq ab$ in the above protocol, then the probability that the dealer succeeds to perform the above is at most $\frac{1}{|K|}$.*

Proof. Suppose there exist two distinct challenges r_1 and r_1' such that if any of them is chosen in Step 3. then D is not disqualified in the next rounds. Step 4. guarantees that honest players have consistent shares of f_β, since we open f_1 and we know f_a is consistent. So there is a well-defined value β shared by f_β. In the same way Step 7 guarantees that $f_{\beta b}$ is consistent, so it defines some value z (which may or may not be βb). Now from Step 4., $r_1 = ra + \beta$ and $r_1' = r'a + \beta$, so from Step 7., we get $(ra + \beta)b + z + rc = 0 = (r'a + \beta)b + z + r'c$ and we conclude that $ab = c$.   □

7   General Adversaries

It is possible to go beyond adaptive security against any dishonest *minority*, by considering *general*, i.e. not necessarily threshold *adversaries* [HM97]. The corruption capability of such an adversary is specified by a family of subsets of the players, where the adversary is restricted to corrupting one of these sets - dishonest minority is clearly a special case. Our results in this paper extend to the general scenario, following ideas developed in [CDM99].

First, by replacing Shamir secret sharing by monotone span program (MSP) secret sharing [KW93] in our VSS, we immediately obtain WSS protocols secure against any Q^2-adversary [HM97], with communication and computation polynomial in the monotone span program complexity of the adversary [CDM99]. A Q^2-adversary is an adversary who is capable of corrupting only subsets of players in a given family of subsets, where no two subsets in the family together cover the full player set.

The reason why the generalized protocol is only a WSS and not a VSS is that for a general linear secret sharing scheme, a qualified subset of shares define uniquely the secret, but NOT necessarily the entire set of shares (in contrast with what is the case for Shamir's threshold scheme).

However, building on the linearity of this WSS and monotone span program secret sharing, we can still construct efficient VSS (with negligible, but non-zero error) secure against any Q^2-adversary.

Roughly speaking, the idea (taken from [CDM99]) is that the dealer will use WSS to commit to his secret and the set of shares. He can then convince the players that this was done correctly. This amounts to showing a number of linear relations on committed values, which is easy by linearity of the WSS. Finally, each commitment to a share is privately opened to the player that is to receive it.

The resulting VSS enables multi-party computation secure against any Q^2-adversary if we base the construction of VSS on a so called MSP with multiplication [CDM99]. Such an MSP always exists, and can be chosen to have size at most twice that of a minimal MSP secure against the adversary. As far as general adversaries are concerned, security against Q^2-adversaries is the maximum attainable level of security.

This construction gives a VSS with complexity $O((k + \log n)nm^3)$ bits, where m is the size of the monotone span program. In some independent work Smith and Stiglic[SS98] present a somewhat similar idea, which however results in a less efficient protocol ($O(k^2(k + \log n)nm^3)$ bits) because they directly apply the ideas from [CDM99] to [Rab94], i.e. replace in [Rab94] Shamir's secret sharing by the monotone span programs with multiplication from [CDM99].

Acknowledgment. We are very grateful to Adam Smith and Anton Stiglic for pointing out an error in the information checking protocols of the almost final version of this paper.

References

[Bea91]   D. Beaver. Secure multiparty protocols and zero-knowledge proof systems tolerating a faulty minority. *Journal of Cryptology*, 4:75–122, 1991.

[BGW88]   M. Ben-Or, S. Goldwasser, and A. Wigderson. Completeness theorems for noncryptographic fault-tolerant distributed computations. In *20th STOC*, pp. 1–10. ACM, 1988.

[BH92]   D. Beaver and S. Haber. Cryptographic protocols provably secure against dynamic adversaries. *Eurocrypt '92*, pp. 307–323. Springer LNCS 658, 1992.

[Can98]   R. Canetti. Security and composition of multiparty cryptographic protocols. Manuscript, to appear, 1998.

[CCD88]   D. Chaum, C. Crepeau, and I. Damgård. Multiparty unconditionally secure protocols. In *20th STOC*, pp. 11–19. ACM, 1988.

[CGMA85]   B. Chor, S. Goldwasser, S. Micali, and B. Awerbuch. Verifiable secret sharing and achieving simultaneity in the presence of faults. In *26th FOCS*, pp. 383–395. IEEE, 1985.

[CDM99]   R. Cramer, I. Damgård, and U. Maurer. General secure multi-party computation from any linear secret-sharing scheme. Manuscript, 1999.

[CFGN96]   Ran Canetti, Uri Feige, Oded Goldreich, and Moni Naor. Adaptively secure multi-party computation. In *28th STOC*, pp. 639–648. ACM, 1996.

[FM88]   P. Feldman and S. Micali. An optimal algorithm for synchronous Byzantine agreement. In *20th STOC*, pp. 148–161. ACM, 1988.

[GL90]   S. Goldwasser and L. Levin. Fair computation of general functions in presence of immoral majority. *Crypto '90*, pp. 77–93. Springer LNCS 537, 1990.

[GMW87]   O. Goldreich, S. Micali, and A. Wigderson. How to play any mental game. In *19th STOC*, pp. 218–229. ACM, 1987.

[GRR98]   R. Gennaro, M. Rabin, and T Rabin. Simplified VSS and fast-track multiparty computations with applications to threshold cryptography. In *17th PODC*, pp. 101–111. ACM, 1998.

[HM97]   M. Hirt and U. Maurer. Complete characterization of adversaries tolerable in general multiparty computations. In *16th PODC*, pp. 25–34. ACM, 1998.

[KW93]   M. Karchmer and A. Wigderson. On span programs. In *Proc. of Structure in Complexity*, pp. 383–395, 1993.

[MR91]   S. Micali and P. Rogaway. Secure computation. *Crypto '91*, pp. 392–404. Springer LNCS 576, 1991.

[MR98]   S. Micali and P. Rogaway. Secure computation: The information theoretic case. Manuscript, to appear, 1998.

[Rab94]   T. Rabin. Robust sharing of secrets when the dealer is honest or faulty. *Journal of the ACM*, 41(6):1089–1109, 1994.

[RB89]   T. Rabin and M. Ben-Or. Verifiable secret sharing and multiparty protocols with honest majority. In *21st STOC*, pp. 73–85. ACM, 1989.

[SS98]   A. Smith and A. Stiglic. Multiparty computations unconditionally secure against Q^2 adversary structures. Manuscript, 1998.

[Yao82]   A.C. Yao. Protocols for secure computations. In *23rd FOCS*, pp. 160–164. IEEE, 1982.

Distributed Pseudo-random Functions and KDCs

Moni Naor*, Benny Pinkas**, and Omer Reingold* * *

Dept. of Computer Science and Applied Math
Weizmann Institute of Science
Rehovot, Israel
{naor, bennyp, reingold}@wisdom.weizmann.ac.il

Abstract. This work describes schemes for distributing between n servers the evaluation of a function f which is an approximation to a random function, such that only authorized subsets of servers are able to compute the function. A user who wants to compute $f(x)$ should send x to the members of an authorized subset and receive information which enables him to compute $f(x)$. We require that such a scheme is **consistent**, i.e. that given an input x all authorized subsets compute the same value $f(x)$.

The solutions we present enable the operation of many servers, preventing bottlenecks or single points of failure. There are also no single entities which can compromise the security of the entire network. The solutions can be used to distribute the operation of a Key Distribution Center (KDC). They are far better than the known partitioning to domains or replication solutions to this problem, and are especially suited to handle users of multicast groups.

1 Introduction

A single server that is responsible for a critical operation is a performance bottleneck and a single point of failure. A common approach for solving this problem is the use of several replicated servers. However this type of solutions degrades the security of the system if the servers should store secrets (e.g. keys) which are required for cryptographic operations. A solution to both the availability and the security problems is to design a system whose security is not affected if a limited number of servers are broken into (see Section 1.2 for a discussion of the availability and security issues for KDCs).

The problem of distributing the evaluation of trapdoor functions for public key cryptography was extensively investigated (see e.g. [18,17,22,42]). However, the problem of distributing the functions needed for private key cryptography,

---

* Research supported by an infrastructure research grant of the Israeli Ministry of Science.
** Research supported by an Eshkol Fellowship of the Israeli Ministry of Science.
* * * Research supported by a Clore Scholars award and an Eshkol Fellowship of the Israeli Ministry of Science.

J. Stern (Ed.): EUROCRYPT'99, LNCS 1592, pp. 327–346, 1999.
© Springer-Verlag Berlin Heidelberg 1999

in particular the distribution of the evaluation of pseudo-random functions, was neglected (an exception is the work of [31]). Threshold evaluation of random-like functions is required for seemingly unrelated applications, for example for secure and efficient metering of web usage [32], for threshold evaluation of the Cramer-Shoup cryptosystem [13], and for the applications we discuss in this paper (in particular, distributed KDCs and long-term repository for encrypted data). These applications require that the protocol for the collective function evaluation does not invovle communication between the parties which evaluate the function. This requirement is not satisfied by most threshold constructions for public key cryptography.

This work describes schemes for distributing between n servers the evaluation of a function f which is an approximation to a random function, such that only authorized subsets of servers are able to compute the function. A user who wants to compute $f(x)$ should send x to the members of an authorized subset and receive information which enables him to compute $f(x)$. We require that such a scheme is consistent, i.e. that given an input x all authorized subsets compute the same value $f(x)$.

Distributed and consistent evaluation of pseudo-random functions is useful for many applications. The consistency property is especially useful for the following three types of applications:

(i) A distributed KDC system (DKDC), in particular for multicast communication. We describe this application in detail in Section 1.2.

(ii) *Long-tem encryption of information*, where a user might want to encrypt personal information and keep the decryption keys safely distributed between many servers (see Section 1.3).

(iii) A realization of a *Random Oracle* or of a beacon [41] that generates randomness which should be shared by remote parties and used in a cryptographic protocol.

We introduce the notion of a *Distributed Pseudo-Random Function (DPRF)*. We describe several constructions of approximations to random functions which are useful for many of the applications of a DPRF. A *threshold DPRF* (depicted in Figure 1) is a system of n servers such that *any* k of them can compute the function f, but breaking into any $k - 1$ servers does not give any information about f (for instance think of a system with $n = 20$ servers and a threshold of $k = 3$). The servers could be distributed across the network, and a party can contact any k of them in order to compute f. If several parties need to compute f for the same input they are not required to contact the same k servers but rather each party can contact a *different* set of k servers (e.g. those to which it has the best communication channels). Furthermore, to reduce the latency of the computation a party can contact the k servers *in parallel*. We also support DPRFs based on general monotone access structures [7,28,3] rather than on threshold ones. There are several scenarios where general access structures might be preferable to threshold access structures (e.g. to allow efficient implementations of quorum systems [38] which enable fast revocation).

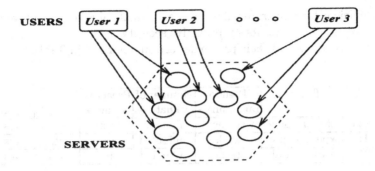

Fig. 1. A Distributed Pseudo-Random Function System.

Our constructions can be further amended to be robust against servers which send incorrect data to users who approach them, (the robustness is based either on error-correcting mechanisms or on proof techniques). The constrcutions can also be further improved to ensure *proactive security* (see [11] and references therein for a general discussion of proactive security), which provides automatic recovery from break-ins: The servers perform some periodic refreshment of their secrets (e.g. once a day), and as a result only an adversary which breaks into k servers in the *same period* can break the security of the system.

1.1 Our Solutions for a Threshold Access Structure

It is unknown how to perform a threshold evaluation of a pseudo-random function without requiring heavy communication between the servers for each given input. Lacking a general construction we describe three different approximations of DPRFs with a threshold access structure.

The first construction generates f as an ℓ-wise independent function. It provides information theoretic security as long as an adversary does not obtain ℓ different values. The scheme is very efficient and requires only multiplications in a *small* finite field (which essentially should only be large enough so that a random element in it can be used as a key for a private-key encryption scheme). The parameter ℓ can therefore be set to be rather large (even several millions).

The second construction is based on a computational assumption: the decisional Diffie-Hellman assumption (see [9]). However the resulting function is only weakly pseudo-random, i.e. it is pseudo-random as long as the inputs on which it is evaluated are pseudo-random. The construction requires a user to compute $O(k)$ exponentiations in order to compute the function's output, and a server should compute only a single exponentiation in order to serve a user. The first two constructions can be easily amended to provide proactive security.

The third construction is based on a monotone CNF formula realizing the threshold k-out-of-n function. This construction computes a full-fledged pseudo-random function and its security depends only on the existence of pseudo-random functions. It can also be adapted to any access structure. Its drawback is that it is only efficient for moderate values of n and small values of k, and we do not know how to enhance it to obtain proactive security.

The constructions and their properties are summarized in Table 1.

	Efficiency	Pseudo randomness	Number of evaluations	Proactive security	Robust.	General access
ℓ-wise ind.	efficient poly	strong	limited	yes	yes	yes
DDH	expensive poly	weak	unlimited	yes	yes	yes
CNF	exponential	strong	unlimited	no	yes	yes

Table 1. A comparison of the threshold schemes.

DPRFs for general access structures: We present constructions of DPRFs based on any monotone access structure. For example, an access structure based on a quorum system allows for fast user revocation by accessing the servers which are members of a single quorum. Our constructions are based either on monotone symmetric branching programs (contact schemes), or on monotone span programs.

1.2   Application to Key Distribution – DKDCs

A Key Distribution Center (KDC): A popular approach for generating common keys between two parties without using public key cryptography is by using a three-party trust model which includes a trusted key distribution center (KDC). In networks which use a KDC there is a dedicated key between the KDC and each of the members of the network. Denote by k_u the key between the KDC and party u. This is the only key that u has to store. Very informally, when two parties (e.g. u and v) wish to communicate, one of them approaches the KDC which then provides a random key, $k_{u,v}$, and sends it to each of the two parties, encrypted with their respective secret keys (i.e. $E_{k_u}(k_{u,v})$ (the encryption of $k_{u,v}$ with the key k_u) is sent to party u, and $E_{k_v}(k_{u,v})$ is sent to v). The parties can now communicate using the key $k_{u,v}$. This approach was initiated by Needham and Schroeder in 1978 [40] and is widely implemented, most notably in the Kerberos system (see e.g. [27]). Bellare and Rogaway [4] give a complexity-theoretic treatment of this model, and present a provably secure protocol for session key distribution based on the existence of pseudo-random functions.

The approach of using a KDC is appealing since each party should only store a single key and when a new party is introduced there is no need to send keys to other parties. However there are various problems in using KDCs, which are due to the fact that a KDC is a single point of failure:

- **Security:** The KDC knows all the keys that are used in the system, and if it is broken into the security of the entire network is compromised.

- **Availability:** (i) The KDC is a performance bottleneck, every party has to communicate with it each time it wishes to retrieve a key. (ii) When the KDC is down or unreachable no party can obtain new keys for starting conversations on the network. (iii) The availability problem is amplified when trying to use a KDC to generate keys for multicast communication (i.e. to be shared by more than two parties), since all the relevant parties have to contact a single KDC.

In order to address these problems the common practice is to use multiple KDCs. However, the known solutions are far from being perfect: **(i)** The security problem is addressed by dividing the network into different *domains* and dedicating a different KDC to each domain. When a KDC is broken into only the domain to which it belongs is compromised. However, the management of inter-domain connections is complicated and a KDC still holds all the secrets of its domain. **(ii)** The availability problem is reduced by replicating the KDC and installing several servers each containing all the information that was previously stored in the KDC. This improves the availability but decreases security: there are multiple sensitive locations and breaking into *any* of these replicated KDCs compromises the security of the network. There is also an additional problem of reliably synchronizing the information that is stored in the different copies.

Multicast communication: The availability problem is relevant to unicast communication between two parties but is even more severe for multicast communication. Multicast communication is sent to a (potentially large) number of parties. Typical applications are the transmission of streams of data (such as video streams) to large groups of recipients, or an interactive multiparty conference. The large (exponential) number of groups in which a party might participate prevents it from storing a key for each potential group. On the other hand, the large number of parties which might require the service of the KDC worsen the availability problem. For example, imagine a source which transmits many video channels over the Internet, with hundreds of thousands of receivers all over the world. A *single* KDC cannot handle requests from all these receivers. Alternatively, consider a multinational company which uses a single KDC for providing keys for virtual meetings of its employees. If some offices are disconnected from the KDC then users in these offices cannot even obtain keys for virtual meetings between themselves.

A Distributed KDC – DKDC: A DPRF can be used to construct a Distributed KDC (DKDC). A DKDC consists of n servers and a user should k of them in order to obtain a key. The servers are responsible for a consistent mapping between key names and key values[1]. Each KDC server should operate as a server in the distributed evaluation of the pseudo-random function f. The key for a certain subset S of users is defined as $k_S = f(S)$. This approach is especially useful for generating keys for multicast groups with many members.

---

[1] Of course, consistency does not prevent groups of users from using different keys at different times (session keys), if this is desired.

Each member might approach a different authorized subset of the KDCs and it is guaranteed that every user obtains the same key. It is also useful to use this construction to generate keys for unicast communication if each of the two parties prefers to access a different subset of KDCs.

Key granting policy: When a user requests a key from a KDC the KDC should decide whether the user is entitled to receive this key. The question of how this decision is made is independent of this work.

One appealing approach is when a group name is derived from the identities of its members and then servers can easily verify whether a user that asks for a key is part of the group of users that use the key. This method is good for "mid-sized" groups. For larger groups the group name can be generated by a method based on hash trees, and then a user can efficiently prove to a server that it is part of a group.

Another appraoch introduces an interesting billing mechanism for multicast transmissions with k-out-of-n DKDCs: the user is required to pay each server $1/k$ of the payment needed for accessing the transmission, and to receive in return the information the server can contribute towards the reconstruction of the decryption key.

1.3 Long-Term Encryption of Information

Suppose one wishes to store encrypted information so that it remains safe for many years. A problem that immediately arises is where to store the keys used for the encryption so that they would not be leaked or lost. Note that the question of storing keys safely arises in many other scenarios, e.g. [8]. One possibility is to use a DPRF as a long term key repository. We add to the system a collection of n servers that act as the servers of the DPRF. These servers are trusted in the sense that no more than k of them become faulty[2]. We should also have some way to specify the policy determining who is allowed to decrypt the file, as the system is likely to be used by many users. We assume that the DPRF has ways to check whether a user is allowed to obtain information with a given policy (this is orthogonal to the issue at hand).

In order for a user to encrypt a file X and decryption policy specified by *who*, it does the following

- Choose an encryption key r for a conventional encryption scheme G and encrypt the file with key r. Let $Y = G_r(X)$.
- Compute $h = H(Y)$ where H is a collision intractable hash function.
- Apply to the DPRF to obtain $s = f(h \circ who)$
- Put Y in the long term storage together with *who* and $s \oplus r$.

  To decrypt an encrypted file Y with policy *who* and encrypted key s':

- Compute $h = H(Y)$.
- Apply the DPRF to obtain $s = f(h \circ who)$
- Decrypt Y with key $s \oplus s'$ to G.

---

[2] In this case the desirability of proactive security is evident since the assumption is that no more than k are broken into at any given period.

Note that we do not require the servers of the DPRF to store anything in addition to their keys. All information related to the file can be stored at the same place. Also note that in order combat changes to the stored information one should use parts of s as an authentication key to Y and r.

1.4 Related Work

DPRF systems   perform multi-party computations. The generic solutions of [25,6,14] for multiparty computations are inefficient for this application (even when applied to the relatively simple pseudo-random functions of [36], see discussion there). In particular, they require communication between the servers which are involved in the evaluation of the function. Their security is also only guaranteed if less than one third (or one half in the passive model) of the servers are corrupted.

There has been a lot of work on designing and implementing KDCs. A good overview of this work can be found in [27] and a formal treatment of the problem is given in [4]. Most of this work was for a trusted party which generates a key "on-the-fly", i.e. where consistency of the key is not required. While this model may be more relevant to unicast it is less applicable when more than two parties are involved.

Naor and Wool [39] considered a different scenario for protecting databases, and when adapted to our scenario their solution is one where the servers are trusted never to reveal their secret keys, but some of them might not have received updates regarding the permissions of users (which is a weaker assumption than regarded in this paper).

Our first two constructions are similar in nature to the constructions of Naor and Pinkas for metering schemes [32]. The problem they considered was to enable a server to *prove* that it served a certain number of clients (a representing application might be to meter the popularity of web sites in order to decide on advertisement fees). In general, not every solution for the metering problem is relevant to the construction of a DPRF (for example, the output of the metering computation should be *unpredictable* whereas the output of a DPRF should be *pseudo-random*). The metering constructions achieve better *robustness* against transmission of corrupt *proof components* than the robustness of our DPRF schemes against corrupt *key components*. On the other hand the metering constructions do not provide proactive security (due to the lack of communication channels between clients in that model) whereas we present very efficient proactive enhancements to the DPRF schemes.

Micali and Sidney [31] showed how to perform a shared evaluation of a pseudo-random function with a non-tight threshold. They provided a lower bound and a non-optimal probabilistic construction which is relevant only for small values of k and n. We describe an deterministic construction for the sharp threshold case which matches their lower bound.

Gong [26] considered a problem related to the DKDC application: a pair of users A and B each have private channels to n servers, and would like to use them to send a secret and authentciated message from A to B (e.g. a key which they will later use). Some of the servers might be corrupt and might change the messages they are asked to deliver (this problem is similar to that considered by Dolev et al [19] since each server is essentially a faulty communication link). Gong's scheme requires A to send through each server a message of length $O(n)$

2   Definitions

2.1   The Model

The following model is used throughout this work.

Setting: We consider a network of many users (clients), which also contains n servers $S_1 \ldots, S_n$. Each user u has a private connection with each of at least k servers (in all but the proactive solutions these channels can be realized using symmetric encryption. A future work [34] describes how to efficiently maintain these channels in the proactive model).

Initialization: At the initialization stage each server S_i receives some secret personal key α_i which it would use in its subsequent operation. It is possible that the values $\{\alpha_i\}_{i=1}^n$ were generated by a central authority from a system key α. If this is the case then α is erased at the end of the initialization stage. Preferably, the servers perform a short joint computation which generates the values $\{\alpha_i\}_{i=1}^n$, such that no coalition C of $k-1$ servers can use its values to learn anything about α_u if $u \notin C$. This prevents even a temporary concentration of the system's secrets at a single location.

Regular operation: A party u that wants to compute $f(x)$, operates as follows:

- It contacts k servers, $S_{i_1}, \ldots, S_{i_k}$, and sends to each of them a message $\langle u, x \rangle$.
- Each server S_i verifies that u is entitled to compute $f(x)$. If so, it computes a function $F(\alpha_i, x)$, and sends the result to u through their private channel.
- u computes $f(x)$ from the answers it received using a function G, namely it computes $f(x) = G(h, F(\alpha_{i_1}, x), \ldots, F(\alpha_{i_k}, x))$.

2.2   Requirements

There are two approaches to approximating random functions: *pseudo-random-ness* and *ℓ-wise independence*. We present approximations to DPRFs which follow both these directions.

Loosely speaking, pseudo-random distributions cannot be efficiently distinguished from uniform distributions. However, pseudo-random distributions have substantially smaller entropy than uniform distributions and are efficiently sampleable. *Pseudo-random function ensembles*, which were introduced in [24], are distributions of functions. These distributions are indistinguishable from the uniform distribution under all (polynomially-bounded) black-box attacks (i.e. the distinguisher can only access the function by adaptively specifying inputs and getting the value of the function on these inputs). Goldreich, Goldwasser, and Micali provided a construction of such functions based on the existence of pseudo-random generators. See [23,29] for further discussions and exact definitions of pseudo-random functions.

We also use ℓ-*wise independent functions*. Their difference from a pseudo-random function is that more than ℓ values of an ℓ-wise independent function are not "random looking" (however, a set of at most ℓ values is completely random rather than pseudo-random).

In a DPRF the ability to evaluate the function is distributed among the servers. The process that is performed by the servers can be defined as k-*out-of-n threshold function evaluation*.

Definition 1 (k-out-of-n threshold evaluation of a pseudo-random func).
Let $\mathcal{F}_m = \{f_\alpha\}$ be a family of pseudo-random functions with security parameter m, keyed by α. A k-out-of-n computation of $\mathcal{F}_m$ is a triple of polynomial time functions $\langle S, F, G \rangle$ (the key sharing, share computation and construction functions), such that

- *For every $f_\alpha \in \mathcal{F}_m$, $S(\alpha) = \langle \alpha_1, \ldots, \alpha_n \rangle$, such that*
- *For every $1 \leq i_1 < \cdots < i_k \leq n$, $G(\langle i_1, F(\alpha_{i_1}, x) \rangle, \ldots, \langle i_k, F(\alpha_{i_k}, x) \rangle) = f_\alpha(x)$. And,*
- *For every $1 \leq i_1 < \cdots < i_{k-1} \leq n$, given $\{\alpha_{i_j}\}_{j=1}^{k-1}$, and given a set Y of polynomially many values (where the inputs in Y were chosen adaptively, possibly depending on $\{\alpha_{i_j}\}_{j=1}^{k-1}$), and the values $\langle f_\alpha(y), \{F(\alpha_i, y)\}_{i=1}^{n} \rangle$ for every $y \in Y$, the restriction of the function f_α to inputs which are not in Y is pseudo-random.*

The definition of k-out-of-n threshold evaluation of an ℓ-wise independent function is similar, except that $\mathcal{F}_m$ is a family of ℓ-wise independent functions, and it is required that given the computation process of any $\ell - 1$ function values, any remaining value is uniformly distributed.

The most important requirement of k-out-of-n threshold function evaluation is that the output of f be consistent. The protocol might be considered as a special case of multi-party computations [25,6,14]. However although it might not be obvious from first reading, our definition includes several efficiency restrictions which do not exist in the definition of multi-party computations and which are actually not satisfied by the constructions of [25,6,14] (their constructions are also for a joint computation by n parties, and are secure only against coalitions of less than $n/2$ or $n/3$ parties. Our requirement is for a joint computation by k parties and security against $k - 1$ servers, where k might be any number up to n). The efficiency requirements, which we explicitly state below, are needed to minimize the communication overhead which is often the most important factor of the system's overhead. The efficiency requirements are:

Communication pattern: In the process of computing $f(x)$ there is no communication between the servers. The only communication is between the servers and the party that computes $f(x)$.

Single round: There is only a single round of communication between the servers and the user. The user can send queries to the servers in parallel, i.e.

there is no need to wait for the answer from one server before sending a query to another server.

Obliviousness: The query to one server does not depend on the identities of the other servers which the user queries. This requirement is important if the user might find (while in the middle of the process of querying the servers) that some of the servers to which it applied are malfunctioning.

Additional requirements can be considered as **security optimizations** to the original definition. They are not obligatory, but improve the quality of a DPRF construction:

Robustness: If a server is controlled by an adversary it might send to the user corrupt information which prevents the user from computing the correct value. It is preferable if the user can identify when such an event happens.

Proactive security (or, Resilience to prolonged attacks): Proactive security enables a system to maintain its overall security even if its components are repeatedly broken into. Systems with proactive security typically use a security parameter k and are secure as long as less than k system components are broken into *in the same time period* (see [11] for a discussion of proactive security).

3   The Threshold Constructions

3.1 ℓ-wise Independence Based on Bivariate Polynomials

The first construction is based on a generalization of the secret sharing scheme of Shamir [44] to bivariate polynomials. It is a threshold construction of an ℓ-wise independent function. The scheme can be used to generate more than ℓ values as long as it is guaranteed that no adversary will get hold of ℓ values. It is not necessarily decided in advance which values will be generated by the scheme.

Setting: The family $\mathcal{F}$ is the collection of all bivariate polynomials $P(x, y)$ over a finite field $\mathcal{H}$, in which the degree of x is $k - 1$ and the degree of y is $\ell - 1$. The key α defines an element $f_\alpha \in \mathcal{F}$ (α consists of the $k\ell$ coefficients of the polynomial). The output of the function is an element in the field $\mathcal{H}$. All the arithmetic operations performed by the scheme are over $\mathcal{H}$.

Initialization: (we describe here an initialization by a central authority, later we also describe how the servers can perform a distributed initialization). The initializer of the system chooses a random key α which defines a random polynomial $P(x, y)$ from $\mathcal{F}$. Each server S_i receives the key $\alpha_i = Q_i(y) = P(i, \cdot)$, which is an $\ell - 1$ degree polynomial in y.

Operation: The value $f(h)$ is defined as $f(h) = P(0, h)$. Consider a user that wishes to compute this value. Say the user approaches server S_i, then it should send him the information $\beta_{i,h} = F(\alpha_i, h) = Q_i(h) = P(i, h)$. After receiving information from k servers $S_{i_1}, \ldots, S_{i_k}$ the user can perform a polynomial interpolation through the points $\{\langle i_j, \beta_{i_j,h} \rangle\}_{j=1}^{k}$ and compute the free coefficient of the polynomial $Q_h(x) = P(\cdot, h)$, namely the value $f(h) = P(0, h)$.

The following points can be easily verified: **(i)** The scheme implements the definition of k-out-of-n evaluation of an ℓ-wise independent function. **(ii)** In a

DKDC application the size of an element in the field $\mathcal{H}$ should be the length of the required key and can therefore be rather small (e.g. 128 bits). The scheme can be therefore used to produce a large number of keys (e.g. $\ell = 10^6$).

Several modifications can enhance the above scheme: (i) Proactive security can be easily obtained, see Section 5. (ii) In order to reduce the complexity of the polynomial interpolation it is possible to use several polynomials of smaller degree and map keys to polynomials at random. (iii) It is possible to perform a distributed initialization of the polynomial P, and then the system's secrets are never held by a single party. The initialization is performed by several servers which each define a bivariate polynomial, and the polynomial used by the system is the sum of these polynomials. Only a coalition of all these servers knows shares of other servers. The initialization uses a new verification protocol we discuss in Section 5.

Robustness: A simple and straightforward procedure to verify that a user is receiving correct information from servers, it to require the user to get shares from $k' > k$ servers and use the error-correction properties of Reed-Solomon codes to construct the correct share (see e.g. [30]).

3.2   Distributed Weak PRFs Based on the DDH Assumption

In this section we describe a different kind of approximation for a DPRF: we show a way to distribute a *weak pseudo-random function* [35,37]. A function f is a weak PRF if it is indistinguishable from a truly random function to a (polynomial-time) observer who gets to see the value of the function on any polynomial number of *uniformly chosen inputs* (instead of any inputs of its choice). The definition of k-out-of-n threshold evaluation of a weak pseudo-random function f is similar to Definition 1. The only difference is that we require that given the computation process of f on any polynomial number of *uniformly chosen inputs*, the value of f on any *additional uniformly chosen input* is indistinguishable from random (this implies that f remains a *weak* pseudo-random function).

The main advantage of a distributed weak PRF compared with distributed ℓ-wise independent function is that the former is secure even when the adversary gets hold of *any polynomial number of values*. However, constructing a distributed weak PRF requires some computational intractability assumption (in particular, the existence of one-way functions). The specific construction described here relies on the decisional version of the Diffie-Hellman assumption (which we denote as the DDH assumption). This construction is rather attractive given its simplicity.

The applicability of weak pseudo-random function: Any distributed weak pseudo-random function f can be transformed to a DPRF by defining $f'(x) = f(RO(x))$, where RO is a random oracle (i.e., a random function that is publicly accessible to all parties as a black-box; see [4]). Therefore, if one postulates the existence of random oracles then the construction we present below can be used for all the applications of DPRFs. However this construction may be applicable even without the use of random oracles. Consider for example the application of

DKDCs for multicast communication. Here there may be several scenarios where a distributed weak pseudo-random function is sufficient. One such scenario is when there exists a *public* mapping H that assigns random names to groups of users. The key of a group can be the value of the distributed function applied to the group's name. It is conceivable that group names are chosen by some trusted party (or by a distributed protocol between several parties), and kept in some (possibly duplicated) publicly available server. In fact, using the specific functions described below is secure as long as some member of the group chooses the group name as g^r and proves that it knows r.

In the scheme we describe below, the user who computes the function f should perform k exponentiations. This overhead is larger than that of a Diffie-Hellman key exchange. However, the overhead is justified even for the DKDC application, since the Diffie-Hellman key exchange protocol cannot be used to solve the availability and the security requirements that underline our solution of a consistent distribution of a KDC (and are especially important for multicast communication).

Related distributed solutions were previously suggested for discrete-log based signatures (e.g. [22]). The novelty in our work is the fact that we prove the *pseudo-randomness* of the evaluated function.

Setting and Assumptions: The scheme is defined for two large primes P and Q such that Q divides $P - 1$, and an element g of order Q in $\mathbb{Z}_P^*$. The values P, Q and g are public and may either be sampled during the initialization or fixed beforehand. We assume that for these values, the decisional version of the Diffie-Hellman assumption (DDH-Assumption) holds. I.e., that given a uniformly distributed pair $\langle g^a, g^b \rangle$, it is infeasible to distinguish between $g^{a \cdot b}$ and a uniformly distributed value g^c with non-negligible advantage. For a survey on the application of the DDH-Assumption and a study of its security see [9].

The functions and their initialization: The family $\mathcal{F}$ is keyed by a uniformly distributed value $\alpha \in \mathbb{Z}_Q^*$. For simplicity, we define the function f_α over $\langle g \rangle$ (where $\langle g \rangle$ denotes the subgroup of $\mathbb{Z}_P^*$ generated by g)[3]. The function f_α is defined by $\forall x \in \langle g \rangle$, $f_\alpha(x) \overset{\text{def}}{=} x^\alpha \bmod P$.

The value α is shared between the servers using the secret sharing scheme of Shamir [44]: The initializer of the system chooses a random polynomial $P(\cdot)$ over $\mathbb{Z}_Q^*$ of degree $k - 1$ such that $P(0) = \alpha$. Each server S_i receives the key $\alpha_i = P(i)$. To facilitate robustness, the initializer also makes the values g^α and $\{g^{\alpha_i}\}_{i=1}^n$ public. It is also possible to let the servers perform a distributed initialization of f.

Operation: Consider a user that wishes to compute $f_\alpha(h)$ and approaches a set of k servers $\{S_i\}_{i \in J}$. Each such server S_i sends to the user the information

---

[3] In fact, one can define f'_α over $\mathbb{Z}_P^*$ by setting $f'_\alpha(x) = f_\alpha(x')$ where $x' = x^{(P-1)/Q} \bmod P$. If f_α is a weak PRF then so is f'_α since: (1) If x is uniform in $\mathbb{Z}_P^*$ then x' is uniform in $\langle g \rangle$. (2) For any $x' \in \langle g \rangle$ one can efficiently compute a uniformly chosen $((P-1)/Q)$-th root of x'. Computing such roots is possible by a generalization of Tonelli's algorithm presented by Adleman, Manders and Miller (see [2] for a survey on this subject).

$\beta_{i,h} = F(\alpha_i, h) = f_{\alpha_i}(h) = h^{\alpha_i}$. After receiving information from the k servers the user can perform a polynomial interpolation through the points $\{\alpha_i\}_{i \in J}$ in the exponent of h. I.e he can compute

$$f_\alpha(h) = h^\alpha = h^{\sum_{i \in J} \lambda_{i,J} \cdot \alpha_i} = \prod_{i \in J} h^{\lambda_{i,J} \cdot \alpha_i} = \prod_{i \in J} f_{\alpha_i}(h)^{\lambda_{i,J}}$$

where all exponentiations are in $\mathbb{Z}_P^*$ and the values $\{\lambda_{i,J}\}$ are the appropriate Lagrange coefficients.

It is easy to verify that querying any k servers for the value $f_{\alpha_i}(h)$ results in the same final value $f_\alpha(h)$. Memory requirements from each server are minimal (i.e. storing a single value in $\mathbb{Z}_Q^*$). In order to serve a user each server should perform a single modular exponentiation in $\mathbb{Z}_P^*$. A user is required to perform k modular exponentiation in $\mathbb{Z}_P^*$.

The security of the scheme is proved by the following theorem.

Theorem 2. *If the DDH-Assumption holds then the above scheme is a k-out-of-n threshold evaluation of a weak pseudo-random function.*

Proof Sketch: For clarity, we ignore at first the issue of corrupted servers and just prove that if the DDH-Assumption holds then $\mathcal{F} = \{f_\alpha\}$ is a family of weak pseudo-random functions. Let D be an efficient algorithm that gets the value of f_α on $q - 1$ uniformly chosen inputs $x_1, \ldots x_{q-1}$ and distinguishes $f_\alpha(x_q)$ from random with advantage ϵ (where x_q is also uniformly distributed). We construct an algorithm A that breaks the DDH-Assumption:

On input $\langle g^\alpha, g^\beta, z \rangle$, the algorithm A first samples random values $\{r_i\}_{i=0}^{q-1}$ (in $\{1, \ldots Q\}$). Then A invokes D and returns its output on the input $\{\langle q_i, f_\alpha(q_i) \rangle\}_{i=0}^{q-1}$ and the additional pair of values $\langle x_q = g^\beta, z \rangle$. Where for each i, $q_i = g^{r_i}$ (and therefore $f_\alpha(q_i) = g^{\alpha \cdot r_i}$ can be evaluated by A). It is easy to verify that the advantage A has in distinguishing between the case that z is uniform in $\langle g \rangle$ and the case the $z = g^{\alpha \cdot \beta}$ is at least ϵ.

We now need to show that no coalition of $k - 1$ corrupt servers $S_{i_1}, \ldots, S_{i_{k-1}}$ can break the threshold scheme. The reason this holds is that such $k - 1$ servers can be simulated by the algorithm D described above. To do so, D samples the secret values of the $k-1$ servers (i.e., $\alpha_{i_1}, \ldots, \alpha_{i_{k-1}}$) by itself. Let P be the degree $k - 1$ polynomial that interpolates these values and α. Define $\alpha_j = P(j)$. D can evaluate every g^{α_j} using interpolation in the exponent of g and can therefore evaluate all the values $f_{\alpha_j}(q_i)$. $\square$

Robustness: Since the values $\{g^{\alpha_i}\}_{i=1}^n$ are public each server can prove the correctness of any answer $f_{\alpha_i}(x) = x^{\alpha_i}$. This can either be done by a zero-knowledge variant of Schnorr's proof for the value of the Diffie-Hellman function [43,15] or by the non-interactive version that uses random-oracles.

It is possible to perform a distributed initialization of the scheme, secure against corrupt servers (even if their only goal is to disrupt the operation of the system rather than to learn keys), and to achieve to achieve proactive security for the scheme.

3.3   DPRFs Based on Any Pseudo-random Function

The following scheme can use any family of pseudo-random functions, but since its overhead for the k-out-of-n access structure is $O(n^{k-1})$ it is useful only if the total number of servers n is moderate and the threshold k is small.

Setting: Define $d = \binom{n}{k-1}$, and define the d subsets $\{G_j\}_{j=1}^d$ as all the subsets of $n - k + 1$ of the n servers.

Let $\mathcal{F}_m$ be a collection of pseudo-random functions with security parameter m. The key α is a d-tuple $\langle a_1, \ldots, a_d \rangle$ of elements from $\{1, \ldots, |\mathcal{F}_m|\}$, and defines a d-tuple $\langle f_{a_1}, \ldots, f_{a_d} \rangle$ of elements from $\mathcal{F}_m$. The function f_α is defined as $f_\alpha(x) = \oplus_{j=1}^d f_{a_j}(x)$.

Initialization: A random key α is chosen. We would like that for every $1 \leq j \leq d$, all the servers in subset G_j would receive the key to the function f_{a_j}. Therefore for every server S_i, $\alpha_i = \{a_j | i \in S_j\}$. Note that the union of the keys of any k servers covers α and is therefore sufficient to compute f_α.

Operation: The DPRF system would provide the value $f_\alpha(h) = \oplus_{j=1}^d f_{a_j}(h)$. When a user approaches a server S_i, and the server approves of the user computing $f(h)$, it should send to the user the information $\{f_{a_j}(h) | a_j \in \alpha_i\}$. I.e., the server should provide to the user the output of all its functions on the input h. After approaching k servers, the user has enough information to compute $f_\alpha(h)$.

For any coalition of $k - 1$ serves there is a subset G_j which does not contain any member of the coalition and thus the coalition members cannot compute f_{a_j}. Therefore it is straightforward to prove that the construction is a k-out-of-n evaluation of a pseudo-random function. The number of functions which each server should be able to compute is $\binom{n-1}{k-1}$, and the total number of functions is $d = \binom{n}{k-1}$. Therefore the scheme cannot be used for systems with a large threshold. However, for a moderate n and a small k the overhead is reasonable (e.g. for $n = 50$ and $k = 4$, $d = 19,600$ and a server should compute $4,606$ functions).

Note that the user receives the value of functions f_{a_j} from more than a single server. Therefore if the user sends to servers the identities of the other servers which it approaches, the communication overhead is reduced if a a simple mapping is used to ensure that the output of each function is sent once. Alternatively the data redundancy can be used to provide robustness against corrupt servers that send incorrect data to users.

Generalization: The scheme can be generalized to any access structure. The construction we used corresponds to a monotone CNF formula which contains all clauses of $n - (k - 1)$ out of n elements. A similar formula can be used to realize any access structure. The total number of pseudo-random functions used is the number of clauses in the monotone CNF formula.

Comparison to previous work: Micali and Sidney [31] considered more general access structures: they defined an (n, t, u)-*resilient collection* (with $t < u < n$) which enables any subset of u (out of n) parties to perform the computation, while no subset of t parties has this ability. We are interested in a sharp threshold, which provides the best security, and therefore require that $k = u = t + 1$.

Micali and Sidney proved a lower bound of $\frac{n!(u-t-1)!}{(n-t)!(u-1)!}$ for the number of functions in an (n, t, u)-resilient collection, and used the probabilistic method to show the existence of a construction which is $\ln \binom{n}{t}$ times larger than the lower bound. Our deterministic construction (for the sharp threshold case) matches their lower bound, and is therefore optimal.

4 General Access Structure KDCs

4.1 Using Monotone Symmetric Branching Programs

We present here generalizations of the threshold schemes to access structures based on *monotone symmetric branching programs*. In Section 4.2 we describe constructions for access structures based on *monotone span programs*. This is a further generalization in that any linear secret sharing scheme can be simulated by a monotone span program of the same size (the converse is also true, i.e. any monotone span program can be simulated by a linear secret sharing scheme of the same size, see [3]). However, the constructions of this section are more efficient (especially for the DH based constructions), as described below.

The application of monotone symmetric branching programs (also called *monotone undirected contact schemes*, and *switching networks*) to secret sharing was suggested by Benaloh and Rudich [7,28,3] and enables to construct a secret sharing scheme for any monotone access structure (the question is the size of the shares). We first present the computational model of monotone symmetric branching programs and then a corresponding DPRF construction.

Monotone symmetric branching programs: Let $G = (V, E)$ be an *undirected* graph, $\psi : E \mapsto \{1, \ldots, n\}$ be a labeling of the edges, and s, t be two special vertices in V. A monotone symmetric branching program is defined as a tuple $\langle G, \psi, s, t \rangle$ and has boolean output. Given an input $x = \{x_1, \ldots, x_n\} \in \{0, 1\}^n$, define G_x as the graph $G_x = (V, E_x)$, where $E_x = \{e \mid e \in E, x_{\psi(e)} = 1\}$. The output of the program is 1 if and only if G_x contains a path from s to t.

A DPRF construction: It is possible to construct DPRFs which are either ℓ-wise independent or weakly pseudo-random, based on monotone symmetric branching programs. A user would have to receive information from a subset of the servers whose characteristic vector corresponds to a "1" output of the monotone symmetric branching program in order to obtain the required value. We present here the ℓ-wise independent construction. Note that the corresponding DH construction is more efficient than with monotone span programs since it requires only multiplications and not exponentiations.

Initialization: A monotone symmetric branching program which realizes the required access structure is constructed. A random polynomial P_s of degree $\ell - 1$ is associated with the node s. The values distributed by the system are defined as $f(h) = P_s(h)$. A random polynomial P_v of degree $\ell - 1$ is associated with any other vertex v, except for the vertex t to which the polynomial $P_t \equiv 0$ is assigned. Every edge $e = (u, v)$ is associated with the polynomial $P_e = P_u - P_v$. Server S_i is given the all the polynomials associated with the edges which are mapped to i (edges e for which $\psi(e) = 1$).

Reconstruction: A user which wants to obtain value $f(h)$ should contact a privileged subset of the servers. Each server S_i which is approached by the user and approves of him evaluating $f(h)$ should provide it with the values $\{P_e(h) | \psi(e) = i\}$. If the user receives information from a privileged subset it can sum the values that correspond to a path from s to t and get $P_s(h)$.

Quorum systems: A Quorum system is a collection of sets (quorums), every two of which intersect (see [38] for a discussion and some novel constructions of quorum systems with optimal load and high availability). A DPRF with an access structure in which every privileged set must contain a quorum has several advantages regarding its maintenance: for example, if a user should not be allowed to compute f it is only required to inform all the servers in a *single* quorum of this restriction, and then every privileged set of servers contains at least one server which will refuse to serve that user. DPRFs with access structures based on the *paths* quorum system [38] can be efficiently realized by the constructions we presented in this section.

Efficiency: The reconstruction of the secret in the Diffie-Hellman variant we presented here requires the user to perform multiplications. It is more efficient than the reconstruction for the monotone span programs based Diffie-Hellman scheme we present in Section 4.2, which requires the user to perform exponentiations.

General prf: Note that a direct use of pseudo-random functions instead of the polynomials or of the Diffie-Hellman construction is insecure. The reason is that an edge (u, v) is associated with a function $f_u - f_v$ and since there is no concise representation for this function which hides f_u and f_v the server which is mapped to the edge should get both functions f_u and f_v. Subsequently, the server can compute $f_u(x)$ or $f_v(x)$ and not just $f_u(x) - f_v(x)$. Therefore a server which is mapped to an edge which touches s has the ability to compute by itself the value of the shared function.

4.2   Using Monotone Span Programs

It is possible to construct DPRFs with access structures which are realized by monotone span programs. Monotone span programs (MSPs) were introduced by Karchmer and Wigderson [28] and their corresponding secret sharing schemes are equivalent to linear secret sharing schemes   in the sense that any secret sharing scheme in one of these classes can be realized by a scheme of the same size in the other class, see [3] for details. Recently MSPs were used by Cramer, Damgard, and Maurer [16] to construct multi-party computation protocols for general monotone sets of subsets of players, any one of which may consist of cheaters. We first present the computational model of monotone span programs and then a DPRF construction.

Monotone span programs: A monotone span program is defined by a triple $\langle K, M, \psi \rangle$ as follows. Let K be a finite field and let M be a matrix with d rows and e columns, and entries in K. The rows of M are labeled by a mapping to server identities, $\psi : \{1, \ldots, d\} \mapsto \{1, \ldots, n\}$. For a subset $A \subset \{1, \ldots, n\}$, define M_A as the matrix consisting of the rows of M which are labeled with $i \in A$, and let d_A be the number of rows in this matrix.

Let $\epsilon = (1, 0, \ldots, 0) \in K^e$ be the *target* vector (ϵ can be replaced by any non-zero vector in K^e). An MSP computes a boolean function $f : \{0, \ldots 1\}^n \mapsto \{0, 1\}$ defined by "$f(x_1, \ldots, x_n) = 1$ if and only if ϵ is in the Image of M_A^t, where $A = \{i | x_i = 1\}$". That is, if there is a linear combination of the d_A rows labeled with an i for which $x_i = 1$, that equals the target vector ϵ. It is known that any monotone boolean function can be computed by an MSP (and the question is what size).

A DPRF construction: The construction is based on the MSP secret sharing scheme. We can achieve either ℓ-wise independence or weak pseudo-randomness. A user would have to receive information from a subset of the servers which corresponds to a "1" output of the MSP in order to obtain the required value. Following we present the DH based construction.

Initialization: An MSP which realizes the required access structure is constructed. All operations are performed over an appropriate field. A vector of random values $\bar{\alpha} = \{\alpha_1, \ldots, \alpha_e\}$ is associated with the columns of M. The function computed by the system is defined as $f(h) = h^{\alpha_1}$.

Server S_i is given the share $\bar{s}_i = M_{\{i\}}\bar{\alpha}$, which is a vector of length $d_{\{i\}}$, the number of rows in $M_{\{i\}}$.

Reconstruction: A user which wants to compute $f(h)$ should contact a privileged subset of the servers. Each server S_i which is approached by the user and approves of computing $f(h)$ should provide him with the values $\{h^\beta | \beta \in \bar{s}_i\}$ (i.e. h raised to the power of each of the coordinates of $\bar{s}_i$). . If the user receives information from a privileged subset then there is a linear combination in the exponents which obtains $f(h) = h^{\alpha_1}$. The user can perform exponentiations and multiplications to compute this combination.

5 Proactive Security

Proactive security enables a system of servers to automatically recover from repeated break-ins while preserving its security. The servers perform a periodical mutual refreshment of their secrets, and security is preserved as long as not too many servers are broken into between two refreshments (see [11] for a survey of proactive security). We can amend our schemes with proactive security while preserving consistency. The value of $f(x)$ computed in two different requests would still be the same, even if several refreshment phases pass between the two requests.

The periodic refreshment requires communication between the servers, which is a new requirement for DPRFs. Alternatively, the refreshment can be controlled by a single secure server which is the only party sending refreshment information to servers. The system is kept secure as long as there is no break-in to this server, but since this server can be highly guarded (e.g. kept off-line at all times except for refreshment phases) this scenario seems reasonable.

We describe very briefly how proactive security is obtained. The periodic refreshment phases employ techniques which are common in proactive refreshments, and a novel method for verifying that the refreshment values sent by each server are indeed correct. In the refreshment of the ℓ-wise independent construction, k servers $S_1, \ldots, S_k$ should each generate a random bivariate polynomial $P_i^t(x, y)$, subject to the constraint $P_i^t(0, \cdot) = 0$. Server S_i sends to each other server S_j the restriction of its polynomial to $x = S_j$, i.e. $P_i^t(S_j, \cdot)$. The new polynomial of each server is the sum of its old polynomial with all the new polynomials it receives.

The servers should run a verification protocol for the values they receive in the refreshment phase, in order to verify that $S_1, \ldots, S_k$ send shares of polynomials of the right degrees which are 0 for $x = 0$. This is essentially a verifiable secret sharing (VSS) protocol. It is possible to use a VSS protocol which is very efficient in both its computation and communication requirements. Very briefly, the verification is done by choosing a random point c, and requiring each S_i to broadcast $P_i^t(\cdot, c)$. Each server should verify that $P_i^t(0, c) = 0$ and that the share it received agrees with this broadcast. Note that unlike the verification protocols of [6,20] this protocol does not require communication between each pair of servers. The random point c can be chosen in a very natural way, it can be defined as a value of the previous polynomial at a point which is only evaluated after the servers send the refreshment values.

Application to distributed initialization: The initialization of the system can be performed in a distributed manner. It is then required to verify that servers that participate in this process do not send incorrect data which would disrupt the operation of the system, i.e. that they send shares of polynomials of the right degrees. This verification can be performed very efficiently using the above protocol and a broadcast channel (note that it is not required to verify that the value of the polynomial is 0 for $x = 0$). The choice of the random point should be done by a distributed protocol which generates several values, where at least one of the values is guaranteed to be random.

Future Work

The most obvious open problem is coming with a construction which has all the properties of a DPRF, i.e. of a function which is strongly pseudo-random and can be evaluated a polynomial number of times. Another interesting line of research is the design of *oblivious* DPRFs, in which the servers do not learn what is the input x for which the user wants to compute $f(x)$. Note that the oblivious polynomial evaluation protocols of [33] are probably too expensive since the number of 1-out-of-2 oblivious transfers is linear in the degree of the polynomial.

Acknowledgments

We thank the anonymous referees for their helpful comments.

References

1. Aho A. V., Hopcroft J. E. and Ullman J. D., **The Design and Analysis of Computer Algorithms**, Addison-Wesley, 1974.
2. Bach E. and Shallit J., **Algorithmic Number Theory**, vol. 1: Efficient Algorithms, The MIT Press, 1996, pp. 155-163.
3. Beimel A., "Secure schemes for secret sharing and key distribution", DSc dissertation, 1996.
4. Bellare M. and Rogaway P., "Provably secure session key distribution — the three party case", *27th Proc. ACM Symp. on Theory of Computing*, (1995), 57–66.
5. Bellare M. and Rogaway P., Random oracles are practical: A paradigm for designing efficient protocols, *Proc. 1st Conf. on Computer and Communications Security*, *ACM*, 1993, pp. 62-73.
6. Ben-Or M., Goldwasser S. and Wigderson A., Completeness theorems for non-cryptographic fault tolerant distributed computation, *20th Proc. ACM Symp. on Theory of Computing*, (1988), 1-9.
7. Benaloh J. and Rudich S., private communication, 1989.
8. Blaze M,, Feigenbaum J. and Naor M., "A Formal Treatment of Remotely Keyed Encryption", *Advances in Cryptology - Eurocrypt'98*, LNCS 1403, Springer-Verlag, 1998, pp. 251–265.
9. D. Boneh, "The Decision Diffie-Hellman Problem", *Proc. of the Third Algorithmic Number Theory Symp.*, LNCS Vol. 1423, Springer-Verlag, pp. 48–63, 1998.
10. Canetti R., Towards realizing random oracles: hash functions that hide all partial information, *CRYPTO '97*, LNCS, Springer, 1997, pp. 455-469.
11. Canetti R., Gennaro R., Herzberg D. and Naor D., "Proactive Security: Long-term protection against break-ins", CryptoBytes.
12. R. Canetti, O. Goldreich and S. Halevi, The random oracle model, revisited, to appear in: *Proc. 30th Ann. ACM Symp. on Theory of Computing*, 1998.
13. Canetti R. and Goldwasser S., "An efficient threshold public-key cryptosystem secure against chosen ciphertext attack", *Advances in Cryptology - Eurocrypt '99*, Springer-Verlag, 1999.
14. Chaum D., Crepeau C. and Damgard I., Multiparty unconditionally secure protocols, *20th Proc. ACM Symp. on Theory of Computing*, (1988), 11-19.
15. Chaum D. and Van Antwerpen H., "Undeniable signatures", *Advances in Cryptology - Crypto '89*, Springer-Verlag LNCS 435, (1990), 212–217.
16. Cramer R., Damgard I. and Maurer U., "General Secure Multi-Party Computation from any Linear Secret-Sharing Scheme", manuscript, 1999.
17. De Santis A., Desmedt Y., Frankel Y., and Yung M., "How to share a function securely", in *Proc. 26th STOC*, 1994, pp. 522-533
18. Desmedt Y. and Frankel Y., "Threshold cryptosystems", *Advances in Cryptology - Crypto '89*, Springer-Verlag LNCS 435, (1989), 307–315.
19. Dolev D., Dwork C., Waarts O. and Yung M., "Perfectly secure message transmission", *JACM* 40(1):17–47, 1993.
20. Feldman P. and S. Micali, "An Optimal Probabilistic Protocol for Synchronous Byzantine Agreement", Siam J. Comp. 26, 1997, pp. 873–933.
21. Gemmell P., "An introduction to threshold cryptography", *CryptoBytes*, Winter 97, 7–12, 1997.
22. Gennaro R., Jarecki S., Krawczyk H. and Rabin T., "Robust threshold DSS signatures", *Adv. in Cryptology - Eurocrypt '96*, Springer-Verlag, LNCS 1070, 354–371.
23. Goldreich O., "Foundations of Cryptography" (fragments of a book), 1995.

24. Goldreich O., Goldwasser S. and Micali S., "How to construct random functions", *J. of the ACM*, Vol. 33, No. 4, (1986), 691–729.
25. Goldreich O., Micali S. and Wigderson A., "How to play any mental game", *19th Proc. ACM Symp. on Theory of Computing*, (1987), 218-229.
26. Gong L., "Increasing availability and security of an authentication service", *IEEE J. SAC*, Vol. 11, No. 5, (1993), 657–662.
27. Kaufmaan C., Perlman R., and Speciner M., **Network Security**, Prentice Hall, 1995.
28. Karchmer M. and Wigderson A., "On span programs", *Proc. 8th Structures in Complexity Theory conf.*, 102–111, 1993.
29. Luby M., **Pseudorandomness and Cryptographic Applications**, Princeton University Press, 1996.
30. McEliece R. J. and Sarwate D. V., "On sharing secrets and Reed-Solomon Codes", *Comm. ACM*, Vol. 24, No. 9, 1981, 583–584.
31. Micali S. and Sidney R., "A simple method for generating and sharing pseudo-random functions, with applications to clipper-like key escrow systems", *Adv. in Cryptology – Crypto '95*, Springer, 185–196.
32. Naor M. and Pinkas B., Secure and efficient metering, *Advances in Cryptology – Eurocrypt '98*, LNCS 1403, Springer-Verlag, 1998, pp. 576–590.
33. M. Naor and B. Pinkas, "Oblivious Transfer and Polynomial Evaluation", *Proc. 31th Ann. ACM Symp. on Theory of Computing*, 1998.
34. Naor M. and Pinkas B., Maintaining secure communication in networks for long terms, manuscript, 1999.
35. Naor M. and O. Reingold, "Synthesizers and their application to the parallel construction of pseudo-random functions", *Proc. 36th IEEE Symp. on Foundations of Computer Science*, 1995, pp. 170-181.
36. Naor M. and Reingold O., "Number-Theoretic constructions of efficient pseudo-random functions", *Proc. 38th IEEE FOCS*, 1997.
37. Naor M. and O. Reingold, "From unpredictability to indistinguishability: A simple construction of pseudo-random functions from MACs", *Advances in Cryptology - CRYPTO '98*, 1998, pp. 267-282.
38. Naor M. and Wool A., "The load, capacity, and availability of quorum systems", *SIAM J. Comput.*, Vol. 27, No. 2, 423-447, April 1998.
39. Naor M. and Wool A., "Access Control and Signatures via Quorum Secret Sharing", *3rd ACM Conf. of Computer and Communication Security*, 1996, 157–168.
40. Needham R. and Schroeder M., "Using encryption for authentication in large networks of computers", *Comm. ACM*, Vol. 21, No. 12, 993–999, December 1978.
41. Rabin M. O., "Transaction Protection by Beacons", *JCSS*, Vol. 27, No. 2, (1983), 256-267.
42. Rabin T., "A Simplified Approach to Threshold and Proactive RSA", *Advances in Cryptology - CRYPTO '98*, 1998, pp. 89-104.
43. Schnorr C. P., Efficient identification and signatures for smart cards, *Proc. Advances in Cryptology - CRYPTO '89*, LNCS, Springer-Verlag, 1990, pp. 239-252.
44. Shamir A., "How to share a secret", *Comm. ACM* Vol. 22, No. 11 (Nov. 1979), 612–613.

Improved Fast Correlation Attacks on Stream Ciphers via Convolutional Codes*

Thomas Johansson and Fredrik Jönsson

Dept. of Information Technology
Lund University, P.O. Box 118, 221 00 Lund, Sweden
{thomas, fredrikj}@it.lth.se

Abstract. This paper describes new methods for fast correlation attacks, based on the theory of convolutional codes. They can be applied to arbitrary LFSR feedback polynomials, in opposite to the previous methods, which mainly focus on feedback polynomials of low weight. The results improve significantly the few previous results for this general case, and are in many cases comparable with corresponding results for low weight feedback polynomials.

Keywords: Stream ciphers, Correlation attacks, Convolutional codes.

1 Introduction

A binary additive stream cipher is a synchronous stream cipher in which the keystream, the plaintext and the ciphertext are sequences of binary digits. The output of the keystream generator, $z_1, z_2, \ldots$ is added bitwise to the plaintext sequence $m_1, m_2, \ldots$, producing the ciphertext $c_1, c_2, \ldots$. Each secret key k as input to the keystream generator corresponds to an output sequence. Since the secret key k is shared between the transmitter and the receiver, the receiver can decrypt, and obtain the message sequence, by adding the output of the keystream generator to the ciphertext, see Figure 1.

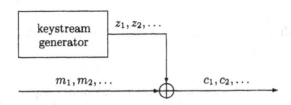

Fig. 1. Principle of binary additive stream ciphers

---

* This work was supported by the Foundation for Strategic Research - PCC under Grant 9706-09.

J. Stern (Ed.): EUROCRYPT'99, LNCS 1592, pp. 347–362, 1999.
© Springer-Verlag Berlin Heidelberg 1999

The goal in stream cipher design is to efficiently produce random-looking sequences that in some sense are "indistinguishable" from truly random sequences. From a cryptanalysis point of view, a good stream cipher should be resistant against a *known-plaintext attack*. In a known-plaintext attack the cryptanalyst is given a plaintext and the corresponding ciphertext, and the task is to determine a key k. For a synchronous stream cipher, this is equivalent to the problem of finding the key k that produced a given keystream $z_1, z_2, \ldots, z_N$. Throughout this paper, we hence assume that a given keystream $z_1, z_2, \ldots, z_N$ is in the cryptanalyst's possession and that cryptanalysis is the problem of restoring the secret key.

In stream cipher design, one usually use linear feedback shift registers, LFSRs, as building blocks in different ways, and the secret key k is often chosen to be the initial state of the LFSRs.

There are several classes of general cryptanalytic attacks against stream ciphers [9]. In our opinion, the most important class of attacks on LFSR-based stream ciphers is *correlation attacks*. Basically, if one can in some way detect a correlation between the known output sequence and the output of one individual LFSR, this can be used in a "divide-and-conquer" attack on the individual LFSR [12,13,7,8]. There is no requirement of structure of any kind for the key generator. The only thing that matters is the fact that, if $u_1, u_2, \ldots$ denotes the output of the particular LFSR, we have a correlation of the form $P(u_i = z_i) \neq 0.5$, see Figure 2.

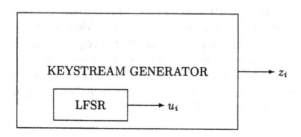

Fig. 2. A sufficient requirement for a correlation attack, $P(u_i = z_i) \neq 0.5$.

A "textbook" methodology for producing random-like sequences from LFSRs is to combine the output of several LFSRs by a nonlinear function f with desired properties. Here f is a binary boolean function in n variables. The purpose is to destroy the linearity of the LFSR sequences and hence provide the resulting sequence with a large linear complexity [9]. This is depicted in Figure 3.

It is worth noticing that there always exists a correlation between the output z_i and either one or a set of M LFSR output symbols $\{u_i^{(i_1)}, u_i^{(i_2)}, \ldots, u_i^{(i_M)}\}$ in the model above. It is well known that if f is a $(M-1)$-resilient (but not M-resilient) function then there is a correlation which can be expressed in the

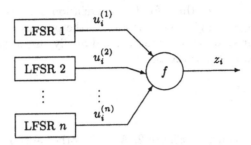

Fig. 3. Principle of nonlinear combination generators

form $P(z_i = u_i^{(i_1)} + u_i^{(i_2)} + \cdots + u_i^{(i_M)}) \neq 0.5$. It is also known that there is a tradeoff between the resiliency and the nonlinearity of f, and hence M must be rather small [12].

Returning to the previously mentioned correlation attacks, the above overview demonstrates that finding a low complexity algorithm that successfully can use the existing correlation in order to determine a part of the secret key can be a very efficient way of attacking such stream ciphers in cryptanalysis. After the initializing ideas of Siegenthaler [12,13], Meier and Staffelbach [7,8] found a very interesting way of exploring the correlation in a fast correlation attack *provided that the feedback polynomial of the LFSR has a very low weight*. This work was followed by several papers, providing minor improvements to the initial results of Meier and Staffelbach, see [10,1,2,11]. For a recent application, see [14]. However, the algorithms that are efficient (good performance and low complexity) still require the feedback polynomial to be of low weight. Due to this requirement, it is today a general advise when constructing stream ciphers that the generator polynomial should not be of low weight.

The problem addressed in this paper is the problem of constructing algorithms achieving the similar performance and similar low complexity as mentioned above *but for any feedback polynomial*. The new algorithms that we propose are based on an interesting observation, namely that one can identify an embedded low-rate convolutional code in the code generated by the LFSR sequences. This embedded convolutional code can then be decoded with low complexity, using the Viterbi algorithm. From the result of the decoding phase, the secret key can be obtained. These algorithms provide a remarkable improvement over previous methods. As a particular example taken from [10], consider a LFSR of length 40 with a weight 17 feedback polynomial, and an observed sequence of length $4 \cdot 10^5$ bits. Let $1 - p$ be the correlation probability. Then the algorithm in [7,8] and the improvement in [10] are successful up to $p \leq 0.104$ and $p \leq 0.122$, respectively, whereas the proposed algorithm is successful up to more than $p \leq 0.4$ with similar computational complexity.

The paper is organized as follows. In Section 2 we give some preliminaries on the decoding model that is used for cryptanalysis, and in Section 3 we shortly review some previous algorithms for fast correlation attacks. In Section 4 we present our new ideas and give a description of the proposed algorithm. In Section 5 the simulation results are presented, and finally, in Section 6 we give some conclusions and possible extensions.

2 Preliminaries

Consider the model shown in Figure 2. As most other authors [13,7,8,10,1], we use the approach of viewing the problem as a decoding problem. Let the LFSR have length l and let the set of possible LFSR sequences be denoted by $\mathcal{L}$. Clearly, $|\mathcal{L}| = 2^l$ and for a fixed length N the truncated sequences from $\mathcal{L}$ is also a linear $[N, l]$ block code [6], referred to as $\mathcal{C}$. Furthermore, the keystream sequence $\mathbf{z} = z_1, z_2, \dots , z_N$ is regarded as the received channel output and the LFSR sequence $\mathbf{u} = u_1, u_2, \dots , u_N$ is regarded as a codeword from $\mathcal{C}$. Due to the correlation between u_i and z_i, we can describe each z_i as the output of the binary symmetric channel, BSC, when u_i was transmitted. The correlation probability $1 - p$, defined by $1 - p = P(u_i = z_i)$, gives p as the crossover probability (error probability) in the BSC. W.l.o.g we can assume $p < 0.5$. This is all shown in Figure 4.

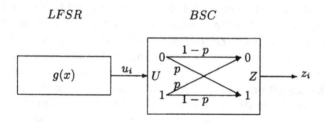

Fig. 4. Model for a correlation attack

The cryptanalyst's problem can be formulated as follows. Given a length N received word $(z_1, z_2, \dots z_N)$ as output of the BSC(p), find the length N codeword from $\mathcal{C}$ that was transmitted.

From simple coding arguments, it can be shown that the length N should be at least around $N_0 = l/(1 - h(p))$ for unique decoding, where $h(p)$ is the binary entropy function. If the length of the output sequence N is modest but allows unique decoding, say $N = N_0 + D$, where D is a constant, the fastest methods for decoding are probabilistic decoding algorithms like Leon or Stern algorithms [5,15].

For received sequences of large length, $N \gg N_0$, *fast correlation attacks* [7,8] are sometimes applicable. These attacks resemble very much the iterative decoding process proposed by Gallager [3] for low-weight parity-check codes.

Due to the fact that the above attacks require the feedback polynomial $g(x)$ (or any multiple of $g(x)$ of modest degree) to have a low weight, one usually refrain from using such feedback polynomials in stream cipher design.

3 Fast Correlation Attacks – An Overview

In [7,8] Meier and Staffelbach presented two algorithms, referred to as A and B, for fast correlation attacks. Instead of an exhaustive search as originally suggested in [13], the algorithms are based on using certain parity check equations created from the feedback polynomial of the LFSR. All different algorithms for fast correlation attacks use two passes. In the first pass the algorithms find a set of suitable parity check equations in the code $\mathcal{C}$ stemming from the LFSR. The second pass uses these parity check equations in a fast decoding algorithm to recover the transmitted codeword and hence the initial state of the LFSR.

The set of parity check equations that was used in [7,8] was created in two separate steps. Let $g(x) = 1 + g_1 x^1 + g_2 x^2 + \ldots + g_l x^l$ be the feedback polynomial, and t the number of taps of the LFSR, i.e., the weight of $g(x)$ (the number of nonzero coefficients) is $t + 1$. Symbol number n of the LFSR sequence, u_n, can then be written as $u_n = g_1 u_{n-1} + g_2 u_{n-2} + \ldots + g_l u_{n-l}$. Since the weight of $g(x)$ is $t + 1$, there are the same number of relations involving a fixed position u_n. Hence, we get in this way $t + 1$ different parity check equations for u_n.

Secondly, using the fact that $g(x)^j = g(x^j)$ for $j = 2^i$, parity check equations are also generated by repeatedly squaring the polynomial $g(x)$. So if $g_0(x) = g(x)$, we create new polynomials by $g_{k+1}(x) = g_k(x)^2$, $k = 1, 2, \ldots$. This squaring is continued until the degree of a polynomial $g_k(x)$ is greater than the length N of the observed keystream. Each of the polynomials $g_k(x)$ are of weight $t + 1$ and hence each gives $t + 1$ new parity check equations for a fixed position u_n.

Combining this squaring technique with shifting the set of equations in time, the same parity check equations are essentially valid in each index position of $\mathbf{u}$. From [7,8] the number of parity check equations, denoted m, that can be found in this way is $m \approx \log(\frac{N}{2l})(t + 1)$, where log uses base 2.

In the second pass, one writes the m equations for position u_n as,

$$u_n + b_1 = 0,$$
$$u_n + b_2 = 0,$$
$$\vdots \tag{1}$$
$$u_n + b_m = 0,$$

where each b_i is the sum of t different positions of $\mathbf{u}$. Applying the same relations above to the keystream we can calculate the following sums,

$$z_n + y_1 = L_1$$
$$z_n + y_2 = L_2$$
$$\vdots$$
$$z_n + y_m = L_m.$$

where y_i is the sum of the positions in the keystream corresponding to the positions in b_i. Assume that h out of the m equations in (1) hold, i.e.,

$$h = |\{i : L_i = 0, 1 \leq i \leq m\}|,$$

when we apply the equations to the keystream. Then it is possible to calculate the probability $p^* = P(u_n = z_n | h$ equation holds) as

$$p^* = \frac{ps^h(1-s)^{m-h}}{ps^h(1-s)^{m-h} + (1-p)(1-s)^h s^{m-h}},$$

where $p = P(z_n = a_n)$, and $s = P(b_i = y_i)$.

Using the parity check equations found above, two different decoding methods were suggested in [7,8]. The first algorithm, called Algorithm A, can shortly be described as follows: First find the equations to each position of the received bit and evaluate the equations. Then calculate the probabilities p^* for each bit in the keystream, select the l positions with highest value of p^*, and calculate a candidate initial state. Finally, find the correct value by checking the correlation between the sequence and the keystream for different small modifications of the candidate initial state.

The second algorithm, called Algorithm B, used another approach. Instead of calculating the probabilities p^* once and then make a hard decision, the probabilities are calculated iteratively. The algorithm uses two parameters p_{thr} and N_{thr}.

1. For all symbols in the keystream, calculate $p*$ and determine the number of positions N_w with $p^* < p_{thr}$.
2. If $N_w < N_{thr}$ repeat step 1 with p replaced by p^*.
3. Complement the bits with $p^* < p_{thr}$ and reset the probabilities to p.
4. If not all equations are satisfied go to step 1.

The performance of the algorithms described above is given in [7,8]. The algorithms above work well when the LFSR contains few taps, but for LFSRs with many taps the algorithms fail. The reason for this failure is that for LFSRs with many taps each parity check equation gives a very small average correction and hence many equations are needed. An improvement was suggested in [10], where a new method for finding parity check equations was suggested. Let u_0 be the initial state of the LFSR. The state after t shifts can be written as $u_t = A^t u_0$, where A is an $l \times l$ matrix that depends of the feedback polynomial. Using powers of the matrix A a set of parity check equations can be found.

Another method of finding parity check equations was suggested in [1]. The idea of this algorithm is to use an algorithm for finding codewords of low weight in a general linear code.

4   New Fast Correlation Attacks Based on Convolutional Codes

The general idea behind the algorithm to be proposed can be described as follows. Looking at the parity check equations as described in (1), they are designed for a second pass that consists of a very simple memoryless decoding algorithm. For a general feedback polynomial, this puts very hard restrictions on the parity check equations that can be used in (1) (weight $\leq t+1$ for a very low t). Our approach considers slightly more advanced decoding algorithms *that include memory*, but still have a low decoding complexity. This allows us to have looser restrictions on the parity check equations that can be used, leading to many more, and more, powerful equations. This work uses the Viterbi algorithm with memory $10 - 16$ as its decoding algorithm. The corresponding restrictions on the parity check equations will be apparent in the sequel.

The proposed algorithm transforms a part of the code C stemming from the LFSR sequences into a convolutional code. The encoder of this convolutional code is created by finding suitable parity check equations from C. Some notation and basic concepts regarding convolutional codes that are frequently used can be found in Appendix A.

The convolutional code will have rate $R = 1/(m + 1)$, where the constant $(m + 1)$ will be determined later. Furthermore, let B be a fixed memory size. In a convolutional encoder with memory B the vector $\mathbf{v}_n$ of codeword symbols at time n is of the form

$$\mathbf{v}_n = u_n G_0 + u_{n-1} G_1 + \ldots u_{n-B} G_B, \tag{2}$$

where in the case $R = 1/(m+1)$ each G_i is a vector of length $(m+1)$. The task in the first pass of the algorithm is to find suitable parity check equations that will determine the vectors $G_i, 0 \leq i \leq m$, defining the convolutional code.

Let us start with the linear code C stemming from the LFSR sequences. There is a corresponding $l \times N$ generator matrix G_{LFSR}. Clearly, $\mathbf{u} = \mathbf{u}_0 G_{LFSR}$, where $\mathbf{u}_0$ is the initial state of the LFSR. The generator matrix is furthermore written on systematic form, i.e., $G_{LFSR} = \left(I_l \ Z \right)$, where I_l is the $l \times l$ identity matrix. Given a generator matrix on this form, the parity check matrix is written as $P_{LFSR} = \left(Z^T \ I_{N-l} \right)$, where each row of P defines a parity check equation in C.

We are now interested in finding parity check equations that involve a current symbol u_n, an arbitrary linear combination of the B previous symbols $u_{n-1}, \ldots, u_{n-B}$, together with at most t other symbols. Clearly, t should be small and we mainly consider $t = 2$.

To find these equations we start by considering the index position $n = B+1$. Introduce the following notation for the generator matrix,

$$G_{LFSR} = \begin{pmatrix} I_{B+1} & Z_{B+1} \\ 0_{l-B-1} & Z_{l-B-1} \end{pmatrix}. \tag{3}$$

Parity check equations for u_{B+1} with weight t outside the first $B + 1$ positions can then be found by finding linear combinations of t columns of Z_{l-B-1} that add to the all zero column vector. This corresponds to the problem of finding weight t codewords in the code dual to Z_{l-B-1}.

For the case $t = 2$ the parity check equations can be found in a very simple way as follows. A parity check equation with $t = 2$ is found if two columns from

G_{LFSR} have the same value when restricted to the last $l - B - 1$ entries (the Z_{l-B-1} part). Hence, we simply put each column of Z_{l-B-1} into one of 2^{l-B-1} different "buckets", sorted according to the value of the last $l-B-1$ entries. Each pair of columns in each bucket will provide us with one parity check equation, provided u_{B+1} is included.

Assume that the above procedure gives us a set of m parity check equations for u_{B+1}, written as

$$u_{B+1} + \sum_{i=1}^{B} c_{i1} u_{B+1-i} + \sum_{i=1}^{\leq t} u_{j_{i1}} = 0,$$
$$u_{B+1} + \sum_{i=1}^{B} c_{i2} u_{B+1-i} + \sum_{i=1}^{\leq t} u_{j_{i2}} = 0,$$
$$\vdots$$
$$u_{B+1} + \sum_{i=1}^{B} c_{im} u_{B+1-i} + \sum_{i=1}^{\leq t} u_{j_{im}} = 0.$$

Now it follows directly from the cyclic structure of the LFSR sequences that *exactly* the same set of parity checks is valid for any index position n simply by shifting all the symbols in time, resulting in

$$u_n + \sum_{i=1}^{B} c_{i1} u_{n-i} + b_1 = 0,$$
$$u_n + \sum_{i=1}^{B} c_{i2} u_{n-i} + b_2 = 0, \qquad (4)$$
$$\vdots$$
$$u_n + \sum_{i=1}^{B} c_{im} u_{n-i} + b_m = 0,$$

where $b_k = \sum_{i=1}^{\leq t} u_{j_{ik}}$, $1 \leq k \leq m$ is the sum of (at most) t positions in **u**.

Using the equations above we next create an $R = 1/(m+1)$ bi-infinite systematic convolutional encoder. Recall that the generator matrix for such a code is of the form

$$\mathbf{G} = \begin{pmatrix} \ddots & \ddots & & \ddots & \\ & G_0 & G_1 & \dots & G_B & \\ & & G_0 & G_1 & \dots & G_B \\ & & & \ddots & \ddots & & \ddots \end{pmatrix}, \qquad (5)$$

where the blank parts are regarded as zeros. Identifying the parity check equations from (4) with the description form of the convolutional code as in (5) gives us

$$\begin{pmatrix} G_0 \\ G_1 \\ \vdots \\ G_B \end{pmatrix} = \begin{pmatrix} 1 & 1 & 1 & \dots & 1 \\ 0 & c_{11} & c_{12} & \dots & c_{1m} \\ 0 & c_{21} & c_{22} & \dots & c_{2m} \\ \vdots & \vdots & \ddots & \ddots & \vdots \\ 0 & c_{B1} & c_{B2} & \dots & c_{Bm} \end{pmatrix}. \qquad (6)$$

For each defined codeword symbol $v_n^{(i)}$ in the convolutional code we have an estimate of that symbol from the transmitted sequence **z**.

Consider $t = 2$. If $v_n^{(i)} = u_n$ (an information bit) then $P(v_n^{(i)} = z_n) = 1 - p$. Otherwise, if $v_n^{(i)} = u_{j_{1i}} + u_{j_{2i}}$ from (4) then $P(v_n^{(i)} = z_{j_{1i}} + z_{j_{2i}}) = (1-p)^2 + p^2$. Using these estimates we can construct a sequence

$$\mathbf{r} = \ldots r_n^{(0)} r_n^{(1)} \ldots r_n^{(m)} r_{n+1}^{(0)} r_{n+1}^{(1)} \ldots r_{n+1}^{(m)} \ldots ,$$

where $r_n^{(0)} = z_n$ and $r_n^{(i)} = z_{j_{1i}} + z_{j_{2i}}$, $1 \le i \le m$, that plays the role of a received sequence for the convolutional code. Then we have from the estimates that $P(v_n^{(0)} = r_n^{(0)}) = 1 - p$ and that $P(v_n^{(i)} = r_n^{(i)}) = (1-p)^2 + p^2$ for $1 \le i \le m$. Next, we enter the decoding phase.

To recover the initial state of the LFSR it is enough to decode l consecutive information bits correctly. Optimal decoding (ML decoding) of convolutional codes uses the Viterbi algorithm to decode.

The original Viterbi algorithm assumes that the convolutional encoder starts in state $\mathbf{0}$. However, in our application there is neither a starting state, nor an ending state. To deal with this we start by assigning the metrics $\log P(\mathbf{s} = z_1, z_2, \ldots, z_B)$ to each state $\mathbf{s}$ in the trellis. We then proceed to decode from $n = B$ as usual. Due to the difference regarding the endpoints, we run the Viterbi algorithm over a number of "dummy" information symbols, before we come to the l information symbols that we try to decode correctly. Similarly, after these l information symbols we continue the Viterbi algorithm over another set of "dummy" information symbols before the algorithm outputs the result. These are well known techniques in Viterbi decoding, and typically one has to decode approximately $4 - 5$ times B "dummy" information symbols, [4], before making the decoding decision. This means that decoding takes place over approximately $J = l + 10B$ information symbols, where the l symbols in the middle are regarded as the l bit sequence that we want to estimate. This estimate from the Viterbi algorithm is then used to provide the corresponding estimate of the initial state of the LFSR. This conclude the general description and we give a detailed summary of the algorithm for $t = 2$.

The Proposed Algorithm ($t = 2$)

Input: The systematic $l \times N$ generator matrix in the form

$$G_{LFSR} = \begin{pmatrix} I_{B+1} & \mathbf{g}_{B+2} & \cdots & \mathbf{g}_J & \mathbf{g}_{J+1} & \cdots & \mathbf{g}_N \end{pmatrix}.$$

1. For $J + 1 \le i, j \le N$ find all pairs of columns $\mathbf{g}_i, \mathbf{g}_j$ such that

$$(\mathbf{g}_i + \mathbf{g}_j)^T = (\underbrace{*, *, \ldots, *}_{B}, 1, \underbrace{0, 0, \ldots, 0}_{l-B-1}),$$

where $*$ means an arbitrary value. Then add

$$(u_{n-B}, u_{n-B-1}, \ldots, u_n, 0, 0, \ldots, 0) \cdot (\mathbf{g}_i + \mathbf{g}_j) + u_{n+i} + u_{n+j} = 0$$

to the set of parity check equations as in (4).

2. From this set, calculate $G_0, G_1, \ldots, G_B$ as in (6).
   Create a received vector $\mathbf{r}$ from $\mathbf{z}$ by $r_n^{(0)} = z_n$ and $r_n^{(i)} = z_{j_{1i}} + z_{j_{2i}}$ for $1 \leq i \leq m$, where j_{1i} and j_{2i} are the indices determined in 1.
3. Let $P(v_n^{(0)} = r_n^{(0)}) = 1 - p$ and $P(v_n^{(i)} = r_n^{(i)}) = (1-p)^2 + p^2$ for $B + 1 \leq n \leq l + 10B$.

Decoding part

4. For each state $\mathbf{s}$, let $\log(P(\mathbf{s} = (z_1, z_2 \cdots, z_B))$ be the initial metric for that state when we start the Viterbi algorithm at $n = B$.
5. Decode the received sequence $\mathbf{r}$ using the Viterbi algorithm from $n = B$ until $n = J$. Output the estimated information sequence $(\hat{u}_{5B+1}, \hat{u}_{5B+2}, \ldots, \hat{u}_{5B+l})$. Finally, calculate the corresponding initial state of the LFSR.

An Illustrating Example

Consider a length 40 LFSR, with feedback polynomial

$$g(x) = 1 + x + x^3 + x^5 + x^9 + x^{11} + x^{12} + x^{17} + x^{19} + $$
$$x^{21} + x^{25} + x^{27} + x^{29} + x^{32} + x^{33} + x^{38} + x^{40}.$$

An observed key sequence $\mathbf{z}$ of length $N = 40000$ is found to be correlated to the LFSR sequence with probability $1 - p = 1 - 0.1$. We want to decode the received sequence $\mathbf{z}$ transmitted over BSC(0.1) using the proposed method with memory $B = 10$.

We start by writing down the generator matrix G_{LFSR}. Then we search for suitable parity check equations by finding all pairs of columns in G_{LFSR} for which the last 29 index positions are all zero. Each such pair gives rise to one parity check equation with $t = 2$. In this case, the following three parity check equations were found

$$u_n + u_{n-1} + u_{n-8} + u_{n-10} + u_{n+4690} + u_{n+23655} = 0,$$
$$u_n + u_{n-2} + u_{n-3} + u_{n-4} + u_{n-7} + u_{n-8} + u_{n+4817} + u_{n+31970} = 0,$$
$$u_n + u_{n-2} + u_{n-3} + u_{n-4} + u_{n-5} + u_{n-9} + u_{n+18080} + u_{n+4626} = 0,$$

which are all valid for $1 \leq n \leq 8030$. We get a fourth codeword symbol by the information symbol u_n itself. Then we can identify

$$\begin{pmatrix} G_0 \\ G_1 \\ \vdots \\ G_B \end{pmatrix} = \begin{pmatrix} 1111 \\ 0100 \\ 0011 \\ \vdots \\ 0100 \end{pmatrix}.$$

Thus, we have created a rate $R = 1/4$ convolutional code having generator matrix

$$G = \begin{pmatrix} \ddots & \ddots & \ddots & \ddots & \ddots & \ddots & \ddots & \ddots & \ddots & \ddots & \ddots \\ & 1111\ 0100\ 0011\ 0011\ 0011\ 0001\ 0000\ 0010\ 0110\ 0001\ 0100 & \\ & \quad 1111\ 0100\ 0011\ 0011\ 0011\ 0001\ 0000\ 0010\ 0110\ 0001\ 0100 & \\ & \ddots & \ddots & \ddots & \ddots & \ddots & \ddots & \ddots & \ddots & \ddots & \ddots \end{pmatrix}.$$

Each r_n in the received sequence $r = r_0 r_1 \ldots$ for the convolutional code is created as

$$r_n^{(0)} = z_n,$$
$$r_n^{(1)} = z_{n+4690} + z_{n+23655},$$
$$r_n^{(2)} = z_{n+4817} + z_{n+31970},$$
$$r_n^{(3)} = z_{n+4626} + z_{n+18080},$$

and $P(v_n^{(0)} = r_n^{(0)}) = 0.9$ and $P(v_n^{(i)} = r_n^{(i)}) = 0.82, 1 \le i \le 3$. Finally we run the Viterbi algorithm, starting in $n = 10$ with all 2^{10} different states $(u_1, u_2, \ldots, u_{10})$. Each state have the initial metric $\log(P(u_1 = z_1)P(u_2 = z_2) \cdots P(u_B = z_B))$. After reaching $n = 140$, we output $(\hat{u}_{51}, \hat{u}_{52}, \ldots, \hat{u}_{90})$.

5   Simulation Results

In this section we present some simulation results for our algorithm. The obtained results are compared with the received results in [7,8,10]. We choose to use exactly the same case as tabulated in [10]. Thus all the simulations are based on a LFSR with length $l = 40$, and a weight 17 feedback polynomial which is

$$g(x) = 1 + x + x^3 + x^5 + x^9 + x^{11} + x^{12} + x^{17} + x^{19} +$$
$$x^{21} + x^{25} + x^{27} + x^{29} + x^{32} + x^{33} + x^{38} + x^{40}.$$

	[7,8]	[10]	Our Algorithm		
N/l	Alg. B	Alg.	$B = 13$	$B = 14$	$B = 15$
10^3	0.092	0.096	0.19	0.22	0.26
10^4	0.104	0.122	0.37	0.39	0.40

Table 1. Maximum p for different algorithms.

In Table 1 the maximum crossover probability p is shown for algorithm B in [7,8], the improvement in [10], and the proposed algorithm. Our results are generated for different sizes of the memory B. As a particular example, we can see that when we have $4 \cdot 10^5$ received symbols the proposed algorithm is successful up to more than $p = 0.4$ for memory $B = 15$, whereas the algorithm in [7,8] and the improvement in [10] are successful only up to a crossover probability of 0.104 and 0.122, respectively. In this case, $B = 15$, the proposed algorithm finds roughly 2300 parity checks and hence the embedded convolutional code is of rate roughly $R = 1/2300$. Also, the decoding takes place over $J = 200$ information symbols. The computational complexity is proportional to $J \cdot m \cdot 2^B$, and in the case $B = 15, M = 2300, J = 200$ the whole attack takes less than one hour on a PC.

Another interesting property to look at is the success rate, i.e., the probability for successful decoding given a channel with crossover probability p. In Figure 5 we plot the success rate as a function of p, when $B = 14$, for 40000 and 400000 received symbols, respectively.

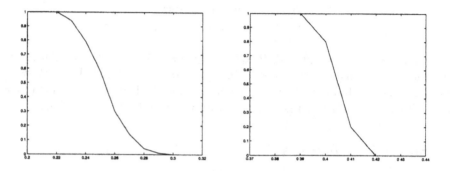

Fig. 5. Success rate for $B = 14$ with $N = 40000$ and $N = 400000$.

Finally, we make a comment regarding the theoretical performance of the proposed algorithm for $t = 2$. For fixed parameters l, B and N, we can determine the expected number of suitable parity checks, i.e., the parameter m. Then one can show that the success rate will be very close to 1 if the rate $R = 1/(m+1)$ is below the *cutoff rate R_0* [4] for the $\text{BSC}(2p(1 - p))$. However, we observe that the simulated results are very close to the capacity C of the $\text{BSC}(2p(1 - p))$, which is $C = 1 - h(2p(1 - p))$.

6   Conclusions

New methods for fast correlation attacks have been proposed, based on identifying an embedded convolutional code in the code $\mathcal{C}$ generated by the LFSR sequences of a fixed length N. The results show a significant improvement compared with previous work regarding general feedback polynomials. We have described the methods using an ordinary convolutional code together with standard Viterbi decoding. There are many different ways to extend these methods that can be considered in future work.

Firstly, we note that by permuting the columns of $\mathcal{C}$ before searching for parity checks, we receive a *time-varying* convolutional code. Secondly, the computational complexity of the Viterbi algorithm is growing exponentially with B, which means that in practice B is bounded to be at most $20 - 30$. But there are several other decoding algorithms, which are not ML, that have a much lower computational complexity. Examples of such algorithms are the M-algorithm (list decoding) and different sequential decoding algorithms [4]. They are promising candidates for improving the performance.

Finally, we also mention the possibility of using iterative decoding. This can roughly be described as follows. Identify several convolutional codes in C that have certain codeword symbols in common. Then decode them using APP (a posteriori probability) decoding algorithms [4] and pass the symbol probabilities to the other decoders. This procedure is iterated until the symbol probabilities have converged to 0 or 1. We believe that this is a very promising approach, and that we might see a further improvement in performance compared to the results in this paper.

References

1. V. Chepyzhov, and B. Smeets, "On a fast correlation attack on certain stream ciphers", In *Advances in Cryptology-EUROCRYPT'91*, Lecture Notes in Computer Science, vol. 547, Springer-Verlag, 1991, pp. 176–185.
2. A. Clark, J. Golic, E. Dawson, "A comparison of fast correlation attacks", *Fast Software Encryption, FSE'96*, Lecture Notes in Computer Science, Springer-Verlag, vol. 1039, 1996, pp. 145–158.
3. R. G. Gallager, *Low-Density Parity-Check Codes*, MIT Press, Cambridge, MA, 1963.
4. R. Johannesson, K. Sh. Zigangirov, *Fundamentals of Convolutional Codes*, IEEE Press, New York, 1999.
5. J. Leon, "A probabilistic algorithm for computing minimum weights of large error-correcting codes", *IEEE Trans. Information Theory*, vol. IT–34, 1988, pp. 1354–1359.
6. F. MacWilliams, N. Sloane, *The Theory of Error Correcting Codes*, North Holland, 1977.
7. W. Meier, and O. Staffelbach, "Fast correlation attacks on stream ciphers", *Advances in Cryptology-EUROCRYPT'88*, Lecture Notes in Computer Science, vol. 330, Springer-Verlag, 1988, pp. 301–314.
8. W. Meier, and O. Staffelbach, "Fast correlation attacks on certain stream ciphers", *Journal of Cryptology*, vol. 1, 1989, pp. 159–176.
9. A. Menezes, P. van Oorschot, S. Vanstone, *Handbook of Applied Cryptography*, CRC Press, 1997.
10. M. Mihaljevic, and J. Golic, "A fast iterative algorithm for a shift register initial state reconstruction given the noisy output sequence", *Advances in Cryptology-AUSCRYPT'90*, Lecture Notes in Computer Science, vol. 453, Springer-Verlag, 1990, pp. 165-175.
11. W. Penzhorn, "Correlation attacks on stream ciphers: Computing low weight parity checks based on error correcting codes", *Fast Software Encryption, FSE'96*, Lecture Notes in Computer Science, vol. 1039, Springer-Verlag, 1996, pp. 159–172.
12. T. Siegenthaler, "Correlation-immunity of nonlinear combining functions for cryptographic applications", *IEEE Trans. on Information Theory*, vol. IT–30, 1984, pp. 776–780.
13. T. Siegenthaler, "Decrypting a class of stream ciphers using ciphertext only", *IEEE Trans. on Computers*, vol. C–34, 1985, pp. 81–85.

14. L. Simpson, E. Dawson, J. Golic, M. Salamasizadeh, "Fast correlation attacks on the multiplexer generator", *Proc. IEEE 1998 International Symposium on Information Theory, ISIT'98*, 1998, p. 270.
15. J. Stern, "A method for finding codewords of small weight," *Coding Theory and Applications*, Springer-Verlag, 1989, pp. 106–113.

A   Convolutional Codes

This section reviews some basic concepts regarding convolutional codes. For a more thorough treatment we refer to [4]. A convolutional code is a linear code where the information symbols and the codeword symbols are treated as infinite sequences. In a general rate $R = b/c$, $b \leq c$ binary convolutional encoder (time-invariant and without feedback) the causal information sequence

$$\mathbf{u} = \mathbf{u_0 u_1} \ldots = u_0^{(0)} u_0^{(1)} \ldots u_0^{(b)} u_1^{(0)} u_1^{(1)} \ldots u_1^{(b)} \ldots$$

is encoded as the causal code sequence

$$\mathbf{v} = \mathbf{v_0 v_1} \ldots = v_0^{(0)} v_0^{(1)} \ldots v_0^{(c)} v_1^{(0)} v_1^{(1)} \ldots v_1^{(c)} \ldots,$$

where

$$\mathbf{v_t} = f(\mathbf{u_t}, \mathbf{u_{t-1}}, \ldots, \mathbf{u_{t-B}}).$$

The function f must be a linear function. Furthermore, the parameter B is called the encoder memory.

In our particular application we only consider convolutional codes for which the rate is of the form $R = 1/c$, i.e., $b = 1$, and thus we now adopt the notation

$$\mathbf{u} = u_0 u_1 \ldots,$$

where $u_i \in \mathbb{F}_2$. Since f is a linear function, it is convenient to write

$$\mathbf{v_t} = u_t G_0 + u_{t-1} G_1 + \cdots + u_{t-B} G_B,$$

where G_i, $0 \leq i \leq B$ is a $1 \times c$ matrix, i.e., a length c vector. Now we can rewrite the expression for the code sequence as

$$\mathbf{v_0 v_1} \ldots = (u_0 u_1 \ldots) \mathbf{G},$$

where

$$\mathbf{G} = \begin{pmatrix} G_0 \, G_1 \ldots G_B & & \\ & G_0 & G_1 & \ldots G_B \\ & & \ddots & \ddots & \ddots \end{pmatrix}, \tag{7}$$

and the blank parts of $\mathbf{G}$ is assumed to be filled with zeros. We call $\mathbf{G}$ the *generator matrix*. The encoder can be illustrated as in Figure 6.

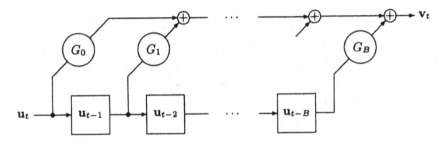

Fig. 6. A general convolutional encoder (without feedback).

The *state* of a system is a description that together with a specification of the present and future inputs, can determine the present and future outputs. From Figure 6 it is easy to see that we can choose the contents of the memory cells at time t as the encoder state σ_t at time t,

$$\sigma_t = u_{t-1}u_{t-2}\ldots u_{t-B}.$$

Thus the encoder has at most 2^B different states at each time instant. We can now consider all possible states σ_t as vertices in a graph and put an edge between two adjacent states σ_t and σ_{t+1} if and only if there is an information symbol u_t such that takes the state from σ_t at time t to σ_{t+1} at time $t+1$. This graph gives rise to a so called *trellis*. The *convolutional code* (or linear trellis code) is the set of all possible codeword sequences (possibly with a predetermined starting and ending state). If we label the edge in the trellis going from σ_t to σ_{t+1} with $v_t = u_t G_0 + u_{t-1}G_1 + \cdots + u_{t-B}G_B$ the set of codeword sequences will correspond to the set of possible paths in the trellis.

Example: Consider the rate $R = 1/2$ convolutional encoder with generator matrix

$$\mathbf{G} = \begin{pmatrix} 11\ 10\ 11 & & \\ & 11\ 10\ 11 & \\ & & \ddots\ \ddots\ \ddots \end{pmatrix}.$$

The encoder can be implemented as in Figure 7, and the corresponding trellis is depicted in Figure 8.

Suppose now that our trellis code is transmitted over the BSC with error probability p. We are interested in determining the most probable codeword from a received sequence $\mathbf{r}$,

$$\mathbf{r} = \mathbf{r}_0\mathbf{r}_1\ldots = r_0^{(0)}r_0^{(1)}\ldots r_0^{(c)}r_1^{(0)}r_1^{(1)}\ldots r_1^{(c)}\ldots.$$

This corresponds to a maximum likelihood decoding problem, ML decoding. The solution to the ML decoding problem for convolutional codes is the famous *Viterbi algorithm*.

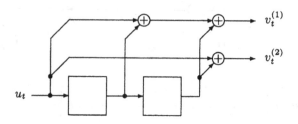

Fig. 7. A rate $R = 1/2$ convolutional encoder.

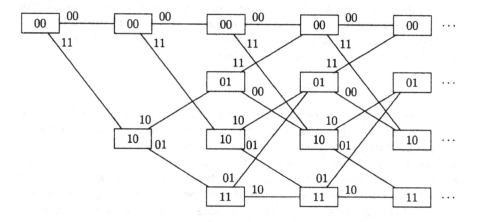

Fig. 8. A binary rate $R = 1/2$ trellis code.

The ML decoder chooses as its estimate $\hat{\mathbf{v}}$ a sequence $\mathbf{v}$ that maximizes $P(\mathbf{r}|\mathbf{v})$. Assuming that the starting and ending state is predetermined to be the zero-state, the ML decoder works as follows. Introduce the Viterbi branch metric, $\mu(\mathbf{r_n}, \mathbf{v_n}) = \sum_i \log P(r_n^{(i)}|v_n^{(i)})$ (One usually introduce a translation and a scaling in order to approximate the metric values with suitable integers [4]).

The Viterbi Algorithm

1. Assign the Viterbi metric to be zero at the initial node, and set $n = 0$.
2. For each node at depth $n+1$: Find for each of its predecessors at depth n the sum of the metric of the predecessor and the branch metric of the connecting branch. Find the maximum sum and assign this metric value to the node. Also, label the node with the shortest path to it.
3. If we have reached the end of the trellis, stop and choose as the estimate $\hat{\mathbf{v}}$ a path to the ending node with largest Viterbi metric; otherwise increment n and go to 2.

Cryptanalysis of an Identification Scheme Based on the Permuted Perceptron Problem

Lars R. Knudsen[1] and Willi Meier[2]

[1] Department of Informatics, University of Bergen, N-5020 Bergen
[2] FH-Aargau, CH-5210 Windisch

Abstract. This paper describes an attack on an identification scheme based on the permuted perceptron problem (PPP) as suggested by Pointcheval. The attack finds the secret key, a vector of n binary elements, in time much faster than estimated by its designer. The basic idea in the attack is to use several applications of a simulated annealing algorithm and combine the outcomes into an improved search. It is left as an open problem to what extent the methods developed in this paper are useful also in other combinatorial problems.

Keywords: Cryptanalysis. Identification Scheme. Perceptron Problem. Simulated Annealing.

1 Introduction

Since the advent of zero-knowledge proofs in 1985 [3], several interactive identification schemes have been proposed. The first protocols, like the Fiat-Shamir scheme [2], were based on number theoretic problems and used arithmetic operations with large numbers. In 1989, Shamir proposed a protocol of a different nature, based on the hardness of an NP-complete problem, the Permuted Kernel Problem, [8]. The distinctive features of this scheme are its use of small integers and its low requirement in memory and processing power. This makes the protocol more suitable for implementations on small processors like smart cards. In the sequel, two new problems (Syndrome Decoding and Constrained Linear Equations) were proposed by Stern, [9], [10]. More recently, Pointcheval presented identification schemes based on the so called Permuted Perceptron Problem (PPP), [5]. PPP is derived from the apparently simpler but still NP-complete Perceptron Problem (PP), which in turn is motivated by the well known Perceptron in Neural Computing. The identification schemes based on (PPP) are attractive as the operations needed are only additions and subtractions of integers of less than a byte. Thus the schemes are particularly well suited for implementation on 8-bit processors. In view of implementations with restricted memory and/or processing power, a precise determination of the security of the identification schemes is required.

The security of Shamir's and Stern's combinatorial schemes have repeatedly been the subject of publications (see e.g. [7] and the references quoted there).

J. Stern (Ed.): EUROCRYPT'99, LNCS 1592, pp. 363–374, 1999.
© Springer-Verlag Berlin Heidelberg 1999

The aim of this paper is to investigate the security of Pointcheval's schemes for the parameter values suggested in [5]. The main conclusion is that the smallest parameter values mentioned in [5], (m=101, n=117), are not secure enough for cryptographic applications. For some technical reasons, all parameters proposed in [5] are odd numbers.

In [5], several attacks against PPP have been tried. The most successful method was a probabilistic search, known as simulated annealing. As any solution for PPP is a solution for PP, simulated annealing is applied for solving PP sufficiently often, until a solution that also satisfies PPP may be found. It is reported in [5], that using this method, no solution for PPP for parameter sizes greater than 71 could be found, even if the search continued for long time. As a consequence of the investigations in [5], the parameters m=101, n=117 were suggested as a secure size for problem PPP.

In this paper, improved search algorithms to solve PP and PPP are developed. In both problems, a solution consists of a vector of size n with values +1 and -1 as entries. Our main aim is to adapt simulated annealing in several ways to directly solve PPP. Along this way, experiments revealed some intrinsic structure in the problem which enables to turn simulated annealing into an iterated search procedure until a solution may be found. Our algorithms turn out to be successful for instances of PPP for parameter sizes as large as 101 or larger. In particular, our methods are able to solve instances of the target case $m = 101, n = 117$. Even if we find a solution only in a fraction of cases for these parameters, our algorithm always identifies a subset of entries of a solution vector which are correct with high probability. Besides test results for solving example instances, our algorithms apply to give bounds for solving average instances: We can generally solve PP and thus also PPP up to 280 times faster than was estimated in [5].

In view of these results it is advisable for secure applications to choose the higher parameter values proposed in [5]. But then the efficiency of the schemes based on PPP compares less favourably to that of the other combinatorial schemes.

It is conceivable that the iterated search method as developed in this paper is also useful in certain other combinatorial and optimization problems.

2   The Permuted Perceptron Problem

We follow the notation in [5] where possible. A vector whose entries have value either +1 or -1 is called ϵ-vector, and similarly for matrices.

If X is a vector of size m, let X_i denote the ith entry in X.

Definition 1. *The Perceptron Problem* **PP**

> *Input   : An $m \times n$ ϵ-matrix, A.*
> *Problem : Find ϵ-vector V of size n, such that*
> $(AV)_i \geq 0$, *for all $i = 1, ..., m$.*

In [5] reference is made to [6], showing that PP is an NP-complete problem. It is possible to design a zero-knowledge identification protocol with every NP-complete problem provided one-way hash functions exist. However such a protocol will be efficient only if the underlying problem is hard already for moderately sized input parameters. Therefore in [5] the following variant of PP is proposed:

Definition 2. *The Permuted Perceptron Problem* **PPP**

> *Input*   : *An $m \times n$ ϵ-matrix, A,*
> *a multiset S of nonnegative numbers of size m*
> *Problem : Find ϵ-vector V of size n, such that*
> $$\{\{(AV)_i | i = \{1, \ldots, m\}\}\} = S.$$

Obviously, a solution for PPP is a solution for PP. In [5] this lead to the conclusion that the Permuted Perceptron Problem is more difficult to solve than the original Perceptron Problem. However, knowledge of the prescribed multiset S may give some hint to a solution of PPP. As will be shown in this paper, this is partially the case.

For cryptographic applications, one needs instances for which a solution is known. To get such instances, a (pseudo-) random ϵ-vector V of size n is chosen which will be a solution of the future instance. Hereafter a random ϵ-matrix A of size $m \times n$ is generated and modified in the following way:

For $i = 1, ..., m$
- If $(AV)_i < 0$, the ith row of A is multiplied by -1.
- If $(AV)_i \geq 0$, the ith row of A remains unchanged.

Finally the multiset $S = \{\{(AV)_i | \{i = 1, ..., m\}\}\}$ is computed. Consequently, (A, S) is an instance of the Permuted Perceptron Problem, with V as a solution.

In an identification protocol, each prover uses a public instance (A, S), and a secret key V. To convince a verifier of his identity, a prover gives him evidence of a solution V to the instance (A, S) by using a zero-knowledge protocol. The description of several such protocols based on PPP is the main subject of [5] and is not detailed here.

2.1   Simulated Annealing for PP and PPP

In [5] several attacks against the two problems PP and PPP have been tried in order to evaluate the security of identification protocols based on PPP. Thereby no structure was found which would enable to solve this problem with obvious methods like Gaussian elimination. The attacks made in [5] are outlined subsequently as far as they are relevant to our investigations:

Let $A = (a_{i,j})$ be an $m \times n$ ϵ-matrix and let $s_j = \sum_{i=1}^{m} a_{i,j}$ for $j = 1, \ldots, n$. In a first attempt to attack PPP, the majority vector M is considered, where $M_j = \text{sign } s_j, j = 1, ..., n$. As in applications m and n are odd, M_j is well defined.

Suppose (A, S) is an instance for PPP with solution V as considered in the previous section, where A is an $m \times n$ - matrix with $n \geq m$. Then in ([5], Theorem 4) it is stated that the vectors M and V are correlated:

$$\#\{j | M_j = V_j, j = 1, ..., n\} \approx 0.8n \tag{1}$$

This leads to a reduced search for a solution V of complexity of order $\binom{n}{0.2n}$. However this number exceeds the common security margin 2^{64} for $n \geq 95$.

Due to the inefficiency of the previous attack, in [5] the method of *simulated annealing* (SA) is proposed. This optimization method simulates a physical cooling process and has been applied for various combinatorial and engineering problems (see e.g., [1], [4]). The idea is to minimize in a probabilistic way an appropriate energy or cost function on a finite space of input variables which has to be provided with a distance measure. SA can only be expected to be efficient if the energy function is roughly continuous, i.e., if the energy difference for neighbouring inputs is bounded by a small number. In [5] the energy function for solving PP is chosen to be

$$E(V) = \frac{1}{2} \sum_{i=1}^{m} (|(AV)_i| - (AV)_i) \tag{2}$$

Figure 1 shows an example of a simulated annealing algorithm for PP and PPP.

Let $\alpha > 1$ and $0 < \beta < 1$. Let $rnd(0, 1)$ be a function which returns a random number between 0 and 1. Choose a candidate vector V' at random.

1. Calculate $E(V')$
2. If $E(V') = 0$ stop
3. $T = \alpha$
4. while $T > 1$ do
   (a) repeat n times
       i. Set $V'' = V'$ and change the sign of one randomly chosen entry of V''
       ii. if $E(V'') < E(V')$ then $V' = V''$
           else if $\exp((E(V') - E(V''))/T) > rnd(0, 1)$ then $V' = V''$
       iii. if $E(V') = 0$ stop
   (b) $T = \beta * T$

Fig. 1. Algorithm for simulated annealing for PP and PPP.

Clearly, $E(V)$ is minimum (i.e. $E(V) = 0$), if and only if the candidate vector V is a solution for PP. For this energy function, SA is successful for instances of size up to 200. On the other hand it is reported in [5] that this way no solution for PPP could be found for instances of size larger than 71, even if the algorithm has been tried for a few months. Hereby, PPP is attacked via solving PP, in the hope of finding a solution for PPP by performing the above SA algorithm sufficiently often. In order to estimate the complexity of such a procedure, the approximate number of solutions for PP and for PPP for average instances has been determined. This suggested that the complexity for solving PPP via PP is maximal if $n \approx m + 16$, in a practically relevant range $100 < m < 200$. As a result, three candidate sizes for instances of PPP are proposed in [5]. The smallest size recommended for the matrix A is $m = 101, n = 117$. It is concluded that solving instances of this size would need 12650 years, corresponding to a complexity of about 2^{64} elementary operations, hence a work load sufficient to guarantee the security of the underlying protocol.

3   Algorithms for Solving PPP

In this section it is shown how to improve the simulated annealing search for a solution of PPP. In general, the success of this search method essentially depends on the choice of a suitable energy function which is deeply connected to the underlying problem. In order to find such an energy function for PPP, let an $m \times n$ ϵ-matrix A and $Y = AV$ be given as before. Determine a histogram vector H over the integers such that $H_i = \#\{Y_j = i | j = 1, \dots, m\}$. With m, n odd, H_i is set only for odd values of i, $1 \leq i \leq n$. In a simulated annealing search let V' denote the candidate for the secret key V, let $Y' = AV'$ and let H_i' denote the histogram vector of Y'. Then an obvious proposal for an energy function would be the distance (in a suitable sense) between the correct and the candidate histogram vector: $E(V) = \sum_{i=1}^{n}(|H_i - H_i'|)$ (where the summation index i is only taken over odd values). For this definition of energy, certainly $E(V) = 0$ if and only if V is a solution for PPP. Experiments have shown however, that this function E is of no use for a search. The reason is that $E(V)$ may change too much even if only a single sign in V is changed. Hence this choice of $E(V)$ is not continuous as is necessary for simulated annealing to work. As a consequence, one may try to combine this function with the function which has already shown to be successful for PP. This motivates an energy function of the form

$$E(V) = g_1 \sum_{i=1}^{m}(|(AV')_i| - (AV')_i) + g_2 \sum_{i=1}^{n}(|H_i - H_i'|). \tag{3}$$

The first part of this function is a multiple of the sum of all negative entries Y_i', the second part is a multiple of the distance between the correct and candidate histogram vector. It is clear that a solution for PPP (and PP) has been found when $E(V) = 0$. The values of g_1 and g_2 provide a weighting of the two sums. It has proved useful to choose $g_1 \geq 30$ and $g_2 = 1$. This introduces additional energy (or penalty) for candidate vectors resulting in negative entries in Y' and such candidate vectors never result in a solution for PPP.

Also, the following equation can be used to increase the probability of success of the PPP-search algorithm.

Corollary 3. *Let $A = (a_{i,j})$ be an $m \times n$ ϵ-matrix. Let $s_j = \sum_{i=1}^{m} a_{i,j}$ for $j = 1, \ldots, n$. If $AV = Y$ then*

$$\sum_{j=1}^{n} s_j \cdot V_j = \sum_{i=1}^{m} Y_i. \tag{4}$$

Note that the righthand side of Eq. 4 is equal to the sum of the elements in the multiset S.

As mentioned in the previous section, the majority vector is in accordance with V, the secret key, for about 80% of the entries. As before, let s_j be the vector computed as the sum of every column of A. By inspection of the majority vector and the solution vector it follows that this fraction increases for larger values of abs(s_j). As an example, for instances of PPP where $m = n = 73$, experiments show that

$$\text{Prob}(M_j = V_j | \text{abs}(s_j) \geq 11) \approx 0.94, \tag{5}$$

whereas for $m = n = 101$ this probability is still 0.92. Moreover for $m = n = 73$ and $m = n = 101$ tests show that on the average abs(s_j) ≥ 11 for 28 respectively 46 entries, or 38% respectively 46% of the entries. Similarly for $m = 101, n = 117$ experiments show that

$$\text{Prob}(M_j = V_j | \text{abs}(s_j) \geq 11) \approx 0.91 \tag{6}$$

and that on the average abs(s_j) ≥ 11 for about 44% of the entries. Note that an attacker knows the vector s, which he can compute from the matrix A. Thus, an attacker can exploit that for large entries of s the corresponding entry in the secret key V is known with a high probability.

3.1   The Search Algorithm

A first attempt for an algorithm is the following. Run the simulated annealing algorithm, Figure 1, t times with the energy function (3). In each run, record the candidate vector V' which gave rise to the lowest value of the energy function. Find the entries, say the set I, which have the same sign in all these t vectors. Run the simulated annealing algorithm again t times, but instead of choosing a random starting vector, let the entries in I have the values of the first t tests; the remaining entries are chosen at random. Record another t vectors and find another set I and repeat the procedure until a solution is found. This algorithm, hereafter called the *PPP-search algorithm*, has proved very successful for PPP with larger values of m and n. The values of α and β play a crucial role for the success of the simulated annealing algorithm. Also, the initial values, that is the values of first t randomly chosen starting vectors, are very important for the further progress of the PPP search algorithm. In the tests reported later in the paper, $\alpha = n$ was chosen.

There are several possible variants of the PPP-search algorithm. The following modifications have been tried with varying success.

1. When in t runs of SA an entry ends up with the same sign t times, this entry is fixed to this value throughout the entire search.
2. Exploit and incorporate Equation 4 and the facts (5) or (6).
3. Repeat the PPP-search algorithm until a sufficiently high number of entries, say u, have the same (correct) sign in t runs of the simulated annealing algorithm. If u is big enough, an exhaustive search for the remaining $n - u$ entries might be possible. Alternatively, one can exhaustively fix a subset of the remaining $n - u$ entries and continue the PPP-search algorithm until a solution is found.

The first variant is faster than the original. Once an entry has the same sign in t consecutive runs of SA, the entry is fixed throughout the remaining search. Clearly, if this entry is correct, this variant will improve the search, but conversely if this entry is incorrect and stays fixed, the search will never find the secret key vector V. It is still possible, though, that the search will find a vector $V' \neq V$, such that $AV' = Y$.

A second variant of the PPP-search algorithm is to choose the entries of the majority vector instead of random values in the starting vector of the first t runs of the simulated annealing algorithm. This has the effect that the set I is larger, however the number of incorrectly assigned entries increases. An alternative is to use the majority vector only for entries where $\text{abs}(s_j)$ has a predetermined high value.

In the following we show how an exhaustive search for a remaining set of entries can be done. First, in the PPP-search algorithm record the frequency of the entries in the candidate vectors V' from SA which gave rise to the lowest value of the energy function. Let W be a vector with n integer entries. After each run of SA set $W = W + V'$. After sufficiently many runs of the above algorithm, tests show that W holds the correct sign for a substantial part of the entries of the correct secret key V. Assume the search has found about $0 < u < n$ correct entries in the candidate solution vector. It turns out that when u is not too small, e.g., if $u > n/4$, the vector W has the correct sign in 85-90% of all entries. That is, let V be the chosen solution vector; if $W_i < 0$ ($W_i > 0$) then with probability $0.85 \ldots 0.90$, $V_i = -1$ ($V_i = 1$). In other words, only 10-15% of the remaining $n - u$ entries are wrong. Thus if $n - u$ is not too big, it is possible to exhaustively try to determine which of these entries are not in accordance with the sign of W. Assume that u entries have been found, and that for n_w entries in W it holds that $W_i \cdot V_i < 0$, whereas $W_i \cdot V_i > 0$ for the remaining entries. Then, if n_w is known, an exhaustive search at this point takes no longer than

$$\binom{n - u}{n_w}$$

steps. An attacker cannot know the exact value of n_w, so the exhaustive search must be repeated for a few values of n_w in the neighbourhood of $0.85 \times n$. Alternatively, one can do an exhaustive search of a subset of the remaining $n - u$ entries. The idea is to fix a subset of entries and continue the PPP-search for a

m	n	Running time	Pointcheval's estimate
101	117	0.3	85
121	137	0.5	130
151	167	1.5	180

Table 1. Running time in seconds for the PP search algorithm (averaged over 50 tests).

number of steps. When the entries in the subset are assigned correct values, the PPP-search will find a solution.

The search can be further improved by incorporating the Equation 4. When doing the exhaustive search for a few entries, one starts by assigning values to the entries, j, for which $abs(s_j)$ are small. After the assignment of a few entries the remaining entries are either forced by Eq. 4 or lead to contradiction. It is assumed that with this improvement the exhaustive search part takes less than $\binom{n-u-1}{n_w}$ operations.

As shown in the next section the PPP-search algorithm with $u = n$ works well on instances of PPP where $m \geq n$. In the case where $m < n$, as recommended by Pointcheval, the algorithm is less successful. This may be due to the fact that for a given matrix A and a multiset S there are several solutions to the problem and the search algorithm is not able to converge to one single solution. This is supported by fact that in the cases where $m > n$ the probability is high that there is only a single solution and the algorithms terminate fast and with a high probability of success.

In the cases where $m < n$ the probability of success can be increased by choosing $u < n$ and performing also the exhaustive search part of the algorithm.

4 Test Results

The computer used for the tests in this section is a Sun Sparc Ultra-1. When running times are given in seconds or minutes, this is the real time it took the tests to succeed when running on a UNIX system with shared resources. Thus, when implementing on a single dedicated machine, the expected time of the algorithms will be lower.

4.1 Results for PP

The PPP-search algorithm can also be used to find solutions for PP by using the energy function (3). For this, PPP-search variant no. 1 choosing $g_1 = 30$ and $g_2 = 0$ in the energy function has proved very successful. Table 1 lists the results of 50 tests on PP with the recommended values of m and n. The running time is taken as the total real time divided by 50. All tests succeeded and found a solution.

m	n	Tests	Solutions	Running time
73	73	50	19	12 min.
81	81	50	11	22 min.
101	101	50	5	84 min.
101	81	50	32	2 min.
121	81	50	43	1 min.

Table 2. Results of the fast search algorithm for PPP. Running time is the average time of all successful tests.

Note that our results listed in Table 1 for PP also improve the bounds determined in [5] for the complexity of average instances of PPP by the same factors. E.g., for $m = 101, n = 117$, our experiments together with the estimates in [5] give a complexity of 2^{56} elementary operations. However, depending on the instance, our experiments with Fast Search as described in the next section show that this complexity may be much lower.

4.2   PPP - Fast Search

When implementing attacks on instances on PPP we found that some solutions were found much faster than others. In this section the results of a series of "fast tests" are given. The attacks do not find the solution of all instances of PPP, but when they find a solution, it is found fast. As mentioned earlier, PPP-search variant no. 1 is faster than the original one. When in t runs of the simulated annealing algorithm an entry holds the same sign in the t output vectors, such an entry is fixed is future tests. In these tests, whenever the search algorithm fixed an entry in the candidate solution vector different from the chosen one, the algorithm aborted. These test results are therefore very pessimistic. It might very well be that when the search algorithm was aborted it was converging towards a solution different from the chosen one. Table 2 lists the results of tests on several instantiations of PPP. The running times are pessimistic, taken as the total real time of all 50 tests divided by the number of correct solutions found. In [5] it was reported that no solutions were found for tests on instances of PPP with $m, n > 71$, even after running several months. Our results show that in about 40% of the cases a solution can be found for $m = n = 73$ in just 12 minutes running time. The cases $m = 101, n = 81$ and $m = 121, n = 81$ are included to illustrate how well the PPP-search algorithm works when $m > n$.

For the target version $m = 101, n = 117$ the PPP-search algorithm variant no. 1 was implemented with $t = 30$, $\alpha = 117$ and $\beta = 0.97$ in the simulated annealing part. In 100 tests the algorithm found a solution in one of the cases. Using the exhaustive search extension, the solution can be found much faster than estimated in [5] in 9 of the tests. Table 3 lists the results of 100 tests. In one test the solution was found using about 2^{31} simple operations. In one other test 72 entries were found when the search stopped. At this point the vector W held the correct sign in 102 of the 117 entries. Thus, an exhaustive search for the

m	n	Tests	Solutions	Complexities (estimates)
101	117	100	9	$2^{31}, 2^{38}, 2^{43}, 2^{45}, \ldots, 2^{52}$

Table 3. The results of a fast PPP-search for $m = 101$, $n = 117$.

m	n	Tests	Solutions	Complexity (estimated)
73	73	20	19	$2^{26}, \ldots, 2^{33}$
101	101	20	14	$2^{30}, 2^{33}, 2^{43}, \ldots, 2^{46}$

Table 4. The results of advanced search algorithms for PPP. Complexities are the total number of simple operations.

remaining 47 entries can be done in time about $\binom{46}{15} \approx 2^{38}$ simple operations. The complexities of other cases are done similarly. The stated complexities are of the exhaustive search for the remaining $n - u$ entries which is greater than the first part using simulated annealing. Therefore in a real-life situation, when attacking single instances, it may be very advantageous to run a more complex simulated annealing part.

4.3 PPP - Advanced Search

In the previous section a solution was found only for a fraction of all instances. Not surprisingly, increasing the complexity of the tests also increases the probability of success. Table 4 lists the results of a series of tests. For the case $m = n = 73$ the PPP-search algorithm was used with $t = 40$ using variant no. 2. In the simulated annealing algorithm $\alpha = 73$ and $\beta = 0.85$ was used. First all entries in the candidate vector for which $abs(s_j) \geq 11$ were fixed $V'_j = M_j$. As mentioned earlier, for these entries these assignments introduce only a few errors. Subsequently, the PPP-search algorithm was run for a certain number of steps. If no solution was found, one of the entries fixed in the beginning was given the opposite sign and the PPP-search algorithm was restarted. After only a small number of restarts a solution was found in 19 of 20 cases. The tests show that a solution for $m = n = 73$ can be found in a few hours with a high probability of success. The total number of simple operations of the attacks varied from 2^{26} to 2^{33}. This attack variant might very well be adapted to the cases $m = n = 101$ and $m = 101, n = 117$.

Variant no. 3 for PPP with $m = n = 101$ and where $t = 40$ was implemented. First the PPP-search algorithm was run with a complexity of maximum 2^{34} simple operations. In 2 of 20 tests a solution was found. For the remaining tests a set of 20 entries which were not found by the search were fixed to the correct value of V and the PPP-search was continued. In 12 of the 18 tests a solution was found in at most 200 runs of the SA algorithm after the assignment of the 20 entries. 200 runs of the SA algorithms in these tests equal about 2^{28} steps. In a real-life situation one should repeat the procedure for all possible values of the 20 entries, making the total complexity $2^{34} + s * 2^{20}$ steps for $s \leq 2^{28}$.

However, the guessing of 20 binary values will succeed after about 2^{19} attempts and furthermore, the 20 entries are correlated to the majority vector, a fact we did not incorporate in these tests.

To measure the success of our algorithms an SA step of relatively low complexity was chosen, such that the first part of the test could be implemented in reasonable time. In a real-life setting when attacking a single instance of the PPP one would choose the parameters such that the complexity of both parts of the algorithm would be roughly equal. With a more complex SA step (higher values of α and β) one can expect more correct entries to be identified by the time of the brute-force assignments of additional entries. For instances of PPP with $m = n = 101$ run the first part of the above attack with a complexity of, say, 2^{40}, whereafter most tests would find a solution shortly after an assignment of 20 correct entries, such that the total complexity would remain around 2^{40} steps. It is conjectured that a solution can be found for a large part of all PPP instances where $m = n = 101$ in time at most 2^{40}.

It is further conjectured that the same variant of the attack is applicable to instances of PPP with $m = 101, n = 117$ where the first part of the above attack has complexities of about $2^{45}, \ldots, 2^{50}$, which also would be the total complexity of the attack.

5   Suggestions for Future Work

As can be seen from the previous sections there are many possible variants of the PPP-search algorithms. The simulated annealing algorithm is very sensitive to the values of the parameters, and a small change in β sometimes produces very different results. As an effect also the behaviour of the PPP-search algorithm changes. In tests on PPP with $m = 31, n = 47$, different sets of solutions were found with different values of the parameters. In the PPP-search algorithm the value of the parameter t is important. For some instances of PPP, solutions are found fast with a low value of t, whereas in other instances a higher value of t seems better. It might be possible also to improve the PPP-search algorithm by using more than one energy function. Either alternate between energy functions from one run of the simulated annealing algorithm to another, or use one energy function in t consecutive runs and another energy function in the next t runs.

It is likely that the limits of our methods as determined in this paper are not optimum and that other variants and/or combinations of the parameters will improve the results.

6   Conclusion

In this paper it was demonstrated that the identification schemes based on the permuted perceptron problem are several orders of magnitudes less secure than previously believed. Therefore it is recommended not to use these schemes with the suggested smallest parameters. As a consequence these identification schemes compare less favorably to other combinatorial schemes. The iterated search methods developed in this paper can be formulated in general terms and thus might be useful also in other combinatorial problems.

References

1. E. Aarts and J. Korst. Simulated Annealing and Boltzmann Machines. Wiley, New York, 1989.
2. A. Fiat, A. Shamir. How to prove yourself: practical solutions of identification and signature problems. In A.M. Odlyzko, editor, *Advances in Cryptology - CRYPTO'86*, LNCS 263, pages 186-194. Springer-Verlag, 1987.
3. S. Goldwasser, S. Micali, C. Rackoff. Knowledge complexity of interactive proof systems. In Proceedings of the 17-th ACM Symposium on the Theory of Computing STOC, ACM, pages 291-304, 1985.
4. M. H. Hassoun. Fundamentals of Artificial Neural Networks. MIT Press, London, England, 1995.
5. D. Pointcheval. A new identification scheme based on the perceptrons problem. In L.C. Guillou and J.-J. Quisquater, editors, *Advances in Cryptology - EURO-CRYPT'95*, LNCS 921, pages 319 - 328. Springer Verlag, 1995.
6. D. Pointcheval. Les réseaux de neurones et leurs applications cryptographiques. Tech. rep. Laboratoire d'Informatique de l'École Normale Supérieure, February 1995. LIENS-95-2.
7. G. Poupard. A Realistic Security Analysis of Identification Schemes based on Combinatorial Problems. European Transactions on Telecommunications, vol. 8, Nr. 5, pages 471-480, 1997.
8. A. Shamir. An efficient identification scheme based on permuted kernels. In G. Brassard, editor, *Advances in Cryptology - CRYPTO'89*, LNCS 435, pages 606-609. Springer-Verlag, 1990.
9. J. Stern. A new identification scheme based on syndrome decoding. In D.R. Stinson, editor, *Advances in Cryptology - CRYPTO'93*, LNCS 773, pages 13-21. Springer-Verlag, 1994.
10. J. Stern. Designing identification schemes with keys of short size. In Y.G. Desmedt, editor, *Advances in Cryptology - CRYPTO'94*, LNCS 839, pages 164-173. Springer-Verlag, 1994.

An Analysis of Exponentiation
Based on Formal Languages

Luke O'Connor

IBM Research Division
Zurich Research Laboratory
Säumerstrasse 4, Rüschlikon
CH-8803, Switzerland
oco@zurich.ibm.com

Abstract. A recoding rule for exponentiation is a method for reducing the cost of the exponentiation a^e by reducing the number of required multiplications. If $w(e)$ is the (hamming) weight of e, and $\bar{e}$ the result of applying the recoding rule A to e, then the purpose is to reduce $w_A(\bar{e})$ as compared to $w(e)$. A well-known example of a recoding rule is to convert a binary exponent into a signed-digit representation in terms of the digits $\{1, \bar{1}, 0\}$ where $\bar{1} = -1$, by recoding runs of 1's. In this paper we show how three recoding rules can be modelled via regular languages to obtain precise information about the resulting weight distributions. In particular we analyse the recoding rules employed by the 2^k-ary, sliding window and optimal signed-digit exponentiation algorithms. We prove that the sliding window method has an expected recoded weight of approximately $n/(k+1)$ for relevant k-bit windows and n-bit exponents, and also that the variance is small. We also prove for the optimal signed digit method that the expected weight is approximately $n/3$ with a variance of $2n/27$. In general the sliding window method provides the best performance, and performs less than 85% of the multiplications required for the other methods for a majority of exponents.

1 Introduction

One of the fundamental operations in cryptography is exponentiation a^e over groups such as $\mathbb{Z}_p^*, \mathbb{Z}_n$, general finite fields, and the group of points on an elliptic curve [21,6,7]. The classical approach to performing this task is the binary method, and the complexity of the exponentiation is usually measured in terms of the number of squarings and multiplications required to determine a^e. Let $e = e_{n-1}e_{n-2}\cdots e_1 e_0$ be an n-bit exponent, $e_i \in \{0,1\}, 0 \le i < n$, and let $w(e) = \sum_{i=0}^{n-1} e_i$ be the weight of e. A simple analysis of the binary method shows that s squarings and $w(e) - 1$ multiplications are required, where s is the index of the most significant bit in e. Many general exponentiation algorithms offer complexity improvements over the binary method include the sliding-window method ([18,4,11] for example), signed-digit representations [20,19,16,13,23], the signed-window method [17], Lempel-Ziv recoding [24], and the string replacement method [8]. The reader is advised to see [18] for a thorough survey.

J. Stern (Ed.): EUROCRYPT'99, LNCS 1592, pp. 375–388, 1999.
© Springer-Verlag Berlin Heidelberg 1999

The common approach of these and other methods is to 'collect' exponent bits according to some rule for reducing the weight of e, hence reducing the number of required multiplications. For example, k consecutive bits are collected to form a single digit in the 2^k-ary method [15], and the binary signed-digit method [8] replaces runs of two or more 1's with just two bits, one signed and one unsigned. We will refer to these and other rules for reducing the weight of e as a *recoding rule*. For a given recoding rule A let $\bar{e}_A = \bar{e}_t\bar{e}_{t-1}\ldots\bar{e}_1\bar{e}_0$ be the result of applying A to e, and let $w_A(\bar{e}) = \sum_{i=0}^{t}[\bar{e}_i \neq 0]$ denote the recoded weight of $\bar{e}_A$. Once the recoding rule is applied, a variant of the b-ary method (b not necessarily equal to 2^k) can be used to complete the exponentiation, potentially after some precomputation has been done. In practice, the exponent recoding and arithmetic operations of the exponentiation are interleaved (see [18] for examples of specific algorithms).

To analyse the computational saving of recoding e according to rule A, we are required to examine the distribution of $w_A(\bar{e})$, and also the cost of any precomputation implied by A. For the 2^k-ary method, $w_A(\bar{e})$ is approximately binomial with parameters $b(n/k, (2^k - 1)/2^k)$, and it is therefore reasonably understood. It is surprising however that in general other recoding methods are discriminated between solely on the basis of $\mathbf{E}[(w_A(\bar{e})]$ and $\max_e w_A(\bar{e})$, the average and worst case weight recodings respectively (see [17,8,23] for such comparisons). We assert that $\mathbf{E}[(w_A(e)]$ and $\max_e w_A(e)$ provide information about the distribution of $w_A(e)$, but without second order statistics, such as the variance, the accuracy and usefulness of this information is uncertain.

In a recoding rule A that produces $\bar{e}_A = \bar{e}_t\bar{e}_{t-1}\ldots\bar{e}_1\bar{e}_0$ from e, often the defining properties of the $\bar{e}_i$ are quite simple, such as $\bar{e}_i = 01^k$ (a run of 1's terminated by a 0, $k \geq 2$) used in signed-digit recoding for example. This reflects the requirement that the recoding rule must be efficient, and also that simple recoding rules can be effective in reducing the cost of exponentiation. For many recoding rules of practical interest, the $\bar{e}_i$ can be represented as elements of a specified regular language [10], implying that the recoding can be performed by an appropriate deterministic finite automata (DFA). For example, the recoding rules presented in [19,17] are analysed in terms of their respective recoding DFAs.

The main contribution in this paper is to propose a framework for analysing the weight distribution of recoding rules which can be described by regular languages. For a recoding rule A, the basis of our analysis is to define a bivariate generating function (bgf) $G_A(x, z) = \sum_{n,m \geq 0} a_{n,m} z^m x^n$ such that

$$\Pr(w_A(\bar{e}) = m \mid \#e = n) = a_{m,n}/2^n,$$

where $\#e$ is the bit length of e. Thus $\Omega_n = \{m \mid a_{m,n} \neq 0, 0 \leq m \leq n\}$ and $\Pr(X_n = m) = a_{m,n}/2^n$, will be the probability space describing the distribution of weights for n-bit exponents recoded according to A. For the binary method (BM), the relevant bgf (derived below) is

$$G_{BM}(x, z) = \frac{1}{1 - (xz + x)} = \frac{1}{1 - x(1 + z)} = \sum_{n \geq 0} x^n \sum_{m \geq 0} \binom{n}{m} z^m \quad (1)$$

which indicates that the weights are distributed binomially, as expected. In general we will derive $G_A(x, z)$ from a A by considering the recoding rules prescribed by A as being performed by a DFA, pass to regular languages, and then enumerate the set of n-bit exponents whose recoded weight is m using standard combinatorial methods (see [22, p.377] or [3, p.342] for example). This analysis technique covers many recoding methods of practical interest, but, for example, does not include the Lempel-Ziv exponentiation method of Yacobi [24], since in this case the $\bar{e}_i$ are produced by the recoding are non-regular (a context-free grammar would be required).

To demonstrate the generality of this approach, we analyse the weight distribution of recoded exponents for the 2^k-ary method (§3), sliding window method (§4) and the optimal signed-digit method (§5). We analyse the 2^k-ary method as it provides an obvious improvement over the binary method, and its analysis is instructive to the bgf approach. The sliding window method was selected since no satisfactory analysis exists (see [11,14] for partial results), and yet it is described as 'the recommended method' for general exponentiation [18, p.617]. We also selected the optimal signed-digit method[8] for analysis since this method and its variants are often suggested for performing elliptic curve scalar multiplication, since group inversion is essentially free [19,16,17,23]. For the 2^k-ary (k, TKM), k-bit sliding window (k, SW), and optimal signed digit (OSD) methods the bgfs for weight are as follows:

$$G_{k,TKM}(x, z) = \frac{z\left(\frac{1-2^k x^k}{1-2x} - \frac{1-x^k}{1-x}\right) + \frac{1-x^k}{1-x}}{1 - x(2^k - 1)x^k - x^k}, \tag{2}$$

$$G_{k,SW}(x, z) = \frac{1 - 2x + zx - zx^k 2^{k-1}}{(1 - x - zx^k 2^{k-1})(1 - 2x)}, \tag{3}$$

$$G_{OSD}(x, z) = \frac{1 - x + xz + -2zx^2 + x^2 z^2}{1 - 2x + x^2 - 2zx^2 + 2zx^3}. \tag{4}$$

Using standard transformations on bgfs we are able to obtain the numerical values of $\mathbf{E}[w_A(\bar{e})]$ and $\mathbf{Var}[w_A(\bar{e})]$ for each bgf from (2) - (4), and thus make comparisons on the number of required multiplications for each method. Since the TKM approximates the binomial distribution $b(n/k, (2^k - 1)/2^k)$, the expectation and variance of $w_{k,TKM}(\bar{e})$ can be approximated accurately. Similar computations for the sliding window method are difficult, but we have been able to show by direct calculation that $\mathbf{E}[w_{k,SW}(\bar{e})] \sim n/(k + 1) + \frac{k(k-1)}{2(k+1)^2}$ for $n \in \{512, 1024\}$, $k \in \{2, 3, \ldots, 6\}$. We currently have no expression for $\mathbf{Var}[w_{k,SW}(\bar{e})]$ but we note that direct calculations show it to be small (for example less than 7 for 6-bit windows on 1024-exponents), and decreasing with k. The expectation and variance the OSD method can be analysed exactly, mainly because there is no window parameter k to complicate the analysis. We prove that $\mathbf{E}[w_{OSD}(\bar{e})] = \frac{n}{3} + \frac{4}{9} - \frac{4(-1)^n}{9 \cdot 2^n}$, and $\mathbf{Var}[w_{OSD}(\bar{e})] = \frac{2n}{27} + \frac{14}{81} + \frac{2n}{27 \cdot 2^n} + o(1)$.

The paper is organised as follows. In §2 we review some concepts of regular languages, and give the principal enumeration theorems. In §3 we derive the bgfs for the binary and 2^k-ary method, demonstrating our method of enumeration. In §4 we analyse the sliding window method, and then in §5 we analyse the optimal signed digit method. Conclusions and open problems are presented in the last section.

2   Regular Expressions and Generating Functions

Regular expressions are defined recursively [10] as follows: if R and S are regular expressions then so is $R+S$ (union), RS (concatenation) and R^* (Kleene closure) where $R^* = \sum_{k \geq 0} R^k = \epsilon + R + RR + RRR + \cdots$. Also let r^k denote the concatenation of r with itself k times, and let $r^+ = r^* - \epsilon$. Over a binary alphabet we will call 1^k a k-run, $k \geq 1$, and any word ω that is a k-run will also be simply referred to as a run.

A regular expression R generates words $\omega = w_1 w_2 \cdots w_n$, $w_i \in \Delta$, and ω is said to have length n, written as $\#\omega = n$. The set of all words generated by the regular expression R, denoted by L_R, is called the regular language generated, or given, by R. Let $L_R^n \subseteq L_R$ denote the set of words in L_R of length $n \geq 0$. We will say that the (ordinary) generating function $G_R(x) = \sum_{n \geq 0} a_n x^n$ enumerates L_R by length if $a_n = \#L_R^n$ for all $n \geq 0$. Let $[x^n]$ be the operator that extracts the coefficient of x^n, so that $[x^n]G_R(x) = a_n$. It is clear that the regular expression $R = (1+0)^*$ generates the language L_R which is the set of all binary strings, and since $|L_R^n| = 2^n$, L_R is enumerated by the geometric series $G_R(x) = 1/(1 - 2x)$. The key property that permits $G_R(x)$ to be derived from R directly is given in the next definition.

Definition 1. A regular expression R is *unambiguous* if there is only one way for R to generate each $\omega \in L_R$.   □

For example $(1 + 0)^*$ is unambiguous, but $(1 + 0 + 10)^*$ is ambiguous since the string $\omega = 10$ can be generated by concatenating 1 and 0, or simply selecting 10. Since it is known that any regular language can be generated by an unambiguous regular expression [22, p.378], the following theorem due to Chomsky and Schutzenberger [5] will be our main enumeration tool.

Theorem 2. Let R and S be unambiguous regular expressions, that are enumerated by the gfs $G_R(x)$ and $G_S(x)$. Then if $R + S$, RS and R^* are also unambiguous, $G_R(x) + G_S(x)$ enumerates $R + S$, $G_R(x)G_S(x)$ enumerates RS, and $1/(1 - G_R(x))$ enumerates R^*.   □

Recall that our goal is to determine the bgf $G_A(x, z) = \sum_{n,m \geq 0} a_{n,m} z^m x^n$ such that $a_{m,n}$ is the number of n-bit exponents recoded to weight m by algorithm A. Fortunately Theorem 2 can also be applied to these bgfs since for the exponent recoding algorithms under consideration there exists a representation of the algorithms in terms of regular expressions for which $w(R+S) = w(R)+w(S)$ and $w(RS) = w(R)w(S)$. We restate this result formally as a corollary to Theorem 2.

Corollary 3. Let R and S be unambiguous regular expressions, that are enumerated by the bgfs $G_R(x, z)$ and $G_S(x, z)$. Then if $R + S$, RS and R^* are also unambiguous, $w(R + S) = w(R) + w(S)$, and $w(RS) = w(R)w(S)$ then $G_R(x, z) + G_S(x, z)$ enumerates $R + S$, $G_R(x, z)G_S(x, z)$ enumerates RS, and $1/(1 - G_R(x, z))$ enumerates R^*.   □

An advantage of using $G_A(x, z)$ for enumeration is that the expectation and variance of $w_A(\bar{e})$ can be directly determined from manipulating $G_A(x, z)$. Using standard operations on bgfs (see for example [22, p.138]) we have that

$$\mathbf{E}[w_A(\bar{e})] = [x^n]\left(\left.\frac{\partial G_A(x/2, z)}{\partial z}\right|_{z=1}\right), \tag{5}$$

$$\mathbf{Var}[w_A(\bar{e})] = [x^n]\left(\left.\frac{\partial^2 G_A(x/2, z)}{\partial^2 z}\right|_{z=1} + \left.\frac{\partial G_A(x/2, z)}{\partial z}\right|_{z=1}\right) \tag{6}$$

$$- \left([x^n]\left(\left.\frac{\partial G_A(x/2, z)}{\partial z}\right|_{z=1}\right)\right)^2,$$

where $[x^n]G(x)$ is the coefficient of x^n in $G(x)$. Thus $\mathbf{E}[w_A(\bar{e})]$ and $\mathbf{Var}[w_A(\bar{e})]$ can be extracted by several differentiations of $G_A(x, z)$ with respect to z, and determining the coefficient of x^n after setting $z = 1$.

3   The Binary and 2^k-ary Methods

As examples of the techniques presented in the previous section, we now derive $G_{BM}(x, z)$ given in (1) for the binary method, and also $G_{k,TKM}(x, z)$ for the 2^k-ary method given in (2). First observe that the binary method processes the exponent bit-by-bit, so the relevant regular expression is $R = (1 + 0)^*$, which clearly generates all binary strings unambiguously. Second, marking $(1 + 0)$ for length and weight gives $zx + x$, and Corollary 3 indicates that R is enumerated by $G_{BM}(x, z) = 1/(1 - (zx + x))$, as shown in (1). Though the 2^k-ary method (TKM) is a natural extension of the binary method, the derivation of $G_{k,TKM}(x, z)$ is more complicated than that of $G_{BM}(x, z)$.

Theorem 4. Let $a_{n,m}$ be the number of binary strings of length n for which the TKM-recoding using k-bit windows has weight m, $0 \leq m < n$. Then

$$G_{k,TKM}(x, z) = \sum_{n,m \geq 0} a_{n,m} x^n z^m = \frac{z\left(\frac{1 - 2^k x^k}{1 - 2x} - \frac{1 - x^k}{1 - x}\right) + \frac{1 - x^k}{1 - x}}{1 - x(2^k - 1)x^k - x^k} \tag{7}$$

Proof. Consider the following regular expression

$$R = R_1^* R_2 = \left((1 + 0)^k\right)^* \left(\epsilon + \sum_{i=1}^{k-1} 1(1 + 0)^i\right).$$

n	k	$E[w_{k,TMK}(\bar{e})]$	$Var[w_{k,TMK}(\bar{e})]$	0.50	0.60	0.75	0.90	0.95	0.99
512	3	149.5	18.8	7	7	9	14	20	44
512	4	120	7.5	4	5	6	9	13	28
512	5	99.6	3.3	3	3	4	6	9	19
512	6	84.4	1.5	2	2	3	4	6	13
1024	3	298.8	37.5	9	10	13	20	28	62
1024	4	240	15	6	7	8	13	18	39
1024	5	198.6	6.2	4	4	5	8	12	25
1024	6	168.3	2.7	3	3	4	6	8	17

Table 1. The 2^k-ary encoding distributions for 512- and 1024-bit exponents. The columns show the value of $\alpha(w_{k,TMK}(\bar{e}),p), p \in \{0.50, 0.60, 0.75, 0.90, 0.95, 0.99\}$.

R_1^* generates all binary k-bit windows repeatedly, while R_2 generates all binary strings of length less than k. R_1 is marked for length and weight as

$$G_{R_1}(x, z) = z(2^k - 1)x^k + x^k \tag{8}$$

which denotes that all windows have length k, and all windows except one (the all-zero window) cost one multiplication in TKM. The marking for R_2 is as follows

$$G_{R_2}(x, z) = z\left(\frac{1 - 2^k x^k}{1 - 2x} - \frac{1 - x^k}{1 - x}\right) + \frac{1 - x^k}{1 - x}. \tag{9}$$

Note that $(1 - 2^k x^k)/(1 - 2x) - (1 - x^k)(1 - x)$ is the number of binary strings of length less than k that are niether empty or all-zero. These strings each cost a multiply in the TKM. The $(1 - x^k)/(1 - x)$ empty or all-zero strings cost no multiplies. The theorem follows from simplifying $G_{R_1}(x, z)G_{R_2}(x, z)$. $\square$

Using (5) and (6), both $E[w_{k,TKM}(\bar{e})]$ and $Var[w_{k,TKM}(\bar{e})]$ can be determined for various values of k and n using a symbolic computation package (we have elected to use Maple [1]). Recall that Chebyshev's inequality bounds the deviation of a random variable X from its mean μ in terms of its variance σ^2: $Pr(|X - \mu| \geq d) \leq \sigma^2/d^2$. Then define $\alpha(X, p)$ as

$$\alpha(X, p) = \min_d \left[\frac{\sigma^2}{d^2} < (1 - p)\right] \tag{10}$$

which states that d is the smallest for which $Pr(|X - \mu| < d) > p$ according to bounds derived by Chebyshev's inequality. Table 1 shows the distribution of TKM recoding weights for various value of k for 512- and 1024-bit exponents, and also the deviations $\alpha(w_{k,TKM}(\bar{e}),p)$ for several probabilities p.

4   The Sliding Window Representation

The sliding-window method [4,11] is a variant of the b-ary method [15], and is the 'recommended method' for general exponentiation [18, p.617]. When $b = 2^k$, the 2^k-ary method can be considered as parsing an exponent e into adjacent k-bit windows, where the window covering the least significant bit may be less than k bits. The idea of the sliding-window method is to select the placement of each k-bit window so that its most and least significant bit are equal to one. The advantage of such a partition over the 2^k-ary method is twofold: first the number of windows is expected to be reduced as runs of zeroes may occur between consecutive windows, and secondly, the amount of precomputation is halved as the windows only represent odd powers. We now derive $G_{k,SW}(x,z)$, the bgf for the sliding window encoding of exponents using k-bit windows.

Theorem 5. Let $a_{n,m}$ be the number of binary strings of length n for which the SW-recoding using k-bit windows has weight m, $0 \le m < n$. Then

$$G_{k,SW}(x,z) = \sum_{n,m\ge 0} a_{n,m}x^n z^m = \frac{1 - 2x + zx - zx^k 2^{k-1}}{(1 - x - zx^k 2^{k-1})(1 - 2x)}. \tag{11}$$

Proof. Consider the following regular expression

$$R = R_1^* R_2 = \left(0 + 10^{k-1} + \sum_{i=0}^{k-2} 1(0+1)^{k-2-i}10^i\right)^* \left(\epsilon + \sum_{i=1}^{k-2} 1(1+0)^i\right).$$

R_1^* generates words of length k that start and end with 1, and also the single word 0. Clearly R_1^* then generates all words corresponding to k-bit windows separated by runs of zeroes. R_2 generates either the empty string or a word beginning with 1, of length less than k, which corresponds to the case where the last there are not $k-1$ bits following the most significant bit of the last window. We now mark R_1 for length and weight: 0 is marked x, 10^{k-1} is marked zx^k meaning it has length k and corresponds to one nonzero digit in the recoding, and $1(0+1)^{k-2-i}10^i$ is similarly marked as $zx^{i+2}(x+x)^{k-2}$. Using the same rules for R_2 we have that

$$G_{R_1}(x,z) = x + zx^k + zx^2 \sum_{i=0}^{k-2} \frac{(x+x)^{k-2}}{(x+x)^i} = x + zx^k + zx^k 2^{k-2}\left(2 - 2^{2-k}\right),$$

$$G_{R_2}(x,z) = 1 + zx\left(\frac{1 - x^{k-1}2^{k-1}}{1 - 2x}\right).$$

The theorem follows from simplifying $G_{R_1}(x,z)G_{R_2}(x,z)$.   □

Using $\alpha(X,p)$ from (10) we can again bound the distribution of weights, which are given in Table 2 for 512-, 768- and 1024-bit exponents. Notice that the expectations are very close to $n/(k+1)$, as previously observed by Hui and Lam

n	k	$n/(k+1)$	$\mathbf{E}[w_{k,SW}(e)]$	$\mathbf{Var}[w_{k,SW}(e)]$	0.50	0.60	0.75	0.90	0.95	0.99
512	4	102.4	102.6	8.3	5	5	6	10	13	29
512	5	85.33	85.6	4.8	4	4	5	7	10	23
512	6	73.14	73.4	3.1	3	3	4	6	8	18
512	7	64	64.3	2.1	3	3	3	5	7	15
1024	4	204.8	205.0	16.5	6	7	9	13	19	41
1024	5	170.67	170.9	9.6	5	5	7	10	14	31
1024	6	146.3	146.6	6.1	4	4	5	8	12	25
1024	7	128	128.3	4.1	3	4	5	7	10	21

Table 2. k,SW encoding distributions for 512- and 1024-bit exponents. The columns show the value of $\alpha(w_{k,SW}(\bar{e}), p)$, $p \in \{0.50, 0.60, 0.75, 0.90, 0.95, 0.99\}$.

[11], and that the variances are quite small. We now consider the case of $k = 5$ explicitly, which is of interest since it is the optimal window size for the 2^k-ary method on 512-bit exponents.

Theorem 6. For a random n-bit exponent and 5-bit windows

$$\mathbf{E}[w_{5,SW}(\bar{e})] \sim \frac{n}{6} + \frac{5}{18}, \quad \mathbf{Var}[w_{5,SW}(\bar{e})] \sim \frac{n}{108} + \frac{35}{324}. \tag{12}$$

Proof. Taking the partial derivative of $G_{k,SW}(x, z)$ with respect to z, setting $k = 5$, and expanding with partial fractions we find that

$$\frac{\partial G_{5,SW}(x, z)}{\partial z}\bigg|_{z=1, k=5} = \frac{1}{6(1 - 2x)^2} + \frac{1}{9(1 - 2x)} + \frac{5 + 3x - 2x^2 - 8x^3}{8x^4 + 4x^3 + 2x^2 + x + 1}$$

$$= \sum_{n \geq 0} \frac{(n + 1)(2x)^n}{6} + \sum_{n \geq 0} \frac{(2x)^n}{9} + \sum_{n \geq 0} O(1.77^n) \tag{13}$$

where 1.77 is the complex root with largest modulus in $x^4 + x^3 + 2x^2 + 4x + 8$, which is the reflected polynomial [9, p.325] of $8x^4 + 4x^3 + 2x^2 + x + 1$. The second derivative with respect to z at $z = 1, k = 5$ has the partial fraction decomposition

$$\frac{\partial^2 G_{5,SW}(x, z)}{\partial^2 z}\bigg|_{z=1, k=5} = \sum_{n \geq 0} \frac{(n + 1)(n + 2)(2x)^n}{36} - \sum_{n \geq 0} \frac{4(n + 1)(2x)^n}{27}$$

$$+ n \cdot \sum_{n \geq 0} O(1.77^n).$$

Thus using (5) and (6), the variance is asymptotic to $n/108 + 35/324$. $\quad\square$

Using similar computations as in Theorem 6 we have verified the following theorem.

Theorem 7. For k in the range $2 \leq k \leq 10$, $\mathbf{E}[w_{k,SW}(\bar{e})] \sim n/(k+1) + \frac{k(k-1)}{2(k+1)^2}$.

We are currently working on extending the proof of the above theorem to all k and n, which involves proving certain terms in the partial fraction expansion of $\mathbf{E}[w_{k,SW}(\bar{e})]$ tend to zero with n. At present we have no expression for $\mathbf{Var}[w_{k,SW}(\bar{e})]$, but note that in general it is small, meaning that the distribution is concentrated around its mean. For example, expanding $G_{5,SW}(x,z)$ directly for 512-bit exponents shows that 99.6% of exponents will be recoded to a weight that lies with ± 6 of $\mathbf{E}[w_{5,SW}(\bar{e})]$. Similarly, 99.998% of 512-bit exponents are recoded to within ± 10 of $\mathbf{E}[w_{5,SW}(\bar{e})]$.

5 Signed-Digit Representations

A signed-digit representation of the number e in base b is of the form $e = \sum_{i=0}^{d} a_i b_i$ where $a_i \in \{0, \pm 1, \pm 2, \ldots, \pm(b-1)\}$, implying that binary numbers are consequently encoded using the digits $\{0, 1, -1 = \bar{1}\}$. In general, the signed-digit representation of a number for a fixed base is not unique, and even the encoding of minimal weight need not be unique. An algorithm for producing minimal a weight signed-digit encoding for a general base b is given by Arno and Wheeler [2].

Working with negative exponents requires group inversions, which can be costly over some groups if the appropriate inverses cannot be precomputed. On the other hand, signed-digit representations are particularly attractive for arithmetic over elliptic curves, since they correspond to addition-subtraction chains, and point addition and subtraction on cryptographic curves have the same cost in terms of group operations [19,16,17,23].

Definition 8. Let $e = \sum_{i=0}^{d} a_i 2_i$, $a_i \in \{0, 1, -1 = \bar{1}\}$ be a minimal weight signed-digit encoding of e. The encoding is called *sparse* if no two consecutive digits a_i, a_{i+1} are both nonzero.

Jedwab and Mitchell [12] prove that sparse encodings are unique and have minimal weight. The algorithm in Figure 1 converts e to a sparse encoding [12] by repeatedly applying the identity $2^{k+1} - 1 = \sum_{i=0}^{k} 2^k$. This guarantees spareness since adjacent bits are encoded as $10 \cdots 0\bar{1}$, and for this reason sparse exponents are also said to be in nonadjacent form [23]. We will refer to exponents recoded according to Figure 1 as *Optimal Signed Digit* encodings, or OSD recodings.

Asymptotic results indicate that the weight of an OSD-encodings approaches $n/3$ for a random n-bit exponent [13,16,2]. It was only recently (1996) that the exact analysis was given by Gollman, Han and Mitchell [8] who proved that the expected weight is $n/3 - 4/9 - \frac{4(-1)^n}{9 \cdot 2^n}$. Previously, Arno and Wheller [2] exhibit a Markov chain P that mimics an OSD-encoding algorithms, whose limiting distribution for the expected number of zeros in the resulting encoding is $\frac{2n}{3}$. We now derive $G_{OSD}(x,z)$ from which we will determine the variance of an OSD-encoding.

$i \leftarrow 0$;
while true
    Find the largest $j > (i+1)$ such that $e' = e_j, e_{j-1}, \ldots, e_i = 01^{j-i}$;
    if there is no such j **then** exit ;
    else replace e' with $10^{j-i-2}\bar{1}$; $i \leftarrow j$;
od

Fig. 1. An algorithm for producing a sparse signed-digit representation of a binary number.

Theorem 9. Let $a_{n,m}$ be the number of binary strings of length n for which the OSD-recoding has hamming weight m, $0 \le m < n$. Then

$$G_{OSD}(x, z) = \sum_{n,m \ge 0} a_{n,m} x^n z^m = \frac{1 - x + xz + -2zx^2 + x^2 z^2}{1 - 2x + x^2 - 2zx^2 + 2zx^3}. \qquad (14)$$

Proof. The proof is based on the following two regular expressions

$$R_1 = 10(10)^*0,$$
$$R_2 = ((10)^+ 11^+ 0 + 11^+ 0)(1^+ 0)^* 0,$$

which describes how bits are propagated between runs of runs separated by at most one 0. Further details are given in the Appendix. □

Using $\alpha(X, p)$ from (10) we can again bound the distribution of weights, which are given in Table 3 for 512-, 768- and 1024-bit exponents.

Theorem 10. For a random n-bit exponent, we have that

$$\mathbf{E}[w_{OSD}(\bar{e})] = \frac{n}{3} + \frac{4}{9} - \frac{4(-1)^n}{9 \cdot 2^n}, \quad \mathbf{Var}[w_{OSD}(\bar{e})] = \frac{2n}{27} + \frac{14}{81} + \frac{2n}{27 \cdot 2^n} + o(1).$$

Proof. The partial fraction decomposition of the derivative of $G_{OSD}(x, z)$ at $z = 1$ is

$$\left. \frac{\partial G_{k,SW}(x, z)}{\partial z} \right|_{z=1} = \frac{1}{3(1 - 2x)^2} + \frac{1}{9(1 - 2x)} + \frac{1}{18(1 + x)} - \frac{1}{2(1 - x)} \qquad (15)$$

giving that $\mathbf{E}[w_{OSD}(\bar{e})] = n/3 + 4/9 - \frac{4(-1)^n}{9 \cdot 2^n}$. The partial fraction decomposition for the second derivative $G_{OSD}(x, z)$ at $z = 1$ is

$$\left. \frac{\partial^2 G_{k,SW}(x, z)}{\partial^2 z} \right|_{z=1} = \frac{2}{9(1 - 2x)^3} + \frac{2}{27(1 + x)^2} - \frac{8}{27(1 - 2x)^2} \qquad (16)$$

for which $[x^n]/2^n$ is $(n+1)(n+2)/9 - 8(n+1)/27 - \frac{2(n+1)}{27 \cdot 2^n}$. Then $\mathbf{Var}[w_{OSD}(\bar{e})]$ is determined directly from (6). □

	$E[w_{OSD}(\bar{e})]$	$Var[w_{OSD}(\bar{e})]$	0.50	0.60	0.75	0.90	0.95	0.99
512	171.1	38.1	9	10	13	20	28	62
768	256.4	57.1	11	12	16	24	34	76
1024	341.7	76.0	13	14	18	28	39	88

Table 3. OSD-encoding distributions for 512-, 768- and 1024-bit exponents. The columns show the value of $\alpha(w_{OSD}(\bar{e}), p)$, $p \in \{0.50, 0.60, 0.75, 0.90, 0.95, 0.99\}$.

6   Comparisons and Conclusions

In this paper we have analysed three recoding rules for improved exponentiation over the binary method. The analysis is thorough in that for the methods considered it is possible to extract both the expectation and variance of the random variable describing the recoded weight. In several of the cases we have derived closed forms for these statistics with respect to the recoding scheme.

It remains to draw comparisons between the three recoding methods. We will only discriminate on the basis of the number of multiplications required by a method, since the number of squaring required by the 2^k-ary and sliding window methods will be similar, and even taking squarings into account, the OSD method is significantly slower. The 2^k-ary method requires $TMK(k) = 2^k + w_{k,TKM}(\bar{e}) - 4$ multiplications [18], and the k-bit sliding window method requires $SW(k) = 2^{k-1} + w_{k,SW}(\bar{e}) - 2$ multiplications. The OSD method requires at least $w_{OSD}(\bar{e}) - 1$, not counting any precomputation.

For 512-bit exponents, the optimal window size is $k = 5$ for both the TKM and SW methods, yielding an average multiplication cost of 127.8 and 100.3 respectively. Thus on average the optimal sliding window method only performs about 78% of the multiplies that the optimal 2^k-ary method performs. From Tables 1 and 2, since $TMK(k) \leq 127.8 + 19 = 146.8$ over 99% of the time, and $SW(k) \geq 100.3 - 23 = 77.3$ over 99% of the time, the optimal sliding window method will perform over 52% of the multiplications required by the optimal 2^k-ary method for most exponents. For the majority of exponents $SW(k) \leq 100.3 + 4 = 104.3$, while the majority of exponents require $TMK(k) \geq 127.8 - 3 = 124.8$ multiplications, meaning that the optimal sliding widow method requires less than 84% of the multiplications required by the optimal 2^k-ary method for a majority of exponents. Further, from Table 3 we find that over 90% of OSD exponents require at least $170 - 20 = 150$ multiplications, implying that the optimal sliding window only performs less than 70% of the multiplication that OSD requires.

Similarly, the optimal sliding window method is superior to the optimal 2^k-ary method for 1024-bit exponents, as it is to the OSD method. In this case the optimal window size for the sliding window method is $k = 6$, while it is $k = 5$ for the 2^k-ary method. Again from Tables 1 and 2, for the majority of exponents $SW(k) \leq 177.3 + 4 = 181.3$, while the majority of exponents have $TMK(k) \geq 226.6 - 4 = 224.6$, meaning that the optimal sliding widow method requires approximately 80% of the multiplications required by the optimal 2^k-ary method for a majority of exponents. With high probability the optimal sliding window method performs at least 60% of the multiplications required by the optimal 2^k-ary method, and with almost certainty performs less than 60% of the multiplications required by the OSD method.

Even more accurate statements can be made if the generating functions are expanded, and probabilities computed directly. In the case of 512-bit exponents and $k = 5$ bit windows, both the sliding window and 2^k-ary methods deviate from their expected weights by more than ± 10 with probability less than 10^{-4}. Further the majority of exponents deviate by less than ± 1 from their expected weight.

OSD recoded exponents tend to a weight of approximately $n/3$ on average. This weight cannot be significantly reduced since smaller weight exponents depend on longer runs of 1's occuring in the original exponent, but a run of length k has probability 2^{-k}. One advantage of the OSD coding is that little space is required for precomputation, and if inverses can be computed quickly then the OSD method may be attactive, say for elliptic curve computations on smart cards.

7 Appendix

Proof of Theorem 9.
Let $e = e_0 e_1 \cdots e_{n-2} e_{n-1}$, $e = \sum_{i=0}^{n-1} e_i 2^i$, be an n-bit exponent, written left-to-right as low order to high order bits. OSD-recoding can be interpreted as initially partitioning an exponent e into blocks

$$e = b_1 0^{j_1} b_2 0^{j_2} \cdots 0^{j_{t-2}} b_{t-1} 0^{j_t} b_t, \tag{17}$$

where $j_d \geq 0$, $1 \leq d \leq t$. Each b_i, $1 \leq i < t$, consists of runs separated by a single zero, where the last run ends in two zeros, which for example might be 100 or 1011100. Also b_t is similar except that the last run is followed by either one or no zeroes. Note that since the b_i and b_{i+1} are separated by at least two zeroes then the recoding of b_i and b_{i+1} according to Figure 1 will be independent.

The regular expression $(1^+0)^*0$ generates words containing runs separated by a single zero, where the last run ends in two zeros. The next step is to determine if the trailing pair of zeros after $b_i, 1 \leq i < t$, is encoded as 10 or 00. In the first case we will say that a *carry* has propagated to the second most significant zero, or more simply, that a carry is present b_i. The main observation is that a carry will be present if and only if b_i contains 110. We define the following two regular expressions to detect the presence of a carry:

$$R_1 = 10(10)^*0,$$
$$R_2 = ((10)^+11^+0 + 11^+0)(1^+0)^*0,$$

Here R_1 generates words with no carry, and R_2 generates words with carry (110 is present). Thus $R_3 = (0 + R_1 + R_2)^*$ generates all blocks in (17) except b_t.

Note that the OSD-encoding of each b_i is length preserving and if $\#b_i = k$ the
it is enumerated as $z^m x^k$ where $m = \#(\text{runs in } b_i) + [110 \text{ is present in } b_i]$. The
gfs for R_1 and R_2 can be derived directly as

$$G_{L_1}(x, z) = \frac{zx^3}{1 - zx^2},$$

$$G_{L_2}(x, z) = \frac{zx^3}{1 - x}\left(1 + \frac{zx^2}{1 - zx^2}\right)\frac{1}{1 - (zx^2)/(1 - x)} \cdot xz.$$

and $G_{L_3}(x, z) = 1/(1 - x - G_{L_1}(x, z) - G_{L_2}(x, z))$. It remains to enumerate block
b_t which is generated by the regular expression $R_4 = (1^+0)^*1^*$. Expanding R_4
so that it can be marked for length and weight we obtain

$$R_4 = (10)^*(\epsilon + 1 + 11) + 110^+(1^+0)^*1^* + (10)^+11^+0(1^+0)^*1^\cdot$$

R_4 is similar to R_3, and $G_{L_4}(x, z)$ is derived in a manner similar to $G_{L_3}(x, z)$.
The theorem follows from simplifying $G_{L_4}(x, z)G_{L_6}(x, z)$.

$\square$

References

1. See the Maple homepage at http://www.maplesoft.com.
2. S. Arno and F. Wheeler. Signed digit representations of minimal hamming weight. *IEEE Transactions on Computers*, 42(8):1007–1010, 1993.
3. E. A. Bender and S. G. Williamson. *Foundations of Applied Combinatorics*. Addison-Wesley Publishing Company, 1991.
4. J. Bos and M. Coster. Addition chain heuristics. *Advances in Cryptology, CRYPTO 89, Lecture Notes in Computer Science, vol. 218, G. Brassard ed., Springer-Verlag*, pages 400–407, 1990.
5. N. Chomsky and P. Schutzenberger. The algebraic theory of context-free languages. In P Braffort and North Holland Hirchberg, D., editors, *Computer programming and formal languages*, pages 118–161, 1963.
6. W. Diffie and M. Hellman. New directions in cryptography. *IEEE Transactions on Information Theory*, 22(6):472–492, 1976.
7. T. ElGamal. A public key cryptosystem and signature system based on discrete logarithms. *IEEE Transactions on Information Theory*, 31(4):473–481, 1985.
8. D. Gollman, Y. Han, and C. Mitchell. Redundant integer representations and fast exponentiation. *Designs, Codes and Cryptography*, 7:135–151, 1996.
9. R. L. Graham, D. E. Knuth, and O. Patshnik. *Concrete Mathematics, A Foundation for Computer Science, First Edition*. Addison Wesley, 1989.
10. J. Hopcroft and J. Ullman. *An Introduction to Automata, Languages and Computation*. Reading, MA: Addison Wesley, 1979.
11. L. Hui and K.-Y. Lam. Fast square-and-multiply exponentiation for RSA. *Electronics Letters*, 30(17):1396–1397, 1994.
12. J. Jedwab and C. Mitchell. Minimum weight modified signed-digit representations and fast exponentiation. *Electronics Letters*, 25:1171–1172, 1989.

13. C. K. Koç. High-radix and bit encoding techniques for modular exponentiation. *International Journal of Computer Mathematics*, 40:139–156, 1991.
14. C. K. Koç. Analysis of sliding window techniques for exponentiation. *Computers and Mathematics with Applications*, 30(10):17–24, 1995.
15. D. E. Knuth. *The Art of Computer Programming : Volume 2, Seminumerical Algorithms*. Addsion Wesley, 1981.
16. N. Koblitz. CM curves with good cryptographic properties. *Advances in Cryptology, CRYPTO 91, Lecture Notes in Computer Science, vol. 576, J. Feigenbaum ed., Springer-Verlag*, pages 279–287, 1992.
17. K. Koyama and T. Tsuruoka. Speeding up elliptic curve cryptosystems using a signed binary window method. In *Advances in Cryptology, CRYPTO 92, Lecture Notes in Computer Science, vol. 740, E. F. Brickell ed., Springer-Verlag*, pages 345–357, 1992.
18. A. Menezes, P. van Oorschot, and S. Vanstone. *Handbook of Applied Cryptography*. CRC press, 1996.
19. F. Morain and J. Olivos. Speeding up the computations on an elliptic curve using addition-subtraction chains. *Theoretical Informatics and Applications*, 24(6):531–544, 1990.
20. G. Reitwiesener. Binary arithmetic. In F. L. Alt, editor, *Advances in Computers*, pages 232–308, 1960.
21. R. L. Rivest, A. Shamir, and L. Adleman. A method for obtaining digital signatures and public key cryptosystems. *Communications of the ACM*, 21(2):120–126, 1978.
22. R Sedgewick and P. Flajolet. *An introduction to the analysis of algorithms*. Addison-Wesley Publishing Company, 1996.
23. J. A. Solinas. An improved algorithm for arithmetic on a family of elliptic curves. *Advances in Cryptology, CRYPTO 97, Lecture Notes in Computer Science, vol. 1294, B. S. Kaliski ed., Springer-Verlag*, pages 357–371, 1997.
24. Y. Yacobi. Exponentiating faster with addition chains. *Advances in Cryptology, EUROCRYPT 90, Lecture Notes in Computer Science, vol. 473, I. B. Damgård ed., Springer-Verlag*, pages 222–229, 1991.

Dealing Necessary and Sufficient Numbers of Cards for Sharing a One-Bit Secret Key

(Extended Abstract)

Takaaki Mizuki[1], Hiroki Shizuya[2], and Takao Nishizeki[3]

[1] Nishizeki Lab., Graduate School of Information Sciences, Tohoku University,
Aoba-yama 05, Aoba-ku, Sendai 980-8579, Japan
`mizuki@nishizeki.ecei.tohoku.ac.jp`
[2] Education Center for Information Processing, Tohoku University,
Kawauchi, Aoba-ku, Sendai 980-8576, Japan
`shizuya@ecip.tohoku.ac.jp`
[3] Graduate School of Information Sciences, Tohoku University,
Aoba-yama 05, Aoba-ku, Sendai 980-8579, Japan
`nishi@ecei.tohoku.ac.jp`

Abstract. Using a random deal of cards to players and a computationally unlimited eavesdropper, all players wish to share a one-bit secret key which is information-theoretically secure from the eavesdropper. This can be done by a protocol to make several pairs of players share one-bit secret keys so that all these pairs form a spanning tree over players. In this paper we obtain a necessary and sufficient condition on the number of cards for the existence of such a protocol. Our condition immediately yields an efficient linear-time algorithm to determine whether there exists a protocol to achieve such a secret key sharing.

1 Introduction

Suppose that there are k (≥ 2) players $P_1, P_2, \cdots, P_k$ and a passive eavesdropper, Eve, whose computational power is unlimited. All players wish to share a common one-bit secret key that is information-theoretically secure from Eve. Let C be a set of d distinct cards which are numbered from 1 to d. All cards in C are randomly dealt to players $P_1, P_2, \cdots, P_k$ and Eve. We call a set of cards dealt to a player or Eve a *hand*. Let $C_i \subseteq C$ be P_i's hand, and let $C_e \subseteq C$ be Eve's hand. We denote this *deal* by $C = (C_1, C_2, \cdots, C_k; C_e)$. Clearly $\{C_1, C_2, \cdots, C_k, C_e\}$ is a partition of set C. We write $c_i = |C_i|$ for each $1 \leq i \leq k$ and $c_e = |C_e|$, where $|A|$ denotes the cardinality of a set A. Note that $c_1, c_2, \cdots, c_k$ and c_e are the sizes of hands held by $P_1, P_2, \cdots, P_k$ and Eve respectively, and that $d = \sum_{i=1}^{k} c_i + c_e$. We call $\gamma = (c_1, c_2, \cdots, c_k; c_e)$ the *signature* of deal C. In this paper we assume that $c_1 \geq c_2 \geq \cdots \geq c_k$; if necessary, we rename the players. The set C and the signature γ are public to all the players and even to Eve, but the cards in the hand of a player or Eve are private to herself, as in the case of usual card games.

We consider a graph called a *key exchange graph*, in which each vertex i represents a player P_i and each edge (i, j) joining vertices i and j represents a

J. Stern (Ed.): EUROCRYPT'99, LNCS 1592, pp. 389–401, 1999.
© Springer-Verlag Berlin Heidelberg 1999

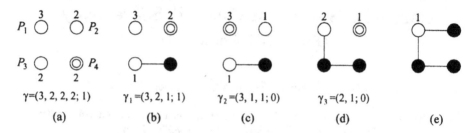

Fig. 1. A generating process of a key exchange graph.

pair of players P_i and P_j sharing a one-bit secret key $r_{ij} \in \{0,1\}$. (See Figure 1.) Refer to [8] for the graph-theoretic terminology. If the key exchange graph is a spanning tree as illustrated in Figure 1(e), then all the players can share a common one-bit secret key $r \in \{0,1\}$ as follows: an arbitrary player chooses a one-bit secret key $r \in \{0,1\}$, and sends it to the rest of the players along the spanning tree; when player P_i sends r to player P_j along an edge (i,j) of the spanning tree, P_i computes the exclusive-or $r \oplus r_{ij}$ of r and r_{ij} and sends it to P_j, and P_j obtains r by computing $(r \oplus r_{ij}) \oplus r_{ij}$.

For the case $k = 2$, Fischer, Paterson and Rackoff give a protocol to form a spanning tree, i.e. a graph having exactly one edge as the key exchange graph by using a random deal of cards [2].

Fischer and Wright [3,6] extend this protocol for any $k \geq 2$, and formalize a class of protocols called "key set protocols," a formal definition of which will be given in the succeeding section. Furthermore they give the so-called SFP protocol as a key set protocol. We say that a key set protocol *works for a signature* γ if the protocol always forms a spanning tree as the key exchange graph for any deal C having the signature γ [2,3,4,5,6]. Let Γ be a set of all signatures, where the number k of players and the total number d of dealt cards are taken over all values. Define sets W and L as follows:

$$W = \{\gamma \in \Gamma \mid \text{there is a key set protocol working for } \gamma\}; \text{ and}$$

$$L = \{\gamma \in \Gamma \mid \text{there is no key set protocol working for } \gamma\}.$$

Thus $\{W, L\}$ is a partition of set Γ. Fischer and Wright show that their SFP protocol works for all $\gamma \in W$ [3,6]. Furthermore they prove that a sufficient condition for $\gamma \in W$ is $c_k \geq 1$ and $c_1 + c_k \geq c_e + k$. They also show that it is a necessary and sufficient condition for the case $k = 2$ [3,6]. However, a simple necessary and sufficient condition for the case $k \geq 3$ has not been known so far [3,6].

Since the SFP protocol works for all $\gamma \in W$, one can determine whether $\gamma \in W$ or not by simulating the SFP protocol for γ. However, it is necessary to simulate the protocol for all "malicious adversaries," and hence the time required by this simulation is exponential in k and such a simulation is impractical.

In this paper for the case $k \geq 3$ we give a simple necessary and sufficient condition on a signature γ for the existence of a key set protocol to work for γ.

Given a signature $\gamma = (c_1, c_2, \cdots, c_k; c_e)$, one can easily determine in time $O(k)$ whether γ satisfies our condition or not. Thus our condition immediately yields an efficient linear-time algorithm for determining whether there exists a key set protocol to work for a given signature γ. Our condition looks in appearance to be similar to the condition for a given degree sequence to be "graphical," and the proof for our condition is complicated as well as those for a degree sequence [1,7,8,10].

2 Preliminaries

In this section we explain the key set protocol formalized by Fischer and Wright, and present known results on this protocol [2,3,6].

We first define some terms. A *key set* $K = \{x, y\}$ consists of two cards x and y, one in C_i, the other in C_j with $i \neq j$, say $x \in C_i$ and $y \in C_j$. We say that a key set $K = \{x, y\}$ is *opaque* if $1 \leq i, j \leq k$ and Eve cannot determine whether $x \in C_i$ or $x \in C_j$ with probability greater than $1/2$. Note that both players P_i and P_j know that $x \in C_i$ and $y \in C_j$. If K is an opaque key set, then P_i and P_j can share a one-bit secret key $r_{ij} \in \{0, 1\}$, using the following rule agreed on before starting the protocol: $r_{ij} = 0$ if $x > y$; $r_{ij} = 1$, otherwise. Since Eve cannot determine whether $r_{ij} = 0$ or $r_{ij} = 1$ with probability greater than $1/2$, the secret key r_{ij} is information-theoretically secure. We say that a card x is *discarded* if all the players agree that x has been removed from someone's hand, that is, $x \notin (\bigcup_{i=1}^{k} C_i) \cup C_e$. We say that a player P_i *drops out* of the protocol if she no longer participates in the protocol. We denote by V the set of indices i of all the players P_i remaining in the protocol. Note that $V = \{1, 2, \cdots, k\}$ before starting a protocol.

The key set protocol has four steps as follows.

1. Choose a player P_s, $s \in V$, as a *proposer* by a certain procedure.
2. The proposer P_s determines in mind two cards x, y. The cards are randomly picked so that x is in her hand and y is not in her hand, i.e. $x \in C_s$ and $y \in (\bigcup_{i \in V - \{s\}} C_i) \cup C_e$. Then P_s proposes $K = \{x, y\}$ as a key set to all the players. (The key set is proposed just as a set. Actually it is sorted in some order, for example in ascending order, so Eve learns nothing about which card belongs to C_s unless Eve holds y.)
3. If there exists a player P_t holding y, then P_t accepts K. Since K is an opaque key set, P_s and P_t can share a one-bit secret key r_{st} that is information-theoretically secure from Eve. (In this case an edge (s, t) is added to the key exchange graph.) Both cards x and y are discarded. Let P_i be either P_s or P_t that holds a smaller hand; if P_s and P_t hold hands of the same size, let P_i be the proposer P_s. P_i discards all her cards and drops out of the protocol. Set $V := V - \{i\}$. Return to step 1.
4. If there exists no player holding y, that is, Eve holds y, then both cards x and y are discarded. Return to step 1. (In this case no new edge is added to the key exchange graph.)

These steps 1–4 are repeated until either exactly one player remains in the protocol or there are not enough cards left to complete step 2 even if two or more players remain. In the first case the key exchange graph becomes a spanning tree. In the second case the protocol fails to form a spanning tree.

We now illustrate the execution of the key set protocol. Let $\gamma = (3, 2, 2, 2; 1)$ be the signature before starting the protocol. Thus there are four players P_1, P_2, P_3, P_4 and Eve; P_1 has a hand of size 3, P_2, P_3 and P_4 have hands of size 2, and Eve has a hand of size 1. At the beginning of the protocol the key exchange graph has four isolated vertices and has no edge, as illustrated in Figure 1(a). In Figure 1 a white circle represents a vertex corresponding to a player remaining in the protocol, and the number attached to a white circle represents the size of the corresponding player's hand. Suppose that P_4 is chosen as a proposer in step 1. In Figure 1 a double white circle represents the vertex corresponding to a proposer. In step 2, P_4 proposes $K = \{x, y\}$ such that $x \in C_4$ and $y \notin C_4$. Assume that $y \in C_3$. Then step 3 is executed, P_3 and P_4 share a one-bit secret key r_{34}, and edge $(3, 4)$ is added to the key exchange graph, as illustrated in Figure 1(b). Since both cards x and y are discarded, the sizes of hands of both P_3 and P_4 decrease by one. Further, since the size of P_3's hand was the same as that of P_4's hand, the proposer P_4 discards all her cards and drops out of the protocol. Thus the resulting signature is $\gamma_1 = (3, 2, 1; 1)$. In Figure 1 a black circle represents a vertex corresponding to a player who has dropped out of the protocol. We now return to step 1. Assume that P_2 is chosen as a proposer and $y \in C_e$. Then step 4 is executed, and the sizes of hands of both P_2 and Eve decrease by one. Thus the resulting signature is $\gamma_2 = (3, 1, 1; 0)$, and no new edge is added to the key exchange graph, as illustrated in Figure 1(c). Since step 4 terminates, we now return to step 1. Assume that P_1 is chosen as a proposer and $y \in C_3$. Then edge $(1, 3)$ is added to the key exchange graph as illustrated in Figure 1(d). Since the size of P_1's hand decreases by one and P_3 drops out of the protocol, the resulting signature is $\gamma_3 = (2, 1; 0)$. We now return to step 1. Assume that P_2 is chosen as a proposer. Then $y \in C_1$ because only P_1 and P_2 remain in the protocol and Eve's hand has already been empty. Thus edge $(1, 2)$ is added to the key exchange graph, and the key exchange graph becomes a spanning tree, as illustrated in Figure 1(e). Thus the protocol terminates. As seen from the example above, during the execution of the key set protocol, each connected component of the key exchange graph always has exactly one vertex (drawn in a white circle) corresponding to a player remaining in the protocol.

Considering various procedures for choosing the proposer P_s in step 1, we obtain the class of *key set protocols*.

First consider the procedure in step 1 for the case $k = 2$. Fischer, Paterson and Rackoff show that, if the procedure always chooses the player with the larger hand as a proposer P_s, then the resulting key set protocol works for any signature $\gamma = (c_1, c_2; c_e)$ such that $c_2 \geq 1$ and $c_1 + c_2 \geq c_e + 2$ [2]. On the other hand, one can easily see that if there exists a key set protocol working for a signature $\gamma = (c_1, c_2; c_e)$ then $c_2 \geq 1$ and $c_1 + c_2 \geq c_e + 2$. Thus the following Theorem 1 holds [3].

Theorem 1. [3] *Let $k = 2$. Then $\gamma \in W$ if and only if $c_2 \geq 1$ and $c_1 + c_2 \geq c_e + 2$.*

Next consider the procedure in step 1 for the case $k \geq 3$. As a key set protocol, Fischer and Wright give the SFP (smallest feasible player) procedure which chooses the "feasible" player with the smallest hand as a proposer P_s [3,6]. Let $\gamma = (c_1, c_2, \cdots, c_k; c_e)$ be the current signature. If $c_e \geq 1$, P_i with $c_i = 1$ were chosen as a proposer, and $y \in C_e$ occurred, then P_i's hand would become empty although she remains in the protocol, and hence the key exchange graph would not become a spanning tree. On the other hand, if $c_e = 0$, then $y \in C_e$ does not occur and hence the procedure appears to be able to choose P_i with $c_i = 1$ as a proposer; however, if $y \in C_j$ and $c_j = 1$, then P_j's hand would become empty and hence the key exchange graph would not become a spanning tree. Thus the procedure can choose P_i with $c_i = 1$ as a proposer only when $c_e = 0$ and $c_j \geq 2$ for every j such that $1 \leq j \leq k$ and $j \neq i$, that is, only when $i = k$ and $c_{k-1} \geq 2$. Remember that $c_1 \geq c_2 \geq \cdots \geq c_k$ is assumed. Hence, we say that player P_i is *feasible* if the following condition (1) or (2) holds.

(1) $c_i \geq 2$.
(2) $c_e = 0$, $c_i = 1$ with $i = k$, and $c_{k-1} \geq 2$.

Thus, if the hands of all the players remaining in the protocol are not empty, i.e. $c_k \geq 1$, and the proposer P_s is feasible, then the hands of all the players remaining in the protocol will not be empty at the beginning of the succeeding execution of steps 1–4.

We define a mapping f from Γ to natural numbers, as follows: $f(\gamma) = i$ if P_i is the feasible player with the smallest hand (ties are broken by selecting the player having the largest index); and $f(\gamma) = 0$ if there is no feasible player. For example, if $\gamma = (4, 3, 2, 2, 1, 1; 3)$, then $f(\gamma) = 4$. If $\gamma = (4, 4, 3, 3, 1; 0)$, then $f(\gamma) = k = 5$ because $c_e = 0$, $c_k = 1$ and $c_{k-1} \geq 2$. If $\gamma = (1, 1, 1; 2)$, then $f(\gamma) = 0$ because there is no feasible player. Hereafter we often denote $f(\gamma)$ simply by f.

From now on let $\gamma = (c_1, c_2, \cdots, c_k; c_e)$. Note that the definition of f immediately implies the following Lemma 2. Lemma 2(a) provides a trivial necessary condition for $\gamma \in W$.

Lemma 2. *The following (a) and (b) hold.*

(a) *If $k \geq 3$ and $\gamma \in W$, then $c_k \geq 1$ and $f(\gamma) \geq 1$ [3].*
(b) *If $c_k \geq 1$, then $c_i = 1$ for every i such that $f(\gamma) + 1 \leq i \leq k$.*

The SFP procedure chooses a proposer P_s as follows:

$$s = \begin{cases} f(\gamma) & \text{if } 1 \leq f(\gamma) \leq k; \\ 1 & \text{if } f(\gamma) = 0. \end{cases}$$

The key set protocol resulting from this procedure is called the *SFP protocol*. The following Theorem 3 has been known on the SFP protocol [3,6].

Theorem 3. [3,6] *Let $\gamma \in \Gamma$. Then there exists a key set protocol working for γ, i.e. $\gamma \in W$, if and only if the SFP protocol works for γ.*

Furthermore the following Lemma 4 is known on a sufficient condition for $\gamma \in W$ [3,6].

Lemma 4. [3,6] *If $c_k \geq 1$ and $c_1 + c_k \geq c_e + k$, then $\gamma \in W$.*

The sufficient condition in Lemma 4 is not a necessary condition in general. For example, $\gamma = (3, 3, 2, 1; 1)$ does not satisfy the condition in Lemma 4, but the SFP protocol works for γ and hence $\gamma \in W$ [3,6]. In this paper we obtain a simple necessary and sufficient condition for $\gamma \in W$ for any $k \geq 3$. As shown later, $\gamma = (3, 3, 2, 1; 1)$ satisfies our necessary and sufficient condition.

3   Necessary and Sufficient Condition

For $k = 3$, we obtain the following Theorem 5 on a necessary and sufficient condition for $\gamma \in W$.

Theorem 5. *Let $k = 3$. Then $\gamma \in W$ if and only if $c_3 \geq 1$ and $c_1 + c_3 \geq c_e + 3$.*

Proof. Given in Section 5.

For $k \geq 4$, we obtain the following Theorem 6 on a necessary and sufficient condition for $\gamma \in W$. Hereafter let $B = \{i \in V \mid c_i = 2\}$, and let $b = \lfloor |B|/2 \rfloor$. Note that, by Lemma 2(a), a trivial necessary condition for $\gamma \in W$ is $c_k \geq 1$ and $f(\gamma) \geq 1$.

Theorem 6. *Let $k \geq 4$, $c_k \geq 1$ and $f \geq 1$. Then $\gamma \in W$ if and only if*

$$\sum_{i=1}^{\tilde{f}} \max\{c_i - h^+, 0\} \geq \tilde{f}, \tag{1}$$

where

$$\bar{f} = f - \delta, \tag{2}$$

$$\tilde{f} = \bar{f} - 2\epsilon, \tag{3}$$

$$h = c_e - c_k + k - \bar{f}, \tag{4}$$

$$h^+ = h + \epsilon, \tag{5}$$

$$\delta = \begin{cases} 0 \text{ if } f = 1; \\ 1 \text{ if } 2 \leq f \leq k - 1; \\ 2 \text{ if } f = k \text{ and } c_{k-1} \geq c_k + 1; \text{ and} \\ 3 \text{ if } f = k \text{ and } c_{k-1} = c_k, \end{cases} \tag{6}$$

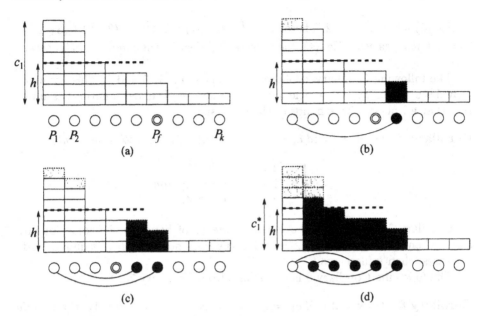

Fig. 2. The evolution of a key exchange graph and the alteration of a signature.

and

$$\epsilon = \begin{cases} \max\{\min\{c_2 - h, b\}, 0\} & if\ 5 \leq f \leq k - 1; \\ \max\{\min\{c_2 - h, b - 1\}, 0\} & if\ 5 \leq f = k\ and\ c_e \geq 1;\ and \\ 0 & otherwise. \end{cases} \qquad (7)$$

Proof. Given in Section 6.

[Remark] Since $c_1 \geq c_2 \geq \cdots \geq c_k$ is assumed, Eq. (1) is equivalent to

$$\sum_{i=1}^{k} \max\{c_i - h^+, 0\} \geq \tilde{f} \qquad (8)$$

where the summation is taken over all i, $1 \leq i \leq k$, although the summation in Eq. (1) is taken over all i, $1 \leq i \leq \tilde{f}$.

Figure 2(a) illustrates Eq. (1); the left hand side of Eq. (1) is equal to the number of cards above the dotted line in Figure 2(a) where the rectangles stacked on a player P_i, $1 \leq i \leq k$, represent the cards of P_i's hand.

As mentioned in Section 2, the SFP protocol works for $\gamma = (3, 3, 2, 1; 1)$, but γ does not satisfy the sufficient condition in Lemma 4 [3,6]. By the definition of f we have $f = f(\gamma) = 3$. Since $k = 4$, we have $2 \leq f = 3 = k - 1$, and hence by Eq. (6) $\delta = 1$. By Eq. (2) $\tilde{f} = 3 - 1 = 2$, and by Eq. (4) $h = 1 - 1 + 4 - 2 = 2$. Since $f = 3 < 5$, by Eq. (7) we have $\epsilon = 0$. Hence by Eq. (3) we have $\tilde{f} = 2 - 0 = 2$ and by Eq. (5) $h^+ = 2 + 0 = 2$. Therefore $\sum_{i=1}^{\tilde{f}} \max\{c_i - h^+, 0\} = (3-2) + (3-2) = 2 = \tilde{f}$. Thus γ satisfies Eq. (1), the necessary and sufficient condition in Theorem 6.

The following Corollary 7 follows from Theorems 1, 5 and 6. This corollary provides a necessary and sufficient condition for $\gamma \in W$ under a natural assumption that all players have hands of the same size.

Corollary 7. *Let $k \geq 2$ and $c_1 = c_2 = \cdots = c_k$. Then $\gamma \in W$ if and only if*

$$c_1 \geq \begin{cases} c_e/2 + 1 & \text{if } k = 2; \\ c_e/2 + 3/2 & \text{if } k = 3; \text{ and} \\ c_e/2 + 2 & \text{if } k \geq 4. \end{cases}$$

Corollary 7 means that the required size c_1 of hands is the same for any $k \geq 4$ when $c_1 = c_2 = \cdots = c_k$. Note that the total number kc_1 of required cards increases when k increases.

The following Corollary 8 is immediate from Corollary 7.

Corollary 8. *Let $k \geq 2$ and $c_1 = c_2 = \cdots = c_k = c_e$. Then $\gamma \in W$ if and only if*

$$c_1 \geq \begin{cases} 2 & \text{if } k = 2; \\ 3 & \text{if } k = 3; \text{ and} \\ 4 & \text{if } k \geq 4. \end{cases}$$

4   Malicious Adversary

In this paper we use a *malicious adversary* in order to prove Theorem 6.

If a key set protocol works for a signature γ, then the key exchange graph must become a spanning tree for any deal C having the signature γ. Hence, whoever has the card y contained in the proposed key set $K = \{x, y\}$, the key exchange graph should become a spanning tree. The malicious adversary determines who holds the card y. Considering a malicious adversary to make it hard for the key exchange graph to become a spanning tree, we obtain a necessary condition for $\gamma \in W$. On the other hand, if under some condition on a signature γ a key set protocol always forms a spanning tree as the key exchange graph for any malicious adversary, then the condition is a sufficient one for $\gamma \in W$.

We use a function $\mathcal{A} : \Gamma \times V \rightarrow V \cup \{e\}$ to represent a malicious adversary, as follows. Remember that Γ is the set of all signatures and that V is the set of indices of all the players remaining in a protocol. Let e be Eve's index. The inputs to the function $\mathcal{A}(\gamma, s)$ are the current signature $\gamma \in \Gamma$ and the index $s \in V$ of a proposer P_s chosen in the protocol. Its output is either the index t of a player P_t remaining in the protocol or the index e of Eve; $\mathcal{A}(\gamma, s) = t \neq e$ means that player P_t holds card y; and $\mathcal{A}(\gamma, s) = e$ means that Eve holds card y.

From now on, we denote by $\gamma = (c_1, c_2, \cdots, c_k; c_e)$ the current signature, and denote by $\gamma'_{(s,\mathcal{A})} = (c'_1, c'_2, \cdots, c'_{k'}; c'_e)$ the resulting signature after executing steps 1–4 under the assumption that P_s proposes a key set $K = \{x, y\}$ and $y \in C_{A(\gamma, s)}$.

The definition of a malicious adversary $\mathcal{A}$ immediately implies the following Lemma 9.

Lemma 9. *Let $k \geq 3$. Then $\gamma \in W$ if and only if there exists a proposer P_s such that $\gamma'_{(s, \mathcal{A})} \in W$ for any malicious adversary $\mathcal{A}$. That is,*

$$\gamma \in W \Longleftrightarrow \exists s \; \forall \mathcal{A} \;\; \gamma'_{(s, \mathcal{A})} \in W,$$

in other words,

$$\gamma \in L \Longleftrightarrow \forall s \; \exists \mathcal{A} \;\; \gamma'_{(s, \mathcal{A})} \in L.$$

From now on let $k \geq 3$. If $f = 0$, then by Lemma 2(a) $\gamma \in L$. On the other hand, if $f \geq 1$, then the index s of the proposer P_s chosen by the SFP procedure satisfies $s = f$. Furthermore, by Theorem 3 the SFP protocol works for all $\gamma \in W$. Thus, if $\gamma \in W$, then $\gamma'_{(f, \mathcal{A})} \in W$ for any malicious adversary $\mathcal{A}$. Hence, the following Corollary 10 immediately follows from Theorem 3.

Corollary 10. *Let $k \geq 3$ and $f(\gamma) \geq 1$. Then $\gamma \in W$ if and only if $\gamma'_{(f, \mathcal{A})} \in W$ for any malicious adversary $\mathcal{A}$. That is,*

$$\gamma \in W \Longleftrightarrow \forall \mathcal{A} \;\; \gamma'_{(f, \mathcal{A})} \in W,$$

in other words,

$$\gamma \in L \Longleftrightarrow \exists \mathcal{A} \;\; \gamma'_{(f, \mathcal{A})} \in L.$$

It follows from the definition of a key set protocol that if two players P_i and P_j hold hands of the same size, that is, $c_i = c_j$, then

$$\forall \mathcal{A} \;\; \gamma'_{(i, \mathcal{A})} \in W \Longleftrightarrow \forall \mathcal{A} \;\; \gamma'_{(j, \mathcal{A})} \in W.$$

Hence, if there exist two or more players P_i with $c_i = c_s$ (including the proposer P_s), then one may assume without loss of generality that P_s has the largest index among all these players. We call it *Assumption 1* for convenience sake. Furthermore, if $\mathcal{A}(\gamma, s) = t \neq e$ and there exist two or more players P_i with $c_i = c_t$ and $i \neq s$ (including P_t), then one may assume without loss of generality that P_t has the largest index among all these players. We call it *Assumption 2* for convenience sake. Under the two assumptions above, $\gamma'_{(s, \mathcal{A})} = (c'_1, c'_2, \cdots, c'_{k'}; c'_e)$ satisfies $c'_1 \geq c'_2 \geq \cdots \geq c'_{k'}$ since γ satisfies $c_1 \geq c_2 \geq \cdots \geq c_k$.

The total size $\sum_{i=1}^{k} c_i$ of all the players' hands decreases by two or more if $\mathcal{A}(\gamma, s) = t \neq e$; it decreases by exactly one if $\mathcal{A}(\gamma, s) = e$. If a key set protocol works for γ, then $\mathcal{A}(\gamma, s) = t \neq e$ occurs $k - 1$ times until the protocol terminates because the key exchange graph becomes a spanning tree having $k - 1$ edges at the end of the protocol. Furthermore $\mathcal{A}(\gamma, s) = e$ would occur c_e times. Hence, if a key set protocol works for γ, then $\sum_{i=1}^{k} c_i \geq 2(k - 1) + c_e = c_e + 2k - 2$. Thus we have the following Lemma 11 as a trivial necessary condition for $\gamma \in W$.

Lemma 11. *If $\gamma \in W$, then $\sum_{i=1}^{k} c_i \geq c_e + 2k - 2$.*

5   Proof of Theorem 5

In this section we give a proof of Theorem 5.

Since Lemma 4 implies the sufficiency of the condition in Theorem 5, we prove its necessity. That is, we show that if $k = 3$ and $\gamma \in W$ then $c_3 \geq 1$ and $c_1 + c_3 \geq c_e + 3$. In order to prove this, we use the following malicious adversary $\mathcal{A}^*$:

$$\mathcal{A}^*(\gamma, s) = \begin{cases} 3 \text{ if } s = 1; \\ 1 \text{ if } s = 2; \\ e \text{ if } s = 3. \end{cases}$$

We first have the following Lemma 12.

Lemma 12. Let $k = 3$, $c_3 \geq 1$ and $c_1 + c_3 \leq c_e + 2$. Then the following (a) or (b) holds.

(a) $\gamma \in L$.
(b) $\gamma'_{(f, \mathcal{A}^*)}$ satisfies $k' = 3$, $c'_3 \geq 1$ and $c'_1 + c'_3 \leq c'_e + 2$.

Proof. Let $k = 3$, $c_3 \geq 1$ and $c_1 + c_3 \leq c_e + 2$. If $f = 0$, then $\gamma \in L$ by Lemma 2(a). Thus one may assume that $1 \leq f \leq 3$. Then there are the following three cases.

Case 1: $f = 1$.
    In this case, by Lemma 2(b), we have $\gamma = (c_1, 1, 1; c_e)$ and hence $c_2 = c_3 = 1$. Thus, by $c_1 + c_3 \leq c_e + 2$ we have $c_1 \leq c_e + 1$. Hence $\sum_{i=1}^{3} c_i \leq (c_e + 1) + 1 + 1 = c_e + 2k - 3$. Therefore $\gamma \in L$ by Lemma 11.

Case 2: $f = 2$.
    In this case, by Lemma 2(b) we have $\gamma = (c_1, c_2, 1; c_e)$. Since $f = 2$, the definition of f implies $c_2 \geq 2$ and $c_e \geq 1$. Furthermore, since $c_3 = 1$ and $c_1 + c_3 \leq c_e + 2$, we have $c_1 \leq c_e + 1$. Since $f = 2$, let P_2 be a proposer P_s. Since $\mathcal{A}^*(\gamma, s) = 1$ for $s = 2$, the size of the hand of P_1 holding card y decreases by one and the proposer P_2 drops out of the protocol, and hence $\gamma'_{(f, \mathcal{A}^*)} = (c_1 - 1, 1; c_e)$. Therefore $c'_1 + c'_2 = (c_1 - 1) + 1 = c_1$. Since $c_1 \leq c_e + 1$ and $c'_e = c_e$, we have $c'_1 + c'_2 \leq c'_e + 1$. Thus by Theorem 1 $\gamma'_{(f, \mathcal{A}^*)} \in L$. Therefore Corollary 10 implies $\gamma \in L$.

Case 3: $f = 3$.
    In this case $c_e \geq 1$; if $c_e = 0$, then by $c_3 \geq 1$ and $c_1 + c_3 \leq c_e + 2 = 2$ we have $c_1 = c_2 = c_3 = 1$, and hence $f = 0$, contrary to $f = 3$. Since $f = 3$, let P_3 be a proposer. Since $\mathcal{A}^*(\gamma, s) = e$ for $s = 3$, the sizes of the hands of both P_3 and Eve decrease by one, and hence $\gamma'_{(f, \mathcal{A}^*)} = (c_1, c_2, c_3 - 1; c_e - 1)$, $k' = 3$ and $c'_e = c_e - 1$. Since P_3 was feasible, we have $c'_3 = c_3 - 1 \geq 1$. Furthermore $c'_1 + c'_3 = c_1 + (c_3 - 1) \leq (c_e + 2) - 1 = c'_e + 2$. Thus (b) holds. ∎

Define the *size* $\text{size}(\gamma)$ of a signature γ as follows: $\text{size}(\gamma) = c_e + k$.
We are now ready to prove the necessity of the condition in Theorem 5.

(Proof for the necessity of the condition in Theorem 5)
Let $k = 3$. We shall show that if $c_3 = 0$ or $c_1 + c_3 \leq c_e + 2$ then $\gamma \in L$. If $c_3 = 0$, then Lemma 2(a) implies $\gamma \in L$. Therefore it suffices to prove the following claim: if $c_3 \geq 1$ and $c_1 + c_3 \leq c_e + 2$ then $\gamma \in L$. We prove the claim by induction on $\text{size}(\gamma) = c_e + k$. Let $c_3 \geq 1$ and $c_1 + c_3 \leq c_e + 2$. Since $k = 3$, $\text{size}(\gamma) \geq 3$.

First consider the case $\text{size}(\gamma) = 3$. In this case, $c_e = 0$, and hence $c_1 + c_3 \leq c_e + 2 = 2$. Thus $c_1 = c_2 = c_3 = 1$, and hence $f = 0$. Therefore by Lemma 2(a) $\gamma \in L$.

Next let $l \geq 4$, and assume inductively that the claim holds when $\text{size}(\gamma) = l - 1$.

Consider any signature γ such that $\text{size}(\gamma) = l$. By Lemma 12, the following (a) or (b) holds:

(a) $\gamma \in L$; and
(b) $\gamma'_{(f,\mathcal{A}^*)}$ satisfies $k' = 3$, $c'_3 \geq 1$ and $c'_1 + c'_3 \leq c'_e + 2$.

Thus one may assume that (b) holds. Then, since $\text{size}(\gamma') = \text{size}(\gamma) - 1 = l - 1$, by the induction hypothesis we have $\gamma'_{(f,\mathcal{A}^*)} \in L$. Therefore Corollary 10 implies $\gamma \in L$. ∎

6   Sketchy Proof of Theorem 6

In this section we outline a proof of Theorem 6.

One can easily prove Theorem 6 for the case $f = 1$ as follows. Let $k \geq 4$, $c_k \geq 1$ and $f = 1$. Then $\delta = \epsilon = 0$ and hence $\hat{f} = \bar{f} = f = 1$. By Lemma 2(b) $c_k = 1$ and hence $h^+ = h = c_e - 1 + k - 1 = c_e + k - 2$. Thus, Eq. (1) is equivalent to $\max\{c_1 - c_e - k + 2, 0\} \geq 1$, and hence equivalent to $c_1 \geq c_e + k - 1$. Therefore Theorem 6 for the case $f = 1$ immediately follows from the following Lemma 13.

Lemma 13. Let $c_k \geq 1$ and $f = 1$. Then $\gamma \in W$ if and only if $c_1 \geq c_e + k - 1$.

Proof. The sufficiency immediately follows from Lemma 4. Therefore it suffices to prove the necessity. Let $c_k \geq 1$, $f = 1$ and $\gamma \in W$. Then by Lemma 11 we have $\sum_{i=1}^{k} c_i \geq c_e + 2k - 2$. On the other hand, since $f = 1$, by Lemma 2(b) $\gamma = (c_1, 1, 1, \cdots, 1; c_e)$ and hence $\sum_{i=1}^{k} c_i = c_1 + k - 1$. Therefore, $c_1 + k - 1 \geq c_e + 2k - 2$ and hence $c_1 \geq c_e + k - 1$. ∎

We then sketch a proof of Theorem 6 for the case $2 \leq f \leq k$. The detail is omitted in this extended abstract. We sketch a proof only for the necessity of the condition in Theorem 6. (One can prove the sufficiency by induction on $\text{size}(\gamma) = c_e + k$.) Let $k \geq 4$, $c_k \geq 1$, $2 \leq f \leq k$ and $\gamma \in W$. Instead of proving Eq. (1) we prove the following equation holds:

$$\sum_{i=1}^{\bar{f}} \max\{c_i - h, 0\} \geq \bar{f}, \tag{9}$$

which is obtained from Eq. (1) by replacing $\tilde{f}$ and h^+ with $\bar{f}$ and h, respectively.

For simplicity, we assume that $\delta = 1$, i.e. $2 \leq f \leq k - 1$. (The proof for $\delta = 2, 3$ is similar.) Then by Lemma 2(b) $c_k = 1$. Furthermore $\bar{f} = f - 1$ and $h = c_e + k - f$. Thus Eq. (9) is equivalent to

$$\sum_{i=1}^{f-1} \max\{c_i - (c_e + k - f), 0\} \geq f - 1. \tag{10}$$

We prove the necessity of Eq. (10). Let $2 \leq f \leq k - 1$. Then the signature is $\gamma = (c_1, c_2, \cdots, c_f, 1, 1, \cdots, 1; c_e)$. That is, there are exactly f feasible players $P_1, P_2, \cdots, P_f$, and each of the remaining $k - f$ players $P_{f+1}, P_{f+2}, \cdots, P_k$ has exactly one card. The key exchange graph has exactly k isolated vertices before starting the protocol, as illustrated in Figure 2(a). In Figure 2, a white rectangle represents a card in players' hands. The SFP protocol chooses the feasible player P_f with the smallest hand as a proposer. Consider a malicious adversary that does not choose Eve and always chooses the player with the largest hand as P_t with $y \in C_t$. Then P_f and the player P_t with the largest hand share a one-bit secret key, the size of P_t's hand decreases by one, P_f drops out of the protocol, and an edge joining two vertices corresponding to these two players is added to the key exchange graph, as illustrated in Figure 2(b). In the example of Figure 2, the size of P_1's hand decreases by one, P_f discards all her cards and drops out of the protocol, and edge $(1, f)$ is added to the key exchange graph. In Figure 2(b), we lightly shade the rectangle corresponding to the card y discarded by $P_t = P_1$, and darkly shade the rectangles corresponding to the cards discarded by P_f who drops out of the protocol. At the next execution of steps 1–4, the proposer is P_{f-1}. By considering the same malicious adversary as above, P_{f-1} and the player with the largest hand share a one-bit secret key as illustrated in Figure 2(c). In Figure 2(b), since P_1 has a hand of the same size as P_2, by Assumption 2 $P_t = P_2$ and hence edge $(2, f - 1)$ is added to the key exchange graph as illustrated in Figure 2(c). Repeat such an operation until P_1 becomes a proposer, i.e. there exists exactly one feasible player as illustrated in Figure 2(d), and let $\gamma^* = (c_1^*, c_2^*, \cdots, c_{k^*}^*; c_e)$ be the resulting signature. Then $k^* = k - f + 1$, $c_2^* = c_3^* = \cdots = c_{k^*}^* = 1$, $f(\gamma^*) = 1$, and the size of Eve's hand remains c_e. By Corollary 10 we have $\gamma^* \in W$. Therefore, by Lemma 13, $c_1^* \geq c_e + k^* - 1 = c_e + k - f = h$. The malicious adversary has chosen $f - 1$ players P_i in total as P_t so far, and hence there are exactly $f - 1$ lightly shaded rectangles in Figure 2(d). The malicious adversary above implies that such a player P_i, $1 \leq i \leq f - 1$, should have a hand of size greater than h when she was chosen by the malicious adversary. Thus there are $f - 1$ or more rectangles above the dotted line in Figure 2(a). Therefore we have $\sum_{i=1}^{f-1} \max\{c_i - h, 0\} \geq f - 1$, and hence Eq. (10) holds.

We have sketched a proof of the necessity of Eq. (9). One can similarly prove the necessity of Eq. (1).

7   Conclusion

In this paper we gave a simple necessary and sufficient condition on signature $\gamma = (c_1, c_2, \cdots, c_k; c_e)$ for the existence of a key set protocol to work for γ. In other words we gave a simple complete characterization of the sets W and L.

Since the SFP protocol works for all $\gamma \in W$ (Theorem 3), one can determine whether $\gamma \in W$ or not by simulating the SFP protocol for γ. However, it is necessary to simulate the protocol for all malicious adversaries, and hence the time required by this simulation is exponential in k and such a simulation is impractical. Clearly one can determine in time $O(k)$ whether our necessary and sufficient condition, i.e. Eq. (1) or (8), holds or not. Thus one can determine in time $O(k)$ whether $\gamma \in W$ or not.

This paper addresses only the class of key set protocols, and hence it still remains open to obtain a necessary and sufficient condition for any (not necessarily key set) protocol to work for γ [5].

An Eulerian circuit is more appropriate as a key exchange graph than a spanning tree if it is necessary to acknowledge the secure key distribution. We have given a protocol to achieve such a key exchange [9].

References

1. T. Asano, "An $O(n \log \log n)$ time algorithm for constructing a graph of maximum connectivity with prescribed degrees," J. Comput. and Syst. Sci., 51, pp. 503–510, 1995.
2. M. J. Fischer, M. S. Paterson and C. Rackoff, "Secret bit transmission using a random deal of cards," DIMACS Series in Discrete Mathematics and Theoretical Computer Science, AMS, 2, pp. 173–181, 1991.
3. M. J. Fischer and R. N. Wright, "An application of game-theoretic techniques to cryptography," DIMACS Series in Discrete Mathematics and Theoretical Computer Science, AMS, 13, pp. 99–118, 1993.
4. M. J. Fischer and R. N. Wright, "An efficient protocol for unconditionally secure secret key exchange," Proceedings of the 4th Annual Symposium on Discrete Algorithms, pp. 475–483, 1993.
5. M. J. Fischer and R. N. Wright, "Bounds on secret key exchange using a random deal of cards," J. Cryptology, 9, pp. 71–99, 1996.
6. M. J. Fischer and R. N. Wright, "Multiparty secret key exchange using a random deal of cards," Proc. Crypto '91, Lecture Notes in Computer Science, Springer-Verlag, 576, pp. 141–155, 1992.
7. S. L. Hakimi, "On realizability of a set of integers as degrees of the vertices of a linear graph. I," J. SIAM Appl. Math., 10, 3, pp. 496–506, 1962.
8. F. Harary, "Graph Theory," Addison-Wesley, Reading, Mass., 1969.
9. T. Mizuki, H. Shizuya and T. Nishizeki, "Eulerian secret key exchange," Proc. COCOON '98, Lecture Notes in Computer Science, Springer, 1449, pp. 349–360, 1998.
10. E. F. Schmeichel and S. L. Hakimi, "On planar graphical degree sequences," SIAM J. Appl. Math., 32, 3, pp. 598–609, 1977.

Computationally Private Information Retrieval with Polylogarithmic Communication

Christian Cachin[1]*, Silvio Micali[2], and Markus Stadler[3]

[1] IBM Zurich Research Laboratory, CH-8803 Rüschlikon, Switzerland,
cachin@acm.org.
[2] Laboratory for Computer Science, MIT, Cambridge, MA 02139, USA.
[3] Crypto AG, P.O. Box 460, CH-6301 Zug, Switzerland,
markus.stadler@acm.org.

Abstract. We present a single-database computationally private information retrieval scheme with polylogarithmic communication complexity. Our construction is based on a new, but reasonable intractability assumption, which we call the Φ-Hiding Assumption (ΦHA): essentially the difficulty of deciding whether a small prime divides $\phi(m)$, where m is a composite integer of unknown factorization.
Keywords: Integer factorization, Euler's function, Φ-hiding assumption, Private information retrieval.

1 Introduction

PRIVATE INFORMATION RETRIEVAL. The notion of *private information retrieval* (PIR for short) was introduced by Chor, Goldreich, Kushilevitz and Sudan [7] and has already received a lot of attention. The study of PIR is motivated by the growing concern about the user's privacy when querying a large commercial database. (The problem was independently studied by Cooper and Birman [8] to implement an anonymous messaging service for mobile users.)

Ideally, the PIR problem consists of devising a communication protocol involving just two parties, the database and the user, each having a secret input. The database's secret input is called the *data string*, an n-bit string $B = b_1 b_2 \cdots b_n$. The user's secret input is an integer i between 1 and n. The protocol should enable the user to learn b_i in a communication-efficient way and at the same time hide i from the database. (The trivial and inefficient solution is having the database send the entire string B to the user.)

INFORMATION-THEORETIC PIRs (WITH DATABASE REPLICATION). Perhaps surprisingly, the original paper [7] shows that the PIR problem is solvable efficiently in an information-theoretic setting if the database does not consist of a single player, but of multiple players, each holding the same data string B, who can communicate with the user but not with each other (a model reminiscent of the multi-prover proof systems of [4]). By saying that this model offers an

---

* Research done at Laboratory for Computer Science, MIT.

J. Stern (Ed.): EUROCRYPT'99, LNCS 1592, pp. 402–414, 1999.
© Springer-Verlag Berlin Heidelberg 1999

information-theoretic solution, we mean that an individual database player cannot learn i at all, no matter how much computation it may perform, as long as it does not collude with other database players.

Several solutions in this model are presented in the paper of Chor *et al.* For example, (1) there are two-database information-theoretic PIRs with $O(n^{1/3})$ communication complexity, and (2) there are $O(\log n)$-database information-theoretic PIRs with polylog(n) communication complexity. In subsequent work, Ambainis gives a construction for k-database information-theoretic PIRs with $O(n^{1/(2k-1)})$ communication complexity [2].

COMPUTATIONAL PIRs (WITH DATABASE REPLICATION). Notice that the latter two information-theoretic PIRs achieve subpolynomial communication complexity, but require more than a constant number of database servers. Chor and Gilboa [6], however, show that it is possible to achieve subpolynomial communication complexity with minimal database replication if one requires only computational privacy of the user input—a theoretically weaker though practically sufficient notion. They give a two-database PIR scheme with communication complexity $O(n^\varepsilon)$ for any $\varepsilon > 0$. Their system makes use of a security parameter k and guarantees that, as long as an individual database performs a polynomial (in k) amount of computation and does not collude with the other one, it learns nothing about the value i.

COMPUTATIONAL PIRs (WITHOUT DATABASE REPLICATION). Though possibly viable, the assumption that the database servers are separated and yet mirror the same database contents may not be too practical. Fortunately, and again surprisingly, Kushilevitz and Ostrovsky [15] show that replication is not needed. Under a well-known number-theoretic assumption, they prove the existence of a *single-database* computational PIR with subpolynomial communication. More precisely, under the quadratic residuosity assumption [13], they exhibit a CPIR protocol between a user and one database with communication complexity $O(n^\varepsilon)$, for any $\varepsilon > 0$, where again n represents the length of the data string. (For brevity, we refer to such a single-database, computational PIR, as a CPIR.)

It should be noted that the CPIR of [15] has an additional communication complexity that is polynomial in the security parameter k, but this additional amount of communication is *de facto* absorbed in the mentioned $O(n^\varepsilon)$ complexity, because for all practical purposes k can be chosen quite small.

This result has raised the question of whether it is possible to construct CPIRs with lower communication complexity.

MAIN RESULT. We provide a positive answer to the above question based on a new but plausible number-theoretic assumption: the Φ *Assumption*, or ΦA for short. The ΦA consists of two parts, the Φ-*Hiding Assumption* (ΦHA) and the Φ-*Sampling Assumption* (ΦSA).

Informally, the ΦHA states that it is computationally intractable to decide whether a given small prime divides $\phi(m)$, where m is a composite integer of unknown factorization. (Recall that ϕ is Euler's totient function, and that computing $\phi(m)$ on input m is as hard as factoring m.) The ΦSA states that it is possible to efficiently find a random composite m such that a given prime p divides $\phi(m)$.

The ΦA is attractively simple and concrete. Finding crisp and plausible assumptions is an important task in the design and analysis of cryptographic protocols, and we believe that the ΦA will prove useful in other contexts and will attract further study. Based on it we prove the following

Main Theorem: Under the ΦA, there is a two-round CPIR whose communication complexity is polylogarithmic in n (and polynomial in the security parameter).

We note that our CPIR is "essentially optimal" in several ways:

Communication complexity. Disregarding the privacy of the user input altogether, in order for the user to obtain the ith bit of an n-bit data string, at least $\log n$ bits have to be communicated between the user and the database in any case.

Computational complexity. Our CPIR is also very efficient from a computational-complexity point of view. Namely, (1) the user runs in time polynomial in $k \log n$ and (2) the database runs in time proportional to n times a polynomial in k. Both properties are close to optimal in our context. The user computational complexity is close to optimal because, as already mentioned, in any scheme achieving sub-linear communication, the user must send at least $\log n$ bits of information, and thus perform at least $\log n$ steps of computation. The database computational complexity is close to optimal because the database must read each bit of its data string in any single-database PIR. (Otherwise, it would know that the user cannot possibly have received any of the unread bits and therefore gain some information about the user input i.)

Round complexity. The round complexity of our CPIR is essentially optimal because, as long as the user can choose his own input i at will in each execution, no single-round CPIR exists[1].

Privacy model. Our CPIR achieves computational privacy. Although information-theoretic privacy is stronger, our scheme is optimal among single-database PIRs since there are no single-database PIRs with information-theoretic privacy (other than sending the entire data string).

---

[1] We do not rule out the possibility of single-round CPIRs in alternative models, for example, in a model where the user always learns the bit in position i in any execution in which the data string has at least i bits.

2 Preliminaries and Definitions

2.1 Notation

INTEGERS. We denote by N the set of natural numbers. Unless otherwise specified, a natural number is presented in its binary expansion whenever given as an input to an algorithm. If $n \in N$, by 1^n we denote the unary expansion of n, that is, the concatenation of n 1's. If $a, b \in N$, we denote that a evenly divides b by writing $a|b$. Let $\mathbb{Z}_m$ be the ring of integers modulo m and $\mathbb{Z}_m^*$ its multiplicative group. The *Euler totient function* of an integer m, denoted by $\phi(m)$, is defined as the number of positive integers $\leq m$ that are relatively prime to m.

STRINGS. If σ and τ are binary strings, we denote σ's length by $|\sigma|$, σ's ith bit by σ_i, and the concatenation of σ and τ by $\sigma \circ \tau$.

COMPUTATION MODELS. By an *algorithm* we mean a (probabilistic) Turing machine. By saying that an algorithm is *efficient* we mean that, for at most but an exponentially small fraction of its random tapes, it runs in fixed polynomial time. By a *k-gate circuit* we mean a finite function computable by an acyclic circuitry with k Boolean gates, where each gate is either a NOT-gate (with one input and one output) or an AND gate (with two binary inputs and one binary output).

PROBABILITY SPACES. (Taken from [5] and [14].) If $A(\cdot)$ is an algorithm, then for any input x, the notation "$A(x)$" refers to the probability space that assigns to the string σ the probability that A, on input x, outputs σ.

If S is a probability space, then "$x \xleftarrow{R} S$" denotes the algorithm which assigns to x an element randomly selected according to S. If F is a finite set, then the notation "$x \xleftarrow{R} F$" denotes the algorithm which assigns to x an element selected according to the probability space whose sample space is F and uniform probability distribution on the sample points.

If $p(\cdot, \cdot, \cdots)$ is a predicate, the notation

$$PROB[x \xleftarrow{R} S; y \xleftarrow{R} T; \cdots : p(x, y, \cdots)]$$

denotes the probability that $p(x, y, \cdots)$ will be true after the ordered execution of the algorithms $x \xleftarrow{R} S$, $y \xleftarrow{R} T, \cdots$.

2.2 Fully Polylogarithmic CPIR

Our proposed CPIR works in only two rounds and achieves both polylogarithmic communication complexity and polylogarithmic user computational complexity. For the sake of simplicity, we formalize only such types of CPIRs below.

Definition: Let $D(\cdot, \cdot, \cdot)$, $Q(\cdot, \cdot, \cdot)$ and $R(\cdot, \cdot, \cdot, \cdot, \cdot)$ be efficient algorithms. We say that (D, Q, R) is a *fully polylogarithmic computationally private information retrieval scheme* (or polylog CPIR for short) if there exist constants $a, b, c, d > 0$ such that,

1. (Correctness) $\forall n$, $\forall$ n-bit strings B, $\forall i \in [1, n]$, and $\forall k$,

$$PROB[(q, s) \overset{R}{\leftarrow} Q(n, i, 1^k) \, ; \, r \overset{R}{\leftarrow} D(B, q, 1^k) : R(n, i, (q, s), r, 1^k) = B_i]$$
$$> 1 - 2^{-ak}$$

2. (Privacy) $\forall n$, $\forall i, j \in [1, n]$, $\forall k$ such that $2^k > n^b$, and $\forall$ 2^{ck}-gate circuits A,

$$\big| PROB[(q, s) \overset{R}{\leftarrow} Q(n, i, 1^k) : A(n, q, 1^k) = 1] -$$
$$PROB[(q, s) \overset{R}{\leftarrow} Q(n, j, 1^k) : A(n, q, 1^k) = 1] \big| \, < \, 2^{-dk}.$$

We call a, b, c, and d the *fundamental constants* (of the CPIR); B the *data string*; D the *database algorithm*; the pair (Q, R) the *user algorithm*; Q the *query generator*; R the *response retriever*; q the *query*; s the *secret* (associated to q); r the *response*; and k the *security parameter*. (Intuitively, query q contains user input i, and response r contains database bit b_i, but both contents are unintelligible without secret s.)

REMARKS.

1. Our correctness constraint slightly generalizes the one of [15]: Whereas there correctness is required to hold with probability 1, we require it to hold with very high probability.
2. As mentioned above, the communication complexity of our CPIR is polylogarithmic in n (the length of the data string) times a polynomial in k (the security parameter). Because k is an independent parameter, it is of course possible to choose it so large that the polynomial dependence on k dominates over the polylogarithmic dependence on n. But choosing k is an overkill since our definition guarantees "an exponential amount of privacy" also when k is only polylogarithmic in n.

2.3 Number Theory

SOME USEFUL SETS. Let us define the sets we need in our assumptions and constructions.

Definition: We denote by $PRIMES_a$ the set of the primes of length a, and by H_a the set of the composite integers that are product of two primes of length a. (For a large, H_a contains the hardest inputs to any known factoring algorithm.)

We say that a *composite integer* m ϕ-*hides a prime* p if $p | \phi(m)$. Denote by $H^b(m)$ the set of b-bit primes p that are ϕ-hidden by m, denote by $\bar{H}^b(m)$ the set $PRIMES_b - H^b(m)$, and denote by H_a^b the set of those $m \in H_a$ (i.e., products of two a-bit primes) that ϕ-hide a b-bit prime.

SOME USEFUL FACTS. Let us state without proof some basic or well-known number-theoretic facts used in constructing our CPIR.

Fact 1: There exists an efficient algorithm that on input a outputs a random prime in $PRIMES_a$.

Fact 2: There exists an efficient algorithm that on input a outputs a random element of H_a.

Fact 3: There exists an efficient algorithm that, on input a b-bit prime p and an integer m together with its integer factorization, outputs whether or not $p \in H^b(m)$.

Fact 4: There exists an efficient algorithm that, on inputs x, p, m, and m's integer factorization, outputs whether or not x has a pth root mod m.

OUR ASSUMPTIONS.

The Φ-Assumption (ΦA):

$\exists e, f, g, h > 0$ such that
- **Φ-Hiding Assumption (ΦHA):** $\forall k > h$ and $\forall\ 2^{ek}$-gate circuits C,

$$PROB[m \overset{R}{\leftarrow} H_{kf}^k \; ; \; p_0 \overset{R}{\leftarrow} H^k(m) \; ; \; p_1 \overset{R}{\leftarrow} \bar{H}^k(m) \; ;$$

$$b \overset{R}{\leftarrow} \{0,1\} : C(m, p_b) = b] \; < \; \frac{1}{2} + 2^{-gk}.$$

- **Φ-Sampling Assumption (ΦSA):** $\forall k > h$, there exists a sampling algorithm $S(\cdot)$ such that for all k-bit primes p, $S(p)$ outputs a random k^f-bit number $m \in H_{kf}^k$ that ϕ-hides p, together with m's integer factorization.

We refer to $e, f, g,$ and h as the *first, second, third,* and *fourth fundamental constant* of the ΦA, respectively.

REMARKS.

1. Revealing a large prime dividing $\phi(m)$ may compromise m's factorization. Namely, if p is a prime $> m^{1/4}$ and $p|\phi(m)$, then one can efficiently factor m on inputs m and p [11,10,9]. Consequently, it is easy to decide whether p divides $\phi(m)$ whenever $p > m^{1/4}$. But nothing similar is known when p is much smaller, and for the ΦHA, it suffices that deciding whether p divides $\phi(m)$ is hard when p is not just a constant fraction shorter than m, but polynomially shorter.
   We further note that if the complexity of factoring is $\Omega(2^{\log m^c})$ for some constant c between 0 and 1, then revealing a prime p dividing $\phi(m)$ cannot possibly compromise m's factorization significantly if $\log p$ is significantly smaller than $(\log m)^c$. Indeed, since p can be represented using at most $\log p$ bits, revealing p cannot contribute more than a speed-up of $2^{\lceil \log p \rceil} \approx p$ for factoring m.
   Note that the ΦHA does not hold for $p = 3$. If $m = Q_1 Q_2$ and $m \equiv 2$ (mod 3), then one can tell that one of Q_1 and Q_2 is congruent to 1 mod 3 and the other is 2 mod 3. In this case, it's obvious that 3 divides $\phi(m) = (Q_1 - 1)(Q_2 - 1)$.

2. The ΦSA is weaker than the well-known and widely accepted Extended Riemann Hypothesis (ERH). Consider the following algorithm $S(\cdot)$:

   Inputs: a k-bit prime p.

   Output: a k^f-bit integer $m \in H^k_{k^f}$ that ϕ-hides p and its integer factorization.

   Code for $S(p)$:

   (a) Repeatedly choose a random $(k^f - k)$-bit integer q_1 until $Q_1 = pq_1 + 1$ is a prime.

   (b) Choose a random k^f-bit prime Q_2.

   (c) Let $m \leftarrow Q_1 \cdot Q_2$ and return m and (Q_1, Q_2).

   Under the ERH, algorithm S finds a suitable m in expected polynomial time in k^f (see Exercise 30 in Chapter 8 of [3]).

3   Our CPIR

3.1   The High-Level Design

At a very high level, the user's query consists of a compact program that contains the user input i in a hidden way. The database runs this program on its data string, and the result of this computation is its response r.

A bit more specifically, this compact program is actually run on the data string in a bit-by-bit fashion. Letting B be the data string, the user sends the database an algorithm A and a k-bit value x_0 (where k is the security parameter), and the database computes a sequence of k-bit values: $x_1 = A(x_0, B_1)$, $x_2 = A(x_1, B_2)$, $\ldots$, $x_n = A(x_{n-1}, B_n)$. The last value x_n is the response r. The user retrieves B_i by evaluating on x_n a predicate R_i, which is hard to guess without the secret key of the user.

This high-level design works essentially because the predicate R_i further enjoys the following properties relative to the sequence of values $x_0, \ldots, x_n$:

1. $R_i(x_0) = 0$;
2. $\forall j = 1, \ldots, i - 1$, $R_i(x_j) = 0$;
3. $R_i(x_i) = 1$ if and only if $B_i = 1$; and
4. $\forall j \geq i$, $R_i(x_{j+1}) = 1$ if and only if $R_i(x_j) = 1$.

It follows by induction that $R_i(x_n) = 1$ if and only if $B_i = 1$.

3.2   The Implementation

To specify our polylog CPIR we must give a database algorithm D and user algorithms Q (query generator) and R (response retriever). These algorithms use two common efficient subroutines T and P that we describe first. Algorithm T could be any probabilistic primality test [17,16], but we let it be a primality *prover* [12,1] so as to gain some advantage in the notation and presentation (at the expense of running time).

BASIC INPUTS.

A number $n \in \mathbb{N}$; an n-bit sequence B; an integer $i \in [1, n]$; and a unary security parameter 1^k such that $k > (\log n)^2$.

PRIMALITY PROVER $T(\cdot)$.

Input: an integer z (in binary).
Output: 1 if z is prime, and 0 if z is composite.
Code for $T(z)$: See [1].

PRIME(-SEQUENCE) GENERATOR $P(\cdot, \cdot, \cdot)$.

Inputs: an integer $a \in [1, n]$; a sequence of k^3 k-bit strings $Y = (y_0, \ldots, y_{k^3-1})$; and 1^k.
Output: a k-bit integer p_a (a prime with overwhelming probability).
Because P is deterministic, for Y and k fixed, it generates a sequence of (probable) primes $p_1, \ldots, p_n$ with $a = 1, \ldots, n$.
Code for $P(a, Y, 1^k)$:
1. $j \leftarrow 0$.
2. $\sigma_{aj} \leftarrow \bar{a} \circ \bar{j}$, where $\bar{a}$ is the $(\log n)$-bit representation of a and $\bar{j}$ the $(k - \log n)$-bit representation of j.
3. $z_j \leftarrow \sum_{l=0}^{k^3-1} y_l \sigma_{aj}{}^l$, where all strings y_l and σ_{aj} are interpreted as elements of $GF(2^k)$ and the operations are in $GF(2^k)$.
4. If $T(z_j) = 1$ or $j = 2^{k-\log n}$, then return $p_a \leftarrow z_j$ and halt; else, $j \leftarrow j+1$ and go to step 2.

QUERY GENERATOR $Q(\cdot, \cdot, \cdot)$.

Inputs: n; an integer $i \in [1, n]$; and 1^k.
Outputs: a query $q = (m, x, Y)$ and a secret s, where m is a k^f-bit composite (f being the second constant of the ΦA), $x \in \mathbb{Z}_m^*$, Y a k^3-long sequence of k-bit strings, and where s consists of m's prime factorization.
Code for $Q(n, i, 1^k)$:
1. Randomly and independently choose $y_0, \ldots, y_{k^3-1} \in \{0, 1\}^k$ and let $Y = (y_0, \ldots, y_{k^3-1})$.
2. $p_i \leftarrow P(i, Y, 1^k)$.
3. Choose a random k^f-bit integer m that ϕ-hides $p_i = P(i, Y, 1^k)$ and let s be its integer factorization.
4. Choose a random $x \in \mathbb{Z}_m^*$.
5. Output the query $q = (m, x, Y)$ and the secret s.

DATABASE ALGORITHM $D(\cdot, \cdot, \cdot)$.

Inputs: B; $q = (m, x, Y)$, a query output by $Q(n, i, 1^k)$; and 1^k.
Output: $r \in \mathbb{Z}_m^*$.

Code for $D(B, q, 1^k)$:
1. $x_0 \leftarrow x$.
2. For $j = 1, \ldots, n$, compute:
   (a) $p_j \leftarrow P(j, Y, 1^k)$.
   (b) $e_j \leftarrow p_j^{B_j}$.
   (c) $x_j \leftarrow x_{j-1}^{e_j} \bmod m$.
3. Output the response $r = x_n$.

RESPONSE RETRIEVER $R(\cdot, \cdot, \cdot, \cdot, \cdot)$:

Inputs: n; i; $(m, x, Y), s)$, an output of $Q(n, i, 1^k)$; $r \in \mathbb{Z}_m^*$, an output of $D(B, (m, x, Y), 1^k)$; and 1^k.
Output: a bit b. (With overwhelming probability, $b = B_i$.)
Code for $R(n, i, (q, s), r, 1^k)$: If r has p_ith roots mod m, then output 1, else output 0.

Theorem: Under the ΦA, (D, Q, R) is a polylog CPIR.

3.3   Proof of the Theorem

RUNNING TIME (SKETCH). Subroutine P is efficient because (on inputs i, Y, and 1^k) its most intensive operation consists, for at most k^3 times, of evaluating once a k-degree polynomial over $GF(2^k)$ and running the primality prover T. Algorithm Q is efficient because subroutines P and T are efficient, because p_i is a k-bit prime with overwhelming probability, and because, under the ΦSA, selecting a random $2k^f$-bit composite $\in H_{k^f}^k$ ϕ-hiding p_i is efficient. (Notice that, because n and i are presented in binary, Q actually runs in time polylogarithmic in n.) Algorithm D is efficient because it performs essentially one exponentiation mod m for each bit of the data string (and thus runs in time polynomial in k and linear in n). Algorithm R is efficient because of Fact 4 and because it has m's factorization (the secret s) available as an input. (R actually runs in time polynomial in k because m's length is polynomial in k.)

CORRECTNESS (SKETCH). Let us start with a quick and dirty analysis of the prime-sequence generator P. Because the elements of Y are randomly and independently selected, in every execution of $P(a, Y, 1^k)$, the values $z_0, \ldots, z_{2^{k-\log n}}$ are k^3-wise independent. Thus with probability lower bounded by $1 - 2^{O(-k^2)}$, at least one of them is prime, and thus p_a is prime. Because the length n of the data string satisfies $n^2 < 2^k$, with probability exponentially (in k) close to 1, *all* possible outputs $p_1, \ldots, p_n$ are primes. Actually, with probability exponentially (in k) close to 1, $p_1, \ldots, p_n$ consists of *random* and *distinct* primes of length k. Observe that the k^f-bit modulus m can ϕ-hide at most a constant number of primes from a set of randomly chosen k-bit primes except with exponentially (in k) small probability. Thus, with probability still exponentially (in k) close to 1, p_i will be the only prime in our sequence to divide $\phi(m)$.

In sum, because it suffices for correctness to hold with exponentially (in k) high probability, we might as well assume that, in every execution of $Q(n, i, 1^k)$, $p_1, \ldots, p_n$ are indeed random, distinct primes of length k, such that only p_i divides $\phi(m)$. Let R_i be the following predicate on $\mathbb{Z}_m^*$:

$$R_i(x) = \begin{cases} 1 & \text{if } x \text{ has a } p_i\text{th root mod } m \\ 0 & \text{otherwise.} \end{cases}$$

The user retrieves b_i by evaluating $R_i(x_n)$. It is easy to check that properties 1–4 of our high-level design hold as promised:

1. $R_i(x_0) = 0$.
   This property follows from the fact that the function $x \to x^{p_j} \bmod m$ on $\mathbb{Z}_m^*$ is 1-to-1 if p_j is relatively prime to $\phi(m)$, and at least p_j-to-1 otherwise. Because p_i is in $\Theta(2^k)$ except with exponentially (in k) small probability, the probability that a random element of $\mathbb{Z}_m^*$ has a p_ith root mod m is also exponentially small (in k). Thus we might as well assume that x_0 has no p_ith roots mod m (remember that correctness should hold only most of the time).[2]

2. $\forall j = 1, \ldots, i - 1, R_i(x_j) = 0$.
   This follows because x_0 has no p_ith roots mod m and because if x has no p_ith roots mod m, for all primes p not dividing $\phi(m)$ also x^p has no p_ith roots mod m. Again because of the size of the primes p_j for $j \neq i$, one can show that except with exponentially small (in k) probability, none of the p_j divides $\phi(m)$.

3. $R_i(x_i) = 1$ if and only if $B_i = 1$.
   If $B_i = 0$, then $x_i = x_{i-1}$. Thus, by property 2 above, x_i has no p_ith roots mod m. If $B_i = 1$, then $x_i = x_{i-1}^{p_i} \bmod m$. Thus, x_i has p_ith roots mod m by construction.

4. $\forall j \geq i, R_i(x_{j+1}) = 1$ if and only if $R_i(x_j) = 1$.
   The "if part" follows from the fact that if x_j has p_ith roots mod m, then there exists a y such that $x_j = y^{p_i} \bmod m$ and therefore also $x_{j+1} = x_j^{p_j} = y^{p_i p_j} = (y^{p_j})^{p_i} \bmod m$ has p_ith roots. For the "only-if part," see the proof of property 2 above.

PRIVACY (SKETCH). Suppose for contradiction that the privacy condition does not hold for (D, Q, R). Then for all $b, c, d > 0$, there exist n, indices i and j ($i \neq j$), $k > \log n^b$, and a 2^{bk}-gate circuit $\widetilde{A}$ (with binary output) such that

$$|\alpha_1 - \alpha_2| \geq \varepsilon$$

for some $\varepsilon > 2^{-dk}$, where

$$\alpha_1 = PROB\big[((m, x, Y), s) \xleftarrow{R} Q(n, i, 1^k) : \widetilde{A}(n, (m, x, Y), 1^k) = 1\big],$$
$$\alpha_2 = PROB\big[((m, x, Y), s) \xleftarrow{R} Q(n, j, 1^k) : \widetilde{A}(n, (m, x, Y), 1^k) = 1\big].$$

---

[2] We choose x_0 at random rather than ensuring that it has no p_ith roots mod m to facilitate proving the privacy constraint.

(Intuitively, $\widetilde{A}$'s advantage ε is always bigger than any exponentially small in k quantity.) Define now the following probability:

$$\beta = PROB\big[m \xleftarrow{R} H_{k'}^{k} \; ; \; x \xleftarrow{R} \mathbb{Z}_m^{*} \; ; \; Y \xleftarrow{R} GF(2^k)^{k^3} : \widetilde{A}(n,(m,x,Y),1^k) = 1\big].$$

(Notice that, in the sequence of experiments defining β, Y still defines a prime p_i and a prime p_j with overwhelming probability, but there is no guarantee that m ϕ-hides either of them.) It follows either $|\alpha_1 - \beta| \geq \varepsilon/2$ or $|\alpha_2 - \beta| \geq \varepsilon/2$. W.l.o.g. assume $|\alpha_1 - \beta| \geq \varepsilon/2$ and also $\alpha_1 - \beta \geq \varepsilon/2$.

We can construct a guessing circuit $\widetilde{C} = \widetilde{C}_{n,i}$ to contradict the ΦHA as follows.

GUESSING CIRCUIT $\widetilde{C}_{n,i}(\cdot,\cdot)$.

Inputs: a number $m \in H_{k'}^{k}$; and a k-bit prime p.
Output: a bit b (indicating whether m ϕ-hides p).
Code for $\widetilde{C}_{n,i}(m,p)$:
   1. Choose k^3 uniformly random k-bit numbers $a_1, \ldots, a_{k^3}$.
   2. Run primality prover T on a_j for $j = 1, \ldots, k^3$ and let j' be the smallest j for which $T(a_j) = 1$. If T returns 0 for all a_j, then $j' \leftarrow k^3$.
   3. Use Lagrange interpolation to find the coefficients $y_0, \ldots, y_{k^3-1}$ of a polynomial $\xi(\sigma)$ over $GF(2^k)$ with degree $k^3 - 1$ such that $\xi(\sigma_{ij}) = a_j$ for $j = 1, \ldots, j'-1, j'+1, \ldots, k^3$ and $\xi(\sigma_{ij'}) = p$, where $\sigma_{ij} \in GF(2^k)$ corresponds to the k-bit string $i \circ j$ as in the prime-sequence generator P. Let $Y = (y_0, \ldots, y_{k^3-1})$.
   4. Choose x at random from $\mathbb{Z}_m^{*}$ and run $\widetilde{A}(n,(m,x,Y),1^k)$. If $\widetilde{A}$ returns 0, then return 1, otherwise (if $\widetilde{A}$ returns 1), then return 0.

Notice that $\widetilde{C}$ can be constructed with a number of gates that is at most polynomially (in k) greater than the number of gates of $\widetilde{A}$.

Above we have defined how $\widetilde{C}$ operates for any $m \in H_{k'}^{k}$ and any $p \in PRIMES_k$. Let us now analyze $\widetilde{C}$'s behavior on the input distribution required by the ΦHA (i.e., when $m \xleftarrow{R} H_{k'}^{k}$ and $p \xleftarrow{R} H^k(m)$ with probability $1/2$ and $p \xleftarrow{R} \bar{H}^k(m)$ with probability $1/2$) and calculate the probability that $\widetilde{C}$ guesses correctly from which distribution p is drawn.

$$
\begin{aligned}
PROB[\widetilde{C} \text{ correct}] \;=\; & \frac{1}{2} \cdot PROB[\widetilde{C} \text{ correct}|p \xleftarrow{R} H^k(m)] \\
& + \frac{1}{2} \cdot PROB[\widetilde{C} \text{ correct}|p \xleftarrow{R} \bar{H}^k(m)] \\
=\; & \frac{1}{2} \cdot PROB[\widetilde{C} = 0|p \xleftarrow{R} H^k(m)] \\
& + \frac{1}{2} \cdot PROB[\widetilde{C} = 1|p \xleftarrow{R} \bar{H}^k(m)].
\end{aligned}
$$

The distribution of the output of $\widetilde{C}$ depends directly on $\widetilde{A}$. If $p \xleftarrow{R} H^k(m)$, then, by construction, $\widetilde{A}$ is run with the same input distribution as in the definition of α_1, except for the case that $\widetilde{C}$ finds no prime among $a_1, \ldots, a_{k^3}$ in step 2 (assume this is not the case for the moment). Let us examine $\tilde{A}$'s input distribution in $\tilde{C}$ when $p \xleftarrow{R} \bar{H}^k(m)$ and compare it to $\tilde{A}$'s input distribution in the definition of β. The experiment leading to β contains three distinct cases for $p_i = P(i, Y, 1^k)$:

1. p_i is composite;
2. $p_i \in H^k(m)$; or
3. $p_i \in \bar{H}^k(m)$.

Note that case 3 is actually how $\tilde{A}$ is called by our $\tilde{C}$ in the ΦHA and occurs with overwhelming probability. Let δ_0 be the probability of case 1, which will be computed below, and assume for the moment that p_i is indeed a random k-bit prime. The probability δ_1 that a random element of $PRIMES_k$ is in $H^k(m)$ is upper bounded by $k^j 2^{-k} = O(2^{-k/2})$. (This is the conditional probability of case 2 above given that p_i is prime.) For $\tilde{C}$, this implies

$$PROB[\tilde{C} = 1 | p \xleftarrow{R} PRIMES_k] \leq PROB[\tilde{C} = 1 | p \xleftarrow{R} \bar{H}^k(m)] + \delta_1.$$

Now consider the case that no prime is detected among $a_1, \ldots, a_{k^3}$ in step 2. Because T is an ideal primality prover, this probability is at most about $(1 - \frac{1}{k})^{k^3}$ and therefore $\delta_0 = O(2^{-k/2})$.

We can now bound $PROB[\tilde{C} \text{ correct}]$ as

$$
\begin{aligned}
PROB[\tilde{C} \text{ correct}] \geq\ & \frac{1}{2} \cdot (1 - \delta_0) \cdot PROB[\tilde{C} = 0 | p \xleftarrow{R} H^k(m)] \\
& + \frac{1}{2} \cdot (1 - \delta_0) \cdot \left(PROB[\tilde{C} = 1 | p \xleftarrow{R} PRIMES_k] - \delta_1 \right) \\
\geq\ & \frac{1}{2} \cdot (1 - \delta_0) \cdot \alpha_1 + \frac{1}{2} \cdot (1 - \delta_0) \cdot (1 - \beta - \delta_1) \\
\geq\ & \frac{1}{2} \cdot (1 + \alpha_1 - \delta_0 - \beta - \delta_0 - \delta_1) \\
\geq\ & \frac{1}{2} + \frac{\varepsilon}{4} - \delta_0 - \frac{\delta_1}{2}.
\end{aligned}
$$

The last inequality follows from the assumption $\alpha_1 - \beta \geq \varepsilon/2$.

To conclude, $\tilde{C}$ distinguishes correctly with probability at least

$$\frac{1}{2} + \frac{\varepsilon}{4} - \delta_0 - \frac{\delta_1}{2}.$$

Intuitively, since δ_1 and δ_0 are exponentially small in k, but ε exceeds any exponentially small quantity, there remains an advantage for $\tilde{C}$ that is not exponentially small and it is clear that $\tilde{C}$ violates the ΦHA. ∎

Acknowledgments

The authors wish to thank Wenbo Mao for interesting comments on an earlier version of the paper.

References

1. L. M. Adleman and M.-D. A. Huang, "Recognizing primes in random polynomial time," in *Proc. 19th Annual ACM Symposium on Theory of Computing (STOC)*, pp. 462–469, 1987.

2. A. Ambainis, "Upper bound on the communication complexity of private information retrieval," in *Proc. 24th ICALP*, vol. 1256 of *Lecture Notes in Computer Science*, Springer, 1997.

3. E. Bach and J. Shallit, *Algorithmic Number Theory*, vol. 1: Efficient Algorithms. Cambridge: MIT Press, 1996.

4. M. Ben-Or, S. Goldwasser, J. Kilian, and A. Wigderson, "Multi prover interactive proofs: How to remove intractability," in *Proc. 20th Annual ACM Symposium on Theory of Computing (STOC)*, pp. 113–131, 1988.

5. M. Blum, A. De Santis, S. Micali, and G. Persiano, "Noninteractive zero-knowledge," *SIAM Journal on Computing*, vol. 20, pp. 1085–1118, Dec. 1991.

6. B. Chor and N. Gilboa, "Computationally private information retrieval," in *Proc. 29th Annual ACM Symposium on Theory of Computing (STOC)*, pp. 304–313, 1997.

7. B. Chor, O. Goldreich, E. Kushilevitz, and M. Sudan, "Private information retrieval," in *Proc. 36th IEEE Symposium on Foundations of Computer Science (FOCS)*, 1995.

8. D. A. Cooper and K. P. Birman, "Preserving privacy in a network of mobile computers," in *Proc. IEEE Symposium on Security and Privacy*, pp. 26–38, 1995.

9. D. Coppersmith, "Finding a small root of a bivariate integer equation; factoring with high bits known," in *Advances in Cryptology: EUROCRYPT '96* (U. Maurer, ed.), vol. 1233 of *Lecture Notes in Computer Science*, Springer, 1996.

10. D. Coppersmith, "Finding a small root of a univariate modular equation," in *Advances in Cryptology: EUROCRYPT '96* (U. Maurer, ed.), vol. 1233 of *Lecture Notes in Computer Science*, Springer, 1996.

11. D. Coppersmith. personal communication, 1998.

12. S. Goldwasser and J. Kilian, "Almost all primes can be quickly certified," in *Proc. 18th Annual ACM Symposium on Theory of Computing (STOC)*, pp. 316–329, 1986.

13. S. Goldwasser and S. Micali, "Probabilistic encryption," *Journal of Computer and System Sciences*, vol. 28, pp. 270–299, 1984.

14. S. Goldwasser, S. Micali, and R. L. Rivest, "A digital signature scheme secure against adaptive chosen-message attacks," *SIAM Journal on Computing*, vol. 17, pp. 281–308, Apr. 1988.

15. E. Kushilevitz and R. Ostrovsky, "Replication is not needed: Single database, computationally-private information retrieval," in *Proc. 38th IEEE Symposium on Foundations of Computer Science (FOCS)*, pp. 364–373, 1997.

16. M. O. Rabin, "How to exchange secrets by oblivious transfer," Tech. Rep. TR-81, Harvard, 1981.

17. R. Solovay and V. Strassen, "A fast monte-carlo test for primality," *SIAM Journal on Computing*, vol. 6, no. 1, pp. 84–85, 1977.

On the Concurrent Composition of Zero-Knowledge Proofs

Ransom Richardson[1] and Joe Kilian[2]

[1] Groove Networks, 800 Cummings Center, Beverly, MA 01915
rrichard@groove.net
[2] NEC Research Institute
joe@research.nj.nec.com

Abstract. We examine the concurrent composition of zero-knowledge proofs. By concurrent composition, we indicate a single prover that is involved in multiple, simultaneous zero-knowledge proofs with one or multiple verifiers. Under this type of composition it is believed that standard zero-knowledge protocols are no longer zero-knowledge. We show that, modulo certain complexity assumptions, any statement in NP has k^ϵ-round proofs and arguments in which one can efficiently simulate any $k^{O(1)}$ concurrent executions of the protocol.

Key Words: Asynchronous Attacks, Zero Knowledge, Black-box Simulation.

1 Introduction

Zero-knowledge proofs [11] and arguments [1] are interactive protocols between a prover (or arguer), P, and a verifier, V, which informally yield no knowledge except for the validity of the assertion. The original formal definition of zero-knowledge considered a very minimal context, and almost immediately, unexpected problems emerged when attempting to apply the notion of zero-knowledge to more practical contexts; the notion of zero-knowledge has been refined accordingly. For example, to make zero-knowledge closed under sequential composition, a number of researchers ([18,20,12]) have proposed a modified definition, known as *auxiliary zero-knowledge*. A still cleaner model, motivated by these issue, is that of *black-box simulation zero-knowledge* [18]; all of the results we will discuss are for this model.

In practice, it is often desirable to run a zero-knowledge proof many times in parallel, so as to lower the error probability without increasing the round complexity. Unfortunately, it is not clear how to efficiently simulate an arbitrary zero-knowledge proof in parallel in polynomial time. Indeed, Goldreich and Krawczyk [10] have shown that for any language L outside of BPP, there is no 3-message protocol for L whose parallel execution can be simulated in black-box zero-knowledge. In their model, the verifier has oracle access to a truly random function; given the existence of cryptographically secure pseudorandom generators, the oracle can be reduced to simply a private string. However, based on

J. Stern (Ed.): EUROCRYPT'99, LNCS 1592, pp. 415–431, 1999.

reasonable computational assumptions, there exist constant-message (indeed, 4 messages suffice) interactive proofs and arguments whose parallel versions remain black-box simulatable.

1.1   Concurrent Repetition

Parallel repetition combines many versions of the same protocol in lock step. When V is supposed to send its ith message, it must send the ith message for all of the parallel runs of the interactive proof. It cannot, for example, delay sending the first message from Game 2 until it has seen the first response in Game 5.

However, in practice, one may wish to engage in many proofs simultaneously and concurrently. For example, one may conceivably give a zero-knowledge proof to establish ones identity whenever one accesses an internet-based service. Different processes may access a number of different services, with no synchronization. This scenario allows for an attack in which a verifier engages in many proofs with the prover, and arbitrarily interleaves the messages in these protocols. Intuitively, the verifier can run some of the protocols ahead in an attempt to gain information that will enable it to attack some of the other protocols.

Beth and Desmedt [3] first discussed such concurrent attacks in the context of identification protocols, and show how to defend against such attacks if parties have precisely synchronized clocks and the adversary is forced to delay its actions.

Dwork, Naor and Sahai [6] consider the role of concurrent attacks on zero-knowledge protocols. They give 4-round zero-knowledge protocols for NP, assuming a weak constraints on the synchrony of weak players: there exist a pair (α, β), where $\alpha \leq \beta$, such that when a good player has observed the passage of β units of time, then every other good player has observed the passage of at least α units of time. Dwork and Sahai [7] reduce (but do not eliminate) the timing constraints required by their defense.

A natural question is defend against arbitrary scheduling without any use of timing. A negative result by Kilian, Petrank and Rackoff [15] extends the Goldreich-Krawczyk result to concurrent attacks, for essentially the same model. They show that for any 4-message proof system for a language L, if one can black-box simulate polynomially many asynchronous proofs, then $L \in BPP$.

1.2   Our Model

Following [6], we consider a malicious verifier V that is allowed to run up to k interactive proofs with the prover P, where k is a free parameter. For our results, k may be replaced with $k^{O(1)}$. Within each proof, V must follow the proper order of the steps, but may arbitrarily interleave steps between different proofs. For example, V may execute the first step of Proof 1, then execute all of Proof 2 in order to obtain an advantage when it executes the second step of Proof 1.

For a given, presumably malicious verifier, V, the simulator is given access to V, but not to the details of its internal state. It is allowed to run V, receiving "requests" for different proofs, and send V responses for these proofs. We assume without loss of generality that V waits for the response before continuing (it never hurts to receive as much information as possible from the prover before sending one's next message). V is allowed to schedule k proofs arbitrarily, subject to the constraint that within each proof the steps are properly ordered.

Following the standard notion of black-box simulatability, the simulator S is allowed to save V's state and rewind V back to a previously saved state. For ease of explication, we do not explicitly state when S is saving V's state, but speak only in terms of rewinding (S may save V's state after every message). Without loss of generality, we assume that V's state includes all of the messages sent to it, though when restored to a previously saved state, no messages are sent since the state was saved are remembered (i.e., we use the reasonable notion of V's "memory"). Given V's initial state, S's interaction with V induces a distribution on V's final state. S's goal is for this distribution to be statistically or computationally indistinguishable from the distribution on V's final state after interacting with P.

Note that in our modeling of the adversary, we are considering *ordering* attacks, but not *timing* attacks [16] in which one uses the actual response time from the prover to obtain information. There are implementation-specific defenses to such attacks [16]; these methods and concerns are orthogonal to our own.

Similarly, we assume that while the verifier can delay a given message M so that other messages are received before M, it cannot delay M so as to make it unclear whether M is actually going to arrive. That is, the prover and simulator can at some point know that no further messages are arriving. Without this stipulation, even a single execution of most protocols seem impossible to simulate: a malicious verifier V might with probability n^{-2C} wait for n^C time-steps before giving its next answer, where C is either ∞ or a large constant unknown to S. This attack forces S to either keep on waiting or risk giving a slightly (but non-negligibly) distorted simulation.

1.3 Results of This Paper

For ease of exposition, we assume the existence of a certain publicly agreed upon bit commitment schemes, both from the prover to the verifier and from the verifier to the prover. We use an unconditionally binding, computationally private bit commitment scheme from the prover to the verifier. We use a computationally binding, unconditionally private bit commitment scheme from the verifier to the prover. The former can be based on one way functions [14,17], and the latter can be based on collision-resistant hash functions [4].

Our main result is a transformation on zero-knowledge protocols for statements in NP. Our transformed protocol for a statement T (or a proof of knowledge) has two parts: an $O(m)$-message *preamble*, for some parameter m and a *main body*. The main body consists of a zero-knowledge proof of knowledge for a witness to a statement T', which is a modified version of T. A witness for T is also a witness for T'. The longer the preamble, the more resistant the resulting argument is to concurrent attacks.

Theorem 1. *Assume the existence of the commitment schemes described above, and a proof system or argument for $T \in NP$ as described above. Let ϵ be an arbitrary positive constant and let $m = k^\epsilon$. The transformed protocol remains a proof of knowledge for T. Furthermore, there exists a polynomial-time black-box simulation for any concurrent attack using at most $k^{O(1)}$ versions of the proof. This simulation achieves computational indistinguishability.*

As we mention in Section 4, there is no need for a public bit-commitment scheme; this convention simply drops some easily handled cases from our simulation and proof.

Quite recently, Rafail Ostrovsky and Giovanni Di Crescenzo have proposed a different solution for defeating concurrent attacks without out timing [19]. Their solution requires a round complexity that is greater than m, an a priori upper bound on the number of attacker; hence, m must be known and bounded in advance. In our solution, $m = k^\epsilon$ is possible, and more to the point m need not really be known in advance, though the larger m is, the longer the simulation takes. However, the result in [19] uses no additional complexity assumptions, and is thus an incomparable result.

1.4   Techniques Used

We use a technique of Feige, Lapidot and Shamir [8] in order to convert witness indistinguishable protocols into zero-knowledge protocols. Instead of proving Theorem T, the prover proves a technically weaker theorem, $T \vee W$, where W is a statement that will fail to hold (or for which the prover will fail to have a witness of) with extremely high probability. However, in the simulation, S obtains a witness for W, and may then act as an ordinary prover. Similarly, we set up our proof system so that the simulator will have a "cheating" witness to the statement being proven.

Discussion Indeed, at first glance it may appear that the method from [8] can be used unchanged. Recall that in the scenario of [8], the world begins with an agreement on a pseudorandom generator $g : \{0,1\}^\ell \to \{0,1\}^{2k}$ and the generation of a random string $R \in \{0,1\}^{2k}$ (for ℓ suitably large). Then, any proof of T is replaced with a proof that T is true or $g^{-1}(R)$ exists. To simulate the world from its creation, the simulator S generates g and $R = g(Q)$ for a random $Q \in \{0,1\}^\ell$. Then S has a witness, (Q), for any statement of the form "T or $g^{-1}(R)$ exists." By an appeal to the witness indistinguishability of the underlying zero-knowledge proof, S is indistinguishable from any other prover for this statement, despite the fact that its witness is quite different than that used by an actual prover.

Space precluded a detailed discussion, but we note that our construction gives a simulatability result that is more "standard" in the zero-knowledge framework. Also, we do not need a common string, guaranteed to be random. Although we, for ease of exposition, assume that a suitable bit commitment scheme has been standardized, we can relax this assumption with only a trivial change to the protocol and no substantive change to the simulator and its proof. On a high level, our methods don't try to "break" or alter the commitment scheme in any way, and thus this scheme can be decided on at the beginning of the protocol.

1.5   Guide to the Rest of the Paper

In Section 2 we describe our transformation and how to simulate it. In Section 3, we analyze the efficiency and efficacy of our simulator. In 4 we discuss some simple extensions of our technique, and some open questions.

2   Transforming the Protocol

2.1   The Protocol

Let T be the statement that P is attempting to prove. We insert an $O(m)$ message preamble to the proof. Instead of simply giving a proof of T, P and V will each randomly choose and commit to m numbers, $p_1, p_2, ...p_m$ and $v_1, v_2, ...v_m$ respectively. P will then prove that either T is true or that for some i $p_i = v_i$.

$V \to P$: Commit to $v_1, v_2, ... v_m$
$P \to V$: Commit to p_1

$V \to P$: Reveal v_1
$P \to V$: Commit to p_2

...

$V \to P$: Reveal v_i
$P \to V$: Commit to p_{i+1}

...

$V \to P$: Reveal v_m

$P \leftrightarrow V$: Zero-Knowledge Proof that $(\exists i$ s.t. $v_i = p_i) \vee (T$ is true$)$

The protocol begins with $m + 1$ message exchanges. First V sends P a commitment to uniformly chosen $v_1, \ldots, v_m \in \{0, 1\}^q$, for some suitably large q. For simplicity, we assume that this commitment is information-theoretically secure. P responds by sending a commitment to p_1. In exchange $i + 1$, for $1 \le i < m$, V reveals v_i and P commits to p_{i+1}. Finally, V reveals v_m. At the conclusion of these exchanges, P responds by giving a zero-knowledge proof that either T is true or that for some i $p_i = v_i$. In the argument model, P gives a statistical zero-knowledge proof that it knows either:

- a witness for T, or
- a witness for a pair (i, REVEAL) such that on seeing REVEAL in the revelation of p_i, V would accept that $p_i = v_i$.

Note that P doesn't reveal which witness it knows, just that it knows one or the other. The general protocols of [13] and [1] may be used for this step (conceivably, more efficient protocols may be designed for useful special cases). The details of this interactive proof (argument) are unimportant.

There are two ways in which P may cause V to accept. Either it proves that T is true or it takes the "easy option" by showing that some $p_i = v_i$. However, regardless of a (possibly malicious) prover $\hat{P}$'s strategy, the easy option will be available with probability at most $m2^{-q}$; by setting q sufficiently large, this option occurs with negligible probability. Hence, the protocol remains a proof (of knowledge) of T.

2.2   Why We Can Simulate the Proof

Since there is so little chance of guessing v_i, P's strategy is to choose p_i at random, or 0^q, and simply proceed with the proof of T. Thus, for the correct prover, the preamble is irrelevant and for a malicious prover, the preamble is not useful. However, the simulator, S, can use the preamble to its advantage. After seeing v_i, it can rewind the conversation to the point where it is required to send p_i, and choose $p_i = v_i$. Because V committed to these v_i in the first message, S need not worry that the v_i change after the rewind as long as it doesn't rewind past the first message of the proof (which it might do while simulating a different proof).

Once S has ensured that for some i $p_i = v_i$, we say it has *solved* the protocol. It can complete the rest of the simulation (of this proof) without any further rewinding. When the actual proof begins, S has an actual witness to the statement being proved, and can therefore proceed according to the algorithm used by the actual prover. Appealing to the witness indistinguishability of the zero-knowledge proof, it is impossible to distinguish whether S used this witness or a witness for T.

2.3   Caveates

We mention three (of many) caveats regarding this approach. First, rewinding a single step in one proof can render irrelevant the simulations of many other proofs; nesting effects can cause exponential blowups in the simulation (as discussed in [6]). However, since S has m places it can rewind in order to fix a proof's simulation, it can choose good times to rewind.

Second, an improper use of rewinding can alter the distribution on the verifiers' questions, rendering the simulation invalid. Our simulation runs in two modes: *normal* and *look-ahead*. The normal mode is a step by step simulation of the k concurrent proofs. A step made in a normal proof is never rewound, facilitating the analysis of the distribution of the verifiers' messages. The look-ahead mode is invoked when the simulator, running in normal mode, is required to commit to p_i to the verifier, for one of the simulated proofs. In look-ahead mode, the simulator will explore many possible simulation paths and return with either the value of v_i, allowing S to solve this proof, or a statement that this is an unsuitable time to solve the proof. Once the look-ahead is complete, the simulator continues the normal-mode simulation. We show that S can use the information obtained in its look-ahead mode yet still maintain a faithful simulation.

We must also take care to avoid malleability attacks [5], where one links a commitment to the value of another parties commitment. For example, the prover might try to commit to the verifier's value, always achieving a match, or the verifier might try to foil the simulation by somehow opening up values different than those committed by the prover. Our assymetric choice of commitment protocols prevents these attacks.

2.4 Preliminaries

Let $v_{i,j}$ and $p_{i,j}$ denote the values of v_i and p_i committed to in the simulation of the jth proof. These values depend greatly on where we are in the simulation. In particular, they may be defined and then undefined when S rewinds the simulation.

Within a simulation path, we number the protocols in order of appearance. Thus, orderings may differ between different paths, but this will not affect our analysis.

During the preamble of a simulated protocol j, the verifier commits to m strings, $v_{1,j}, \ldots, v_{m,j}$. By a standard argument, the probability that $v_{i,j}$ is successfully revealed to be different values at different times after being committed to is negligible. Thus, we'll speak of the "value" of $v_{i,j}$. However, if the simulator rewinds past the point where the verifier committed to $v_{1,j}, \ldots, v_{m,j}$, these values become undefined.

At some point in the protocol, the simulated verifier will send a string that is supposed to reveal $v_{i,j}$. This string will either actually reveal this unique value or fail to reveal any value. Note that in the actual protocol, P aborts in the latter case.

During a path in the simulation, we say that a simulated protocol j is *solved* if, for some i, $v_{i,j}$ has been determined and $p_{i,j}$ has not yet been sent. We say that a simulated protocol j is *aborted* if the verifier fails to reveal $v_{i,j}$ when scheduled to do so. Note that rewinding and choosing a new path can change whether a simulated protocol is solved or is aborted.

If protocol j has been solved, the simulator simulates the prover's messages as follows. If the prover is supposed to send $p_{i,j}$ then it sends $v_{i,j}$ if it is known and an arbitrary string otherwise. During the main body of the proof, the simulator has a witness to the statement being proved, and acts according to the algorithm used by an honest prover. In particular, no rewinding is ever needed.

2.5 The Simulator

We let k_0 be a constant set to the initial value of k, which denotes the number of concurrent proofs. We show that if the number of message exchanges in the preamble is $m = k_0{}^\epsilon$ for any $\epsilon > 0$ then the above protocol can be simulated in time $k^{O(1/\epsilon)}$. This section describes the look-ahead procedure used by the simulator and how the simulator works in normal mode.

Look-Ahead Mode An *n-proof look-ahead* is a procedure used by S to gather information about the messages V is likely to send in the future. In the look-ahead phase, the simulation is allowed to proceed until certain events occur that cause it to be (prematurely) ended. The limited duration of the look-ahead makes it much more efficient than a full simulation, and indeed it is called many times during the simulation.

The n-proof look-ahead is called when S is required to commit to some $p_{i,j}$. The main simulator runs many ($100k_0{}^2$, to be precise) look-ahead simulations; we call these *threads*. We first describe one of one such thread, then describe how to use the results from many threads.

In the course of the simulation, the simulator is required to commit to strings $p_{a,b}$ and to engage in the main body of proofs. Along the way, it receives the values of strings $v_{a,b}$. A particular run of the n-proof look-ahead terminates when either $v_{i,j}$ has been revealed or the $n+1^{st}$ new proof, which started since the look-ahead began, is seen. The former case means that the mission is accomplished: S can set $p_{i,j} = v_{i,j}$. The latter case means that the simulation is proceeding too far and risks becoming too complicated; it may not be cost effective to keep waiting for $v_{i,j}$ to be revealed.

We differentiate between the protocols $1, \ldots, z$ that have already begun and the protocols $z + 1, \ldots, z + n$ that begin during the look-ahead simulation. The look-ahead simulator recursively starts a normal-mode simulator (described later). The normal-mode simulator requires a parameter specifying the maximum number of simultaneous proofs; this parameter is set to n. All messages and requests related to proofs $z + 1, \ldots, z + n$ are forwarded to this recursive simulation.

The normal-mode simulation has the property that, with all but negligible probability, by the time $p_{m,a}$ has been committed to, proof a will have been solved (see lemma 8). The look-ahead mode is less careful about solving proofs which began before the look-ahead. In this mode, S may commit to $p_{m,a}$ for an unsolved proof, and subsequently be unable to enter the main body of the proof. However, S will only get stuck if $v_{m,a}$ is revealed. Whenever this happens, S aborts the look-ahead and rewinds. We note that this rewinding will take S to before it committed to $p_{m,a}$, so proof a is now solved. Since less than k_0 proofs can begin before any given look-ahead, only k_0 of the look-ahead simulation paths will be aborted (the effect of these aborted paths is dealt with in lemma 7). Because these simulation paths are always rewound they will not affect the distribution of the normal-mode simulation.

We formally describe the n-proof look-ahead simulation by a case analysis of how the simulator responds to various messages. Only the first three cases are related to the purpose of the look-ahead; the rest are simply to keep the simulation going in a faithful fashion. By convention, the simulation takes as its first message the message being handled at the time it was called.

$V \rightarrow S$: Valid revelation of $v_{i,j}$.
$S \rightarrow V$: Terminate the simulation. *(proof j has been solved)*

$V \rightarrow S$: A commitment to $v_{1,a}, \ldots, v_{m,a}$, for $a = z + n + 1$.
$S \rightarrow V$: Terminate the simulation. *(look-ahead is finished)*

$V \rightarrow S$: Invalid revelation of $v_{i,j}$.
$S \rightarrow V$: Terminate the simulation. *(no chance of recovering $v_{i,j}$)*

$V \rightarrow S$: Any message related to protocol a, for $z < a \leq z + n$.
$S \rightarrow V$: Forward the message to the recursive simulation.
(We assume $1 \leq a \leq z$ in the remaining)

$V \rightarrow S$: Any message related to a solved proof a.
$S \rightarrow V$: Answer according to the standard fashion for solved proofs.

$V \rightarrow S$: Valid revelation of $v_{b,a}$, where a is unsolved and $b < m$.
$S \rightarrow V$: Commit to an arbitrary (random) value of $p_{b+1,a}$.

$V \rightarrow S$: Invalid message related to proof a
$S \rightarrow V$: Sign-off message from simulated prover for proof a.

$V \rightarrow S$: The value of $v_{m,a}$ for an unsolved proof a.
$S \rightarrow V$: Abort this line of the simulation. *(the simulator cannot simulate the main body of the proof of an unsolved proof)* However, note that proof a is now solved at the point where the look-ahead simulation began, which allows us to bound how often this bad case occurs.

Combining the Result of the Look-Aheads. Each run of the n-proof look-ahead simulation either returns a solution to the proof $(v_{i,j})$ or announces a failure to do so. To give a faithful simulation, the simulator must flip coins (e.g., when committing to $p_{b,a}$. Thus, there is a probability distribution on these results. In the normal-mode simulation, whenever the look-ahead simulation is invoked it is in fact invoked $100k_0{}^2$ times. If a solution is found in any of these invocations, the proof is solved. Otherwise, we will argue that with high (but not overwhelming) probability, at least n proofs in the actual (normal-mode) simulation will be started before $v_{i,j}$ is revealed.

Normal Mode Simulation Our simulator, working in normal mode, services the requests for up to k asynchronous proofs; the parameter k will be changed during recursive calls. Valid responses from S can take the following form:

- S signs off due to an invalid message.
- S engages in the main body of a proof.
- S commits to some $p_{i,j}$.

Handling invalid messages is trivial. Once S has solved a proof a, it can easily engage in the main body of a proof as a prover, since it has a witness for this proof. We will ensure that with all but negligible probability, S will always have solved a proof before it enters into the main body.

When S must commit to some $p_{i,j}$, it runs the n-proof look-ahead procedure $100k_0^2$ times, where $n = \lceil 2k/m \rceil$, in an attempt to recover $v_{i,j}$. If it succeeds it commits to $p_{i,j} = v_{i,j}$; otherwise, it commits to an arbitrary value of $p_{i,j}$.

As before, we describe the behavior of S by its response to various messages.

V $\rightarrow$ S: Any message related to a solved proof a.
S $\rightarrow$ V: Answer according to the standard fashion for solved proofs.

V $\rightarrow$ S: A commitment to $v_{1,a}, \dots, v_{m,a}$.
S $\rightarrow$ V: Invoke the $n = \lceil 2k/m \rceil$-proof look-ahead simulation $100k_0^2$ times. If $v_{1,a}$ is recovered, set $p_{1,a} = v_{1,a}$, else set $p_{1,a}$ arbitrarily. Commit to $p_{1,a}$.

V $\rightarrow$ S: valid revelation of $v_{b,a}$, $b < m$.
S $\rightarrow$ V: Invoke the $n = 2k/m$-proof look-ahead simulation $100k_0^2$ times. If $v_{b+1,a}$ is recovered, set $p_{b+1,a} = v_{b+1,a}$, else set $p_{b+1,a}$ arbitrarily. Commit to $p_{b+1,a}$.

V $\rightarrow$ S: Any invalid message related to proof a
S $\rightarrow$ V: Sign-off message from simulated prover for proof a.

3   Analysis of the Simulation

Theorem 2. *The simulator, S, described in Section 2 is a black box simulator for the protocol in Section 2 that runs in time $k^{O(1/\epsilon)}$ on k non-synchronized proofs when $m = k^\epsilon$ for $\epsilon > 0$.*

Proof. (Sketch) This theorem will follow from the following lemmata. Lemma 3 and Lemma 4 show that it runs in time $k^{O(1/\epsilon)}$. Lemma 6 shows that the simulator produces a valid output as long as it never gets stuck. Lemma 8 shows that the chance of getting stuck is negligibly small in m and the security parameter for the bit commitment schemes.   □

3.1   Bounding the Running Time

We assume that a simulator can handle a single message and give a response in *unit time*. We note that since we consider the main body of the proof to be a single message, we consider that proof to be given in unit time. A more precise (and more cumbersome) statement is that the running time is $k^{O(1/\epsilon)}$ times the amount of time it takes to perform the main body of the proof.

Lemma 3. *The running time of the simulator is bounded by the function*

$$t(k) = 100mk_0^3 t(\lceil \frac{2k}{m} \rceil) + k_0^{O(1)}.$$

Proof. (Sketch) First we note that each look-ahead thread begins a recursive simulation that handles up to $\lceil \frac{2k}{m} \rceil$ further proofs. This takes time bounded by $t(\lceil \frac{2k}{m} \rceil)$. Each look-ahead is repeated up to $100k_0^2$ times and S may attempt to solve each of the k_0 games by performing these look-aheads in m different places. This results in the coefficient of $100mk_0^3$. Each look-ahead thread also handles messages from previous game, whether solved or not. This takes unit time for each message. The number of messages, games and look-aheads are all polynomial in k_0. So the cost of this in all look-aheads is bounded by the $k_0^{O(1)}$ term.   □

Lemma 4. *The recurrence* $t(k) = 100mk_0^3t(\lceil \frac{2k}{m} \rceil) + k_0^{O(1)}$ *is* $k_0^{O(1/\epsilon)}$ *when* $m = k_0^\epsilon$.

Proof. (Sketch) At each recursive step, k is divided by $k_0^\epsilon/2$. Thus the total depth of the recursion is $O(1/\epsilon)$. Both the coefficient of for the recursive term and the cost at each level of the recursion are bounded by $k_0^{O(1)}$. Therefore, the total cost is $k_0^{O(1/\epsilon)}$.   □

3.2 The Simulation Is Valid

Note that (S, V) doesn't just simulate the conversation, it implicitly simulates the internal state of V - that is, the state of the "black-box" V that S is interacting with. We can consider the conversation generated thus far to be part of V's internal state. Thus, the process of S interacting with V constitutes a sequence of transformation on V's state. We can similarly consider the interaction of P and V to be a sequence of transformations on V's state.

We say that S becomes *stuck* if it enters the main body of a proof that hasn't been solved. We designate all other moves made by the simulator as *safe*. Lemma 6 says that S will produce a valid simulation as long as S only performs safe moves.

Lemma 5. *Any sequence of safe operations performs the identical (up to computational indistinguishability) transformations on V as the corresponding operations performed by P.*

Proof. (Sketch) Whenever S interacts in a solved proof, it has a witness for the statement to be proven. Due to the witness indistinguishability of the zero-knowledge proof in the main body, and the security of the bit commitment scheme used by the prover, all actions taken by S are computationally indistinguishable from those taken by any other prover.

When S commits to $p_{i,j}$, it may first launch into many recursive subsimulations involving many backtrackings. However, at the end of all these subsimulations, S restores V to its initial state, chooses a value for $p_{i,j}$ and commits to $p_{i,j}$. The value of $p_{i,j}$ depends on the results of these subsimulations; its distribution may be completely different from that generated by P (indeed, whenever a proof is solved, it's distribution is quite different). However, the distribution of messages sent for the commitment is the same (up to computational indistinguishability), regardless of this value.

Finally, by inspection, S responds to any illegal messages the same way as does P.   □

Note that the notion of a corresponding operations makes sense, because neither S nor P can control which *type* of operation it must make in response to V. Here, we are appealing to the computational indistinguishability of the commitment scheme from the prover to the verifier and the witness indistinguishability of the zero-knowledge protocols.

Now, given a particular configuration of V, S may run many simulations, due to the look-ahead mode. However, these simulations are ultimately thrown away. If one ignores all simulation threads arising from *further* recursive calls to the look-ahead mode (a currently active look-ahead mode may be continued), one obtains a unique sequence of transformations on V. This holds regardless of whether the particular configuration of V is encountered in normal mode or look-ahead mode. We call this sequence the *main line* of the evolution of V.

The main line of V from its starting configuration constitutes the simulation of the proofs. The main line from the beginning of a look-ahead thread goes to the point where S has finished this line or has been forced to discontinue the simulation (or get stuck).

Lemma 6. *Consider the evolution of V's configuration along its main line. Assuming that S never gets stuck, this evolution will be indistinguishable from the corresponding evolution obtained by interacting with P.*

Proof. (Sketch) The evolution of V consists of it sending messages to S $\{P\}$ and then having S $\{P\}$ perform operations. As long as S doesn't get stuck, all of its operations will be safe, and by Lemma 5, their effect on V will be indistinguishable from the effect of the corresponding operations performed by P.

It remains to be shown that the evolution of V when it generates its next message in the simulation is faithful to that in the actual protocol. Note that this is a nontrivial statement: S could conceivable run V many times and pick a path in which V sends a message that is amenable to S. However, by inspection of the simulation algorithms, S never selects which path to follow based on what V says. Indeed, it's selection process is completely rigid: Paths taken in look-ahead mode are ultimately not pursued; the main line path is pursued, at least locally (it may be thrown away later if it is part of a larger look-ahead path). Along any mainline path, S obtains V's message exactly once, by running V in the normal matter. Hence, V's internal evolution is also identical to that obtained by interacting with P.   □

3.3   Bounding the Probability of Getting Stuck

Lemma 6 implies that the main line from a configuration of V is indeed simulated correctly, as far as it goes (since in look-ahead mode, a simulation is typically ended prematurely) and as long as S doesn't get stuck. We now show that S gets stuck along any main line with negligible probability.

While in look-ahead mode, S never becomes stuck on a proof which began before the look-ahead, since it simply aborts the thread if it is about to become stuck. From then on that proof is solved, so the number of times S aborts is limited. This strategy cannot be employed in normal mode (at least on the top level of the recursion) since such stopping would constitute a failure to finish the overall simulation.

Recall that for any proof started in normal mode, the simulator tries m times to solve some proof a, by going into look-ahead mode in order to determine $v_{i,a}$ for each i. We characterize the various outcomes of this attempt.

- (complete success) The look-ahead recovers $v_{i,a}$, solving proof a.
- (win by forfeit) During the main line, the next message from V regarding proof a is ill formed (does not reveal $v_{i,a}$ when it should have).
- (honorable failure) The look-ahead fails to solve proof a, and during the main line more than $2k/m$ new proofs are begun before V reveals $v_{i,a}$.
- (dishonorable failure) The look-ahead fails to solve proof a, then during the main line, at most $2k/m$ new proofs are begun, after which V then sends a correct revelation of $v_{i,a}$.

Clearly, a single complete success or a win by forfeit will cause the game to be solved. We must show that with high probability, one of the m attempts will result in a complete success or a win by forfeit.

We next observe that an honorable failure can happen at most $m/2$ times. Since the normal-mode only handles k games, $m/2$ honorable failures result in more than $(2k/m) \cdot (m/2) = k$ new games, a contradiction. Thus, it remains to bound the probability of $m/2$ dishonorable failures.

We will prove that the chance of getting stuck at any level in the recursion is negligibly small by using induction on k (the number of proofs in a call to the simulator). The base case, $k = 1$, is when the simulator is solving a single proof. The look-aheads will never encounter another proof and as a result can never become stuck. The following lemma will be needed to complete the inductive step.

Lemma 7. *During any attempt to solve proof a the probability of a dishonorable failure is at most $1/10$ as long as the chance of getting stuck in a look-ahead is negligibly small.*

Proof. (Sketch) S attempts to solve proof a by performing $100k_0{}^2$ look-aheads after being asked to commit to $p_{i,a}$. We note that since these look-aheads have a negligibly small chance of getting stuck, lemma 6 implies that they give a valid sampling of the possiblie paths of the conversation. In order for a dishonorable failure to happen V must not reveal $v_{i,a}$ during any of those look-aheads but then reveal it when S continues on in normal mode. We may assume that the chance of V revealing $v_{i,a}$ is at least $p = 1/10$.

Now we must show that the chance of S not learning $v_{i,a}$ in any of the look-aheads is smaller than p. We must remember that some of the look-aheads could have been aborted if V revealed $v_{m,b}$ for some unsolved proof b. But each time a look-ahead is aborted we solve proof b. So the maximum number of times the look-ahead is aborted is $k_0 - 1$. Thus we need to show that the chance of seeing $v_{i,a}$ at least k_0 times is greater than p.

It is easy to verify that for any $b < a/3$, $\binom{a}{b-1} \le \binom{a}{b}/2$. Therefore the chance of seeing $v_{i,a}$ at most k_0 times is at most twice the cost of seeing it exactly k_0 times. This cost is less than

$$\binom{100k_0^2}{k_0} p^{k_0}(1-p)^{100k_0^2 - k_0}$$

$$< 100^{k_0} k_0^{2k_0} (1/10)^{k_0}(9/10)^{99k_0^2}$$

Which is dominated by the $(9/10)^{99k_0^2}$ and therefore (much) less than $1/10$.

Note that the above argument glossed over the fact that the safe steps are only computationally close to "real" steps. By a standard argument, this does not affect the analysis by more than a negligible amount.   □

Lemma 8. *The chance of S getting stuck is negligibly small in m and the security parameter of the bit commitment scheme.*

Proof. (Sketch) We use induction on k. In the base case of $k = 1$ the simulator will trivially never get stuck because there are no proofs to get stuck on. By induction we may assume that all look-aheads (which have smaller values of k) get stuck with negligibly small probability. The total number of look-aheads must be polynomial because the total running time of S is polynomial (see Lemma 3 and Lemma 4). Therefore the total chance of getting stuck in any look-ahead is also negligibly small. By lemma 7 the chance of each dishonorable failure is less than $1/10$. We note that for S to get stuck during a proof it must have had at least $m/2$ dishonorable failures. The chance of this is less than $\binom{m}{m/2}(1/10)^{m/2} < 2^{m/2}(1/10)^{m/2} < 2^{-m/2}$. There are a total of at most k proofs on which the simulator can get stuck, so the total chance of getting stuck is negligibly small.

Note that as with the previous argument, the above argument glossed over the fact that the safe steps are only computationally close to "real" steps. For this reason, the probability of getting stuck is negligibly small, not exponentially small.   □

4   Extensions and Open Questions

4.1   Extensions

Our proof assumes that k is known. In the case where k is unknown, S may start by assuming that $k = 1$ and double k and restart the interaction each time it discovers that k is larger than it assumed. It is easy to verify that this has no effect on the output distribution and that the total running time is still polynomial.

We do not really need to have globally agreed upon commitment schemes. The modification is to add two messages to the protocol in which each party specifies the commitment scheme that should be used to commit to it (first the verifier, then the prover). The property we desire is that the commitments are unconditionally guaranteed to be zero-knowledge, regardless of how it is specified (illegal specifications are treated as invalid messages). Thus, a party can use the other party's bit commitment system without any loss of its security. The party specifying the protocol has no obvious reason to make it easy to break, but this is not enforced. Such bit commitment schemes are easily constructed based on [4] and [14,17]. Due to space limitations, details are omitted from this manuscript.

We also note that there is essentially no reason why our construction doesn't work even if the k proofs are different.

4.2   Open Questions

It is unknown whether there is a perfect simulation for non-synchronized composition of zero-knowledge proofs. We note that in the case when V always follows the protocol and successfully reveals v_i we can modify the simulator so that it never gets stuck. We do this by having the simulator look-ahead from the point it is forced to commit to p_m until v_m is revealed if the proof has not yet been solved. If V is required to reveal v_m this is always successful. Then instead of being stuck, S is just in a bad case which may take longer to simulate, but it is still possible to do in polynomial time. Assuming the existence of a perfect commitment scheme, this non-cheating verifier allows us to provide a perfect simulation.

It would also be useful to show that it is possible to simulate non-synchronized composition with a constant number of message exchanges in the preamble. Again, assuming that the verifier always reveals v_i, our protocol can be modified so that it runs in time $k^{O(\log k)}$ with a constant number of messages. This is interesting because it shows that the techniques used in [15] to show that any four message protocol takes time $2^{O(k)}$ to simulate can not be extended to any constant round proof.

As mentioned in the introduction, we do not address timing issues in the verifier's attack. Even modeling what zero-knowledge should mean in this context, in a way that is both useful and possible, is an interesting open question.

Finally, it is paradoxical that such seemingly meaningless alterations in the protocol can restore zero-knowledge. Intuitively, it seems implausible that the protocol has been made more secure in practice. Ideally, one would like to have a notion of security that is more or less invariant under such transformations. The notions of witness hiding and witness indistinguishable protocols are good steps in this direction.

5 Acknowledgments

Kilian would like to thank Cynthia Dwork, Uri Feige, Moni Naor, Amit Sahai and Erez Petrank for many illuminating conversations on this subject.

References

1. G. Brassard, D. Chaum, C. Crépeau. Minimum Disclosure Proofs of Knowledge. Journal of Computer and System Sciences, Vol. 37, 1988, pp. 156–189.
2. C. Brassard, C. Crepeau and M. Yung, "Constant-Round Perfect Zero-Knowledge Computationally Convincing Protocols", *Theoretical Computer Science*, Vol. 84, 1991, pp. 23-52.
3. T. Beth and Y. Desmedt. Identification tokens - or: Solving the chess grandmaster problem. In A. J. Menezes and S. A. Vanstone, editors, *Proc. CRYPTO 90*, pages 169–177. Springer-Verlag, 1991. Lecture Notes in Computer Science No. 537.
4. Damgård, Torben P. Pedersen, and Birgit Pfitzmann. On the existence of statistically hiding bit commitment schemes and fail-stop signatures. In Douglas R. Stinson, editor, *Proc. CRYPTO 93*, pages 250–265. Springer, 1994. Lecture Notes in Computer Science No. 773.
5. D. Dolev, C. Dwork, and M. Naor. Non-malleable cryptography. In ACM, editor, *Proceedings of the twenty third annual ACM Symposium on Theory of Computing, New Orleans, Louisiana, May 6-8, 1991*, pages 542–552, 1109 Spring Street, Suite 300, Silver Spring, MD 20910, USA, 1991. IEEE Computer Society Press.
6. Cynthia Dwork, Moni Naor, and Amit Sahai. Concurrent zero knowledge. In *Proceedings of the 30th Annual ACM Symposium on Theory of Computing (STOC-98)*, pages 409–418, New York, May23–26 1998. ACM Press.
7. C. Dwork and A. Sahai. Concurrent zero-knowledge: Reducing the need for timing constraints. *Lecture Notes in Computer Science*, 1462, 1998.
8. U. Feige, D. Lapidot, and A. Shamir. Multiple non-interactive, zero-knowledge proofs based on a single random string. In *Proc. 31st Ann. IEEE Symp. on Foundations of Computer Science*, pages 308–317, 1990.
9. U. Feige and A. Shamir, "Zero Knowledge Proofs of Knowledge in Two Rounds", *Advances in Cryptology – Crypto 89 proceedings*, pp. 526-544, 1990.
10. O. Goldreich, H. Krawczyk. On the Composition of Zero-Knowledge Proof Systems. SIAM J. on Computing, Vol. 25, No.1, pp. 169-192, 1996
11. S. Goldwasser, S. Micali, C. Rackoff. The Knowledge Complexity of Interactive Proofs. Proc. 17th STOC, 1985, pp. 291-304.
12. S. Goldwasser, S. Micali, C. Rackoff. The Knowledge Complexity of Interactive Proof Systems. SIAM J. on Computing, Vol. 17, 2(1988), pp. 281-308.
13. S. Goldwasser, S. Micali, A. Wigderson. Proofs that Yield Nothing But their Validity or All Languages in NP have Zero-Knowledge Proofs. J. of the ACM, Vol. 38, No. 3, July 1991, pp. 691-729.
14. Johan Hastad, Russell Impagliazzo, Leonid A. Levin, and Michael Luby. Construction of a pseudo-random generator from any one-way function. Technical Report TR-91-068, International Computer Science Institute, Berkeley, CA, December 1991.
15. Kilian, Petrank, and Rackoff. Lower bounds for zero knowledge on the internet. In *FOCS: IEEE Symposium on Foundations of Computer Science (FOCS)*, 1998.

16. Paul C. Kocher. Timing attacks on implementations of Diffie-Hellman, RSA, DSS, and other systems. In Neal Koblitz, editor, *Advances in Cryptology—CRYPTO '96*, volume 1109 of *Lecture Notes in Computer Science*, pages 104–113. Springer-Verlag, 18–22 August 1996.

17. Moni Naor. Bit commitment using pseudo-randomness. In *Advances in Cryptology: CRYPTO '89*, pages 128–137, Berlin, August 1990. Springer.

18. Y. Oren. On the cunning powers of cheating verifiers: Some observations about zero knowledge proofs. In Ashok K. Chandra, editor, *Proceedings of the 28th Annual Symposium on Foundations of Computer Science*, pages 462–471, Los Angeles, CA, October 1987. IEEE Computer Society Press.

19. R. Ostrovsky and G. Di Crescenzo. Personal Communication, September 15, 1998.

20. M. Tompa and H. Woll. Random self-reducibility and zero-knowledge interactive proofs of possession of information. In *Proc. 28th Ann. IEEE Symp. on Foundations of Computer Science*, pages 472–482, 1987.

Pseudorandom Function Tribe Ensembles Based on One-Way Permutations: Improvements and Applications

Marc Fischlin

Fachbereich Mathematik (AG 7.2)
Johann Wolfgang Goethe-Universität Frankfurt am Main
Postfach 111932
60054 Frankfurt/Main, Germany
marc@mi.informatik.uni-frankfurt.de
http://www.mi.informatik.uni-frankfurt.de/

Abstract. Pseudorandom function tribe ensembles are pseudorandom function ensembles that have an additional collision resistance property: almost all functions have disjoint ranges. We present an alternative to the construction of pseudorandom function tribe ensembles based on one-way permutations given by Canetti, Micciancio and Reingold [7]. Our approach yields two different but related solutions: One construction is somewhat theoretic, but conceptually simple and therefore gives an easier proof that one-way permutations suffice to construct pseudorandom function tribe ensembles. The other, slightly more complicated solution provides a practical construction; it starts with an arbitrary pseudorandom function ensemble and assimilates the one-way permutation to this ensemble. Therefore, the second solution inherits important characteristics of the underlying pseudorandom function ensemble: it is almost as efficient and if the starting pseudorandom function ensemble is invertible then so is the derived tribe ensemble. We also show that the latter solution yields so-called committing private-key encryption schemes. i.e., where each ciphertext corresponds to exactly one plaintext — independently of the choice of the secret key or the random bits used in the encryption process.

1 Introduction

In [7] Canetti, Micciancio and Reingold introduce the concept of pseudorandom function tribe ensembles. Informally, such tribe ensembles consists of pseudorandom functions that have an independent public key in addition to the secret key. Though this public key, called the tribe key, is independent of the secret key, it guarantees that any image/preimage pair commits to the secret key. More specifically, for a random tribe key t there do not exist secret keys $k \neq k'$ and a value x such that the functions determined by the keys k, t resp. k', t map x to the same value (except with exponentially small probability, where the probability is taken over the choice of t). Canetti et al. [7] use such pseudorandom function tribe ensembles to construct perfectly one-way probabilistic hash functions.

J. Stern (Ed.): EUROCRYPT'99, LNCS 1592, pp. 432–445, 1999.

In contrast to ordinary one-way functions, such perfectly one-way probabilistic hash functions hide all partial information about the preimage (secrecy), yet finding a hash value together with distinct preimages is infeasible (collision resistance). In [3] Canetti presents perfectly one-way hash functions based on a specific number-theoretic assumption, namely the Decisional-Diffie-Hellman assumption. Generalizing this result, Canetti, Micciancio and Reingold [7] show that perfectly one-way functions can be constructed from any cryptographic hash function (achieving secrecy statistically and collision resistance computationally) or from any pseudorandom function tribe ensembles (with computational secrecy and statistical collision resistance). In the latter case, the pseudorandomness of the tribe ensemble provides secrecy and collision resistance follows from the property of the tribe key. Canetti et al. [7] also prove that PRF tribe ensembles exist if one-way permutations exist. Their construction is a modification of the GGM-tree design of PRF ensembles [10] combined with a generalization of the Goldreich-Levin hardcore predicate [12]. A sketch of this construction is given in Appendix A. Here, we take a different approach which consists of two elementary and independent steps. First, we show that any one-way permutation suffices to construct a PRF ensemble such that for distinct secret keys k, k' the functions determined by k and k' map 1^n to different values. We call such ensembles fixed-value-key-binding as the key is determined by the function value for 1^n or, using a minor modification, for any other fixed value instead of 1^n. Second, we prove that fixed-value-key-binding PRF ensembles yield PRF tribe enembles. After presenting a conceptually simple construction of fixed-value-key-binding ensembles based on the GGM-tree design to the authors of [7], they pointed out an improvement that led to the more practical solution which does not necessarily involve the GGM-construction. Instead it works with a every PRF ensemble by assimilating the one-way permutation to the given ensemble. This yields a fixed-value-key-binding PRF ensemble and, in turn, a PRF tribe ensemble which is almost as efficient as the starting PRF ensemble. Moreover, if the functions of the ordinary ensemble are invertible then so are the functions of the tribe ensemble. From a theoretical and practical point of view this gives us the best of both worlds: As for the theory, we obtain a simple proof that the existence of one-way permutations implies the existence of PRF tribe ensembles. For practical purposes, we present a construction where pseudorandomness is slightly harder to prove, but which has nice properties. In both cases, the second step deriving the tribe ensemble from the fixed-value-key-binding ensemble is identical. We give an outline of this part. It is a reminiscent of Naor's statistically-binding bit commitment scheme [17]. There, the receiver sends a random $3n$-bit string A to the committing party who applies a pseudorandom generator $G : \{0,1\}^n \rightarrow \{0,1\}^{3n}$ to a random value $r \in \{0,1\}^n$ and returns $G(r) \oplus A$ to commit to 1 resp. $G(r)$ to commit to 0. The receiver cannot distinguish both cases with significant advantage because of the pseudorandomness of the generator's output. On the other hand, to open a commitment ambiguously the sender has to find r, r' such that $G(r) = G(r') \oplus A$. But $\#\{G(r) \oplus G(r') \mid r, r'\} \leq 2^{2n}$, hence $A \in \{G(r) \oplus G(r') \mid r, r'\}$ with probability at most 2^{-n} (over the choice of A).

This means that the commitment cannot be opened ambiguously with probability at least $1-2^{-n}$. We adopt this idea to define our PRF tribe ensemble. Given a fixed-value-key-binding PRF ensemble we define an appropriate fixed-value-key-binding PRF ensemble F^{stretch} with functions f_k^{stretch} that stretch the input to a sufficiently large output. We then show that there exists a value I_k (depending on the secret key k) and a function $\text{XOR}(t, I_k)$ of the tribe key t and I_k such that from the key-binding property it follows that for different keys k, k' and random t the value $\text{XOR}(t, I_k) \oplus \text{XOR}(t, I_{k'})$ is a uniformly distributed string having the same length as the output of f_k^{stretch}.[1] In other words, $\text{XOR}(t, I_k)$ is an xor universal hash function [8] with argument I_k and description t. Define the functions f_k^t of the PRF tribe ensemble by $f_k^t(x) = f_k^{\text{stretch}}(x) \oplus \text{XOR}(t, I_k)$. A collision $f_k^t(x) = f_{k'}^t(x)$ for $x, k \neq k'$ implies

$$f_k^{\text{stretch}}(x) \oplus f_{k'}^{\text{stretch}}(x) = \text{XOR}(t, I_k) \oplus \text{XOR}(t, I_{k'})$$

Since the output length of the functions in F^{stretch} is much bigger than the input length and as $\text{XOR}(t, I_k) \oplus \text{XOR}(t, I_{k'})$ is a random string for random t, collision resistance of the tribe ensemble is obtained as in Naor's bit commitment scheme. Additionally, we will show that the pseudorandomness of the tribe ensemble follows from the pseudorandomness of F^{stretch}.

Finally, based on our PRF tribe ensemble, we present a committing private-key encryption scheme, i.e., such that one cannot later open an encryption ambiguously by pretending to have used a different secret key. Secure committing *public*-key encryption systems can be derived for example from trapdoor permutations using the Goldreich-Levin hardcore predicate. In fact, constructing the opposite, public-key schemes that allow to open encryptions ambiguously, is a very interesting problem, because such schemes yield multiparty protocols secure against adaptive adversaries [5,6,4]. Given an arbitrary fixed-value-key-binding PRF ensemble we present a straightforward solution for a committing private-key system. Unfortunately, this scheme allows to deduce if two encryptions have been generated with the same secret key; a drawback which schemes based on PRF ensembles usually do not have. Therefore, we present another committing system that does not have this disadvantage, and prove that this scheme is secure against chosen ciphertext and plaintext attacks or, equivalent, non-malleable.

2   Preliminaries

For sake of self-containment, we briefly recall basic definitions of pseudorandom functions, pseudorandom generators, etc. See [11] for the underlying intuition. At the end of this section, we repeat the GGM-construction and the definition of pseudorandom function tribe ensembles. We present all definitions for uniform adversaries only; replacing the term "polynomial-time algorithm" by "polynomial circuit family" one easily obtains the nonuniform counterpart.

---

[1] Actually, this string will be uniformly distributed in a sufficiently large subset of the binary strings of the output length.

A function $\delta(n)$ is called *negligible in n* if $\delta(n) < 1/p(n)$ for any positive polynomial $p(n)$ and all sufficiently large n. A polynomial-time computable function f is *one-way* if for any probabilistic polynomial-time algorithm A the probability $\text{Prob}\big[A(1^n, f(x)) \in f^{-1}(x)\big]$ that A outputs a preimage of $f(x)$ for random $x \in \{0,1\}^n$ is negligible in n. A one-way function f is a *one-way permutation* if f permutes $\{0,1\}^n$ for every n. A *hardcore predicate* of a one-way function f is a polynomial-time computable predicate B such that for any probabilistic polynomial-time algorithm A it holds that $\text{Prob}[A(1^n, f(x)) = B(x)]$ for random $x \in \{0,1\}^n$ is negligible in n. According to a result of Goldreich and Levin [12] every one-way function can be modified to have a hardcore predicate. A polynomial-time computable function G is a *pseudorandom generator* if there exists some function $\ell(n)$ such that $\ell(n) > n$ and $G(x) \in \{0,1\}^{\ell(n)}$ for all $x \in \{0,1\}^n$ and all n, and such that for any probabilistic polynomial-time algorithm D the advantage $|\text{Prob}[D(G(x)) = 1] - \text{Prob}[D(y) = 1]|$ is neglible in n, where x is chosen at random from $\{0,1\}^n$ resp. y from $\{0,1\}^{\ell(n)}$. Pseudorandom generators exist if and only if one-way functions exist [14]. A function ensemble with key space $K = \{K_n\}_{n \in \mathbb{N}}$, input length $\text{in}(n)$ and output length $\text{out}(n)$ is a sequence $F = \{F^{(n)}\}_{n \in \mathbb{N}}$ of function families $F^{(n)} = \{f_k\}_{k \in K_n}$ such that for any $k \in K_n$ the function f_k maps bit strings of length $\text{in}(n)$ to bit strings of length $\text{out}(n)$. A function ensemble is *polynomial-time computable* if the length of the keys of $K = \{K_n\}$ and $\text{in}(n)$ are bounded by some polynomial in n and if there exists a polynomial-time algorithm Eval such that $\text{Eval}(k, x) = f_k(x)$ for all n, $k \in K_n$ and $x \in \{0,1\}^{\text{in}(n)}$. In the sequel we denote by $\mathcal{R} = \{\mathcal{R}^{(n)}\}_{n \in \mathbb{N}}$ the function ensemble that contains all functions $g : \{0,1\}^{\text{in}(n)} \to \{0,1\}^{\text{out}(n)}$; here $\text{in}(n)$ and $\text{out}(n)$ and therefore the key space of $\mathcal{R}^{(n)}$ will be understood from the context. A polynomial-time computable function ensemble F (with key space K and input/output length $\text{in}(n)$ and $\text{out}(n)$) is a *pseudorandom function ensemble* (PRF ensemble) if for any probabilistic polynomial-time algorithm D, called the distinguisher, the advantage $|\text{Prob}\big[D^f(1^n) = 1\big] - \text{Prob}[D^g(1^n) = 1]|$ is negligible, where f is chosen at random from $F^{(n)}$ (by selecting a random key from K_n) and g is a random function of $\mathcal{R}^{(n)}$ (where each function in $\mathcal{R}^{(n)}$ has input/output length $\text{in}(n)$ and $\text{out}(n)$). A PRF ensemble F with key space K and input/output length $\text{in}(n) = \text{out}(n)$ is called a *pseudorandom permutation ensemble* (PRP ensemble) if f_k is a permutation for any key $k \in K_n$ and the advantage $|\text{Prob}\big[D^f(1^n) = 1\big] - \text{Prob}[D^g(1^n) = 1]|$ is negligible for any probabilistic polynomial-time algorithm D, where f is a random function of $F^{(n)}$ and g is a random permutation with input/output length $\text{in}(n) = \text{out}(n)$. A PRP ensemble F is said to be a *strong* PRP ensemble if it even holds that $|\text{Prob}[D^{f,f^{-1}}(1^n) = 1] - \text{Prob}[D^{g,g^{-1}}(1^n) = 1]|$ is negligible for any probabilistic polynomial-time algorithm D.

Pseudorandom function ensembles can be constructed from any pseudorandom generator via the GGM-tree design [10]. Let G denote a length-doubling pseudorandom generator, i.e., with output length $\ell(n) = 2n$; such generators can be constructed from any pseudorandom generators by modifying the output length. Let $G^0(x)$ resp. $G^1(x)$ denote the left and right half of $G(x)$ and define the function ensemble F with key space $K_n = \{0,1\}^n$ and input/output length $\text{in}(n) = \text{out}(n) = n$ by $f_k(x) = G^{x_n}(\cdots G^{x_2}(G^{x_1}(k))\cdots)$. Here, $x_1,\ldots,x_n \in \{0,1\}$ and $x = x_1;\cdots;x_n$ is the concatenation of $x_1,\ldots,x_n$. The function f_k can be described by a binary tree of depth n where the root is labeled with k and each left (right) child of a node v is labeled with $G^0(\text{label}(v))$ resp. $G^1(\text{label}(v))$. A value $x \in \{0,1\}^n$ then determines a path from the root to some leaf and the function value $f_k(x)$ equals the label of this leaf. Goldreich et al. [10] prove that the derived ensemble F is pseudorandom.

A PRF tribe function ensemble with key space $K = \{K_n\}_{n\in\mathbb{N}}$ and tribe key space $T = \{T_n\}_{n\in\mathbb{N}}$ is a function ensemble $F = \{\{F_t^{(n)}\}_{t\in T_n}\}_{n\in\mathbb{N}}$ of function families $F_t^{(n)} = \{f_k^t\}_{k\in K_n}$ such that $\{F_{t_n}^{(n)}\}_{n\in\mathbb{N}}$ is a PRF ensemble for any sequence $\{t_n\}_{n\in\mathbb{N}}$, $t_n \in T_n$ of tribe keys, and such that for a randomly chosen tribe key $t \in T_n$ the probability that there exist $x \in \{0,1\}^{\text{in}(n)}$, $k,k' \in K_n$ with $k \neq k'$ and $f_k^t(x) = f_{k'}^t(x)$ is at most 2^{-n}. The latter property is called (statistical) collision resistance.

3   Constructing PRF Tribe Ensembles

We first show how to construct an PRF ensemble F^{bind} such that $f_k^{\text{bind}}(1^n) \neq f_{k'}^{\text{bind}}(1^n)$ for keys $k \neq k'$. Put differently, the function value at 1^n commits to the key. We therefore say that this ensemble *binds the key* (for a fixed value) because once we have seen the value at 1^n one cannot later pretend to have used another key. Obviously, we can also take any other fixed value x_0 instead of 1^n by setting $f_k^*(x) = f_k^{\text{bind}}(x \oplus x_0 \oplus 1^n)$. We then use such a fixed-value-key-binding PRF ensemble to derive a pseudorandom function (with tribe key t) where $f_k^t(x) \neq f_{k'}^t(x)$ for any $x, k \neq k'$ with probability $1 - 2^{-n}$ over the choice of t. This is achieved by using Naor's idea as explained in the introduction. We can even modify the construction to obtain a *key-binding-and-invertible* pseudorandom function that binds the key and can be efficiently inverted given the secret key. Particularly, this implies that $f_k^t(x) \neq f_{k'}^t(x')$ for $(k,x) \neq (k',x')$ with probability $1 - 2^{-n}$, i.e., the function binds the key *and* the preimage with high probability. This somewhat weaker property can also be derived extending the universal hash function $\text{XOR}(t, I_k)$ to take arguments x and I_k instead of I_k. We discuss this construction at the end of the section. However, it is not clear that this solution is efficiently invertible using the key, a requirement that we need in Section 4 applying our construction to private-key encryption.

3.1   A Fixed-Value-Key-Binding PRF Ensemble

Clearly, a pseudorandom function ensemble with $f_k(1^n) \neq f_{k'}(1^n)$ for $k \neq k'$ can be derived via the GGM-construction using a length-doubling pseudorandom generator G which is one-to-one on the right half. In this case, the function value at 1^n is $G^1(\cdots G^1(k))$ and since G^1 is one-to-one this yields different values for different keys. According to a result by Yao [23] such a pseudorandom generator

G where G^1 is one-to-one can be constructed from any one-way permutation g by setting

$$G(x) = B(x); B(g(x)); \cdots; B(g^{|x|-1}(x)); g^{|x|}(x)$$

Here, $g^i(x) = g(g^{i-1}(x))$ and $g^1(x) = g(x)$ and B denotes some hardcore predicate of g. Obviously, $G^1(x) = g^{|x|}(x)$ is one-to-one (in fact, it is a permutation).

Another construction of fixed-value-key-binding ensembles was proposed by the authors of [7] after presenting the GGM-based approach to them. The advantage is that we use the underlying pseudorandom function as a black box and merely add the length-doubling generator G (with G^1 being one-to-one) on. Particularly, instead of using the GGM-construction one can start with any PRF ensemble. For instance, more efficient constructions of PRF ensembles based on synthesizers [18] resp. on the Decisional-Diffie-Hellman assumption [19] suffice. In practice, one can also use appropriate candidates like the forthcoming AES.

So let F^{start} be an arbitrary PRF ensemble (the starting point). For simplicity, we suppose that each function f_k^{start} of $F^{\text{start},(n)}$ maps n bits to n bits and that the key length equals n, too. We discuss below how to patch other cases. Set $k^b = G^b(k)$ for $b \in \{0, 1\}$ and define the functions of $F^{\text{bind},(n)}$ by

$$f_k^{\text{bind}}(x) = \begin{cases} k^1 & \text{if } x = 1^n \\ f_{k^0}^{\text{start}}(x) & \text{else} \end{cases}$$

Proposition 1. F^{bind} *is a fixed-value-key-binding PRF ensemble.*

Proof. (Sketch) The proof follows by standard hybrid techniques. Given a distinguisher D^{bind} that distinguishes F^{bind} and $\mathcal{R}$ with advantage $\delta(n)$ for infinitely many n, we either obtain an algorithm that distinguishes the output of G from random bits with advantage $\delta(n)/2$ infinitely often or we derive an algorithm that distinguishes F^{start} and $\mathcal{R}$ with advantage $\delta(n)/2$ for infinitely many n. Obviously, F^{bind} binds the key for the fixed value 1^n because G^1 is one-to-one. $\square$

Though F^{start} might be a pseudorandom permutation ensemble, F^{bind} does not inherit this property in general. However, a slight modification works: swapping the values that map to k^1 and $f_{k^0}^{\text{start}}(1^n)$ we let

$$f_k^{\text{bind}}(x) = \begin{cases} k^1 & \text{if } x = 1^n \\ f_{k^0}^{\text{start}}(1^n) & \text{if } f_{k^0}^{\text{start}}(x) = k^1 \text{ and } x \neq 1^n \\ f_{k^0}^{\text{start}}(x) & \text{else} \end{cases} \qquad (1)$$

It is easy to see that f_k^{bind} is a permutation if $f_{k^0}^{\text{start}}$ is. Moreover, the inverse of f_k^{bind} is efficiently computable (given the key k) if $f_{k^0}^{\text{start}}$ has this property. We remark that every PRF ensemble can be turned into a PRP ensemble [15]; see [20] for recent results. Yet, using the Luby-Rackoff transformation, the key length of the derived permutation grows. This can be handled by stretching the output length of the generator G accordingly; it suffices that G is one-to-one on the bits that replace the output at 1^n. In particular, if the output length of $f_{k^0}^{\text{start}}$ is smaller than right half of $G(k)$ then we can first stretch the output of $f_{k^0}^{\text{start}}$ at the cost of decreasing the input length slightly. We will use this technique in the next section, too, so we omit further details here. The proof that the ensemble F^{bind} defined by equation (1) is pseudorandom is similar to the proof of Proposition 1. It is also easy to show that F^{bind} is a *strong* PRP ensemble if F^{start} is.

Proposition 2. *If F^{start} is a [strong] PRP ensemble then F^{bind} as defined in equation (1) is a fixed-value-key-binding [strong] PRP ensemble.*

We remark that once the key is generated (by evaluating the pseudorandom generator) computing $f_k^{\text{bind}}(x)$ is as fast as computing $f_k^{\text{start}}(x)$. Particularly, f_k^{start} may be any fast practical pseudorandom function candidate. In contrast, using the GGM-based approach we have to apply n times a pseudorandom generator which is one-to-one on the right half, e.g., based on a number-theoretic one-way permutation like RSA.

3.2 PRF Tribe Ensembles from Key-Binding PRF Ensembles

Let F^{bind} be a fixed-value-key-binding PRF ensemble (for the value 1^n). In another intermediary step we define a PRF ensemble F^{stretch} that has input length $n-3$, but stretches the output length to $5n$. Define the functions f_k^{stretch} : $\{0,1\}^{n-3} \to \{0,1\}^{5n}$ by

$$f_k^{\text{stretch}}(x) = f_k^{\text{bind}}(x000); \cdots ; f_k^{\text{bind}}(x011); f_k^{\text{bind}}(x111)$$

Obviously, F^{stretch} is a PRF ensemble if F^{bind} is. Also note that computing f_k^{stretch} takes at most five evaluations of f_k^{bind}; but due to the common prefix one might not need to carry out all evaluations of f_k^{bind} from scratch and save time.

Now we are able to define our tribe ensemble F of functions $f_k^t : \{0,1\}^{n-3} \to \{0,1\}^{5n}$. The tribe key $t = (t_1, \ldots, t_n)$ consists of n uniformly and independently chosen values $t_i \in \{0,1\}^{4n} \times \{0^n\}$, i.e., t_i is a random $4n$-bit string filled up with 0-bits. Denote

$$I_k = f_k^{\text{bind}}(1^n) = \text{rightmost } n \text{ bits of } f_k^{\text{stretch}}(1^{n-3})$$

and let

$$\text{XOR}(t, I_k) = \bigoplus_{i\text{-th bit}(I_k)=1} t_i$$

Then we set

$$f_k^t(x) = f_k^{\text{stretch}}(x) \oplus \text{XOR}(t, I_k) \tag{2}$$

Note that once k and t are chosen, $\text{XOR}(t, I_k)$ is also fixed. Therefore, evaluating f_k^t at some point x is quasi as efficient as computing $f_k^{\text{stretch}}(x)$. The proof that F

is pseudorandom *for any sequence of tribe keys* is given below. We stress that the pseudorandomness of F does not depend on the random choice of the tribe key. See the discussion in [7]. Also note that if f_k^{bind} is one-to-one (e.g., a permutation) then $f_k^{\text{stretch}}(x) \neq f_k^{\text{stretch}}(x')$ resp. $f_k^t(x) \neq f_k^t(x')$ for $x \neq x'$.

Proposition 3. *F is a PRF ensemble for any sequence of tribe keys.*

Proof. (Sketch) The proof follows by standard simulation arguments. Given an adversary D that distinguishes a random function of F and a randomly chosen function from the ensemble $\mathcal{R}$ we obtain a distinguisher D^{stretch} that distinguishes F^{stretch} and $\mathcal{R}$ with the same advantage. Note that both D and D^{stretch} are given an arbitrary tribe key t as input. For a function $f : \{0,1\}^{n-3} \rightarrow \{0,1\}^{5n}$ let $f^{\text{sim}}(x) = f(x) \oplus \text{XOR}(t, I)$, where I denotes the rightmost n bits of $f(1^{n-3})$. D^{stretch} simulates D by answering all oracle queries x with $f^{\text{sim}}(x)$, where the underlying oracle f of D^{stretch} is chosen from $F^{\text{stretch},(n)}$ or $\mathcal{R}^{(n)}$. If f is chosen at random from F^{stretch} then f^{sim} is a random function of F. Assume that f is a random function of $\mathcal{R}^{(n)}$. It is easy to see that in this case any value $f^{\text{sim}}(x)$ is distributed independently of the other function values. Hence, it suffices to show that $\text{Prob}_f \left[f^{\text{sim}}(x) = y \right] = 2^{-5n}$ for any x, y. This is clear for $x \neq 1^{n-3}$. Consider the case $x = 1^{n-3}$. The rightmost n bits of $f(x)$ are random bits and the rightmost n bits of $\text{XOR}(t, I)$ equal 0^n. Hence, with probability 2^{-n} we have equality on these bits. The leftmost $4n$ bits of $f(x)$ are random bits that are independent of the other n bits. Therefore, the probability that these bits equal the leftmost $4n$ bits of $y \oplus \text{XOR}(t, I)$ is 2^{-4n} and both probabilities multiply due to the independence.   □

Recall that a PRF tribe ensemble is collision-resistant (in a statistical sense) if there do not exist x and k, k' such that $k \neq k'$ and $f_k^t(x) = f_{k'}^t(x)$ except with exponentially small probability (over the random choice of the tribe key). In our case, we have $I_k \neq I_{k'}$ for $k \neq k'$ and a collision

$$f_k^t(x) = f_k^{\text{stretch}}(x) \oplus \text{XOR}(t, I_k) = f_{k'}^{\text{stretch}}(x) \oplus \text{XOR}(t, I_{k'}) = f_{k'}^t(x)$$

implies

$$f_k^{\text{stretch}}(x) \oplus f_{k'}^{\text{stretch}}(x) = \text{XOR}(t, I_k) \oplus \text{XOR}(t, I_{k'}) = \text{XOR}(t, I_k \oplus I_{k'})$$

Because $I_k \oplus I_{k'} \neq 0^n$, the value $\text{XOR}(t, I_k \oplus I_{k'})$ is uniformly distributed in $\{0,1\}^{4n} \times \{0^n\}$ for fixed $x, k \neq k'$ and random t. By the union bound we conclude that

$$\text{Prob}_t \left[\exists x, k \neq k' \text{ s.t. } f_k^t(x) = f_{k'}^t(x) \right] \leq 2^{3n-3} \cdot 2^{-4n} \leq 2^{-n}$$

Thus we obtain:

Theorem 4. *The ensemble F defined by equation (2) is a PRF tribe ensemble.*

Clearly, we can lower the error probability of the collision resistance. For example, to achieve an error of 2^{-4n} we extend f_k^{stretch} to $8n$ bits output and choose the t_i's at random from $\{0,1\}^{7n} \times \{0^n\}$. If, in addition to an extended output length of at least $6n$ bits, we use a pseudorandom permutation F^{start} then we derive a pseudorandom function tribe ensemble F such that $f_k^t(x) \neq f_{k'}^t(x')$ for $(k,x) \neq (k',x')$ with probability at least $1 - 2^{-n}$ (taken over the choice of the tribe key only) and which is efficiently invertible given the secret key (for all possible tribe keys); to invert a value $y = f_k^t(x)$ invert the rightmost n bits of y under the starting pseudorandom function to obtain $x111$ and therefore x (note that the rightmost n bits of $\text{XOR}(t, I_k)$ equal 0^n). We call such an ensembles key-binding-and-invertible. Observe that the key-and-preimage-binding property alone can be achieved by taking output length $8n$ bits, choosing $2n - 3$ strings t_i from $\{0,1\}^{7n} \times \{0^n\}$ and letting $\text{XOR}(t, I_k, x)$ be the exclusive-or of the t_i's for which the i-th bit of $I_k; x$ equals 1.

4   Committing and Key-Hiding Private-Key Encryption

A well-known private-key encryption scheme based on PRF ensembles is given by $\text{Enc}_k(m, r) = (r, f_k(r) \oplus m)$, where k is the secret key, m is the message and r is chosen at random. To decrypt a pair (r, c) compute $m = \text{Dec}_k(r, c) = f_k(r) \oplus c$. This encryption scheme is not comitting in general, i.e., for an encryption (r, c) there might exist (k, m), (k', m') with $m \neq m'$ and $\text{Enc}_k(m, r) = (r, c) = \text{Enc}_{k'}(m', r)$. Conversely, we call a cryptosystem committing if for each ciphertext c there exists a unique message m such that c must have been derived by applying the encryption algorithm to m — this holds independently of the choice of the secret key and the coin tosses used during the encryption process.

Before presenting the formal definition of committing schemes we sketch the definition of a private-key cryptosystem. A private-key encryption scheme is a triple (KGen, Enc, Dec) of probabilistic polynomial-time algorithms such that

- KGen on input 1^n generates a random key k,
- Enc on input 1^n, key k, message m (of some appropriate length) and randomness r outputs a ciphertext $c = \text{Enc}(1^n, k, m, r)$,
- $\text{Dec}(1^n, k, \text{Enc}(1^n, k, m, r)) = m$.

Wlog. we assume that 1^n is recoverable from k and therefore write $\text{Enc}(k, m, r)$ or $\text{Enc}_k(m, r)$ instead of $\text{Enc}(1^n, k, m, r)$. Similarly for Dec.

Definition 5 (Committing Private-Key Encryption Scheme). *A private-key encryption scheme* (KGen, Enc, Dec) *is called committing if for any key k, message m, randomness r and encryption $c = \text{Enc}_k(m, r)$ there do not exist k', m', r' such that $m \neq m'$ and $\text{Enc}_{k'}(m', r') = \text{Enc}_k(m, r)$.*

Using a fixed-value-key-binding PRF ensemble the obvious solution $\text{Enc}_k(m, r) = (f_k(1^n), r, f_k(r) \oplus m)$ works. The drawback of this solution is that an eavesdropper knows whenever the parties change the secret key. In some settings hiding this fact might be crucial. For instance, if one party sends the new secret key by encrypting it with the current one, then breaking this encryption by an exhaustive search makes all the following messages visible to the adversary. Applying the key-and-preimage-binding PRF tribe ensemble of Section 3 we can over-come this disadvantage. But before presenting our committing and key-hiding scheme we formalize the notion of a key-hiding scheme. Let $(\mathsf{KGen}, \mathsf{Enc}, \mathsf{Dec})$ be a private-key encryption scheme and D be a probabilistic polynomial-time algo-rithm. We consider two experiments. In the first experiment, we independently execute $\mathsf{KGen}(1^n)$ twice to obtain two keys k, k'. D is given 1^n as input and is allowed to query the probabilistic oracles Enc_k and $\mathsf{Enc}_{k'}$ in the following way: In the first part, D is allowed to obtain encryptions of messages of its choice by querying the oracle Enc_k. Then it passes a message SWITCH to the oracle Enc_k. It continues to query for messages of its choice, but this time the answers are given by the second oracle $\mathsf{Enc}_{k'}$. Finally, D outputs a bit, denoted $D^{\mathsf{Enc}_k, \mathsf{Enc}_{k'}}(1^n)$, and stops. The second experiment differs only in the way the oracles are ini-tialized. This time we let $k' = k$, i.e., we do not change the keys. Denote by $D^{\mathsf{Enc}_k, \mathsf{Enc}_k}(1^n)$ the output.

Definition 6 (Key-Hiding Private-Key Encryption Scheme). *A private-key encryption scheme* $(\mathsf{KGen}, \mathsf{Enc}, \mathsf{Dec})$ *is said to be key-hiding if for any prob-abilistic polynomial-time algorithm* D *the value* $\left| \mathrm{Prob}\left[D^{\mathsf{Enc}_k, \mathsf{Enc}_{k'}}(1^n) = 1 \right] - \mathrm{Prob}\left[D^{\mathsf{Enc}_k, \mathsf{Enc}_k}(1^n) = 1 \right] \right|$ *is negligible in* n.

Actually, every secure[2] scheme should "hide" the key, i.e., it should not reveal the key. Otherwise it can be easily broken. However, Definition 6 demands even more. For instance, an encryption scheme where each encryption leaks the Hamming weight of the key with some probability that is not negligible does not hide the key as defined above. Yet, the scheme may be secure.

We remark that we do not grant D access to the decryption oracles Dec_k and $\mathsf{Dec}_{k'}$, respectively. Otherwise D could distinguish both cases easily: D encrypts some message m with the first oracle, sends SWITCH and tries to decrypt with the second decryption oracle; this only yields m again if the keys have not changed.

We define the committing and key-hiding encryption scheme $(\mathsf{KGen}^{\mathsf{com}}, \mathsf{Enc}^{\mathsf{com}}, \mathsf{Dec}^{\mathsf{com}})$. Let F be a PRF tribe ensemble derived by the technique of Section 3.2 from a key-and-preimage-binding ensemble F^{bind}. We assume that some trusted party chooses a random tribe key t and publishes it or sends it to the partici-pating parties, respectively. Hence, we do not achieve the committing property of Definition 5 perfectly, but only with exponentially small error probability. Abusing notations we will also call this derived scheme committing. Algorithm $\mathsf{KGen}^{\mathsf{com}}(1^n)$ selects a random $k \in K_n$. Let $\mathsf{Enc}_k^{\mathsf{com}}(m, r) = (f_k^t(r), r \oplus m)$ where $m, r \in \{0,1\}^{n-3}$. To decrypt a pair (y, c) compute r from the rightmost n bits of y by applying the inverse of f_k^{bind}. Finally, recover m by $m = r \oplus c$.

Proposition 7. *The encryption scheme* $(\mathsf{KGen}^{\mathsf{com}}, \mathsf{Enc}^{\mathsf{com}}, \mathsf{Dec}^{\mathsf{com}})$ *is a commit-ting and key-hiding encryption scheme.*

Proof. (Sketch) It remains to show that the scheme is key-hiding. But this follows directly from the pseudorandomness of F^{bind}. $\square$

---

[2] Here, security does not refer to any formal definition. It is used in a rather liberal sense.

It is quite easy to see that this scheme is polynomially secure as defined in [13]. We sketch this and other security notions in Appendix B. In fact, it is not hard to show either that it is even secure against lunchtime attacks [21].

Proposition 8. *The scheme* $(\mathsf{KGen}^{\mathsf{com}}, \mathsf{Enc}^{\mathsf{com}}, \mathsf{Dec}^{\mathsf{com}})$ *is a committing and key-hiding private-key encryption which is secure against lunchtime attacks.*

The proof is omitted from this extended abstract.

The encryption scheme can be easily broken with a chosen ciphertext and plaintext attacks (see [22] or Appendix B) because given a ciphertext (y, c) the adversary can query the decryption oracle for $(y, c \oplus 1^{|c|})$ and easily recover m from the answer. Using an idea of Bellare and Rogaway [2] we can turn the scheme above into an encryption scheme $(\mathsf{KGen}^{\mathsf{ccp}}, \mathsf{Enc}^{\mathsf{ccp}}, \mathsf{Dec}^{\mathsf{ccp}})$ which is secure against chosen ciphertext and plaintext attacks. To do so, we let

$$\mathsf{Enc}_k^{\mathsf{ccp}}(r, m) = (f_k^t(r; m), r \oplus m)$$

for $m, r \in \{0, 1\}^{n/2-1}$. Defining $\mathsf{Dec}^{\mathsf{ccp}}$ is straightforward. Loosely speaking, appending m to the argument r of the pseudorandom function serves as a proof that one knows the values r, m explicitely. Again, the formal proof is omitted.

Proposition 9. *The committing and key-hiding private-key encryption scheme* $(\mathsf{KGen}^{\mathsf{ccp}}, \mathsf{Enc}^{\mathsf{ccp}}, \mathsf{Dec}^{\mathsf{ccp}})$ *is secure against chosen ciphertext and plaintext attacks.*

Recently, Dolev et al. [9] showed that (semantic) security against chosen ciphertext and plaintext attacks implies non-malleability. Hence, our scheme is non-malleable as well.

Acknowledgements

We are grateful to Ran Canetti, Daniele Micciancio and Omer Reingold for discussing our ideas and proposing the more practical construction of F^{bind} in Section 3.1. We also thank the anonymous referees of Eurocrypt'99 for their comments.

References

1. M.BELLARE, A.DESAI, E.JOKIPII, P.ROGAWAY: A Concrete Security Treatment of Symmetric Encryption, *Proceedings of the 38th IEEE Symposium on Foundations of Computer Science (FOCS), pp. 394–403, 1997.*

2. M.BELLARE, P.ROGAWAY: Random Oracles are Practical: A Paradigm for Designing Efficient Protocols, *First ACM Conference on Computer and Communications Security*, 1993.

3. R.CANETTI: Towards Realizing Random Oracles: Hash Functions that Hide All Partial Information, *Crypto '97, Lecture Notes in Computer Science, Vol. 1294, Springer-Verlag, pp. 455–469, 1997.*

4. R.CANETTI, C.DWORK, M.NAOR. R.OSTROVSKY: Deniable Encryption, *Crypto '97, Lecture Notes in Computer Science, Vol. 1294, Springer-Verlag, pp. 90–104,* 1997.
5. R.CANETTI, U.FEIGE, O.GOLDREICH, M.NAOR: Adaptively Secure Multi-Party Computation, *Proceedings of the 28th Annual ACM Symposium on the Theory of Computing (STOC), pp. 639–648,* 1996.
6. R.CANETTI, R.GENNARO: Incoercible Multiparty Computation, *Proceedings of the 37th IEEE Symposium on Foundations of Computer Science (FOCS), pp. 504–513,* 1996.
7. R.CANETTI, D.MICCIANCIO, O.REINGOLD: Perfectly One-Way Probabilistic Hash Functions, *Proceedings of the 30th Annual ACM Symposium on the Theory of Computing (STOC),* 1998.
8. L.CARTER, M.WEGMAN: Universal Classes of Hash Functions, *Journal of Computer and System Science, vol. 18, pp. 143–154,* 1979.
9. D.DOLEV, C.DWORK, M.NAOR: Non-Malleable Cryptography, *submitted journal version; a preliminary version appeared in Proceedings of the 23rd Annual ACM Symposium on the Theory of Computing (STOC) in 1991,* 1999.
10. S.GOLDWASSER, O.GOLDREICH, S.MICALI: How to Construct Random Functions, *Journal of ACM, vol. 33, pp. 792–807,* 1986.
11. O.GOLDREICH: Foundations of Cryptography (Fragments of a Book), *Department of Computer Science and Applied Mathematics, Weizmann Institute of Science, Rehovot, Israel,* 1995.
12. O.GOLDREICH, L.LEVIN: A Hardcore Predicate for All One-Way Functions, *Proceedings of the 21st Annual ACM Symposium on the Theory of Computing (STOC), pp. 25–32,* 1989.
13. S.GOLDWASSER, S.MICALI: Probabilistic Encryption, *Journal of Computer and System Science, Vol. 28, pp. 270–299,* 1984.
14. J.HASTAD, R.IMPAGLIAZZO, L.LEVIN, M.LUBY: Construction of a Pseudorandom Generator from any One-Way Function, *to appear in SIAM Journal on Computing, preliminary versions in STOC'89 and STOC'90,* 1989/90.
15. M.LUBY, C.RACKOFF: How to Construct Pseudorandom Permutations from Pseudorandom Functions, *SIAM Journal on Computing, Vol. 17, pp. 373–386,* 1988.
16. S.MICALI, C.RACKOFF, B.SLOAN: The Notion of Security for Probabilistic Cryptosystems, *SIAM Journal on Computing,* 1988.
17. M.NAOR: Bit Commitment Using Pseudo-Randomness, *Journal of Cryptology, vol. 4, pp. 151–158,* 1991.
18. M.NAOR, O.REINGOLD: Synthesizers and Their Application to the Parallel Construction of Pseudorandom Functions, *Proceedings of the 36th IEEE Symposium on Foundations of Computer Science (FOCS), pp. 170–181,* 1995.
19. M.NAOR, O.REINGOLD: Number-Theoretic Constructions of Efficient Pseudorandom Functions, *Proceedings of the 38th IEEE Symposium on Foundations of Computer Science (FOCS), pp. 458–467,* 1997.
20. M.NAOR, O.REINGOLD: On the Construction of Pseudorandom Permutations: Luby-Rackoff Revisited, *Journal of Cryptology, vol. 12, no. 1, pp. 29–66,* 1999.
21. M.NAOR, M.YUNG: Public-Key Cryptosystems Provably Secure Against Chosen Ciphertext Attacks, *Proceedings of the 20th Annual ACM Symposium on the Theory of Computing (STOC), pp. 427–437,* 1990.
22. C.RACKOFF, D.SIMON: Non-Interactive Zero-Knowledge Proof of Knowledge and Chosen Ciphertext Attacks, *Crypto '91, Lecture Notes in Computer Science, Vol. 576, Springer-Verlag, pp. 433–444,* 1991.

23. A.C.Yao: Theory and Application of Trapdoor Functions, *Proceedings of the 23rd IEEE Symposium on Foundations of Computer Science (FOCS), pp. 80–91, 1982.*

A   The CMR PRF Tribe Ensemble — In a Nutshell

We sketch the construction of PRF tribe ensembles from one-way permutations given in [7]. See their paper for discussions and proofs. Let g' be a one-way permutation over $\{0,1\}^{6n}$ and assume that $g'(x;r) = g(x);r$ for $x, r \in \{0,1\}^{3n}$. Furthermore, we can assume wlog. that g has no cycles of length less than $12n$. Let p be a non-constant polynomial over $GF[2^{6n}]$ and define a hardcore predicate $B_p : \{0,1\}^{6n} \rightarrow \{0,1\}$ of g' by the inner product $B_p(x;r) = p(x) \cdot r$ of $p(x), r \in \{0,1\}^{3n}$. Then, for any polynomial p, we construct a length-doubling pseudorandom generator by

$$G_p(x;r) = B_p(x;r); B_p(g(x);r); \cdots ; B_p(g^{6n-1}(x);r); g^{6n}(x); r$$

Denote by $G_p^0(x;r)$ and $G_p^1(x;r)$ the left and right half of $G_p(x;r)$. Additionally, we let $G : \{0,1\}^n \rightarrow \{0,1\}^{6n}$ denote an arbitrary pseudorandom generator which is one-to-one on the right half.

The tribe key t consists of n random, non-constant polynomials $p_1, \ldots, p_n$ of degree less than $6n$. Then let

$$f_k^t(x) = G_{p_n}^{x_n}(\cdots G_{p_1}^{x_1}(G(k)))$$

That is, f_k^t is a GGM-tree using pseudorandom generators based on the modified Goldreich-Levin hardcore predicate.

B   Security Notions of Private-Key Encryption Schemes

In this section we recall the notions of polynomial security [13], security against lunchtimes attacks [21] resp. against chosen ciphertext and plaintext attacks [22]. See [1] for further security definitions for symmetric schemes. We refer the reader to [9] for a definition of non-malleable schemes, a notion that turned out to be equivalent to security against chosen ciphertext and plaintext attacks.

Consider the following attack on a private-key cryptosystem. Let (F, D) be a pair of probabilistic polynomial-time algorithms. First, a secret key k is chosen according to $KGen(1^n)$ and kept secret from F and D. Then the message finder F gets the input 1^n and outputs two messages m_0, m_1. Let $b \in \{0,1\}$ be a fixed bit. A ciphertext $c = Enc_k(m_b, r)$ for randomness r is generated. Now D is given input 1^n, m_0, m_1 and c and is supposed to predict b, i.e., to distinguish encryptions of m_0 and m_1. Let $\delta_{F,D}^b(n)$ denote the probability that D outputs 1 if m_b is encrypted. The probability is taken over all random choices, including the internal coin tosses of F and D.

An encryption scheme is *polynomially secure* if D cannot distinguish an encryption of m_0 from an encryption of m_1 significantly. More formally, it is polynomially secure if for all (probabilistic polynomial-time) adversary pairs (F, D) the value $|\delta_{F,D}^0(n) - \delta_{F,D}^1(n)|$ is negligible in n.

A lunchtime attack is similar to the aforementioned attack, but F is also allowed to adaptively query the encryption/decryption oracle for plaintexts and ciphertexts of its choice before outputting m_0, m_1, and D is given the history of this query/answer sequence as additional input. An encryption scheme is *secure against lunchtime attacks* if it still holds that $|\delta_{F,D}^0(n) - \delta_{F,D}^1(n)|$ is negligible in n for all (probabilistic polynomial-time) adversary pairs (F, D).

A chosen ciphertext and plaintext attack is a lunchtime attack where D is also allowed to adaptively query the encryption/decryption oracle — though D is of course not allowed to decipher the challenge c. Again, an encryption scheme is *secure against chosen ciphertext and plaintext attacks* if $|\delta_{F,D}^0(n) - \delta_{F,D}^1(n)|$ is negligible in n for all (probabilistic polynomial-time) adversary pairs (F, D).

Secure Communication in Broadcast Channels: The Answer to Franklin and Wright's Question*

Yongge Wang[1] and Yvo Desmedt[1,2]

[1] Department of EE & CS, University of Wisconsin – Milwaukee,
P.O. Box 784, WI 53201 Milwaukee, USA,
{wang,desmedt}@cs.uwm.edu
[2] The Center of Cryptography, Computer and Network Security
CEAS, University of Wisconsin – Milwaukee, and
Dept. of Mathematics, Royal Holloway, University of London, UK

Abstract. Problems of secure communication and computation have been studied extensively in network models. Goldreich, Goldwasser, and Linial, Franklin and Yung, and Franklin and Wright have initiated the study of secure communication and secure computation in multi-recipient (broadcast) models. A "broadcast channel" (such as ethernet) enables one processor to send the same message—simultaneously and privately— to a fixed subset of processors. In their Eurocrypt '98 paper, Franklin and Wright have shown that if there are n broadcast lines between a sender and a receiver and there are at most t malicious (Byzantine style) processors, then the condition $n > t$ is necessary and sufficient for achieving efficient probabilisticly reliable and probabilisticly private communication. They also showed that if $n > \lceil 3t/2 \rceil$ then there is an efficient protocol to achieve probabilisticly reliable and perfectly private communication. And they left open the question whether there exists an efficient protocol to achieve probabilisticly reliable and perfectly private communication when $\lceil 3t/2 \rceil \geq n > t$. In this paper, by using a different authentication scheme, we will answer this question affirmatively and study related problems.

Keywords: Network security, Privacy, Perfect secrecy, Reliability.

1 Introduction

If two parties are connected by a private and authenticated channel, then secure communication between them is guaranteed. However, in most cases, many parties are only indirectly connected, as elements of an incomplete network of private and authenticated channels. In other words they need to use intermediate or internal nodes. Achieving participants cooperation in the presence of faults is

---

* Research supported by DARPA F30602-97-1-0205. However the views and conclusions contained in this paper are those of the authors and should not be interpreted as necessarily representing the official policies or endorsements, either expressed or implied, of the Defense Advance Research Projects Agency (DARPA), the Air Force, of the US Government.

J. Stern (Ed.): EUROCRYPT'99, LNCS 1592, pp. 446–458, 1999.
© Springer-Verlag Berlin Heidelberg 1999

a major problem in distributed networks. The interplay of network connectivity and secure communication have been studied extensively (see, e.g., [1,4,6,7,12]). For example, Dolev [6] and Dolev et al. [7] showed that, in the case of t Byzantine faults, reliable communication is achievable only if the systems's network is $2k+1$ connected. Hadzilacos [12] has shown that even in the absence of malicious failures connectivity $t + 1$ is required to achieve reliable communication in the presence of t faulty participants.

Goldreich, Goldwasser, and Linial [11], Franklin and Yung [9], and Franklin and Wright [8] have initiated the study of secure communication and secure computation in *multi-recipient (broadcast)* models. A "broadcast channel" (such as ethernet) enables one participant to send the same message—simultaneously and privately—to a fixed subset of participants. Franklin and Yung [9] have given a necessary and sufficient condition for individuals to exchange private messages in broadcast models in the presence of passive adversaries (passive gossipers). For the case of active Byzantine adversaries, many results have been presented by Franklin and Wright [8]. Note that Goldreich, Goldwasser, and Linial [11] have also studied the fault-tolerant computation in the public broadcast model in the presence of active Byzantine adversaries.

There are many examples of broadcast channels. A simple example is a local area network like an Ethernet bus or a token ring. Another example is a shared cryptographic key. By publishing an encrypted message, a participant initiates a broadcast to the subset of participants that is able to decrypt it.

We will abstract away the concrete network structures and consider multicast graphs. Specifically, a multicast graph is just a graph $G(V, E)$. A vertex $A \in V$ is called a neighbor of another vertex $B \in V$ if there there is an edge $(A, B) \in E$. In a multicast graph, we assume that any message sent by a node A will be received identically by all its neighbors, whether or not A is faulty, and all parties outside of A's neighbor learn nothing about the content of the message. The neighbor networks have been studied by Franklin and Yung in [9]. They have also studied the more general notion of hypergraphs, which we do not need.

As Franklin and Wright [8] have pointed out, unlike the simple channel model, it is not possible to directly apply protocols over multicast lines to disjoint paths in a general multicast graph, since disjoint paths may have common neighbors. Franklin and Wright have shown that in certain cases the change from simple channel to broadcast channel hurts the adversary more than it helps, because the adversary suffers from the restriction that an incorrect transmission from a faulty processor will always be received identically by all of its neighbors.

It was shown [8] that if there are n broadcast lines (that is, n paths with disjoint neighborhoods) between a sender and a receiver and there are at most t malicious (Byzantine style) processors, then the condition $n > t$ is necessary and sufficient for achieving efficient probabilisticly reliable and probabilisticly private communication. They also showed that there is an efficient protocol to achieve probabilisticly reliable and perfectly private communication when $n > \lceil 3t/2 \rceil$, and there is an exponential bit complexity protocol for achieving probabilisticly reliable and perfectly private communication when $\lceil 3t/2 \rceil \geq n > t$. However, they left open the question whether there exists an efficient protocol to achieve probabilisticly reliable and perfectly private communication when $\lceil 3t/2 \rceil \geq n > t$. In this paper, by using a different authentication scheme, we will answer this question affirmatively and study related problems. We will also show that it is **NP**-complete to decide whether a multicast graph has n disjoint broadcast lines (that is, n paths with disjoint neighborhoods).

Note that, similar as in Franklin and Wright [8], we will only consider the scenario when the underlying graph is known to all nodes. For the scenario that the graph is unknown, the protocols may be completely different, see Burmester, Desmedt, and Kabatianski [2].

2   Models

Throughout this paper, n denotes the number of multicast lines and t denotes the number of faults under the control of the adversary. We write $|S|$ to denote the number of elements in the set S. We write $x \in_R S$ to indicate that x is chosen with respect to the uniform distribution on S. Let $\mathbf{F}$ be a finite field, and let $a, b, M \in \mathbf{F}$. We define $\mathrm{auth}(M, a, b) = aM + b$ (following [10,13,14]). In this paper, we will also use a multiple authentication scheme. That is, for $a, b, c, d, M \in \mathbf{F}$, let $\mathrm{bauth}(M, a, b, c, d) = aM^3 + bM^2 + cM + d$. Note that the main advantage of the function $\mathrm{bauth}()$ is that each authentication key (a, b, c, d) can be used to authenticate three different messages M_0, M_1, and M_2 without revealing any information of the authentication key. While for the function $\mathrm{auth}()$ each authentication key (a, b) can only be used to authenticate one message (that is, it is a kind of one-time pad) (see Simmons [15]). Note that den Boer [5] used similar polynomials to construct one-time authentication schemes.

Theorem 1. *Let (a, b, c, d) be chosen uniformly from $\mathbf{F}^4$, $M_i \in \mathbf{F}$ for $i = 0, 1, 2$, and $s_i = \mathrm{bauth}(M_i, a, b, c, d)$ for $i = 0, 1, 2$ be the authentication code of M_i respectively. Then, for any $a_0, b_0, c_0, d_0 \in \mathbf{F}$,*

$$\Pr[a = a_0 | view_0] = \Pr[b = b_0 | view_0] = \Pr[c = c_0 | view_0] = \Pr[d = d_0 | view_0] = \frac{1}{|\mathbf{F}|}$$

where $view_0 = (M_0, s_0, M_1, s_1, M_2, s_2)$

Proof. By the condition, we have the following three equations with four unknowns:
$$M_0^3 a + M_0^2 b + M_0 c + d = s_0$$
$$M_1^3 a + M_1^2 b + M_1 c + d = s_1$$
$$M_2^3 a + M_2^2 b + M_2 c + d = s_2.$$

Since the coefficient matrix of the above equations is a so-called Vandermonde matrix, no value of a can be ruled out. That is, every a is equally likely given the values $(M_0, s_0, M_1, s_1, M_2, s_2)$. (A similar argument applies for b, or c or d.) This completes the proof of the theorem. $\square$

Following Franklin and Wright [8], we consider multicast as our only communication primitive. A message that is multicast by any node in a multicast neighbor network is received by all its neighbors with privacy (that is, non-neighbors learn nothing about what was sent) and authentication (that is, neighbors are guaranteed to receive the value that was multicast and to know which neighbor multicast it). In our models, we assume that all nodes in the multicast graph know the complete protocol specification and the complete structure of the multicast graph. In a message transmission protocol, the sender A starts with a message M^A drawn from a message space $\mathcal{M}$ with respect to a certain probability distribution. At the end of the protocol, the receiver B outputs a message M^B. We consider a synchronous system in which messages are sent via multicast in rounds. During each round of the protocol, each node receives any messages that were multicast by its neighbors at the end of the previous round, flips coins and perform local computations, and then possibly multicast a message. We will also assume that the message space $\mathcal{M}$ is a representable subset of the finite field $\mathbf{F}$.

Generally there are two kinds of adversaries. A passive adversary (or gossiper adversary) is an adversary who can only observe the traffics through t internal nodes. An active adversary (or Byzantine adversary) is an adversary with unlimited computational power who can control t internal nodes. That is, an active adversary will not only listen to the traffics through the controlled nodes, but also control the message sent by those controlled nodes. Both kinds of adversaries are assumed to know the complete protocol specification, message space, and the complete structure of the multicast graph. At the start of the protocol, the adversary chooses the t faulty nodes. A passive adversary can view the behavior (coin flips, computations, message received) of all the faulty nodes. An active adversary can view all the behavior of the faulty nodes and, in addition, control the message that they multicast. We allow for the strongest adversary. (An alternative interpretation is that t nodes are collaborating adversaries.)

For any execution of the protocol, let adv be the adversary's view of the entire protocol. We write $adv(M, r)$ to denote the adversary's view when $M^A = M$ and when the sequence of coin flips used by the adversary is r.

Definition 2. *(see Franklin and Wright [8])*

1. *A message transmission protocol is δ-reliable if, with probability at least $1 - \delta$, B terminates with $M^B = M^A$. The probability is over the choices of M^A and the coin flips of all nodes.*
2. *A message transmission protocol is ε-private if, for every two messages M_0, M_1 and every r, $\sum_c |\Pr[adv(M_0, r) = c] - \Pr[adv(M_1, r) = c]| \leq 2\varepsilon$. The probabilities are taken over the coin flips of the honest parties, and the sum is over all possible values of the adversary's view.*
3. *A message transmission protocol is perfectly private if it is 0-private.*
4. *A message transmission protocol is (ε, δ)-secure if it is ε-private and δ-reliable.*

5. *An (ε, δ)-secure message transmission protocol is efficient if its round complexity and bit complexity are polynomial in the size of the network, $\log \frac{1}{\varepsilon}$ (if $\varepsilon > 0$) and $\log \frac{1}{\delta}$ (if $\delta > 0$).*

3   Background: Reliable Communication over Neighbor Networks

In this section, we review Franklin and Wright's Eurocrypt '98 protocols for reliable communication over multicast lines. The reader familiar with these protocols can skip this section. For two vertices A and B in a multicast graph $G(V, E)$, we say that A and B are connected by n neighborhood (except A and B) disjoint *lines* if there are n lines $p_1, \ldots, p_n \subseteq V$ with the following properties:

- For each $j \leq n$, the j-th line p_j is a sequence of $m_j + 2$ nodes $A = X_{0,j}$, $X_{1,j}, \ldots, X_{m+1,j} = B$ where $X_{i,j}$ is a neighbor of $X_{i+1,j}$.
- For each i_1, i_2, j_1, and j_2 with $j_1 \neq j_2$, the only possible common neighbors of X_{i_1,j_1} and X_{i_2,j_2} are A and B.

If there is no ambiguity we drop the "except A and B."

Without loss of generality, in this section we assume that party A (the message transmitter) and party B (the message recipient) are connected by n neighborhood disjoint *lines*, and we assume that $m_1 = m_2 = \ldots = m_n$.

Basic Propagation Protocol (Franklin and Wright [8]) In this protocol, A tries to propagate a value s^A to B.

- In round 1, A multicast s^A.
- In round ρ for $2 \leq \rho \leq m + 1$, each $X_{\rho-1,j}(1 \leq j \leq n)$ expects to receive a single element from $X_{\rho-2,j}$. Let $u_{\rho-1,j}$ be this value if a value was in fact received, or a publicly known default element otherwise. At the end of round ρ, $X_{\rho-1,j}$ multicast $u_{\rho-1,j}$.
- In round $m + 2$, B receives a single element from each $X_{m,j}$, or substitutes the default element. Let s_j^B be the value received or substituted on line j.

From now on when a party substitutes the default element, we just say that the party substitutes.

Full Distribution Protocol (Franklin and Wright [8]) In this protocol, each internal node $X_{i,j}$ tries to transmit an element $s_{i,j}$ to both A and B.

- In round 1, each $X_{i,j}(1 \leq i \leq m, 1 \leq j \leq n)$ multicast $s_{i,j}$ to (in particular) $X_{i-1,j}$ and $X_{i+1,j}$.
- In round ρ for $2 \leq \rho \leq m + 1$:
  - For $1 \leq j \leq n$ and $\rho \leq i \leq m$, each $X_{i,j}$ expects to be the intended recipient of an element from $X_{i-1,j}$ (initiated by $X_{i-\rho+1,j}$). Let $u_{i,j}$ be the received value or a default value if none is received.

- For $1 \leq j \leq n$ and $1 \leq i \leq m - \rho + 1$, $X_{i,j}$ expects to be the intended recipient of an element from $X_{i+1,j}$ (initiated by $X_{i+\rho-1,j}$). Let $v_{i,j}$ be the received or default value.
- For $1 \leq j \leq n$, B expects to be the intended recipient on the j-th line of a single element (initiated by $X_{m-\rho+2,j}$). Let $s^B_{m-\rho+2,j}$ be the received or default value.
- For $1 \leq j \leq n$, A expects to be the intended recipient on the j-th line of a single element (initiated by $X_{\rho-1,j}$). Let $s^A_{\rho-1,j}$ be the received or default value.
- $X_{i,j}$ multicasts $u_{i,j}$ to $X_{i+1,j}$ if $\rho \leq i \leq m$, and $v_{i,j}$ to $X_{i-1,j}$ if $1 \leq i \leq m - \rho + 1$.

Fact 3. *(Franklin and Wright [8]) If there are no faults on the j-th line, then $s^A_{i,j} = s^B_{i,j}$ for all $1 \leq i \leq m$. Further, if $X_{i,j}$ is the only fault on the j-th line, then $s^A_{i,j} = s^B_{i,j}$.*

Reliable Transmission Protocol (Franklin and Wright [8]) In this protocol, A tries to reliably transmit a message M^A to B.

- The nodes on all the n lines execute an instance of the Full Distribution Protocol, which takes place during rounds 1 through $m+1$. The element that $X_{i,j}$ initiates is $(a_{i,j}, b_{i,j})$ which is randomly chosen from $\mathbf{F}^2$. Let $(a^A_{i,j}, b^A_{i,j})$ and $(a^B_{i,j}, b^B_{i,j})$ be the values that A and B receive or substitute as the element initiated by $X_{i,j}$.
- The nodes on all the n lines execute an instance of the Basic Propagation Protocol from A to B, which takes place during rounds $m+2$ through $2m+3$. The element that A initiates is $\{(i,j, M^A, \mathrm{auth}(M^A, a^A_{i,j}, b^A_{i,j})) : 1 \leq i \leq m, 1 \leq j \leq n\}$. In round $2m+3$, B receives or substitutes $\{(i, j, M^B_{i,j,k}, u^B_{i,j,k}) : 1 \leq i \leq m, 1 \leq j \leq n\}$ on the k-th line, $1 \leq k \leq n$.
- Let $r_k(M) = \{j : \exists i(M = M^B_{i,j,k} \& u^B_{i,j,k} = \mathrm{auth}(M^B_{i,j,k}, a^B_{i,j}, b^B_{i,j}))\}$. B outputs M^B that maximizes $\max_k |r_k(M^B)|$.

Theorem 4. *(Franklin and Wright [8]) If $\delta > 0$, $n > t$, and $|\mathbf{F}| > mn^2/\delta$, then the Reliable Transmission Protocol is an efficient δ-reliable message transmission protocol.*

4   Reliable and Private Communication over Neighbor Networks

4.1   Survey of Franklin-Wright's Results

As in the previous section, we assume that party A (the message transmitter) and party B (the message recipient) are connected by n neighborhood disjoint *lines*. Franklin and Wright showed the following results regarding to privacy in broadcast networks:

1. If $n > t$, $\delta > 0$ and $\varepsilon > 0$, then there is an efficient (ε, δ)-secure message transmission protocol between A and B.
2. If $n > \lceil 3t/2 \rceil$ and $\delta > 0$, then there is an efficient $(0, \delta)$-secure message transmission protocol between A and B, that is, a δ-reliable and perfect private message transmission protocol.
3. If $t < n \leq \lceil 3t/2 \rceil$ and $\delta > 0$, then there is an exponential bit complexity $(0, \delta)$-secure message transmission protocol between A and B.

4.2   The Franklin-Wright's Open Problem

They left open the question whether it is possible to efficiently achieve perfect privacy when $t < n \leq \lceil 3t/2 \rceil$. That is, does there exist a polynomial time $(0, \delta)$-secure message transmission protocol between A and B when $t < n \leq \lceil 3t/2 \rceil$? We give an affirmative answer to this question.

4.3   The Solution

Intuitively, our protocol proceeds as follows. First, using the Full Distribution Protocol from the preceding section, each internal node $X_{i,j}$ transmits a random authentication key $(a_{i,j}, b_{i,j}, c_{i,j}, d_{i,j}) \in_R \mathbf{F}^4$ to both A and B. Secondly, using the Basic Propagation Protocol, B transmits to A a random $r \in_R \mathbf{F}$ authenticated by the keys in $\{(a_{i,j}, b_{i,j}, c_{i,j}, d_{i,j}) : 1 \leq i \leq m, 1 \leq j \leq n\}$. Thirdly, for each $1 \leq j \leq n$, A decides whether A and B agree on at least one authentication key on the j-th line. Let

$$K^A = \{(i_j, j) : (a^A_{i_j,j}, b^A_{i_j,j}, c^A_{i_j,j}, d^A_{i_j,j})$$
$$\text{is the first key agreed upon by } A \text{ and } B \text{ on the } j\text{-th line}\}.$$

Lastly, A encrypts the message M^A using the sum of the pads $a^A_{i_j,j}$ $((i_j, j) \in K^A)$ and, using the Basic Propagation Protocol, transmits to B the set K^A and the ciphertext authenticated by the keys in $\{(a^A_{i,j}, b^A_{i,j}, c^A_{i,j}, d^A_{i,j}) : 1 \leq i \leq m, 1 \leq j \leq n\}$. Lastly, B decrypts the message.

Perfectly Private Transmission Protocol

- The nodes on all the n lines execute an instance of the Full Distribution Protocol, which takes place during rounds 1 through $m + 1$. The element that $X_{i,j}$ initiates is $(a_{i,j}, b_{i,j}, c_{i,j}, d_{i,j})$ which is randomly chosen from $\mathbf{F}^4$. Let $(a^A_{i,j}, b^A_{i,j}, c^A_{i,j}, d^A_{i,j})$ and $(a^B_{i,j}, b^B_{i,j}, c^B_{i,j}, d^B_{i,j})$ be the values that A and B receive or substitute as the element initiated by $X_{i,j}$.
- The nodes on all the n lines execute an instance of the Basic Propagation Protocol from B to A, which takes place during rounds $m+2$ through $2m+3$. The element that B initiates is $\{(i, j, r^B, \text{bauth}(r^B, a^B_{i,j}, b^B_{i,j}, c^B_{i,j}, d^B_{i,j})) : 1 \leq i \leq m, 1 \leq j \leq n\}$, where $r^B \in_R \mathbf{F}$. In round $2m+3$, A receives or substitutes $\{(i, j, r^A_{i,j,k}, u^A_{i,j,k}) : 1 \leq i \leq m, 1 \leq j \leq n\}$ on the k-th line, $1 \leq k \leq n$.

- Let $r_k(r) = \{j : \exists i(r = r_{i,j,k}^A \& u_{i,j,k}^A = \text{bauth}(r_{i,j,k}^A, a_{i,j}^A, b_{i,j}^A, c_{i,j}^A, d_{i,j}^A))\}$, r^A be the message that maximizes $|r_{k^A}(r^A)| = \max_k |r_k(r^A)|$, and let $K^A = \{(i_j, j) : j \in r_{k^A}(r^A), \forall (0 < i < i_j)(u_{i,j,k^A}^A \neq \text{bauth}(r^A, a_{i,j}^A, b_{i,j}^A, c_{i,j}^A, d_{i,j}^A))\}$. A computes $z^A = M^A + \sum_{(i_j,j) \in K^A} a_{i_j,j}^A$.

- In rounds $2m + 4$ through $3m + 5$, the nodes on all the n lines execute an instance of the Basic Propagation Protocol from A to B. The element that A initiates is $\{(i, j, z^A, K^A, \text{bauth}(\langle z^A, K^A \rangle, a_{i,j}^A, b_{i,j}^A, c_{i,j}^A, d_{i,j}^A)) : 1 \leq i \leq m, 1 \leq j \leq n\}$, where $\langle z^A, K^A \rangle$ denotes the concatenation of z^A and K^A (without loss of generality, we assume that prefix-free codes are used so that we can uniquely recover z^A and K^A from $\langle z^A, K^A \rangle$). In round $3m + 5$, B receives or substitutes $\{(i, j, z_{i,j,k}^B, K_{i,j,k}^B, u_{i,j,k}^B) : 1 \leq i \leq m, 1 \leq j \leq n\}$ on the k-th line, $1 \leq k \leq n$.

- $R_k(\langle z, K \rangle) = \{j : \exists i(\langle z, K \rangle = \langle z_{i,j,k}^B, K_{i,j,k}^B \rangle \& u_{i,j,k}^B = \text{bauth}(\langle z_{i,j,k}^B, K_{i,j,k}^B \rangle, a_{i,j}^B, b_{i,j}^B, c_{i,j}^B, d_{i,j}^B))\}$, and let $\langle z^B, K^B \rangle$ be the message that maximizes the following: $|R_{k^B}(\langle z^B, K^B \rangle)| = \max_k |R_k(\langle z^B, K^B \rangle)|$. B outputs $M^B = z^B - \sum_{(i_j,j) \in K^B} a_{i_j,j}^B$.

The Perfectly Private Transmission Protocol provides efficient $(0, \delta)$-secure message transmission provided that the field $\mathbf{F}$ used by bauth() satisfies $|\mathbf{F}| \geq \frac{2(3n+mn^2)}{\delta}$. Since reliable communication is not possible when $t \geq n$, this protocol provides matching upper and lower bounds for perfect privacy and probabilistic reliability.

Theorem 5. *If $\delta > 0$, $n > t$, and $|\mathbf{F}| > 2(3n + mn^2)/\delta$, then the Perfectly Private Transmission Protocol is an efficient $(0, \delta)$-secure message transmission protocol.*

Proof. Let w_0 denote the number of lines with no faults, w_1 the number with exactly one fault, and w_+ the number with two or more faults. Then since $n > t$, it follows that $w_0 > w_+$. By Fact 3, $|K^A| \geq w_0 + w_1 > w_+ + w_1$. Whence there is a $(i_{j^*}, j^*) \in K^A$ such that the j^*-th line is a non-faulty line, and $a_{i_{j^*},j^*}^A = a_{i_{j^*},j^*}$, $b_{i_{j^*},j^*}^A = b_{i_{j^*},j^*}$, $c_{i_{j^*},j^*}^A = c_{i_{j^*},j^*}$, and $d_{i_{j^*},j^*}^A = d_{i_{j^*},j^*}$. By Theorem 1, the adversary gets no information about $a_{i_{j^*},j^*}^A$ given the view adv_{M^A}, where adv_{M^A} consists of the following information:

1. $\{(i, j, r^B, \text{bauth}(r^B, a_{i,j}^B, b_{i,j}^B, c_{i,j}^B, d_{i,j}^B)) : 1 \leq i \leq m, 1 \leq j \leq n\}$;
2. $\{(i, j, z^A, K^A, \text{bauth}(\langle z^A, K^A \rangle, a_{i,j}^A, b_{i,j}^A, c_{i,j}^A, d_{i,j}^A)) : 1 \leq i \leq m, 1 \leq j \leq n\}$; and
3. at most one randomly guessed (by the adversary) correct authenticator of some random message.

It should be noted that the above item 3 in the adversary's view adv_{M^A} is important for the following reasons: with non zero probability the first transmission from B to A may fail (i.e. in rounds $m + 2$ through $2m + 3$). That is, the adversary may create a bogus $(r^B)'$ (which is different from r^B) and guess the value $\text{bauth}((r^B)', a_{i_{j^*},j^*}^B, b_{i_{j^*},j^*}^B, c_{i_{j^*},j^*}^B, d_{i_{j^*},j^*}^B)$ correctly. Then at the end of round $2m + 3$, A may choose $r^A = (r^B)'$. The consequence is that there may be an item $(i_{j'}, j') \in K^A$ such that

$$(a^A_{i_{j'},j'}, b^A_{i_{j'},j'}, c^A_{i_{j'},j'}, d^A_{i_{j'},j'}) \neq (a^B_{i_{j'},j'}, b^B_{i_{j'},j'}, c^B_{i_{j'},j'}, d^B_{i_{j'},j'}).$$

It is easy for the adversary to decide whether such kind of item exists in K^A. When such an item exists, the adversary knows that he has guessed a correct authenticator of the message $(r^B)'$.

Since $z^A = M^A + a^A_{i_{j^*},j^*} + \sum_{(i_j,j) \in K^A, j \neq j^*} a^A_{i,j}$, we have that every M^A is equally likely given adv_{M^A}. Since this is the only relevant information about M^A in adv, we have that $\Pr[adv(M_0, r) = c] = \Pr[adv(M_1, r) = c]$ for every pair of messages M_0 and M_1, adversary's coin flips r, and the possible view c. It follows that $\sum_c |\Pr[adv(M_0, r) = c] - \Pr[adv(M_1, r) = c]| = 0$.

We now prove reliability. Let

$$K^{AB} = \{(i_j, j) : \exists i((a^A_{i,j}, b^A_{i,j}, c^A_{i,j}, d^A_{i,j}) = (a^B_{i,j}, b^B_{i,j}, c^B_{i,j}, d^B_{i,j})) \text{ and}$$
$$\forall (0 < i < i_j)((a^A_{i,j}, b^A_{i,j}, c^A_{i,j}, d^A_{i,j}) \neq (a^B_{i,j}, b^B_{i,j}, c^B_{i,j}, d^B_{i,j}))\}.$$

It follows from the use of bauth() that the probability that there exists a k and $r' \neq r^B$ with $r_k(r') > w_1 + w_+$ is less than or equal to the probability that at least one fault node guesses a correct authenticator of r', which is again less than $mn^2/|\mathbf{F}|$ (see Franklin and Wright [8]). That is, the first transmission from B to A (i.e. in rounds $m + 2$ through $2m + 3$) succeeds with the probability at least $1 - mn^2/|\mathbf{F}|$. Let FTR denote the event that the first transmission from B to A succeeds. Now assume that $r^A = r^B$ and $u^A_{i,j,k_A} = \text{bauth}(r^A, a^A_{i,j}, b^A_{i,j}, c^A_{i,j}, d^A_{i,j}))$. Then

$$a^B_{i,j}(r^B)^3 + b^B_{i,j}(r^B)^2 + c^B_{i,j}r^B + d^B_{i,j} = a^A_{i,j}(r^A)^3 + b^A_{i,j}(r^A)^2 + c^A_{i,j}r^A + d^A_{i,j}$$

which implies that r^B is a solution of the equation

$$(a^B_{i,j} - a^A_{i,j})(r^B)^3 + (b^B_{i,j} - b^A_{i,j})(r^B)^2 + (c^B_{i,j} - c^A_{i,j})r^B + (d^B_{i,j} - d^A_{i,j}) = 0 \quad (1)$$

Since $a^A_{i,j}, b^A_{i,j}, c^A_{i,j}, d^A_{i,j}, a^B_{i,j}, b^B_{i,j}, c^B_{i,j}$, and $d^B_{i,j}$ are fixed before the random choice of r^B, and the equation (1) has at most three solutions, it follows that for any fixed $(i_j, j) \in K^A$,

$$\Pr[(a^A_{i_j,j}, b^A_{i_j,j}, c^A_{i_j,j}, d^A_{i_j,j}) \neq (a^B_{i_j,j}, b^B_{i_j,j}, c^B_{i_j,j}, d^B_{i_j,j})|\text{FTR}] \leq 3/|\mathbf{F}|. \quad (2)$$

Then, by the relation (2),

$$\Pr[K^A = K^{AB}|\text{FTR}]$$
$$\geq 1 - \sum_{(i_j,j) \in K^A} \Pr[(a^A_{i_j,j}, b^A_{i_j,j}, c^A_{i_j,j}, d^A_{i_j,j}) \neq (a^B_{i_j,j}, b^B_{i_j,j}, c^B_{i_j,j}, d^B_{i_j,j})|\text{FTR}]$$
$$\geq 1 - \frac{3n}{|\mathbf{F}|}.$$

Whence we have

$$\Pr[K^A = K^{AB}] = \Pr[K^A = K^{AB}|\text{FTR}] \cdot \Pr[\text{FTR}]$$
$$\geq \left(1 - \frac{mn^2}{|\mathbf{F}|}\right)\left(1 - \frac{3n}{|\mathbf{F}|}\right)$$
$$\geq 1 - \frac{3n + mn^2}{|\mathbf{F}|}.$$

A similar analysis shows that the probability that $K^B \neq K^{AB}$ or $z^B \neq z^A$ is less than $\frac{3n + mn^2}{|\mathbf{F}|}$. Hence our protocol is reliable with the probability

$$\Pr[K^A = K^{AB}] \cdot \Pr[K^B = K^{AB}] \geq \left(1 - \frac{3n + mn^2}{|\mathbf{F}|}\right)^2 \geq 1 - \frac{2(3n + mn^2)}{|\mathbf{F}|}.$$

Since $|\mathbf{F}| > 2(3n + mn^2)/\delta$, it follows that $\Pr[M^B = M^A] > 1 - \delta$. $\qquad\square$

Remark: Note that in rounds $2m+4$ through $3m+5$ of our Perfectly Private Transmission Protocol, the information K^A is transmitted explicitly. Indeed, this is not necessary. We can omit the transmission of K^A. Then at the end of round $3m+5$, using the same method that A used to compute the set K^A at the end of round $2m+3$, B can compute K^B (which equals to K^A with high probability). If K^A is not transmitted explicitly, then we can also use the authentication code $bauth(M, a, b, c) = aM^2 + bM + c$ instead of $bauth(M, a, b, c, d)$, since even the adversary guesses a correct authentication code on a random $(r^B)'$, he has no idea whether he has succeed. For this modification, the proof for the corresponding Theorem 5 remains the same.

5   Weak Connectivity

In a more general setting of multicast graph, there is a channel from each node to its neighbor nodes. We say that two nodes A and B of a multicast graph is *strongly t-connected* (which was implicitly introduced by Franklin and Wright [8]) if there are t neighborhoods (except A and B) disjoint paths connecting A and B. Franklin and Wright [8] have observed that the multicast lines protocol can be simulated on any strongly $t + 1$-connected multicast graph. That is, if A and B are strongly $t + 1$-connected, then our results in the previous section shows that $(0, \delta)$-secure message transmission between A and B are possible. In the following, we show that this condition is not necessary.

Franklin and Yung [9] define that two nodes A and B in a multicast graph $G(V, E)$ are *weakly t-connected* if for any set $V_1 \subseteq V \setminus \{A, B\}$ with $|V_1| < t$, the removal of $neighbor(V_1)$ and all incident edges from $G(V, E)$ does not disconnect A and B, where $neighbor(V_1) = V_1 \cup \{v \in V \mid \exists u \in V_1 : (u, v) \in E\} \setminus \{A, B\}$. Franklin and Yung [9] show that it is co**NP** hard to decide whether a given graph is weakly t-connected.

Let A and B be two nodes on a multicast graph $G(V, E)$ and $t < n$. We say that A and B are *weakly (n, t)-connected* if there are n vertex disjoint paths

$p_1, \ldots, p_n$ between A and B and, for any vertex set $T \subseteq (V \setminus \{A, B\})$ with $|T| \leq t$, there exists an i ($1 \leq i \leq n$) such that all vertices on p_i have no neighbor in T. Obviously, if two vertices are weakly (n, t)-connected then they are weakly $t + 1$-connected.

Theorem 6. *If A and B are weakly (n, t)-connected for some $t < n$, then the Perfectly Private Transmission Protocol in the previous section is an efficient $(0, \delta)$-secure message transmission between A and B.*

Proof. It follows straightforward from the proof of Theorem 5. $\square$

Franklin and Yung [9] show that, in the context of a t-passive adversary, weak $t + 1$-connectivity is necessary and sufficient for achieving private communications. Theorem 6 provides a sufficient condition for achieving perfect privacy and probabilistic reliability against a t-active adversary in a general multi-cast graph. It is an open question whether the condition in Theorem 6 is also necessary.

It is easily observed that strong $t + 1$-connectivity implies weak $(t + 1, t)$-connectivity. The following example shows that (n, t)-weak connectivity does not imply strong $t + 1$-connectivity.

Example 7. Let $G(V, E)$ be the graph defined by $V = \{A, B\} \cup \{v_{i,j} : i, j = 1, 2, 3\}$ and $E = \{(A, v_{i,1}) : i = 1, 2, 3\} \cup \{(v_{i,j}, v_{i,j+1}) : i = 1, 2, 3; j = 1, 2\} \cup \{(v_{i,3}, B) : 1 = 1, 2, 3\} \cup \{(v_{1,1}, v_{2,1}), (v_{2,2}, v_{3,2}), (v_{3,3}, v_{1,3})\}$. Then it is straightforward to show that A and B are weakly $(3, 1)$-connected but not strongly 2-connected in G.

Theorem 6 shows that, for at most one malicious node, efficient $(0, \delta)$-secure message transmission between A and B is possible in the multicast graph defined in Example 7. Note that this multicast graph is only strongly 1-connected, and so Franklin-Wright's results have no bearing on this example.

Similarly, for any $n > 2$ the following example gives a graph G and two vertices A and B such that A and B are weakly $(n, 1)$-connected but not weakly 3-connected.

Example 8. Let $G(V, E)$ be the graph defined by $V = \{A, B\} \cup \{v_{i,j} : i = 1, \ldots n; j = 1, 2\}$ and $E = \{(A, v_{i,1}) : i = 1, \ldots, n\} \cup \{(v_{i,1}, v_{i,2} : i = 1, \ldots, n\} \cup \{(v_{i,2}, B) : 1 = 1, \ldots, n\} \cup \{(v_{1,2}, v_{i,2}) : i = 2, \ldots, \lfloor \frac{n}{2} \rfloor\} \cup \{(v_{\lfloor \frac{n}{2} \rfloor+1,2}, v_{i,2}) : i = \lfloor \frac{n}{2} \rfloor + 2, \ldots, n\}$. Then it is straightforward to show that A and B are weakly $(n, 1)$-connected but not weakly 3-connected in G.

Then Theorem 6 shows that, for at most one malicious node, efficient $(0, \delta)$-secure message transmission between A and B is possible in the graph G defined in Example 8. The result by Franklin and Yung [9] shows that secure message transmission between A and B is impossible in this graph when there are two malicious nodes. However, if $n > 2t + 1$ and we use non-broadcast channels, then secure message transmission is possible between A and B against t malicious nodes (see, e.g., Dolev, Dwork, Waarts, and Yung [7]). It follows that in certain cases broadcast *helps* adversaries "more", which contrasts with Franklin and Wright's result [8] that in certain cases broadcast *hurts* adversaries "more".

We close our paper by showing that it is **NP**-hard to decide whether a given multicast graph is strongly k-connected.

Theorem 9. *It is* **NP**-*complete to decide whether a given multicast graph is strongly k-connected.*

Proof. It is clear that the specified problem is in **NP**. Whence it suffices to reduce the following **NP**-complete problem IS (Independent Set) to our problem. A similar (but not identical) reduction for a different problem has appeared in Burmester, Desmedt, and Wang [3]. The independent set problem is:

Instance: A graph $G(V, E)$ and a number k.
Question: Does there exist a node set $V_1 \subseteq V$ of size k such that any two nodes in V_1 are not connected by an edge in E?

The input $G(V_G, E_G)$, to IS, consists of a set of vertices $V_G = \{v_1, \dots, v_n\}$ and a set of edges E_G. In the following we construct a multicast graph $f(G) = MG(V; E)$ and two nodes $A, B \in V$ such that there is an independent set of size k in G if and only if A and B are strongly k-connected.

Let $V = \{A, B\} \cup \{u_{i,j} : i, j = 1, \dots n\} \cup \{u_i : i = 1, \dots, n\}$, and E be the set of the following edges.

1. For each pair $i, j = 1, \dots, n$, there is an edge $(A, u_{i,j}) \in E$.
2. For each pair $i, j = 1, \dots, n$: if there is exists an edge $(v_i, v_j) \in E_G$, then there are four edges $(u_{i,j}, u_i)$, $(u_{i,j}, u_j)$, $(u_{j,i}, u_i)$, and $(u_{j,i}, u_j)$ in E.
3. For each i, there is an edge $(u_i, B) \in E$.

It is clear that two paths P_1 and P_2 connecting A and B which go through u_i and u_j respectively are node disjoint and have no common neighborhoods (except A and B) if and only if there is no edge (v_i, v_j) in E_G. Hence there is an independent set of size k in G if and only if A and B are strongly k-connected.
□

Similarly, we can define the corresponding problem for weak (n, t)-connectivity as follows:

Instance: A graph $G(V, E)$ and two number $n > k$.
Question: Is G weakly (n, t)-connected?

Using a reduction from the **NP**-complete problem "Vertex Cover", a similar argument as in the proof of Theorem 9 can be used to show that the above problem is coNP-hard (the details are omitted). Indeed, it is straightforward to show that the above problem belongs to Σ_2^p (that is, the second level of the polynomial time hierarchy). It remains open whether this problem is coNP-complete, or Σ_2^p-complete, or neither.

Acknowledgment

The authors thank Matt Franklin (Xerox) and Rebecca Wright (AT&T) for informing us, after the paper had been accepted, that Donald Beaver has found a different method to address the Franklin-Wright open problem.

References

1. M. Ben-Or, S. Goldwasser, and A. Wigderson. Completeness theorems for non-cryptographic fault-tolerant distributed computing. In: *Proc. ACM STOC, '88*, pages 1–10, ACM Press, 1988.
2. M. Burmester, Y. Desmedt, and G. Kabatianski. Trust and Security: A New Look at the Byzantine Generals Problem. *DIMACS Series in Discrete Mathematics and Theoretical Computer Science* **38**, pages 75–83, American Mathematical Society, 1998.
3. M. Burmester, Y. Desmedt, and Y. Wang. Using approximation hardness to achieve dependable computation. In: *Proc. of the Second International Conference on Randomization and Approximation Techniques in Computer Science*, LNCS 1518, pages 172–186, Springer Verlag, 1998.
4. D. Chaum, C. Crepeau, and I. Damgard. Multiparty unconditional secure protocols. In: *Proc. ACM STOC, '88*, pages 11–19, ACM Press, 1988.
5. B. den Boer. A simple and key-economical unconditional authentication scheme. *Journal of Computer Security*, 2:65–71, 1993.
6. D. Dolev. The Byzantine generals strike again. *J. of Algorithms*, 3:14–30, 1982.
7. D. Dolev, C. Dwork, O. Waarts, and M. Yung. Perfectly secure message transmission. *J. of the ACM*, **40**(1):17–47, 1993.
8. M. Franklin and N. Wright. Secure communication in minimal connectivity models. In: *Advances in Cryptology, Proc. of Euro Crypt '98*, LNCS 1403, pages 346–360, Springer Verlag, 1998.
9. M. Franklin and M. Yung. Secure hypergraphs: privacy from partial broadcast. In: *Proc. ACM STOC, '95*, pages 36–44, ACM Press, 1995.
10. E. Gilbert, F. MacWilliams, and N. Sloane. Codes which detect deception. *The BELL System Technical Journal*, **53**(3):405–424, 1974.
11. O. Goldreich, S. Goldwasser, and N. Linial. Fault-tolerant computation in the full information model. *SIAM J. Comput.* **27**(2):506–544, 1998.
12. V. Hadzilacos. *Issues of Fault Tolerance in Concurrent Computations*. PhD thesis, Harvard University, Cambridge, MA, 1984.
13. T. Rabin. Robust sharing of secrets when the dealer is honest or faulty. *J. of the ACM*, **41**(6):1089–1109, 1994.
14. T. Rabin and M. Ben-Or. Verifiable secret sharing and multiparty protocols with honest majority. In: *Proc. ACM STOC, '89*, pages 73–85, ACM Press, 1989.
15. G. J. Simmons. A survey of information authentication. In: *Contemporay Cryptology, The Science of Information Integrity*, pages 379–419. IEEE Press, 1992.

Efficient Communication-Storage Tradeoffs for Multicast Encryption

Ran Canetti[1], Tal Malkin[2*], and Kobbi Nissim[3]

[1] IBM T. J. Watson Research Center, Yorktown Height, NY, 10598,
`canetti@watson.ibm.com`
[2] Laboratory for Computer Science, Massachusetts Institute of Technology,
545 Technology Square, Cambridge, MA 02139,
`tal@theory.lcs.mit.edu`
[3] Dept. of Computer Science and Applied Math, Weizmann Institute of Science,
Rehovot 76100, Israel,
`kobbi@wisdom.weizmann.ac.il`

Abstract. We consider re-keying protocols for secure multicasting in a dynamic multicast group with a center. There is a variety of different scenarios using multicast, presenting a wide range of efficiency requirements with respect to several parameters. We give an upper bound on the tradeoff between storage and communication parameters. In particular, we suggest an improvement of the schemes by Wallner *et al.* and Wong *et al.* [13,14] with sub-linear center storage, without a significant loss in other parameters.

Correctly selecting the parameters of our scheme we can efficiently accommodate a wide range of scenarios. This is demonstrated by Applying the protocol to some known benchmark scenarios.

We also show lower bounds on the tradeoff between communication and user storage, and show that our scheme is almost optimal with respect to these lower bounds.

1 Introduction

Multicast communication (and, in particular, IP multicast routing) is an attractive method for delivery of data to multiple recipients. The motivation for multicast communication is its efficiency – multicast group users get the same message simultaneously, hence the reduction of both sender and network resources. A wide range of applications benefit from efficient multicast: interest groups, file and real-time information update, video multi-party conferences, on-line games and pay TV are few examples.

Securing multicast communication is non-trivial and poses a number of challenges, ranging from algorithmic problems, through system and communication design, to secure implementation. (See overview in [5,4].) The main security concerns are typically *access control* — making sure that only legitimate members

---

* Supported by DARPA grant DABT63-96-C-0018. Part of this work was done while the author was visiting the IBM T.J. Watson Research Center.

of a multicast group have access to the multicast group communication, *source authentication* — verifying that received multicasted data is unmodified and originates with the claimed source, and *maintaining availability* — protecting against denial-of-service and clogging attacks.

This paper focuses on providing access control for multicast communication. The standard technique to this end is to maintain a common key that is known to all the multicast group members, but is *unknown* to non-members. All group communication is then encrypted using the shared key. (We remark that long-term secrecy is typically not a concern for multicast communication; encryption is used mainly for obtaining short-term access control.) The main problem here is *key management* — how to maintain the invariant that all group members, and only them, have access to the group key in a group with dynamic membership. We limit ourselves to the case where there is a centralized *group controller* (or, *group center*) who handles the task of key management. Whenever a member joins or leaves the group, the group key needs to be changed and the new key needs to be let known to all members.

We concentrate on efficient schemes for this *re-keying* problem. In particular, we show a *tradeoff* between communication and storage parameters for the group controller and members, and provide nearly optimal upper and lower bound for some of these parameters. Our protocol is parameterized in terms of the tradeoff, allowing different choices of parameters to result in a variety of scheme performances. This makes the protocol suitable for different applications and scenarios. The works of [13,14] on efficient re-keying schemes are the starting point for this work.

1.1 Security of Re-keying Schemes

A standard security requirement from the data encryption mechanism is *semantic security* [8] of the group communication. Assuming the usage of appropriate (semantically secure) encryption schemes, this requirement reduces to the semantic security of the group session key k_s, shared by the group members. I.e. it is required that an adversary cannot distinguish the real session key from a random key.

If the only operation allowed is joining new users to the group, the re-keying problem is solved by simply giving the session key k_s to the new users. If backward privacy is also required (i.e. new users should not have access to past messages), then a new session key k_s^{new} may be selected and given to the new users, and $E_{k_s}(k_s^{new})$ is multicasted. (Alternatively, the new key can be locally computed as a pseudorandom function of the old key.)

Removing users from the group requires the change of k_s (and possibly other data) to guarantee the semantic security of the new key against any coalition of removed users. It is stressed that security is required against *any* coalition of removed users. In particular, we do not assume any limit on the size or structure of the coalition.

To be able to focus on the re-keying problem we assume authenticated and reliable communication, or more specifically that the messages sent by the group center arrive at their destination and messages are not modified, generated, or replayed by an adversary. These concerns should be addressed separately.

1.2 Efficiency of Re-keying Schemes

Efficiency of multicast re-keying schemes is measured by several parameters: (i) communication complexity, (ii) user storage and (iii) center storage and (iv) time complexity. In this paper we concentrate on the communication and storage complexity measures (of course, without letting the time complexity be infeasible).

Communication complexity is probably the most important measure, as it is the biggest bottleneck in current applications. (Indeed, reducing communication is the main motivation for using multicast technology.)

Reducing the *center storage* enables small memory in the security module (which is responsible for key management). This module is typically separate from the module(s) handling group membership; this latter task typically requires special handling of each member upon joining and leaving the group, and is left out of scope of this work. The module separation can be either logical or physical. Furthermore, for large groups the membership module may consist of several disparate components handling different regions, while the key management module remains centralized. Also, the performance and latency requirement from the key-management module may be more stringent.

Using our scheme, the center storage may indeed be sub-linear, thereby improving on the the best previously known schemes [13,14], without a significant change in other parameters. E.g. with current technology, for a million users multicast group, our reduction enables a security module with all its storage in fast cache memory, making it considerably more efficient.

The motivation for reducing *user storage* stems from applications in which the users are low-end, and have severe memory restrictions (e.g. when the multicast group consists of cable TV viewers and the user module resides in the cable converter unit).

Since there is a large number of potential multicast scenarios it seems unlikely that a single solution will fit all scenarios. This motivates a *tradeoff* between efficiency parameters. Simple solutions suggest that such a tradeoff exists: (i) One extreme is a center that shares, in addition to the session key, a distinct symmetric key with each user. When a user is removed, the center sends new symmetric keys and a new session key to each of the users separately. Thus, user storage is minimal but the communication costs are proportional to the number of group users. (ii) An opposite extreme is having a key for every possible subset of users, where every potential user gets all the keys for the subsets that contain her. Whenever a user is removed, the session key is set to the key of the remaining subset of users. The length of the re-keying message of this solution is optimal (it suffices to declare each removed user), but the number of keys held by each user is clearly prohibitive (at least 2^{n-1} keys, where n is the group size).

Our goal is to study the tradeoff between communication and storage, and construct schemes which are flexible enough to fit a variety of scenarios, in a way that is provably optimal (or close to optimal).

We achieve this goal with respect to the tradeoff between communication and user storage. For the tradeoff between communication and center storage, our upper bound is better than all previously known schemes. Proving a lower bound on the latter tradeoff remains an intriguing open problem.

1.3   Summary of Results

We give an upper bound on the tradeoff between user storage, center storage and communication, and a lower bound relating user storage and the minimal communication. The gap between the bounds is at most logarithmic in the size of the group. Moreover, for a natural class of protocols, including all currently known ones, the gap is closed, namely our scheme is optimal for this class. Thus, our upper bound is nearly optimal with respect to our lower bound, in a strong sense. Our upper bounds are based on the re-keying schemes of Wallner et al. and Wong et al. [13,14], with improvements of [4] and McGrew and Sherman [10]. These schemes communicate $\log n$ encrypted keys per update, and require linear center storage ($2n - 1$ keys), and logarithmic user storage ($\log n$ keys).

Upper Bound   We give an upper bound (i.e. a protocol) which allows trading center storage with communication, with the restriction that communication is lower bounded as a function of user storage. Specifically, for a group of n users with user storage of $b + 1$ keys, the communication is $O(bn^{1/b} - b)$ encrypted keys. Center storage multiplied by communication length is roughly $O(n)$.

One instance yields $O(\log n)$ communication, $O(\log n)$ user storage, $O(\frac{n}{\log n})$ center storage. This is the first scheme with center storage sub-linear in n. Other instances are suitable for different applications, as we demonstrate by applying our scheme to benchmark scenarios.

In practice, re-keying protocols may be used in "batch mode", where the center does not immediately perform updates, but rather waits until several updates accumulate and perform all of them at once. (This is acceptable for most applications.) Doing this allows in many cases (such as in our scheme) significant savings in the communication. However this paper focuses on updates one-by-one, as this is the worst case scenario.

Lower Bounds   We first give a lower bound on the communication of re-keying protocols as a function of user storage. We prove that if each user holds at most $b + 1$ keys, the communication costs are at least $n^{1/b}$ encrypted messages.

We further consider the class of *structure preserving protocols* (to which currently known schemes belong [13,4]). Intuitively, structure preserving protocols are those that maintain the property of "u_1 knows m keys which u_2 doesn't" across updates. That is, if user u_1 holds m keys which are not known to user u_2, then after deleting a user $u_3 \notin \{u_1, u_2\}$ and performing the necessary updates, u_1 still holds about m keys not known to u_2. For structure preserving protocols, we show a tight (up to small constant factors) lower bound of $bn^{1/b} - b$ messages (matching our upper bound protocol).

The lower bound is for algorithms that use a "generic" key encryption mechanisms. Formally, we assume a "black-box encryption service" that is the only means of encryption (i.e., the algorithm should provide perfect secrecy in the idealized model). Consequently, the implication of the lower bounds is that in order to achieve more efficient protocols than ours one would have to use specific properties of a particular encryption system, such as exploit algebraic properties of the keys used.

1.4   Related Work

A different approach to solving the problem of allowing only legitimate users to access multicasted data is put forward by Fiat and Naor [6]. In their formalization, a center uses a broadcast channel to transmit messages to a group of users $\mathcal{U}$. There are two *pre-specified* sets: (i) collection $S \subseteq 2^{\mathcal{U}}$ of legal subsets of recipients, and: (ii) collection $\mathcal{C} \subseteq 2^{\mathcal{U}}$ of possible "bad" coalitions. The goal is to enable the center to communicate data secretly to a given set $S \in S$ of users, while preventing any coalition from $\mathcal{C} - S$ to gather information on the data. Any such mechanism can be used to establish a group key and thus provides a solution to the re-keying problem.

The [6] solution is radically different than ones discussed here. In particular, it allows encrypting multicast communication even without requiring all users to have a single common key; in addition, joining and leaving of members does not necessarily require *any* action by the other members. However, their solution assumes in a critical way some bound on the size or structure of the coalition of adversarial non-members. This work considers schemes where no such assumptions are made.

There have been some works in broadcast encryption models that consider lower bounds on storage and communication, and show that both cannot be simultaneously low. Luby and Staddon [9] allow arbitrary coalitions, but restrict the possible subsets of recipients to be all sets of certain size $n - m$. In this model they study the tradeoff between the number of keys held by each user, and the number of transmissions needed for establishing a new broadcast key. They assumed a security model that allows translating the problem to a combinatorial (set theoretic) problem. Their lower bound states that either the number of transmissions is very high, or the number of keys held by every user is high.

Blundo, Frota Mattos and Stinson [2] and Stinson and Trung [12] study communication storage tradeoff in a model of *unconditionally secure* broadcast encryption [2] by providing some upper and lower bounds for key pre-distribution schemes (e.g. [3,11]) and broadcast encryption. This model further differs from ours in that information theoretic security is required, and storage and communication are measured in terms of amount of secret *information* stored by each user, and the broadcast *information rate*.

Organization In Section 2 we describe our communication and encryption model. The upper bound scheme is described in Section 3. Finally, we prove lower bounds on the tradeoff between user storage and communication in Section 4.

2 Preliminaries

Let $\mathcal{U}$ denote the universe of all possible users[1], and GC denote the group center. We consider a set $M = \{u_1, \ldots, u_n\} \subseteq \mathcal{U}$, called the *multicast group* (for simplicity, $GC \notin M$). A session key k_s is initially shared by all users in M and by GC (and is not known to any user $v \notin M$). In addition, other information may be known to the users and the center. We abstract away the details of the initialization phase by which the users get their initial information. In particular we may assume that each user joining M has an authenticated secure unicast channel with the center GC for the purpose of initialization. (In practice this may be obtained by using a public key system.) After the initialization phase, and throughout the lifetime of the system, the only means of communication with group members is via a multicast channel on which the group center may broadcast messages that will be heard by all users in $\mathcal{U}$. Our goal is to securely update the session key when the group M changes, so that all users in the group, and only them, know the session key at any given time.

A *multicast protocol* specifies an algorithm by which the center may update the session key (and possibly other information) for the following two update operations on M:

- *remove(U)* where $U \subseteq M$. The result is the removal of users in U from the multicast group: $M^{new} = M \setminus U$.
- *join(U)* where $U \subseteq \mathcal{U}$. The result is the joining of users in U to the multicast group: $M^{new} = M \cup U$.

Since the worst case for the re-keying protocol is when $|U| = 1$, from now on we assume $|U| = 1$ and measure the efficiency of our protocols accordingly. In our description we focus on the removal of users from the multicast group, since dealing with joining users is much simpler and can be done with virtually no communication overhead.

Since we do not want to consider specific private key encryption and their particular properties, we concentrate on a general key-based model, where the cryptographic details are abstracted away. This is modeled by a publically available black-box pair E, D, such that E given as inputs a key k and a message m outputs a *random* ciphertext $c = E(k, m)$; given a ciphertext c and a key k, the decryption algorithm D outputs the plaintext m. (We assume that the encryption is deterministic; that is, two applications with the same message and key will result in the same ciphertext. Probabilistic encryption can be built upon E, D in straightforward ways.) This model guarantees that, when multicasting a message encrypted with a key k, any user holding k will be able to decrypt, and any coalition of users that does not hold k gains no information from hearing the ciphertext. To formalize our requirement that all encryption and decryption is being done via the black-box pair E, D, we let the adversary be computationally *unbounded*. A lower bound in our model means that any scheme which beats the

---

[1] There is no need to a-priori have an explicit representation of $\mathcal{U}$. For example, $\mathcal{U}$ may be the set of all users connected to the Internet.

bound must be based on a particular encryption scheme and its particular (We remark that, although this model is formalized with the lower bounds in mind, our re-keying schemes can be proven secure even in this model.)

Multicast Encryption Protocols We define the model of key-based multicast as follows. Let l be a security parameter, and let the number of users n be polynomial in l. Let $K \subset \{0,1\}^l$ be a set of *keys*. Each user $u_i \in M$ holds a subset $K(u_i) \subseteq K$ of keys. In particular, there is a "session key" $k_s \in K$ such that every $u \in M$ holds k_s. For a set of users $U \subseteq M$ we define $K(U) = \bigcup_{u \in U} K(u)$. We say that a set $U \subseteq M$ holds a key $k \in K$ if $k \in K(U)$.

In response to a request for update operation the group center (following a given protocol) sends a multicast message that results in changed group keys (and possible other keys). For a key $k \in K$ and a string $m \in \{0,1\}^l$, the group center GC may send over the broadcast channel the ciphertext $E_k(m)$. Users holding k may decrypt and obtain m. After all the ciphertexts for an update have been broadcasted by the center, the users who can decrypt ciphertexts do so, and follow the protocol specification to update their keys. The new total set of keys is denoted by K^{new}.

For the definition of security, we consider an adaptive adversary who may, repeatedly and in an arbitrary order, submit update (remove/join) operations to the center for subsets of his choice, and break into users $u \in \mathcal{U}$ of his choice (thereby getting all of u's information).

We say that a multicast system is *secure* if for any adversary, after any sequence of operations as above, if the adversary has not broken into any user who was in the multicast group while a key k_s was the session key, then the adversary has no advantage in distinguishing k_s from a random key. Note that this definition implies *backward security* as well, since the adversary is not allowed to learn any information about a previous session key, unless he broke into a user who legitimately belonged to the group at the time that key was used). We also do not put a restriction on the number of users the adversary may break into.

Finally, by convention, when performing a *remove(U)* operation, all keys in $K(U)$ are removed from K^{new} (since we require arbitrary resilience, it can be shown that there is no advantage in using a key of a removed user to broadcast a message, and thus these keys may be removed). In particular, k_s is also removed, and thus a new key must resume the special role of a session key k_s^{new}.

The *communication complexity* of an update operation is measured by the number of ciphertexts that need to be broadcasted by the center per update (for the worst case choice of update), and is denoted by $c(n)$ for a group of size n. The *storage* is measured by the number of keys that need to be stored.

3 A Re-keying Scheme

We start by describing two schemes that our construction will be built upon. The first (described in Section 3.1) is a simple scheme achieving minimal (constant) storage for the center and each user, but highly inefficient (linear) communication complexity. The second (described in Section 3.2) is a widely used scheme by Wallner *et al.* and Wong *et al.* [13,14] (with an improvement of [4]), which we call the basic tree scheme. This scheme achieves logarithmic communication complexity and logarithmic storage for each user, but linear storage for the center. We then show (in Section 3.4) how the basic tree scheme can be generalized and combined with the minimal storage scheme, so as to achieve an improved scheme with a tradeoff between the parameters. As a special case, we get a reduction of the center storage in the tree scheme by a logarithmic factor.

3.1   A Minimal Storage Scheme

We describe a simple scheme, which requires the smallest possible amount of storage – two keys for the center and each user[2], but is very communication intensive, requiring $(n-1)$ ciphertext sent per removal of a user. We will later use this scheme as a building block in our construction.

In this scheme each user u holds the session key k_s, and a unique symmetric key k_u not known to any other user. The center should be able to generate the keys of all users, which is possible by holding a single secret key r, an index to a pseudo-random function f_r [7] (which can be constructed from the same black-box used for encryption). The keys can be generated by applying the function to the user's index, namely $k_u = f_r(u)$.

When a group of users U is removed from the group, the center chooses a new session key k_s^{new}, and sends it to each user, by broadcasting the ciphers $E_{k_u}(k_s^{new})$ for all $u \in M^{new} = M \setminus U$.

The security of this scheme is based on the security of the encryption scheme and pseudo-random function. The parameters are summarized in Table 1.

3.2   The Basic Tree Scheme

We describe the scheme by Wallner *et al.* and Wong *et al.* [13,14] (with the improvement of [4]). For a detailed description, we refer the reader to [13,14,4].

The group center creates a balanced binary tree with at least n leaves and assigns a l-bit random key to every node. Let k_ϵ denote the key assigned with the tree root v_ϵ. Denote the left and right children of node v_σ by $v_{\sigma 0}, v_{\sigma 1}$ and their assigned keys by $k_{\sigma 0}, k_{\sigma 1}$ respectively (i.e. the left and right children of the node indexed by σ are indexed by σ concatenated with 0 or 1 respectively). Every user in M is assigned a leaf and is given the $\log n + 1$ keys assigned to nodes on the path from the root to this leaf. Since k_ϵ is known to all group members it is used as the session key: $k_s = k_\epsilon$.

Notation Let $\sigma \in \{0,1\}^*$. Denote by σ^i the string resulting by erasing the i rightmost bits of σ. Denote by $flip(\sigma)$ the string resulting by flipping the rightmost bit of σ.

Let $G : \{0,1\}^l \to \{0,1\}^{2l}$ be a pseudo random generator that doubles the size of its input [15,1]. Let $G_L(x), G_R(x)$ be the left and right halves of $G(x)$ respectively. Upon removal of a user u_σ, The group center chooses a random number $r_{\sigma 1} \in_R \{0,1\}^k$. For $i = 1, \ldots, \log n$ the group center sets $k_{\sigma^i}^{new}$ to $G_L(r_{\sigma^i})$, sets $r_{\sigma^{i+1}}$ to $G_R(r_{\sigma^i})$ and broadcasts $E_{k_{flip(\sigma^{i-1})}}(r_{\sigma^i})$.

---

[2] This is minimal by Corollary 3 in the next section.

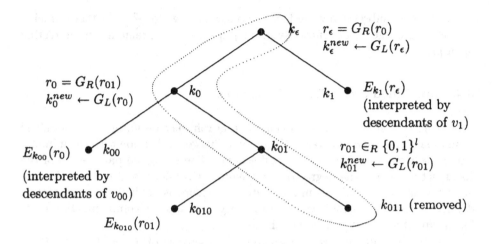

$$r_\epsilon = G_R(r_0)$$
$$k_\epsilon^{new} \leftarrow G_L(r_\epsilon)$$

$$r_0 = G_R(r_{01})$$
$$k_0^{new} \leftarrow G_L(r_0)$$

$$E_{k_1}(r_\epsilon)$$
(interpreted by descendants of v_1)

$$r_{01} \in_R \{0,1\}^l$$
$$k_{01}^{new} \leftarrow G_L(r_{01})$$

$$E_{k_{00}}(r_0)$$
(interpreted by descendants of v_{00})

$$E_{k_{010}}(r_{01})$$

k_{011} (removed)

Fig. 1. The basic tree scheme actions when holder of k_{011} is removed. (The figure shows only tree nodes that affected by the removal.)

E.g. if u_{011} is removed (see Figure 1), r_{01} is chosen at random, k_{01}^{new} is set to $G_L(r_{01})$, r_0 is set to $G_R(r_{01})$ and r_{01} is broadcasted encrypted with k_{010}. Then, k_0^{new} is set to $G_L(r_0)$, r_ϵ is set to $G_R(r_0)$ and r_0 is broadcasted encrypted with k_{00}. Finally, the new session key $k_s^{new} = k_\epsilon^{new}$ is set to $G_L(r_\epsilon)$ and r_ϵ is broadcasted encrypted with k_1. Now, every user can compute the changed keys on his root-to-leaf path.

The basic tree scheme parameters appear in Table 1.

	minimal storage scheme	basic tree scheme
user storage	2	$\log n + 1$
center storage	2	$2n - 1$
communication	$(n - 1)$	$\log n$

Table 1. Parameters of the basic schemes.

3.3   On the Storage Requirements of the Group Center

On first glance, reducing the center storage requirements in the tree scheme may proceed as follows. Instead of having the center keep all keys on the tree, the keys may be generated from a single key, say by applying a pseudo-random function, and the center will keep only this secret key. However, this idea does not seem to work, since when an update occurs, the center will have to change the secret key, requiring changing the entire tree, thus bringing the communication to linear.[3]

---

[3] Alternatively, the secret key may stay the same, but some counter be changed for every update. However, this is only useful if we require threshold security (requiring

In the next subsection we reduce the center storage to $\frac{n}{\log n}$. Further reducing the center storage, or alternatively proving it impossible, remains an interesting open problem.

3.4 Combined a-ary Tradeoff Scheme

The basic tree shown in the previous paragraph may be naturally generalized from binary trees to a-ary trees. We combine this generalization with the minimal storage scheme to create our tradeoff scheme. There are two parameters of the construction (i) a - the degree of the tree internal nodes, and (ii) m - the size of user subsets to which the minimal storage scheme is applied. The parameters determine the number of keys given to every user and the communication costs for an update operation. Details follow.

Divide the multicast group users to disjoint subsets of size m: $U_1, \ldots, U_{n/m}$, $\cup_{i=1}^{n/m} U_i = M$. The group center constructs an a-ary tree of height $b = \lceil \log_a \lceil \frac{n}{m} \rceil \rceil$ (i.e. the tree has at least n/m leaves). Assign subset U_i with the ith leaf of the tree. As in the basic tree scheme, a random key is assigned with each tree node.

For $m = 1$, $U_i = \{u_i\}$, the scheme is a simple generalization of the basic tree scheme to a-ary trees. For $m > 1$, we combine the basic tree scheme and the minimal storage scheme as follows. Every user $u \in U_i$ is given the b keys assigned to the nodes on the path from the root to the ith leaf. The center holds all these keys, as well as secret keys r_i for each leaf i (r_i's are not known to any user). r_i is used as the seed for the minimal storage scheme between the group center and U_i, namely r_i is used for generating a unique private key for every $u \in U_i$. Whenever a user $u \in U_i$ is removed, the keys on the path from the ith leaf to the root are changed. The center sends to every user in $U_i \setminus \{u\}$ the new key for the ith leaf as in the minimal storage scheme, and then sends the ciphertexts necessary to update the path to the root as in the basic tree scheme.

The security of this scheme follows from the security of the minimal storage scheme and the basic tree scheme (based on the security of the pseudorandom function). The parameters of the scheme appear in Table 2.

	general m, a	Example 1	Example 2
user storage	$\log_a(\frac{n}{m}) + 1$	$O(\log n)$	2
center storage	$\frac{n}{m} \cdot \frac{a}{a-1}$	$O(\frac{n}{\log n})$	$n^{1/2} + 1$
communication	$m - 1 + (a-1)\log_a(\frac{n}{m})$	$O(\log n)$	$2n^{1/2} - 2$

Table 2. Parameters of the tradeoff scheme. Note that setting $a = m = n$ gives the minimal storage scheme and setting $m = 1$, $a = 2$ gives the basic tree scheme. In Example 1, $a = 2$, $m = O(\log n)$, in Example 2, $a = m = n^{1/2}$.

---

storage which is linear in the size of the coalition). For the strong notion of security against arbitrary coalitions, this would again require linear storage from the center.

Denote the center storage by s_{GC}, the user storage by $b+1$ (i.e. $b = \log_a(\frac{n}{m})$, or equivalently $a = \frac{n}{m}^{1/b}$) and the communication by $c = c(n)$. The tradeoff scheme allows trading center storage and communication costs, subject to the restriction that communication costs are lower bounded as a function of user storage. Specifically:

Theorem 1. *There exist secure multicast encryption protocols such that*

1. $s_{GC} \cdot c = \Theta(n)$.
2. $c = \Theta(bn^{\frac{1}{b}})$.

These bounds follow from the parameters of our scheme in Table 2.

Thus, our scheme is flexible enough to deal with a large range of applications, adjusting the parameters accordingly (see, for exampele, [4,5] for a discussion and two very different benchmark scenarios).

In particular, it follows that using our scheme the center storage may be reduced by a factor of $\log n$ with respect to the storage in [13,14,4]. Further reduction in the center storage is achieved by noticing that the center need not hold an *explicit* representation of keys, instead it can hold a shorter representation from which it is possible to compute the keys efficiently. Consider, for instance, the case where the group center holds a secret key r to a pseudo-random function $f_r : \{0,1\}^l \to \{0,1\}^l$, and a counter cnt which is initially set to zero. Set $m > 2$. When a user in U_i is removed, the center uses $r_i = f_r(cnt)$, stores cnt in the leaf corresponding to U_i and advances cnt. All the nodes on the path from the ith leaf to the root store a pointer to leaf i. This way, the center may compute any key in the tree via one application of f_r and $O(\log_a \frac{n}{m})$ applications of the pseudo random generator G.

As an example, consider a group with a million users using DES (7-bytes keys). In the basic construction, the needed center memory is $2 \cdot 10^6 \cdot 7 = 14$Mbytes. Using our construction with a 4-bytes counter reduces the center memory to $2 \cdot 10^6 \cdot 4/20 = 400$Kbytes, which is small enough to be put in a fast cache memory.

4 Lower Bounds

In this section we describe lower bounds on the amount of storage and the communication complexity per update (both measured in units of l bits, namely the key size), and the relation between the two. We begin by observing simple lower bounds on the user storage and the number of keys in the system.

Lemma 2. *For any secure multicast encryption protocol, $\forall U \subseteq M \; \exists k \in K$ such that $k \in K(U)$ but $\forall v \in M \setminus U$, $k \notin K(v)$ (every subset of users has a key which does not belong to any other user outside the subset).*

Proof. Assume for contradiction that there exists a subset $U \subseteq M$ such that $\forall k \in K(U)$, $k \in K(M \setminus U)$. That is, every key held by users in U is also held by some user in $M \setminus U$. It follows that any multicast message which is understood by someone in U is also understood by the coalition $M \setminus U$. Consider the operation $remove(M \setminus U)$ (whether done by removing the users one by one, or a more general removal of the whole subset). By the above, there is no way to provide U with a new session key that is not known to the coalition $M \setminus U$, and thus this update operation cannot be performed securely, yielding a contradiction. □

Corollary 3. *For any secure multicast encryption protocol,*

1. *Every user $u \in M$ must hold at least two keys: a unique key k_u known only to u and GC, and the session key k_s.*
2. *The total number of keys in the system is $|K| \geq n + 1$.*

We now turn to prove lower bounds regarding the tradeoff between communication and user storage. Consider any given secure multicast encryption protocol. Recall that n denotes the number of users in the multicast group M, and $c(n)$ the denotes the maximal communication complexity required for re-keying following a deletion of a user from the group. We let $b(n) + 1$ denote the maximal number of keys, including the session key, held by any user in M (for convenience, we sometimes omit the argument n from the notation of b). We will prove bounds on the relation between $b(n)$ and $c(n)$.

We start with the special case of $b(n) = 1$, namely for a system where each user holds only one key in addition to the session key. This case will be used in the following general theorems.

Lemma 4. *If the maximal number of keys held by each user is $b(n) + 1 = 2$, then the re-keying communication costs satisfy $c(n) \geq n - 1$.*

Proof. Since each user u holds at most two keys, by Corollary 3 these must be the session key k_s and a unique key k_u known only to u. When a user is removed, the other $n - 1$ users must be notified in order to establish the new session key. But since k_s is known to the removed user it cannot be used, forcing the center to use the unique keys k_u for each user who stays in the group, requiring one message per user, for a total of $n - 1$ messages. □

The minimal storage scheme presented in Section 3.1 matches the above lower bound.

Theorem 5. *Let $b(n) + 1$ be the maximal number of keys held by any user in M. Then, the re-keying communication costs satisfy $c(n) \geq n^{1/b(n)} - 1$.*

Proof. The proof is by induction on b. The base case, $b(n) = 1$, is proved in Lemma 4. For $b(n) > 1$, denote by t_k the number of users holding key k. Denote by k_{max} a key other than the session key, such that $t = t_{k_{max}}$ is maximal.

On one hand, consider the set of t users holding the key k_{max}. By the induction hypothesis there exists a user holding k_{max} whose removal incurs re-keying communication costs at least $t^{\frac{1}{b-1}} - 1$, even if only the t users holding k_{max} are considered. On the other hand, when removing any user, the communication must be $c(n) \geq \frac{n}{t}$, since each message is an encryption under some key k which is understood by at most t users. It follows that the re-keying communication complexity is at least

$$c(n) \geq \max(t^{\frac{1}{b-1}} - 1, \frac{n}{t}) \geq \max(t^{\frac{1}{b-1}}, \frac{n}{t}) - 1 \geq n^{1/b} - 1$$

where the last inequality holds for any $1 \leq t \leq n$. □

For constant b the above bound is tight (upto a constant factor), and agrees with the scheme in Section 3. Otherwise, there is an $O(b)$ (and at most $O(\log n)$) gap between the above lower bound and the upper bound in Section 3.

In the following we consider a class of *structure preserving* re-keying protocols, defined below, that includes our protocol in Section 3 as well as the other known protocols. We show a tight lower bound (matching our upper bound) for this class, which is $c(n) \geq b n^{1/b}$. For the special case $b(n) = 2$ this bound holds even for protocols that are not structure preserving, and we find it useful to prove it in the following lemma. The proof follows the direction of the proof of Theorem 5 above with a more careful analysis.

Lemma 6. *If the maximal number of keys held by each user is $b(n) + 1 = 3$, then the re-keying communication costs satisfy $c(n) \geq 2\sqrt{n} - 2$.*

Proof. Each user u holds at most 3 keys, which by Corollary 3 must include the session key k_s, a unique key k_u, and a possible additional key. As before, let t denote the number of users holding a key k_{max} other than the session key, which is held by the maximal number of users. Consider the operation of removing one of the users holding k_{max}. All other $t - 1$ users holding k_{max} can only receive messages encrypted by their unique key, since the other two keys they are holding, k_{max} and k_s, were known to the removed user. This requires $t - 1$ messages. Since these messages are sent using unique keys, they do not give any information to the $n - t$ users not holding k_{max}, and thus additional messages should be sent to those users, requiring at least $\frac{n-t}{t}$ encryptions. Altogether,

$$c(n) \geq t - 1 + \frac{n - t}{t} = t + \frac{n}{t} - 2 \geq 2\sqrt{n} - 2$$

where the last inequality holds for any $1 \leq t \leq n$. □

An instance of tradeoff scheme (Example 2 in Table 2) matches the above lower bound.

Definition 7. *A protocol is* structure preserving *if $\forall U \subseteq M$ and $\forall v, v' \in M$ ($v \neq v'$), if there exists $k \in K$ such that $\forall u \in U$, $k \in K(u)$ but $k \notin K(v)$, then after the operation $remove(v')$ there exists $k' \in K^{new}$ such that $\forall u \in U \setminus v'$, $k' \in K^{new}(u)$ but $k' \notin K^{new}(v)$.*

Intuitively, structure preserving protocols are those that maintain the property of "the set U has advantage over the user v" across updates, for any subset U and user v. That is, if there is a set of users U all sharing a key k, and a user v which does not have this key, then after removing another user v' (whether $v' \in U$ or not), the users U still holds some key k' that v does not hold.

Theorem 8. *For structure preserving protocols, the re-keying communication costs satisfy $c(n) \geq bn^{1/b} - b$, where $b + 1$ denotes the maximal number of keys held by any user in M.*

Proof. The proof is by induction on b (using a stronger induction hypothesis described below). The base case $b = 1$ follows from Lemma 4. We have also proved the case $b = 2$ in the proof of Lemma 6, and in fact we use here the same idea as in the proof of Lemma 6. However, the difference is that for $b = 2$, the messages sent to the $t - 1$ users holding k_{max} cannot be interpreted by anyone who does not hold k_max (since they are sent using unique keys), and thus they can be simply added to the messages sent to the users that do not hold k_max. In contrast, for $b > 2$, this is not necessarily true: some keys can be shared both by users holding k_{max} and users that do not hold k_max. Here we use the fact that the protocols is structure preserving and count the $t - 1$ messages needed to update k_max which cannot be interpreted by users that do not hold k_max. Details follow.

We start by describing a process for selecting a user to be removed: we choose a maximal subset holding some key, then choose a maximal subset of this subset holding another key, and so on, going to smaller and smaller subset until we reach a single user. More formally, denote by $k_{max}^{b+1} = k_s$ (the session key), and $U_{max}^{b+1} = M$ (the entire multicast group). For $i = b, b - 1, \ldots, 1$ let $k_{max}^i \notin \{k_{max}^{i+1} \ldots, k_{max}^{b+1}\}$ be a key that is held by a maximal number of users. Let U_{max}^i be the set of users holding k_{max}^i. At the end of the process $U_{max}^1 = \{u\}$ is a singleton, since k_{max}^1 is the unique key of a user u. Select to remove u.

Lemma 9. *When removing a user according to the selection process described above, the communication re-keying costs satisfy $c(n) \geq t_2 + \frac{t_3}{t_2} + \cdots + \frac{t_{b+1}}{t_b} - b$, where $t_i = |U_{max}^i|$ (in particular, $t_{b+1} = n$).*

We prove the claim by induction on b. For $b = 1$ we simply need to prove $c(n) \geq t_2 - 1$ where $t_2 = n$, which follows from Lemma 4. For $b \geq 2$, let u be the user to be removed according to the selection process above. Consider the set U_{max}^b, which is a maximal-size set of users holding a key $k_{max}^b \neq k_s$. Since the protocol is structure preserving, after removing u there should be a key k' which is held by every user in $U_{max}^b \setminus \{u\}$, but not by any other user. Because of the way u was chosen, if $|U_{max}^b \setminus \{u\}| = t_b - 1 > 1$ then no such key k' unknown to u exists before the update, because otherwise the next maximal subset would be chosen as $U_{max}^{b-1} = U_{max}^b \setminus \{u\}$, and u would not be selected. Therefore, the center needs to send messages to generate this key. By the induction hypothesis, this requires communication of at least $t_2 + \frac{t_3}{t_2} + \cdots + \frac{t_b}{t_{b-1}} - (b-1)$, which cannot be interpreted by any user outside of U_{max}^b. Adding to it the communication costs for these outside users (in order to establish a new session key), sums up to

$$c(n) \geq t_2 + \frac{t_3}{t_2} + \cdots + \frac{t_b}{t_{b-1}} - (b-1) + \frac{n - t_b}{t_b} = t_2 + \frac{t_3}{t_2} + \cdots + \frac{n}{t_b} - b$$

as needed.

The only cases which we did not handle are those where $U^b_{max} \setminus \{u\}$ is small (empty or a singleton). If $U^b_{max} \setminus \{u\} = \phi$, by the maximality of U^b_{max}, each user holds only the session key and a unique key, and the bound of Lemma 4 can be applied. If it a singleton, any key other than the session key is held by at most two users, which implies that a message sent to the user in $U^b_{max} \setminus \{u\}$ (in order to update the session key) is encrypted by the unique key and cannot be interpreted by other users, thus the same calculation as above holds.

Thus, we have proved the claim. The theorem follows by observing that

$$t_2 + \frac{t_3}{t_2} + \cdots + \frac{n}{t_b} \geq bn^{1/b}$$

which can be proven by induction on b. Thus, $c(n) \geq bn^{1/b} - b$, and the proof is complete. □

Acknowledgments

We thank Moni Naor for pointing out the improvement using counters described at the end of Section 3.4.

References

1. M. Blum and S. Micali, How to generate cryptographically strong sequences of pseudorandom bits, SIAM J. Comput. **13** (1984), no. 4, 850–864.
2. C. Blundo, L. A. Frota Mattos and D. R. Stinson, Trade-offs between communication and storage in unconditionally secure schemes for broadcast encryption and interactive key distribution, in *Advances in cryptology—CRYPTO '96 (Santa Barbara, CA)*, 387–400, Lecture Notes in Comput. Sci., 1109, Springer, Berlin.
3. C. Blundo, A. De Santis, A. Herzberg, S. Kutten, U. Vaccaro and M. Yung, Perfectly secure key distribution in dynamic conferences, in *Advances in cryptology—CRYPTO '92*, 471–486, Lecture Notes in Comput. Sci., 740, Springer, Berlin.
4. R. Canetti, J. Garay, G. Itkis, D. Micciancio, M. Naor and B. Pinkas, Multicast Security: A Taxonomy and Efficient Authentication, Infocomm 1999.
5. R. Canetti and B. Pinkas, A Taxonomy of Multicast Security Issues, Internet draft <draft-canetti-secure-multicast-taxonomy-00.txt>, ftp://ftp.ietf.org/internet-drafts/draft-canetti-secure-multicast-taxonomy-00.txt.
6. A. Fiat and M. Naor, Broadcast Encryption, in *Advances in cryptology—CRYPTO '93 (Santa Barbara, CA)*, 480–491, Lecture Notes in Comput. Sci., 773, Springer, Berlin.
7. O. Goldreich, S. Goldwasser, and S. Micali. How to Construct Random Functions. *JACM*, Vol. 33, No. 4, pages 792–807, 1986.
8. S. Goldwasser and S. Micali, Probabilistic encryption, J. Comput. System Sci. **28** (1984), no. 2, 270–299.
9. M. Luby and J. Staddon, Combinatorial Bounds for Broadcast Encryption, in K. Nyberg, editor, *Advances in Cryptology—EUROCRYPT '98 (Espoo, Finland)*, 512–526, Lecture Notes In Comput. Sci., 1403, Springer, Berlin.
10. McGrew D. A., and Sherman A. T., Key Establishment in Large Dynamic Groups using One-way Function Trees. Manuscript, 1998.

11. D.R. Stinson, On some methods for unconditionally secure key distribution and broadcast encryption, to appear in Designs, Codes and Cryptography.
12. D.R. Stinson and T. van Trung, Some new results on key distribution patterns and broadcast encryption, to appear in Designs, Codes and Cryptography.
13. D. M. Wallner, E. J. Harder and R. C. Agee, Key Management for Multicast: Issues and Architectures, Internet draft <draft-wallner-key-arch-01.txt>, ftp://ftp.ietf.org/internet-drafts/draft-wallner-key-arch-01.txt.
14. C. K. Wong, M. Gouda and S. S. Lam, "Secure Group Communication Using Key Graphs", SIGCOMM '98. Also, University of Texas at Austin, Computer Science Technical report TR 97-23.
15. A. C. Yao, Theory and applications of trapdoor functions, in *23rd annual symposium on foundations of computer science (Chicago, Ill., 1982)*, 80–91, IEEE, New York.

Author Index

Springer
and the
environment

At Springer we firmly believe that an international science publisher has a special obligation to the environment, and our corporate policies consistently reflect this conviction.
We also expect our business partners – paper mills, printers, packaging manufacturers, etc. – to commit themselves to using materials and production processes that do not harm the environment. The paper in this book is made from low- or no-chlorine pulp and is acid free, in conformance with international standards for paper permanency.

Lecture Notes in Computer Science

For information about Vols. 1–1505
please contact your bookseller or Springer-Verlag

9 783540 658894